Robust Design of Microelectronics Assemblies Against Mechanical Shock, Temperature and Moisture

Related titles

Reliability characterisation of electrical and electronic systems
(ISBN 978-1-78242-221-1)

Modeling, characterization and production of nanomaterials: Electronics, photonics and energy applications
(ISBN 978-1-78242-228-0)

Optofluidics, sensors and actuators in microstructured optical fibres
(ISBN 978-1-78242-329-4)

Woodhead Publishing Series in Electronic and Optical Materials: Number 81

Robust Design of Microelectronics Assemblies Against Mechanical Shock, Temperature and Moisture

E.-H. Wong and Y.-W. Mai

ELSEVIER

AMSTERDAM • BOSTON • CAMBRIDGE • HEIDELBERG
LONDON • NEW YORK • OXFORD • PARIS • SAN DIEGO
SAN FRANCISCO • SINGAPORE • SYDNEY • TOKYO
Woodhead Publishing is an imprint of Elsevier

WP
WOODHEAD PUBLISHING

Woodhead Publishing is an imprint of Elsevier
80 High Street, Sawston, Cambridge, CB22 3HJ, UK
225 Wyman Street, Waltham, MA 02451, USA
Langford Lane, Kidlington, OX5 1GB, UK

British Library Cataloguing in Publication Data
A catalogue record for this book is available from the British Library

Library of Congress Control Number: 2015931788

ISBN 978-1-84569-528-6 (print)
ISBN 978-0-85709-911-2 (online)

For information on all Woodhead Publishing publications visit our website at http://store.elsevier.com/

Contents

Woodhead Publishing Series in Electronic and Optical Materials

1 **Circuit analysis**
J. E. Whitehouse
2 **Signal processing in electronic communications: For engineers and mathematicians**
M. J. Chapman, D. P. Goodall and N. C. Steele
3 **Pattern recognition and image processing**
D. Luo
4 **Digital filters and signal processing in electronic engineering: Theory, applications, architecture, code**
S. M. Bozic and R. J. Chance
5 **Cable engineering for local area networks**
B. J. Elliott
6 **Designing a structured cabling system to ISO 11801: Cross-referenced to European CENELEC and American Standards**
Second edition
B. J. Elliott
7 **Microscopy techniques for materials science**
A. Clarke and C. Eberhardt
8 **Materials for energy conversion devices**
Edited by C. C. Sorrell, J. Nowotny and S. Sugihara
9 **Digital image processing: Mathematical and computational methods**
Second edition
J. M. Blackledge
10 **Nanolithography and patterning techniques in microelectronics**
Edited by D. Bucknall
11 **Digital signal processing: Mathematical and computational methods, software development and applications**
Second edition
J. M. Blackledge
12 **Handbook of advanced dielectric, piezoelectric and ferroelectric materials: Synthesis, properties and applications**
Edited by Z.-G. Ye
13 **Materials for fuel cells**
Edited by M. Gasik
14 **Solid-state hydrogen storage: Materials and chemistry**
Edited by G. Walker
15 **Laser cooling of solids**
S. V. Petrushkin and V. V. Samartsev

16 **Polymer electrolytes: Fundamentals and applications**
Edited by C. A. C. Sequeira and D. A. F. Santos
17 **Advanced piezoelectric materials: Science and technology**
Edited by K. Uchino
18 **Optical switches: Materials and design**
Edited by S. J. Chua and B. Li
19 **Advanced adhesives in electronics: Materials, properties and applications**
Edited by M. O. Alam and C. Bailey
20 **Thin film growth: Physics, materials science and applications**
Edited by Z. Cao
21 **Electromigration in thin films and electronic devices: Materials and reliability**
Edited by C.-U. Kim
22 ***In situ* characterization of thin film growth**
Edited by G. Koster and G. Rijnders
23 **Silicon-germanium (SiGe) nanostructures: Production, properties and applications in electronics**
Edited by Y. Shiraki and N. Usami
24 **High-temperature superconductors**
Edited by X. G. Qiu
25 **Introduction to the physics of nanoelectronics**
S. G. Tan and M. B. A. Jalil
26 **Printed films: Materials science and applications in sensors, electronics and photonics**
Edited by M. Prudenziati and J. Hormadaly
27 **Laser growth and processing of photonic devices**
Edited by N. A. Vainos
28 **Quantum optics with semiconductor nanostructures**
Edited by F. Jahnke
29 **Ultrasonic transducers: Materials and design for sensors, actuators and medical applications**
Edited by K. Nakamura
30 **Waste electrical and electronic equipment (WEEE) handbook**
Edited by V. Goodship and A. Stevels
31 **Applications of ATILA FEM software to smart materials: Case studies in designing devices**
Edited by K. Uchino and J.-C. Debus
32 **MEMS for automotive and aerospace applications**
Edited by M. Kraft and N. M. White
33 **Semiconductor lasers: Fundamentals and applications**
Edited by A. Baranov and E. Tournie
34 **Handbook of terahertz technology for imaging, sensing and communications**
Edited by D. Saeedkia
35 **Handbook of solid-state lasers: Materials, systems and applications**
Edited by B. Denker and E. Shklovsky
36 **Organic light-emitting diodes (OLEDs): Materials, devices and applications**
Edited by A. Buckley
37 **Lasers for medical applications: Diagnostics, therapy and surgery**
Edited by H. Jelínková
38 **Semiconductor gas sensors**
Edited by R. Jaaniso and O. K. Tan

39 **Handbook of organic materials for optical and (opto)electronic devices: Properties and applications**
Edited by O. Ostroverkhova
40 **Metallic films for electronic, optical and magnetic applications: Structure, processing and properties**
Edited by K. Barmak and K. Coffey
41 **Handbook of laser welding technologies**
Edited by S. Katayama
42 **Nanolithography: The art of fabricating nanoelectronic and nanophotonic devices and systems**
Edited by M. Feldman
43 **Laser spectroscopy for sensing: Fundamentals, techniques and applications**
Edited by M. Baudelet
44 **Chalcogenide glasses: Preparation, properties and applications**
Edited by J.-L. Adam and X. Zhang
45 **Handbook of MEMS for wireless and mobile applications**
Edited by D. Uttamchandani
46 **Subsea optics and imaging**
Edited by J. Watson and O. Zielinski
47 **Carbon nanotubes and graphene for photonic applications**
Edited by S. Yamashita, Y. Saito and J. H. Choi
48 **Optical biomimetics: Materials and applications**
Edited by M. Large
49 **Optical thin films and coatings**
Edited by A. Piegari and F. Flory
50 **Computer design of diffractive optics**
Edited by V. A. Soifer
51 **Smart sensors and MEMS: Intelligent devices and microsystems for industrial applications**
Edited by S. Nihtianov and A. Luque
52 **Fundamentals of femtosecond optics**
S. A. Kozlov and V. V. Samartsev
53 **Nanostructured semiconductor oxides for the next generation of electronics and functional devices: Properties and applications**
S. Zhuiykov
54 **Nitride semiconductor light-emitting diodes (LEDs): Materials, technologies and applications**
Edited by J. J. Huang, H. C. Kuo and S. C. Shen
55 **Sensor technologies for civil infrastructures**
Volume 1: Sensing hardware and data collection methods for performance assessment
Edited by M. Wang, J. Lynch and H. Sohn
56 **Sensor technologies for civil infrastructures**
Volume 2: Applications in structural health monitoring
Edited by M. Wang, J. Lynch and H. Sohn
57 **Graphene: Properties, preparation, characterisation and devices**
Edited by V. Skákalová and A. B. Kaiser

58 **Silicon-on-insulator (SOI) technology**
Edited by O. Kononchuk and B.-Y. Nguyen
59 **Biological identification: DNA amplification and sequencing, optical sensing, lab-on-chip and portable systems**
Edited by R. P. Schaudies
60 **High performance silicon imaging: Fundamentals and applications of CMOS and CCD sensors**
Edited by D. Durini
61 **Nanosensors for chemical and biological applications: Sensing with nanotubes, nanowires and nanoparticles**
Edited by K. C. Honeychurch
62 **Composite magnetoelectrics: Materials, structures, and applications**
G. Srinivasan, S. Priya and N. Sun
63 **Quantum information processing with diamond: Principles and applications**
Edited by S. Prawer and I. Aharonovich
64 **Advances in non-volatile memory and storage technology**
Edited by Y. Nishi
65 **Laser surface engineering: Processes and applications**
Edited by J. Lawrence, C. Dowding, D. Waugh and J. Griffiths
66 **Power ultrasonics: Applications of high-intensity ultrasound**
Edited by J. A. Gallego-Juárez and K. F. Graff
67 **Advances in delay-tolerant networks (DTNs): Architectures, routing and challenges**
Edited by J. J. P. C. Rodrigues
68 **Handbook of flexible organic electronics: Materials, manufacturing and applications**
Edited by S. Logothetidis
69 **Machine-to-machine (M2M) communications: Architecture, performance and applications**
Edited by C. Anton-Haro and M. Dohler
70 **Ecological design of smart home networks: Technologies, social impact and sustainability**
Edited by N. Saito and D. Menga
71 **Industrial tomography: Systems and applications**
Edited by M. Wang
72 **Vehicular communications and networks: Architectures, protocols, operation and deployment**
Edited by W. Chen
73 **Modeling, characterization and production of nanomaterials: Electronics, photonics and energy applications**
Edited by V. Tewary and Y. Zhang
74 **Reliability characterisation of electrical and electronic systems**
Edited by J. Swingler
75 **Handbook of industrial wireless sensor networks: Monitoring, control and automation**
Edited by R. Budampati and S. Kolavennu
76 **Epitaxial growth of complex metal oxides**
Edited by G. Koster, M. Huijben and G. Rijnders
77 **Semiconductor nanowires: Materials, synthesis, characterization and applications**
Edited by J. Arbiol and Q. Xiong

78 **Superconductors in the power grid**
Edited by C. Rey
79 **Optofluidics, sensors and actuators in microstructured optical fibres**
Edited by S. Pissadakis
80 **Magnetic nano- and microwires: Design, synthesis, properties and applications**
Edited by M. Vázquez
81 **Robust design of microelectronics assemblies against mechanical shock, temperature and moisture**
E.-H. Wong and Y.-W. Mai
82 **Biomimetic technologies**
Edited by T. D. Ngo
83 **Directed self-assembly of block co-polymers for nano-manufacturing**
Edited by R. Gronheid and P. Nealey

Foreword

The reliability of electronic/photonic components, modules, and systems has been a critical issue in the microelectronics industry for decades. Although tremendous efforts and resources have been spent on this subject in the past, it continues to be a very active research area in academia and industry due to the endless demands for higher performance and better reliability from applications and markets. The main drivers include more complicated operating environments, more mobile consumer electronics, and tighter government legislation on lead-free electronic devices. I have been studying microelectronics reliability for almost 20 years. I am very pleased to see Dr E.-H. Wong and Prof. Y.-W. Mai, who are my long-time friends and colleagues in the same research area, publish their book on *Robust Design of Microelectronics Assemblies Against Mechanical Shock, Temperature, and Moisture*. This is a timely technical monograph on its intended subject for both scientific researchers and engineering practitioners. The readers of this book will benefit from the fundamental theories and the case studies provided by the authors. Through the introduction and propagation of this book, I look forward to seeing more robust microelectronic products with better reliability in the market.

S.W. Ricky Lee, PhD
Fellow of IEEE, ASME, IMAPS
Professor of Mechanical and Aerospace Engineering
Hong Kong University of Science and Technology
Junior Past-President of IEEE Components
Packaging and Manufacturing Technology Society

Preface

Unlike the cautious design practices of aircraft, the automotive industry, buildings, bridges, and machinery, microelectronic assemblies are designed to the very limit of the strength of their constituting materials. Failures of microelectronic assemblies are not uncommon and they offer good learning opportunities. Scientists and engineers in the microelectronic assembly community are presented with numerous opportunities to test the limit of their designs and learn from the failure analysis, aided by test structures built into miniaturised electrical circuitries to provide continuous monitoring of structural damage as it occurs. At the same time, being a relatively young engineering field, the design of microelectronic assemblies for robustness can tap into the vast wealth of knowledge accumulated in the other traditional fields of engineering.

The first idea for this book was conceived soon after EHW completed his PhD at the University of Sydney in 2006. Sizable materials for this book were the result of research activities performed while EHW was with the Institute of Microelectronics, Singapore. While the intent of this book is primarily for the benefit of practicing engineers, it is also a valuable textbook for graduate students who are interested in the detailed elaboration given to the theories and the derivation of analytical solutions.

The presentation of this book reflects the philosophy of the authors in addressing engineering problems: that it is essential to develop a profound understanding of the science behind a problem; examine critically the current practices, be ready to challenge them and, whenever possible, develop more precise solutions; the best solution is the one that is consistent with the science and can be easily understood and used by the practicing engineers.

This book offers a number of improved solutions to existing problems in the microelectronics assembly industry. Chapter 2 presents an improved analytical solution for describing thermoelastic stresses in the bonding layer and in discrete solder joints. Chapter 3 provides a unified equation that presents a simple and consistent method of modelling the creep-fatigue of solder joints. Chapter 4 summarizes a number of advanced techniques for characterizing moisture diffusivity. Chapter 5 offers advanced techniques for modelling moisture diffusion in microelectronic assemblies using the thermal-mass modelling analogy. Chapters 6 to 12 present a comprehensive approach to investigating a new damage driver – the drop impact. In the process, new testing techniques are established, new material characteristics are generated, and new analytical solutions are developed. Chapter 7 describes the development of micro-impacting and the high speed cyclic bending test techniques; the former has been incorporated into an industry test standard, while the latter has been shown to offer faster testing. Chapter 8 and Chapter 9 describe the generation of fatigue strain-life characteristics of solder joints and the in situ measurement of crack-growth in solder joints, respectively. Both experiments were performed at high cyclic frequency. It is noted that such experimental techniques and data are scarce in the literature. Chapter 10 offers a series of analytical solutions for the dynamics of mechanical shock. Chapter 11 presents a sophisticated analytical solution for describing the stresses in the bonding layer and in discrete solder joints when microelectronic assemblies are subject to bending. Chapter 12 describes the generation of the stress–strain characteristic of solder joints at high strain rates.

The authors would like to express their gratitude to their many collaborators who helped make this publication possible. In particular, EHW would like to thank his former management (or managers) Thiam-Beng Lim for the conducive research environment and for being a mentor in life, and Mahadevan Iyer for his continued support. Much of this research would not have been possible without the support of colleagues in the Institute of Microelectronics; notably, Ranjan Rajoo, Cheryl Selvanayagam, and most notably, Simon Seah, whose contribution in the 'drop-impact induced damage' of electronic assemblies is especially profound. The materials on 'vapour pressure modeling' in Chapter 5 are attributed to the brilliance of K.H. Lee and K.M. Lim and the dedication of Sau-Wee Koh, all from the National University of Singapore. The experimental data for Chapter 6 was generated under the collaborative project between Institute of Microelectronics, National University of Singapore, and the University of Cambridge.

The experimental data for Chapters 7–9 and 12 were generated in collaboration with industry partners; sincere thanks are due to the supports of Jo Caers (Philips), William van Driel (NXP, Philips), Yi-Shao Lai (ASE), N. Owens (Freescale), and Daisuke Watanabe (Nihon Superior). The experimental work for Chapter 7 would not have been possible without the unyielding support of Kuo-Tsing Tsai from Instron, Singapore, who was instrumental in the development of the micro-impact tester and the high-speed cyclic bend tester.

Finally, the authors would like to dedicate this book to K.H. Lee, who was a mentor and a friend to EHW, and whose demise was a great loss to the scientific community and to those who knew him.

E.-H. Wong and Y.-W. Mai
Christchurch and Sydney
December, 2014

Introduction

1

1.1 Introduction to microelectronic packaging

A typical semiconductor wafer is made out of extremely pure silicon that is grown into monocrystalline cylindrical ingots up to 450 mm in diameter. The semiconductor is doped with either boron or phosphorus, which is added to the molten intrinsic material in precise amounts in order to change the silicon crystal into n-type or p-type extrinsic semiconductor. These ingots are then sliced into wafers to predefined thicknesses and polished to flat and parallel surfaces. Using photolithography, multiple layers of integrated circuits (ICs) are embedded on one of the surfaces of the wafer.

The process of converting the semiconductor wafer into individual components that are electrically functional and ready to be assembled onto a printed circuit board (PCB) is known as IC packaging (Tummala, Rymaszewski, & Klopfenstein, 1997); the individual components are referred to as IC components or microelectronic assembly. Figure 1.1 illustrates a typical IC packaging process flow that consists of (1) singulating the IC wafer into multiples of IC chips, each embedded with identical ICs; (2) attaching the IC chip onto the interposer, which may be a single layer of metal (lead frame) or a multilayered organic substrate; (3) electrically interconnecting the IC chip to the interposer using either wire bond or solder joints; (4) encapsulating the IC chip for mechanical protection; and (5) applying solder to the interposer, preparing it for mounting onto the PCB.

IC packaging has the important functions of (1) providing paths for external electrical current to power the ICs on the chip; (2) providing paths for transmission of electrical signals between the ICs on the chip and between the IC chip and the PCB; (3) removing heat generated by the ICs; and (4) providing mechanical support and protection to the delicate ICs.

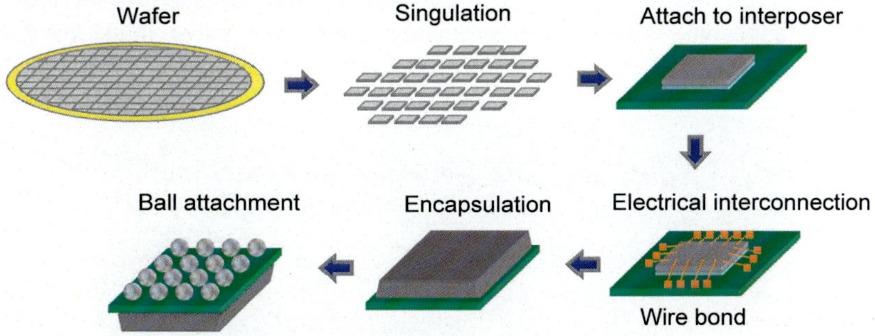

Figure 1.1 Integrated circuit packaging process flow.

Robust Design of Microelectronics Assemblies Against Mechanical Shock, Temperature and Moisture
http://dx.doi.org/10.1016/B978-1-84569-528-6.00001-0

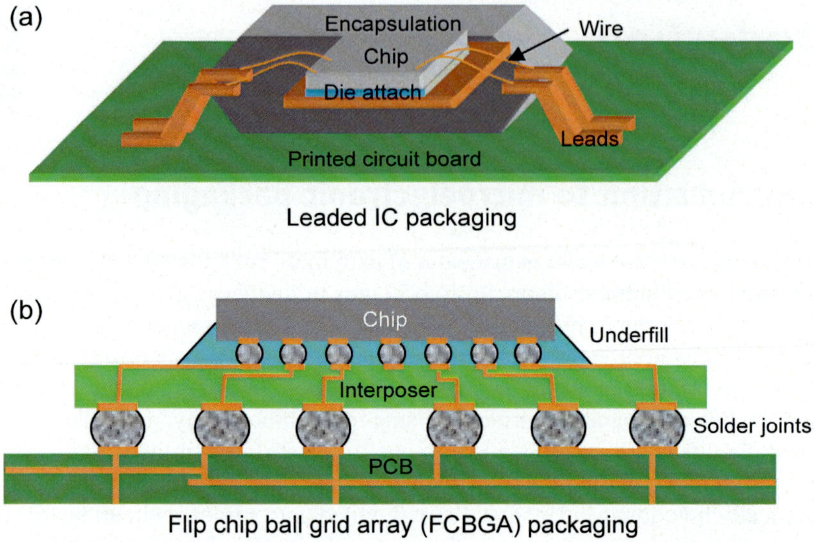

Figure 1.2 Schematics of (a) leaded integrated circuit (IC) and (b) area array IC packaging.

Figure 1.2 shows the schematics of two major classes of IC packaging: leaded packaging and ball grid array (BGA) packaging. Figure 1.3 shows a picture of a PCB assembly with multiple BGA packages. The interposers used for the leaded and the BGA packaging are leadframe and multilayered substrate, respectively. The former is made of copper alloy, while the latter is made of multiple layers of glass-reinforced composite impregnated with thermal setting plastic and separated by planes of copper traces. This shall henceforth be referred to as 'substrate'. In view of the very different sophistication of the fabrication equipment and processes used in fabricating the IC wafer and the PCB, the resolution and accuracy of the inputs/outputs (I/Os) on the IC chip are at least an order finer than that on the PCB. The interposer serves to 'fan out' the I/Os on the IC chip to a larger dimension and pitch to match the I/Os on the PCB.

The IC chip comes with I/Os along the periphery of the chip. For IC packaging that uses a leadframe interposer, the IC chip is first attached to the interposal using adhesive (the process is referred to as 'die attach' and the adhesive is generally referred to as die-attached adhesive) with its active surface facing up; wires made of gold or

Figure 1.3 Ball grid array packages on the board assembly of a mobile device.

aluminium are used to interconnect the individual I/O along the periphery of the IC chip to the corresponding I/O (lead) along the periphery of the leadframe. This is known as wire-bond interconnection. Encapsulation is applied to provide protection to the ICs and the I/Os. The leads of the leadframe are shaped to correspond with the metal pads on the PCB. The leads are coated with a layer of solder preparing them for assembly to the corresponding metal pads on the PCB (Figure 1.2(a)). The peripheral layout of the leads limits the number of I/Os possible for the leaded packaging.

For IC packaging that uses a substrate interposer, the I/Os on the IC chip may be interconnected to those on the interposer through wire bonding, as in the case of the leadframe interposer, or through solder joining. In the latter, the I/Os along the periphery of the IC chip must first be rerouted to form area array of I/Os on the active surface of the chip, thus allowing greater numbers of I/Os. Individual I/O is then deposited (via screen printing) with a tiny volume of solder; this is followed by raising the temperature slightly above the melting temperature of the solder to reflow it and form a metallurgical bond with the under-bump metallisation on the I/O, forming a solder bump. The interposer is prepared with a corresponding area array of I/Os on its front. When interconnecting the IC chip with the interposer, the IC chip is 'flipped' such that the active surface is facing down and the I/Os on the IC chip are aligned and in physical contact with the corresponding I/Os on the interposal through the solder bumps. The IC chip and the interposal are then heated to reflow the solder bumps to form solder joints that interconnect the I/Os on the IC chip to the interposer. The space between the solder joints is then 'underfilled' with nonconductive adhesive to protect the ICs and also serve as a structural reinforcement to the solder joints (Figure 1.2(b)). The I/Os on the front of the interposer are routed to the back of the interposer and arranged in an area array. Solder balls are reflowed onto these I/Os forming a BGA. For this reason, IC packaging using substrate interposers is generally referred to as BGA packaging. The area array arrangement is capable of handling a higher number of I/Os for the same I/O pitch. IC components are assembled onto the PCB to form PCB assembly. In assembling to PCB, the solder balls are reflowed to form solder joints that interconnect the I/Os on the IC component to the PCB. These are generally referred to as board-level solder joints. The PCB and the microelectronic assemblies are collectively referred to as electronic assemblies.

In view of the worse compliance of solder joints compared to the leads of the leadframe interposer, the solder joints of the BGA packaging experience higher stress/strain whenever there is mismatched deformation between the IC component and the PCB. The trend toward higher density of I/Os leads to reduced feature size of the solder joints, rendering the solder joints increasingly vulnerable.

The drive toward ever more functionality without increasing the size of electronic products can only be supported by ever increasing densities of the transistors, but more critically, the ever increasing density of the I/Os. This has been achieved through reducing the pitch between the I/Os and by transitioning from peripheral to BGA interconnections. Beyond these, the industry has resorted to advanced microelectronic packaging, including:

Wafer-level packaging: the footprint of an IC packaging is typically larger than that of an IC chip. By using silicon crystal as the interposal for redistribution of circuitries, the footprint of the IC packaging becomes the same size as the IC chip – a truly chip-scale packaging. The redistribution, encapsulation, and solder bumping processes are carried out at the wafer level; hence, the term wafer-level packaging. The final process involves singulating the wafer-level assembly into individual IC components.

3D packaging: there is a limit to increasing the area density of the I/Os; this leads to the 3D packaging in which the I/Os are added along the *z*-axis. The simplest 3D-packaging technology is *package-on-package* (Tummala, 2008) in which multiple IC packages are stacked on top of each other and are interconnected through the interposers using solder joints. The next simplest technology is *stacked die* in which multiple chips are stacked on each other using adhesive and are interconnected to the common interposer. The numerous wires in the stacked die present significant electrical inductance resulting in transmission losses and delay. By remedying the inductance from the numerous wires, the *through silicon via* (TSV) technology, in which the stacked dies are interconnected using via that runs through the thickness of the silicon chip (Yoon et al., 2011), offers excellent electrical performance. The combination of TSV and wafer-level packaging leads to TSV wafer-level packaging.

1.2 Introduction to robust design

1.2.1 Robust design and challenges

The quest for lighter, stronger, safer and more durable designs has been propelling the advancements of engineering and technologies and improving the quality of our lives. These advancements are reflected in the changing faces of our modern world in terms of taller buildings, longer bridges, larger airplanes, faster trains, stronger machine tools, and not the least, cheaper, lighter and more compact electronic products that are enabled by the cheaper, higher-density and robust microelectronics and assemblies.

The concept of robust designs is not new. Conventional robust designs are stuffed with layers of excessive fats, known as safety factors, or more cynically, ignorance factors. However, robust designs that are relying on excessive fats are inefficient and uncompetitive. Modern planes are ever lighter, larger and capable of flying further; modern buildings are ever taller, and bridges ever longer. Fats have been trimmed, yet these modern structures are no less safe. All these are attributable to the continuing advancement in predictive robust design.

Predictive robust design is described by a simple inequality:

$$\text{Failure driving force}\begin{cases} < \text{material resistance (robust)} \\ > \text{material resistance (unsafe)} \end{cases} \tag{1.1}$$

On the left is the failure driving force, and on the right the material resistance. A design (structure) is safe and robust if the failure driving force is smaller than the material resistance and unsafe otherwise. The critical factor in predictive robust design is to define as precisely as possible the magnitudes of the failure drive force (or forces) and the material resistance.

Equation (1.1) is deceptively simple. Many forces could induce failure in a design (structure): temperature, moisture, magnetic and electromagnetic forces, electromigration, electrochemical, radiation and mechanical loadings. These forces could act singly or in a combination. In the latter, the resultant forces would not be a simple summation of the combined forces. The definition of material resistance is much more complex than the failure driving forces. Many forms of material resistance have been proposed: principal stress, principal strain, von Mises stress, maximum shear stress, maximum shear strain, fracture toughness (of different modes), interfacial fracture toughness (of different modes), stress range, strain range, creep strain, etc. Moreover, the material resistance is a function of environmental factors such as temperature, humidity, chemicals, radiation, etc.; and at small feature size, such as microelectronics and microelectronic assemblies, the material resistance can also be a function of the geometrical size of the structure.

1.2.2 Unique features of microelectronic assemblies

The predictive robust design of microelectronic assemblies has some unique features:

1. the assemblies are made of layered structures and are particularly susceptible to delamination;
2. compared to the conventional structures, microelectronic assemblies are very small in feature size, which implies: (1) the material resistance obtained from a large specimen may be invalid; and (2) conventional practice of damage monitoring by strain gauging the critical structural element is infeasible;
3. structural health of constituting elements can be very conveniently monitored electrically, which offers a unique opportunity that is not available to other structures for studying damages; and
4. there is no mandatory safety factor in the robust design of microelectronic assemblies for commercial electronic products.

Failures of real-life structures such as buildings, bridges, spacecraft and airplanes have disastrous consequences; failures are rare as these structures are protected by layers of 'fats'. Many scientists and engineers in these professions have never seen failure. By contrast, failures of microelectronic assemblies are common. Failure offers a wonderful learning opportunity. Thus, scientists and engineers in the microelectronic assembly community are blessed with numerous opportunities to test the limits of their designs, analyse failure and learn from failure. This is helped by the ease of continuing monitoring of structural damage.

1.2.3 Principal forces of damage for microelectronic assemblies

1.2.3.1 Temperature

The PCB and the microelectronic assemblies undergo thermal excursion during the assembly processes, the accelerated test, and while in service. The PCB and the microelectronic assemblies are made of multilayer structures made of different materials with different coefficients of thermal expansion. The individual layers undergo different magnitudes of thermal expansion during the temperature excursion. The mismatch in the expansion that leads to internal stresses and failure along the bonding interfaces is not uncommon. More common than interfacial delamination is the creep fatigue of solder joints due to the combined action of creep, which is induced by the elevated temperature, and fatigue, which is induced by the cyclical mismatched thermal expansions.

The investigation of analytical equations for describing the interfacial stresses in a multilayer structure arising from mismatched thermal expansion has a long history, but the first investigation on electronic packaging was probably in the 1970s. The creep fatigue of solder joints has been a subject of keen investigation since then.

1.2.3.2 Moisture

Since the transition from ceramic packaging to plastic packaging in the 1980s, the microelectronic assembly has become susceptible to the ingression of moisture during storage. During the subsequent reflow of solder, the moisture turns into high-pressure vapour within a delamination, resulting in violent cracking of the IC component. This is frequently referred to as 'popcorn cracking'. Hygroscopic swelling accompanies the absorption of moisture; differential hygroscopic swelling between the constituting materials of the microelectronic assembly results in interfacial stresses. Absorption of moisture is usually associated with reduction in adhesive bond strength, possibly due to the degradation of the hydrogen bond. More subtle effects of moisture absorption are the change in the instantaneous and relaxation modulus of the polymeric materials that leads to plasticising and hence a redistribution of stresses in the microelectronic assembly. Less violent, but no less damaging, is increasing the dielectric constant of the dielectric layers in the microelectronic assembly, resulting in increased delay in signal propagation. Similarly, the presence of moisture increases the optical index of optical coupling adhesive leading to increased transmission losses.

1.2.3.3 Drop impact

Since the emergence of affordable cell phones in the 1980s, mobile electronic devices have become essential tools for communication and entertainment. The widespread use of portable electronic devices (PED) has given rise to a new reliability issue for electronic packaging – that of the robustness of electronic assemblies within a PED experiencing drop impact. The miniaturisation of the product housing has resulted in its reduced capacity to absorb and dissipate the energy of the impact before transferring the residual energy to the electronic components within, resulting in high magnitude of deceleration of the PCB assembly. The inertia of the PCB assembly during deceleration leads to its bending; and the differential flexural deformation between the IC package and the PCB leads to severe deformation of the interconnecting solder joints.

The miniaturisation of electronic devices was accompanied by miniaturisation of the electrical interconnections and accompanied fragility. But it was not until the early 2000s, concurrent with the introduction of lead-free solders, that widespread failures of solder joints during drop impact were reported.

1.2.4 An overview introduction to robust design against drop impact

Compared to robust designs against temperature and moisture, robust design against drop impact is relatively recent and has received much less attention. In reality, the subject is much more complicated to deal with and deserves much more discussions. While the robustness of the microelectronic assemblies against temperature and moisture could be established independent of the eventual electronic devices that the electronic assemblies will be attached to, the robustness of the microelectronic assemblies depends heavily on the mechanical design of the PED. The propensity for brittle fracture of the solder joints demands a more sophisticated method for assessing the manufacturing quality of the solder joints. While the electronic packaging community has accumulated vast experience in modelling thermal stresses, including the advanced topic of viscoplasticity and the associated constitutive equation such as Anan's equation, the community is much less familiar with transient dynamics and the rate-dependent constitutive equation. Similarly, while the community is familiar with the reduced fatigue resistance of solders due to creep, it is relatively alien to the reduced fatigue resistance of solders at increased strain rate. For these reasons, the major content of this book is dedicated to the robust design of microelectronic assemblies against drop impact, and this section is dedicated to providing an overview introduction to this subject.

A typical drop impact of a PED is shown in the high-speed video sequence of Figure 1.4, where the device impacts upon a rigid surface after a brief period of free fall under gravity. Upon impact, a large portion of the kinetic energy is converted into elastic vibrational energy of the mechanical elements of the device, including the housing, the liquid crystal display and the PCB assembly. Figure 1.5 shows an example of PCB assembly, which is made up of electronic components, viz the IC components and the passive components, assembled onto a PCB using solder joints. In case the strain energies experienced by the mechanical elements exceed their material strength, then some of these strain energies would be converted into plastic work and even work of fracture.

The failure of LCD was the main concern in the early days of PED, however, the recent concern has been with the failure of the solder joints that interconnect the IC components to the PCB. The increased risk of failure of the interconnecting solder joints to the drop impact of PED is attributable to two factors. The first is the

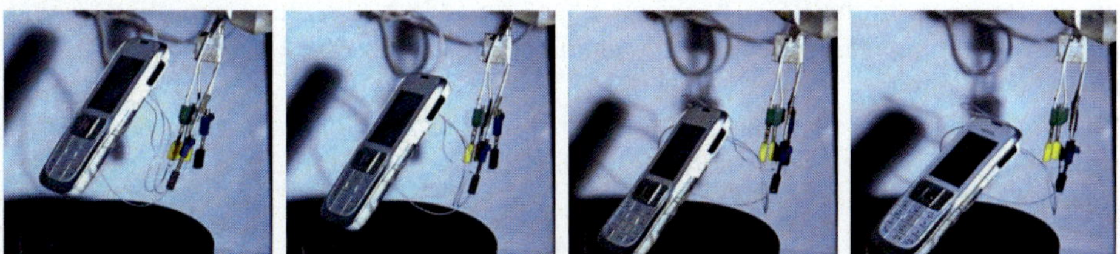

Figure 1.4 High-speed camera sequence of drop impact of a mobile phone. The high degree of dynamic flexing is discernible even in the exterior housing of the phone.

Figure 1.5 An example of printed circuit board assembly.

miniaturisation of PED despite their increased functionality. A chain of consequences follows this, starting with the miniaturisation of the product housing, which results in its reduced capacity to absorb and dissipate the energy of the impact before transferring the residual energy to the electronic components within. The PCB assembly has to be minimised to fit the miniaturised product housing. Conversely, the number of electronic components has to be increased to support the increased product functionality, and these components have to be assembled onto the reduced footprint of the PCB. This is only possible by drastically reducing the footprint of the IC components. This necessitates the switch from the leaded IC package, where interconnections are positioned along the perimeter of the IC package, to the area-array IC package, where the interconnections are distributed over the entire face of the IC package. This reduces the footprint of the IC component on the PCB for the same number of interconnections and for the same spacing (pitch) between the interconnections. Compared to the lead in the leaded packages, the solder joints in the BGA packages have considerably reduced compliances and are therefore more susceptible to the mismatched deformation between the PCB and the IC component.

The second factor contributing to the drop-impact problem is the recent industry-wide adoption of lead-free solders, a move driven by the 2006 Restriction of Hazardous Substances Directive (RoHS) that bans the use of well-established tin–lead solders in consumer electronics. Several lead-free solder joints that performed well under thermocyclic loading were found to be extremely fragile under the high strain rates associated with drop-impact loading. Figure 1.6 contrasts

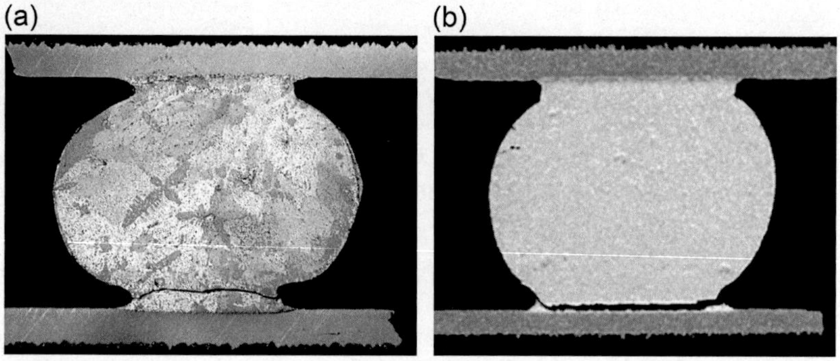

Figure 1.6 Cross-sections of solder ball joints damaged by drop impact. (a) Sn−37Pb ball on Cu pad with crack in bulk solder. (b) Lead-free Sn−1.0Ag−0.1Cu ball on Cu pad with crack along bonding interface.

(a) the crack that runs in the ductile bulk solder of the tin−lead solder joint with (b) the crack that runs along the brittle intermetallic compound (IMC) interface of the lead-free solder and the metal pad. The transition from ductile fracture in the tin−lead solder to brittle fracture in the lead-free solder is attributable to the higher flow stress of the lead-free solders that has resulted in a more rigid solder joint which has in turn elevated the stresses in the solder joint; brittle fracture occurs when the elevated stresses exceed the fracture strength of the IMC.

Future generations of PED are expected to be increasingly fragile against drop impact due to the simple fact that the dimensions of the solder joint will continue to shrink, resulting in reduced ability to accommodate the differential bending between the IC component and the PCB, in order to support the increased density of interconnections that is required to support the ever-increasing thirst for increased functionality of the PED.

The failure of solder joints in the drop impact of PED was not expected and caught the original product producers and the supply chain – the product assemblers, the PCB assemblers, the IC component manufacturers, the PCB manufacturers and the solder materials suppliers – by surprise. The biggest challenge was to find a common test method that would allow qualification of the design of the IC components and the solder joints, which includes the shape and size of solder joints, the solder alloys, and the surface finishes on the metal pads on which the solder joint is formed. The obvious solution is to drop test the PCB assembly in the particular PED to which the PCB assembly is to be attached. However, such a test is not possible for the simple reason that IC components and the associated solder joints are not designed for a specific PED but are intended for generic PED, including those products that are yet to be designed. Therefore, a realistic qualification test should be one that is independent of the design of the portable product. The establishment of such a test requires an understanding of the exact physics of failure of the solder joints interconnecting the IC component to the PCB. The suspected physics were: (1) overstress of solder joints due to the inertia force of the IC component; (2) overstress of solder joints due to differential bending between the IC component and the PCB; and (3) impact stress

Figure 1.7 High-speed video sequence of the JEDEC board-level drop test, showing the dynamic bending of the board assembly after impact.

induced in the solder joints due to knocking of the IC component against the product housing and against other components.

An industry-wide effort led by Amkor led to the July 2003 publication of the JEDEC standard, JESD22-b111 "board level drop test method of components for handheld electronic products". The test attempted to duplicate the drop-impact nature of a portable product by subjecting the PCB assembly to multiple mechanical shocks induced by drop impacting the holding fixture against a rigid target. A high-speed video sequence of the JEDEC test, showing dynamic bending of the board assembly subjected to a mechanical shock, is illustrated in Figure 1.7. The test reproduces (though inadequately) the high strain rate experienced by the solder joints in a drop impact of the PED. The standardisation of test board and test methodology offered by the JESD22-b111 test standard has facilitated industry production of IC components and solder joints destined for mobile applications.

It is worth highlighting that the lack of quantitative correlation between the JESD22-b111 test condition and the product-level drop test limits the applicability of this JEDEC test to evaluating and comparing the designs of IC components and the associated solder joints, but not qualifying the designs of IC component and solder joints. That is, an IC component and solder joint system that passes the JESD22-b111 test condition may not survive a product-level drop test and vice versa.

Probably the main criticism of the JESD22-b111 test method is the high magnitude of the generated noise and ground shock and the poor controllability of the prescribed acceleration shock applied to the bolting sites on the PCB. These limitations have catalysed significant research efforts aimed at improving the test standard, one of which is the high-speed cyclic bend test.

The intrinsic fracture strength of the solder-pad IMC depends on the solder alloy and the preservative finish on the copper pad. However, the actual fracture strength of the IMC varies significantly with manufacturing process. The control on manufacturing process that is required to produce IMC of consistent fracture strength is significantly more stringent than that required to produce bulk solder of consistent creep-fatigue strength. While it is possible to use a board-level test method such as JEDEC standard JESD22-B111 for assuring the manufacturing quality of the IC components and solder joints, it would be extravagantly expensive. A great deal of research effort had been directed at developing a quality assurance test method that does not require the assembly of an IC component to a PCB – a component-level test method in a nutshell. In a component-level test, a single solder ball that has been reflowed onto the metal pad of an IC component is mechanically detached

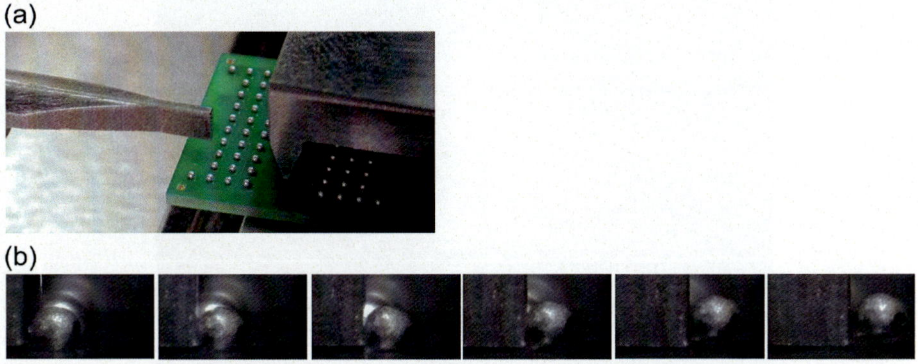

Figure 1.8 Component-level high-speed shear. (a) Shearing blade next to ball-on-substrate samples (Instron MicroImpact tester). (b) High-speed video sequence of test.

from the IC component at relatively high speed to simulate the strain rate experienced by the solder joints in a product-level drop test. The force and energy of separation are registered and the brittle/ductile mode of fracture are analysed to provide a qualitative assessment of the manufacturing quality of the joining between the solder ball and the IC component. Despite the initial apprehension with the component-level test methods and in interpreting the test results, the effort has led to a new JEDEC test standard, JESD22-B115 'ball pull test', and a revised JEDEC test standard, JESD22-B117A 'ball shear test', credited to Keith Newman of Sun Microsystems. A high-speed ball shear test set-up and a series of high-speed images of solder ball shearing are shown in Figure 1.8.

The establishment of the board-level and the component level test standards/methods were important milestones for the industry; however, while the board-level test could be used to evaluate the robustness of a design, further understanding of the mechanics of drop impact is needed to establish a predictive robust design capability. The fundamentals of predictive robust design is the quantification of failure driving force and the material resistance. A priori quantification of the failure driving force is its definition, which requires an answer to the question: 'in the drop impact of a PED, do the board level solder joints fail through overstress in a single bending cycle or through fatigue after multiple bending cycles?'

The evidence of fatigue as the mechanics of failure for solder joint is presented in Figure 1.9 that shows fatigue striations on the fracture surface of a solder joint in a PCB assembly of a PED that has experienced several drop impacts. The parallel striations are a classic indicator of cyclic fatigue, where each striation indicates the incremental advance of the crack for one cycle of load – bending of PCB in this case. The number of drops of the PED and the associated number of bending of the PCB required to cause complete electrical failure of the solder joints is much less than the number of temperature cycles as specified in JESD22-B104D to cause complete electrical failure. This is attributed partly to the higher amplitude of stress experienced by the solder joints for each drop impact of the PED and partly to the more rapid crack propagation rate of the brittle fracture of the IMC.

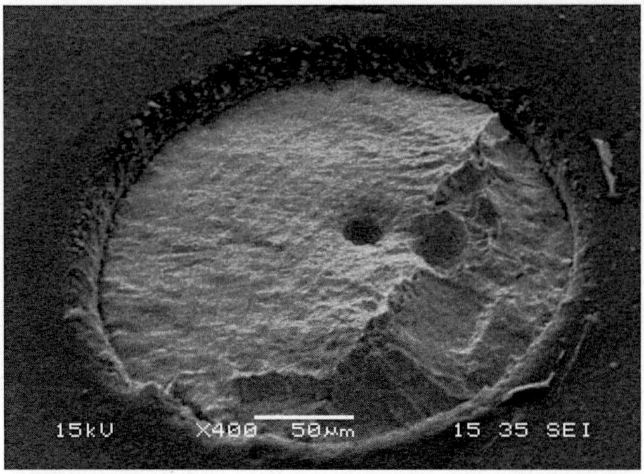

Figure 1.9 Parallel striations (crack propagation is from the left to right) on the fracture surface of a joint that has failed under drop impact.

Equation (1.1) may now be written more specifically for the drop-impact failure of solder joints as:

$$\text{Fatigue driving force} \begin{cases} < \text{Fatigue resistance of solder joints (robust)} \\ > \text{Fatigue resistance of solder joints (unsafe)} \end{cases} \quad (1.2)$$

The fatigue resistance of solder joints on the PCB assembly may be expressed in the conventional form of a power law equation:

$$\Delta\sigma = C_o N^{\beta_o} \quad (1.3)$$

where $\Delta\sigma$ is the peeling stress range in the solder joints. The minute feature size of solder joints denies experimental instrumentation, and it is experimentally much more convenient to measure the fibre strain of the PCB; the fatigue resistance of solder joints may be expressed as

$$\Delta\varepsilon_{\text{pcb}} = C_o N^{\beta_o} \quad (1.4)$$

A relevant stress-life relation must be established at a strain rate that is representative of that experienced by the solder joints in a typical drop impact of PED. In contrast to creep fatigue, the fatigue life of solder joints decreases with increasing strain rate as shown in Figure 1.10. More insights into the fatigue resistance of solder joints at high strain rate may be gained by studying the growth characteristics of fatigue cracks. Through the use of a high-sensitivity voltmeter and the specially designed test board, the fatigue crack growth in a critical solder joint could be monitored through the change

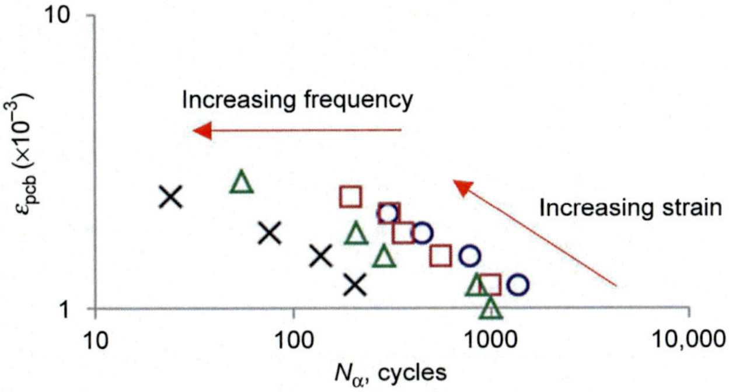

Figure 1.10 ε_{pcb}–N_α characteristics versus bending frequency of SAC101 OSP solder joint.

in electrical resistance. Figure 1.11 shows the fatigue crack growth characteristic of the SnPb_OSP solder joint.

The fibre strain range on the PCB, $\Delta\varepsilon_{pcb}$, may be treated as the indirect fatigue driving force as suggested in Eqn (1.4). An analytical equation has been derived for $\Delta\varepsilon_{pcb}$ as a function of mechanical shock. The peeling stress range in the solder joints, $\Delta\sigma$, is the direct fatigue driving force as suggested in Eqn (1.3). An analytical equation has been established between $\Delta\sigma$ and $\Delta\varepsilon_{pcb}$ assuming a linear system, which facilitates the robust design analysis of solder joints against drop impact. In practice, the solder joints are highly nonlinear, and their flow stress increases with increasing strain rate; hence, the magnitude of $\Delta\sigma$ increases with increasing strain rate. Accurate modelling

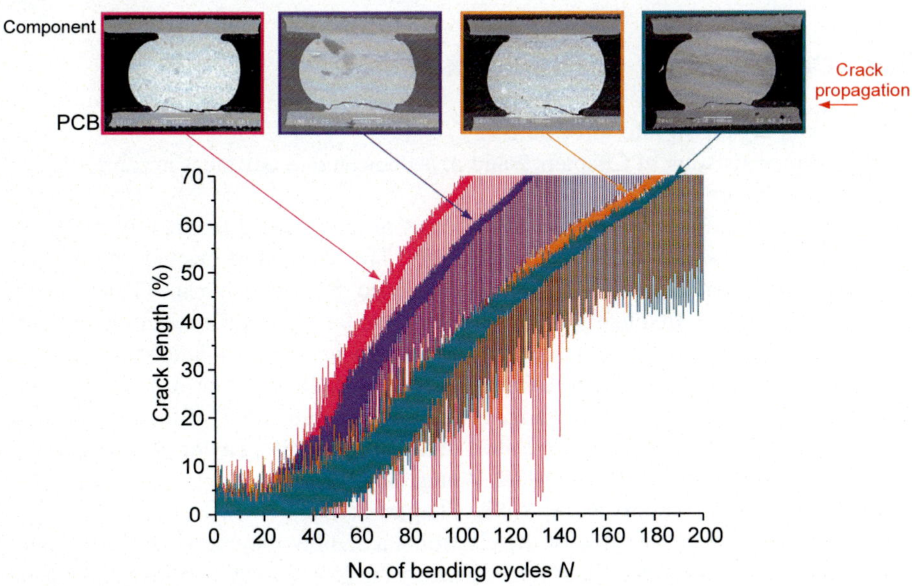

Figure 1.11 Crack propagation characteristics of SnPb_OSP solder joint.

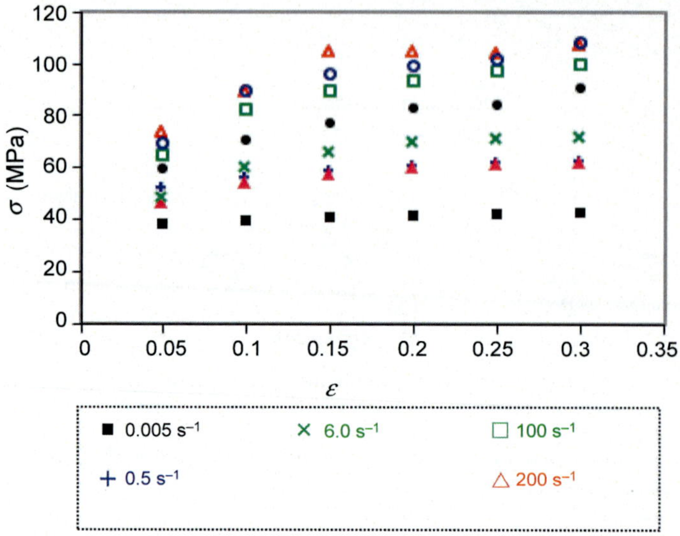

Figure 1.12 Stress—strain characteristics of Sn—37Pb solder expressed in curves of initial strain.

of the magnitude of $\Delta\sigma$ using numerical analysis requires knowledge of the stress—strain relations of solder joints over the relevant range of strain rates as shown in Figure 1.12. Such information is not readily available and must be characterised using specialised experimental techniques.

1.3 Organisation of the book

This book is organised into three parts:

Part One, consisting of Chapters 2 and 3, is dedicated to advances in robust design against temperature.

Chapter 2 describes advances in the analysis of mismatched thermal expansion in microelectronic assembly and in PCB assembly. The analysis of mismatched thermal expansion of layered structure was first reported by Timoshenko (Timoshenko, 1925). In view of the frequently observed interfacial failure of microelectronic assembly, there has been extensive interest in formulating analytical solutions for thermal stresses in layer structure (Chen, Cheng, & Geerhardt, 1982; Chen & Nelson, 1979; Heinrich, Shakya, & Lee, 1997; Jiang, Huang, & Chandra, 1997; Kuo, 1989; Lee & Jasiuk, 1991; Pao & Eisele, 1991; Suhir, 1986, 1988, 1989; Taylor & Yuan, 1962; Wen & Basaran, 2004; Willams, 1985; Wong, Lim, & Mai, 2009), culminating in a few dedicated books (Suhir et al., 1988, Chapter 5; Lau, 1993; Kelly, 1999; Wild, 1973). The layered construction of the electronic assembly may be viewed as a sandwich structure; for example, IC chip bonded to interposer or IC component bonded to PCB. In most of the published articles, these were modelled as a beam that is adhesively bonded to another beam. These analytical solutions have consistently found shearing stress to be the predominant stress, while the peeling stress is of negligible magnitude; this does not agree with numerical analysis that suggests the peeling stress to be the dominant stress (Wong, Kim, et al., 2009). The deviation is attributable to the overly simplified interfacial shear-stress function assumed for the beam–adhesive interface and the overly simplified transverse-stress function assumed for the adhesive layer. This deviation has severely limited the applicability of the existing analytical solutions for robust design. While numerical analysis may be used for robust design analysis, it has a limited range of validity and suffers from stress/strain singularity. An accurate analytical solution of the peeling stress is thus essential. *Chapter 2* introduces an advanced analytical solution in which the interfacial shearing stress is prescribed with a more sophisticated stress function, while the transverse stress in the adhesive layer is prescribed with a stress function that is consistent with the theory of elasticity. This has led to analytical solutions of the interfacial shearing and peeling stresses that are consistent with the numerical solution. Design analysis using the advanced solution has led to many useful insights for robust design of microelectronic assembly against thermally induced failure.

Chapter 3 is concerned with advances in the creep-fatigue modelling of solder joints. Even more extensively researched and published than interfacial stresses is the creep fatigue of solder joints that has led to a number of dedicated books on the topic (Engelmaier, 1989; Frear, Burchett, & Morgan, 1993; Kelly, 1999; Lau, 1991, 1993; Liu & Liu, 2011; Perkins & Sitaraman, 2008; Madenci, Guven, & Kilic, 2002; Suhir et al., 1988, Chapter 5; Tu, 2007; Wild, 1973). The main subject of interest has been the life prediction of solder joints. A number of life prediction models have been proposed and a few of them are heavily used by the IC packaging industry. These models typically come with a number of fitting parameters. In view of the highly empirical nature of the models, the fitting parameters tend to have a limited validity window; for example, different sets of parameters are needed for different time–temperature profiles of temperature cycling even for the same solder joints. It is a daunting task to characterise the fitting parameters for different temperature–time profiles for the many lead-free solder alloys.

The dependence of fitting parameters on the temperature–time profile is indeed evidence of incorrectness of a life prediction model. Metals are known to exhibit a well-defined time–temperature creep characteristic cummulating in a time–temperature transformation relation. Thus, the contribution of creep damage in creep fatigue can be captured through a 'creep function' that is independent of the time–temperature profile of temperature cycling. Based on this concept, a unified equation for creep fatigue has recently been proposed (Wong & Mai, 2014).

Part Two, consisting of Chapters 4 and 5, is dedicated to advances in robust design against moisture.

Chapter 4 focusses on the techniques of characterising solubility, diffusivity and hygroscopic swelling. Unlike temperature-related properties, moisture-related properties of microelectronic materials are difficult to come by and techniques for their characterisations are much less established. The characterisations of solubility and diffusivity are complicated by the presence of non-asymptotic saturation, believed to be attributable to chemisorption. Organic laminated substrate exhibits transversely anisotropic diffusivity, whose characterisation has so far eluded published standards. This is also true for hygroscopic swelling.

Chapter 5 is concerned with advances on diffusion modelling. Fundamental to the analysis of moisture-induced failure in microelectronic packaging is knowledge of the distribution of moisture within the microelectronic assembly. This can be analysed through transient-diffusion modelling. A diffusion problem can be analysed just as a heat-conduction problem using heat-diffusion analogy (Crank, 1980). Through the introduction of 'wetness fraction', defined as $w = C/C_{sat}$, which overcomes the discontinuity of concentration, moisture diffusion in multimaterial microelectronic assembly can be modelled using the heat-conduction module of commercial finite element software (Wong, Teo, & Lim, 1998). At its conceptualisation, fundamental proof of the continuity of 'wetness' was not presented. Thus, a fundamental proof of the continuity of wetness based on the continuity of chemical potential is given in this chapter. The fact that C_{sat}, the saturate concentration of moisture in an absorbent, is a function of temperature renders wetness discontinuous whenever a microelectronic assembly experiences a temperature variation with time. The continuity of wetness can be overcome using the 'internal source' technique.

The last section of *Chapter 5* discusses advances in the modelling of vapour pressure in the delamination of a microelectronic assembly due to the rapid rise of temperature during solder reflow. With the 'wetness' defined, the upper-bound estimation of the vapour pressure in a cavity (delamination) in a microelectronic assembly can be readily established using a simple equation. A more elaborate solution requires a coupled analysis of moisture diffusion and structural deformation.

Part Three, consisting of Chapters 6—12, is dedicated to robust design against drop impact, whose organisation is presented in Figure 1.13.

Chapter 6 describes the physics and mechanics of electrical failure when a PED experiences a drop impact. The ever-higher performance of PEDs is accompanied by the ever-delicate product housing and the ever-smaller feature size of microelectronic assembly. The delicate housing lacks the ability to dissipate the kinetic energy of impact, transferring most of the kinetic energy to the ever-delicate microelectronic and PCB assemblies within the product housing. Electrical failure was almost inevitable, and this was triggered in the early 2000s. In order to understand the relation between thc design of a PED housing and the layout of the PCB assembly within the housing, extensive product-level drop-impact testing was carried out on more than 20 PEDs at various impact orientations. The PCB assembly within the PED was instrumented to extract the resultant acceleration and strain. *Section 6.1* reports the drop-test experiments, the instrumentation and the resultant acceleration. The resultant high magnitude of impact force sets up high magnitude of acceleration and hence high magnitude of inertial force in the PCB assembly, especially in the interconnections (more specifically, the solder joints) of the microelectronic assembly of large mass; the inertia of the PCB imparts differential flexural deformation between the PCB and the surface-mounted microelectronic assembly, putting severe strain on the solder joints that interconnect the microelectronic assembly to the PCB; the close packing of components within the housing leads to knocking of the components against the

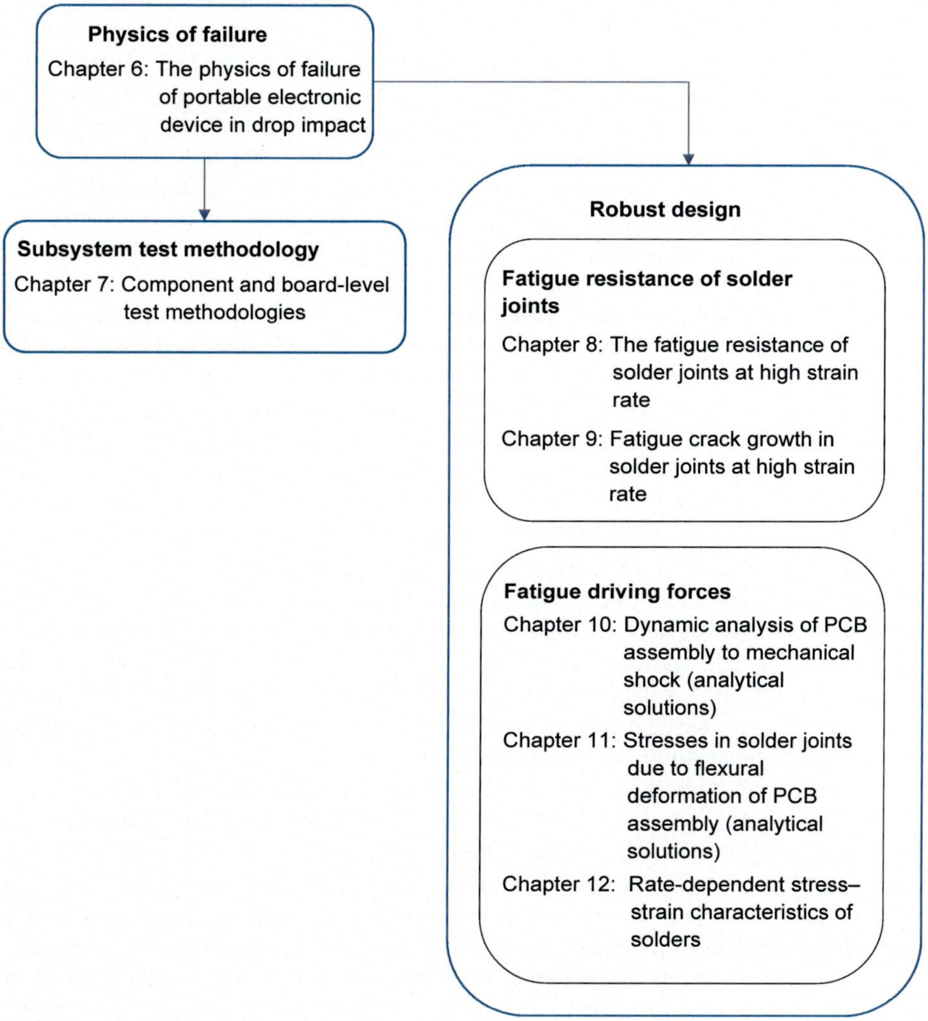

Figure 1.13 Organisation of Part Three.

housing and the inter-knocking of these components, thus inducing high magnitude of impact force, which in turn causes local bending of the PCB assembly. *Section 6.2* investigates these failure mechanisms using a combination of experiments and simple analytical equations.

Chapter 7 is dedicated to investigating the subsystem and the sub-subsystem test methods for microelectronic assembly. In early 2000, the electrical failure of mobile phones upon drop impact caught the microelectronic assembly and the PCB assembly industries by surprise and evolved into a small crisis: how should a microelectronic assembly provider evaluate its component for drop-impact application? It was decided that testing of microelectronic assembly should be independent of the design of the PED since the microelectronic assembly industry has no control over the design of the product. The subsystem test must be capable of replicating the key failure mechanism observed in a product-level drop test. A number of subsystem test (generally referred to as board-level test) methods have since been reported (Goyal, Buratynski, & Elko, 2000; Hirata, Yoshida, & Morisaki, 2001; JEDEC Standard JESD22-B111, 2003; Kim et al., 2001; Marjamäki, Mattila, & Kivilahti, 2006; Mishiro et al., 2002; Nakado, Yaguchi, Terasaki, & Yamamoto, 2006; Peng et al., 2001; Qiang et al., 2002; Reiff & Bradley, 2005; Seah, Wong, Mai, Rajoo, & Lim, 2006; Sogo & Hara, 2001; Varghese & Dasgupta, 2003; Yaguchi, Yamada, & Yamamoto, 2003; Zhu, 2001). The JEDEC standard JESD22-B111 (2003) involves mounting the microelectronic assemblies of interest onto a specified PCB and mounting the PCB assembly onto a rigid fixture and dropping the fixture against a target to a specified acceleration–time pulse. This is technically a board-level drop-shock test (BLDST). While the JEDEC BLDST is credited for standardising the test standard among the industry, it is susceptible to inconsistency (because of the difficulty of reproducing the acceleration–time pulse) and is not necessarily the most efficient test procedure. Fundamentally, the BLDST sets up flexural oscillations of the PCB assembly, and the resulting differential bending between the microelectronic assembly and the PCB leads to fatigue deformation of the interconnecting solder joints at high strain rate. The mechanics of failure can theoretically be replicated by bending the PCB assembly statically and cyclically at high speed (Seah et al., 2006) – a high-speed cyclic bending test (HSCBT), which is far more consistent and efficient. *Section 7.1* describes the design of a specialised high-speed cyclic bend tester and reports the benchmarking of the HSCBT against the JEDEC BLDST.

The subsystem board-level test has been used to evaluate the design of the microelectronic assembly, which includes the layout of the solder joints, the dimensions of the solder joints, the selection of the solder alloy and the metal pad finishing on the microelectronic assembly and the PCB, etc. Once the design has been selected, it is uneconomical to use the system test as a quality test method. A sub-subsystem test method that evaluates the quality of solder joints without assembling the microelectronic assembly onto a board is highly desirable; this has been generally referred to as a component-level test method. The component-level test must be capable of replicating the experience of solder joints in a subsystem board-level test, namely, the brittle-prompt failure of solder joints at high strain rate. A number of component-level test methods were proposed (Date, Shoji, Fujiyoshi, Sato, & Tu, 2004a,b; Lai, Chang, & Yeh, 2007; Newman, 2005; Ou, Xu, Tu, & Alam, 2005; Shoji, Yamarmoto, Kajiwara, Morita, Sato, & Date, 2002; Song et al., 2007; Song, Lee, Newman, Clark, & Sykes, 2007; Song, Lee, Newman, Sykes, & Clark, 2007; Valotaet, Losavioa, Renarda, & Vicenzob, 2006; Wong et al., 2005, 2006, 2008; Yeh, Lai, Chang, & Chen, 2005, 2007; Zhao, Caers, de Vries, Wong, & Rajoo, 2007), and the ball-impact shear method has been incorporated into the JEDEC standard JESD22-B117A (2006). *Section 7.2* presents a review of these test methods, including a comprehensive benchmarking of the ball-impact shear test method with the subsystem board-level test method.

The robust design of a mechanical structure, such as a microelectronic assembly, depends on the tussle between the material resistance to failure and the forces that are driving the materials to failure. The board-level solder joints have been identified as the weakest link in the drop impact of PEDs, and low-cycle fatigue has been identified as the mode of failure. *Chapters 8* and *9* are dedicated to the fatigue resistance of solder joints, while *Chapters 10* and *11* are focused on the failure driving forces in the solder joints.

Chapter 8 is concerned with establishing the fatigue resistance of solder joints through a strain-life equation. Solder joints, which are formed by solder alloys and the metal pad finishing, that are robust against creep fatigue may not be robust against high-strain-rate fatigue. The strain-life characteristic of a combination of leaded and lead-free solder alloys and pad finishing at elevated strain rate have been characterised, which not only facilitates the selection of material system for solder joints but also enables quantitative fatigue life design. The strain life of solder joints is a function of strain rate and temperature — at increasing strain rate and decreasing temperature, the fatigue failure mode of the solder joints transits from ductile failure of the bulk solder to brittle failure of IMC, accompanied by decreasing fatigue life (Wong et al., 2009). The effect of strain rate has been incorporated as a rate function into the strain-life equation, similar to the creep function in creep fatigue, to form a unified equation that facilitates design analysis (Wong, Seah, Caers, & Lai, 2014).

Chapter 9 investigates the fatigue resistance of solder joints through fatigue crack growth. While the expression of fatigue resistance in the form of strain-life facilitates fatigue design, the study of fatigue crack growth in solder joints provides valuable insights that could aid robust design of solder joints. Through the use of a sophisticatedly designed test vehicle and delicate instrumentation, the electronic resistance across a critical solder joint could be monitored to nano-ohm resolution, which translates to micrometer resolution for fatigue crack growth. Fatigue crack in the solder joint was observed to have formed within a single board-level drop shock without a period of crack initiation. When cycled at constant strain range at elevated strain rate, the fatigue crack path in most solder joints tends to cut through a mixture of bulk solder and IMC, resulting in mix-mode fracture, though the ductile—brittle mixity differs between solder joints of different material system. Significantly lower fatigue crack growth rate (da/dN) was observed when the crack was propagating within the ductile bulk solder; solder joints with higher fatigue resistance generally have a longer fatigue crack propagation path within the bulk solder. The fatigue resistance of a solder joint can be significantly improved by deflecting the fatigue crack into the bulk solder, away from the IMC.

The rate of fatigue crack growth (da/dN) increases at small crack length and peaks before declining gradually. Analysing numerically using 3-D fracture mechanics, the magnitude of J-integral around the crack tip has been found to correspond with the observed rate of fatigue crack growth. The declining J-integral at longer fatigue crack length is attributable to the increased compliance of the solder joint. Thus, fatigue resistance of a solder joint can be improved by increasing the compliance of critical solder joints.

Chapter 10 establishes analytically the dynamic response of the PCB assembly due to a board-level drop shock, with the ultimate intention of establishing the maximum magnitude of PCB strain, which is used in Chapter 11 for evaluating the resultant stresses (failure driving force) on solder joints. The PCB was first modelled as a spring–mass system, then as a beam supported at its two ends and subjected to a half-sine shock. The analytical solution for damped vibration was provided. The analytical solutions provide many insights into the shock response of the PCB assembly such as that specified in the JEDEC standard JESD22-B111. For example, the PCB assembly would exhibit transient response followed by harmonic vibration upon being subjected to a shock; however, with the choice of appropriate frequency ratio (the ratio of the natural frequency of the PCB assembly to the shock frequency), the PCB assembly could be made to exhibit almost only transient response without harmonic response. Probably more useful is the insight that the magnitude of response of the test board in the JEDEC standard JESD22-B111 increases linearly with the shock impulse – the area under the acceleration–time curve, independent of the shape of the shock pulse.

Given a prescribed force–time or acceleration–time profile, the boundary condition for a transient dynamic problem is fully described; such problems can be modelled numerically and (for a simple flexible system such as a PCB assembly) even analytically. This is not the case for an impact event with an unspecified impulse. The resultant force time or acceleration time is a function of many parameters. Such problems are usually treated numerically. For reason of comprehensiveness, Chapter 10 provides an introduction to the numerical algorithms for transient dynamics used in the commercial finite element software.

Chapter 11 focusses on establishing analytically the stresses acting on the solder joints interconnecting the microelectronic assembly to the PCB when a PCB assembly undergoes flexural deformation. The differential flexural deformation between the microelectronic assembly and the PCB has to be taken up by the interconnecting solder joints leading to high magnitude of stresses in the solder joints. The analytical solution of stresses in the solder joints is an extension of the advanced model presented in *Chapter 2* for the modelling of interfacial stresses due to mismatched thermal expansion, with an additional term that describes the bending of solder joints. Design analysis using the advanced solution has led to many useful insights for robust design of solder joints against mechanically induced PCB bending. For example, design analysis using the analytical solution has suggested the peeling stress in a critical solder joint can be reduced by increasing its cross-sectional area and its flexural compliance. By reducing the volume of the critical solder joint during solder printing, and hence shaping it into hourglass, it has been shown experimentally that its fatigue life at high strain rate has been doubled.

Chapter 12 develops the dynamic stress–strain characteristics of solder materials. The solder joints experience extensive plastic deformation during the flexural deformation of the PCB assembly in the JEDEC BLDST and in the HSCBT. The plastic strain of metals is highly rate dependent and its evaluation requires good description of the stress–strain characteristic at appropriate strain rates. This chapter describes the experimental characterisation of a number of leaded and lead-free solders using the drop-weight tester. Due to the reducing velocity of the drop weight and the reducing length of test specimen with increasing deformation, it is impossible to maintain a constant strain rate in the specimen during the characterisation; analytical equations have been established that enable the transformation of engineering strain rate to true strain rate. Using regression analysis, the rate-dependent stress–strain equations of the leaded and lead-free solders have been established.

References

Chen, D., Cheng, S., & Geerhardt, T. D. (1982). Thermal stresses in laminated beams. *Journal of Thermal Stresses, 5*, 67–84.

Chen, W. T., & Nelson, C. W. (1979). Thermal stresses in bolted joints. *IBM Journal of Research and Development, 23*, 178–188.

Crank, J. (1980). *The mathematics of diffusion* (2nd ed.). Oxford Science Publications.

Date, M., Shoji, T., Fujiyoshi, M., Sato, K., & Tu, K. N. (2004a). Ductile-to-brittle transition in Sn–Zn solder joints measured by impact test. *Scripta Material, 51*, 641–645.

Date, M., Shoji, T., Fujiyoshi, M., Sato, K., & Tu, K. N. (2004b). Impact reliability of solder joints. In *Proc. 54th electronic components and technology conference* (pp. 668–674).

Engelmaier, W. (1989). Performance considerations: thermal–mechanical effects. In *Section 6: Soldering and mounting technology, electronic materials handbook* (Vol. 1, p. 740). Materials Park, OH: ASM International Packaging.

Frear, D. R., Burchett, S. N., & Morgan, H. S. (1993). *Mechanics of solder alloy interconnects*. Chapman and Hall.

Goyal, S., Buratynski, E. B., & Elko, G. W. (2000). Role of shock response spectrum in electronic product suspension design. *International Journal of Microcircuits and Electronic Packaging, 23*(2), 182–189.

Heinrich, S. M., Shakya, S., & Lee, P. S. (1997). Improved analytical estimate of global CTE mismatch displacement in area-array solder joints. *ASME Transactions Journal of Electronic Packaging, 119*, 218–227.

Hirata, I., Yoshida, Y., & Morisaki, I. (2001). Drop-simulation of electronic boards mounted with CSPs. In *Proc. 1st international conference for electronic packaging* (pp. 422–426).

JEDEC Standard JESD22-B111. (July 2003). *Board level drop test method of components for handheld electronic products*.

JEDEC Standard JESD22-B117A. (October 2006). *Solder ball shear*.

Jiang, Z. Q., Huang, Y., & Chandra, A. (1997). Thermal stresses in layered electronic assemblies. *ASME Journal of Electronic Packaging, 119*, 1127–1132.

Kelly, G. (1999). *The simulation of thermomechanically induced stress in plastic encapsulated IC packages*. Kluwer Academic Publishers.

Kim, J.-Y., Kang, I. S., Park, M. G., Kim, J. H., Cho, S. J., Park, I. S., et al. (2001). Characterization of wafer level package for mobile phone application. In *Proc. 51st electronic components and technology conference* (pp. 1398–1401).

Kuo, A. Y. (1989). Thermal stress at the edge of a bi-metallic thermostat. *ASME Journal of Applied Mechanics, 56*, 585–589.

Lai, Y.-S., Chang, H.-C., & Yeh, C.-L. (2007). Evaluation of solder joint strengths under ball impact test. *Microelectronics Reliability, 47*(12), 2179–2187.

Lau, J. H. (Ed.). (1991). *Solder joint reliability: Theory and applications*. New York: Van Nostrand Reinhold.

Lau, J. H. (Ed.). (1993). *Thermal stress and strain in microelectronics packaging*. Van Nostrand Reinhold.

Lee, M., & Jasiuk, I. (1991). Asymptotic expansions for the thermal stresses in bonded semi-infinite biomaterial strips. *ASME Journal of Electronic Packaging, 113*, 173–177.

Liu, S., & Liu, Y. (2011). *Ch 14: Power and thermal cycling, solder joint fatigue life, modeling and simulation for microelectronic packaging assembly: Manufacturing, reliability and testing*. Wiley online.

Madenci, E., Guven, I., & Kilic, B. (2002). *Fatigue life prediction of solder joints in electronic packages with ANSYS*. Kluwer Academic Publishers.

Marjamäki, P., Mattila, T. T., & Kivilahti, J. K. (2006). A comparative study of the failure mechanisms encountered in drop and large amplitude vibration tests. In *Proc. 56th electronic components and technology conference* (pp. 95–101).

Mishiro, K., Ishikawa, S., Abe, M., Kumai, T., Higashiguchi, Y., & Tsubone, K. (2002). Effect of the drop impact on BGA/CSP package reliability. *Microelectronics Reliability, 42*, 77–82.

Nakado, K., Yaguchi, A., Terasaki, T., & Yamamoto, K. (2006). Fatigue strength of BGA type solder joints between package and printed wiring board of portable device. *JSME, Series A, 49*(2), 220–228.

Newman, K. (2005). BGA brittle fracture – alternative solder joint integrity test methods. In *Proc. 55th electronic components and technology conference* (pp. 1194–1201).

Ou, S., Xu, Y., Tu, K. N., & Alam, M. O. (2005). Micro-impact test on lead-free BGA balls on Au/electrolytic Ni/Cu bond pad. In *Proc. 55th electronic components and technology conference* (pp. 467–471).

Pao, Y. H., & Eisele, E. (1991). Interfacial shear and peel stresses in multi-layered thin stacks subjected to uniform thermal loading. *ASME Journal of Electronic Packaging, 113*, 164–172.

Peng, H., Johnson, R. W., Flowers, G. T., Ricketts, A.-G., Yeager, E. K., Konarski, M. M., et al. (2001). Underfilling fine pitch BGAs. *IEEE Transactions on Electronics Packaging Manufacturing, 24*(4), 293–299.

Perkins, A. E., & Sitaraman, S. K. (2008). *Solder joint reliability – prediction for multiple environment*. Springer.

Qiang, Y. U., Kikuchi, H., Ikeda, S., Shiratori, M., Kakino, M., & Fujiwara, N. (2002). Dynamic behaviour of electronics package and impact reliability of BGA solder joints. In *Proc. inter society conf. on thermal phenomena* (pp. 953–960).

Reiff, D., & Bradley, E. (2005). A novel mechanical shock test method to evaluate lead-free BGA solder joint reliability. In *Proc. 55th electronic components and technology conference* (pp. 1519–1525).

Seah, S. K. W., Wong, E. H., Mai, Y.-W., Rajoo, R., & Lim, C. T. (2006). High-speed bend test method and failure prediction for drop impact reliability. In *Proc. 56th electronic components and technology conference* (pp. 1003–1008).

Shoji, T., Yamarmoto, K., Kajiwara, R., Morita, T., Sato, K., & Date, M. (2002). Study of interface reaction phenomena between Pb-free solder and Cu. In *Proc. 16th JIEP annual meeting, (in Japanese)* (pp. 97–98).

Sogo, T., & hara, S. (2001). Estimation of fall impact strength for BGA solder joints. In *Proc. 1st international conference for electronic packaging* (pp. 369–373).

Song, F. B., Lee, S. W. R., Newman, K., Clark, S., & Sykes, B. (2007). Comparison of joint strength and fracture energy of lead-free solder balls in high speed ball shear/pull tests and their correlation with board level drop test. In *Proc. 9th electronic packaging technology conference* (pp. 450–458).

Song, F. B., Lee, S. W. R., Newman, K., Reynolds, H., Sykes, B., & Clark, S. (2007). Effect of thermal aging on high speed ball shear and pull tests of SnAgCu lead-free solder balls. In *Proc. 9th electronic packaging technology conference* (pp. 463–470).

Song, F. B., Ricky Lee, S. W., Newman, K., Sykes, B., & Clark, S. (2007). High-speed solder ball shear and pull tests vs. board level mechanical drop tests: correlation of failure mode and loading speed. In *Proc. 57th electronic component technology conference* (pp. 1504–1513).

Suhir, E. (1986). Stresses in bi-metal thermostats. *ASME Journal of Applied Mechanics, 53*, 657–660.

Suhir, E. (1988). An approximate analysis of stresses in multilayered elastic thin films. *ASME Journal of Applied Mechanics, 55*, 143–148.

Suhir, E. (1988). Thermal stress failures in microelectronic components – review and extension. In A. Bar-Cohen, & A. D. Kraus (Eds.), *Advances in thermal modeling of electronic components and systems* (Vol. 1, pp. 337–412). New York: Hemisphere.

Suhir, E. (1989). Interfacial stresses in bimetal thermostats. *ASME Journal of Applied Mechanics, 56*, 595–600.

Taylor, T. C., & Yuan, F. L. (1962). Thermal stress and fracture in shear-constrained semiconductor device structures. *IRE Transactions on Electron Devices, 9*, 303–308.

Timoshenko, S. (1925). Analysis of bi-metal thermostats. *Journal of Optical Society of America, 11*, 233–255.

Tu, K.-N. (2007). *Solder joint technology: Materials, properties, and reliability*. Springer.

Tummala, R. (2008). *System on package: Miniaturization of the entire system*. McGraw-Hill.

Tummala, R., Rymaszewski, E. J., & Klopfenstein, A. G. (1997). *Microelectronics packaging handbook* (2nd ed.). Springer.

Valotaet, A. T., Losavioa, A., Renarda, L., & Vicenzob, A. (2006). High speed pull test characterization of BGA solder joints. In *Proc. 7th international conference on thermal, mechanical and multiphysics simulation and experiments in micro-electronics and micro-systems* (pp. 1–6). http://dx.doi.org/10.1109/ESIME.2006.1644005.

Varghese, J., & Dasgupta, A. (2003). Test methodology for impact testing of portable electronic products. In *Proc. 2003 ASME international mechanical engineering congress & exposition, IMECE2003–41844*.

Wen, Y., & Basaran, C. (2004). An analytical model for thermal stress analysis of multi-layered microelectronic packaging. *Mechanics of Materials, 36*, 369–385.

Wild, R. N. (1973). *Some fatigue properties of solders and solder joints*. IBM Federal Systems Division, Electronics Systems Center. http://books.google.co.nz/books/about/Some_Fatigue_Properties_of_Solders_and_S.html?id=Gi7xtgAACAAJ&redir_esc=y.

Willams, H. E. (1985). Asymptotic analysis of the thermal stresses in a two-layer composite with an adhesive layer. *Journal of Thermal Stresses, 8*, 183–203.

Wong, E. H., Lim, K. M., & Mai, Y.-W. (2009). Analytical solutions for PCB assembly subjected to mismatched thermal expansion. *IEEE Transactions on Advanced Packaging, 32*(3), 602–611.

Wong, E. H., & Mai, Y.-W. (2014). A unified equation for creep fatigue. *International Journal of Fatigue, 68*, 186–194.

Wong, E. H., Mai, Y.-W., Rajoo, R., Tsai, K. T., Liu, F., Seah, S. K. W., et al. (2006). Micro impact characterisation of solder joint for drop impact application. In *Proc. 56th electronic component technology conference* (pp. 64–71).

Wong, E. H., Rajoo, R., Mai, Y.-W., Seah, S. K. W., Tsai, K. T., & Yap, L. M. (2005). Drop impact: fundamentals and impact characterisation of solder joints. In *Proc. 55th electronic component technology conference* (pp. 1202–1209).

Wong, E. H., Rajoo, R., Seah, S. K. W., Selvanayagam, C. S., van Driel, W. D., Caers, J. F. J. M., et al. (2008). Correlation studies for component level ball impact shear test and board level drop test. *Microelectronic Reliability, 48*, 1069—1078.

Wong, E. H., Seah, S. K. W., Caers, J. F. J. M., & Lai, Y.-S. (2014). Frequency-dependent strain-life characteristics of Sn1.0Ag0.1Cu solder on cupper pad at high cyclic frequency. *International Journal of Fatigue, 59*, 43—49.

Wong, E. H., Seah, S. K. W., van Driel, W. D., Caers, J. F. J. M., Owens, N., & Lai, Y.-S. (2009). Advances in the drop-impact reliability of solder joints for mobile applications. *Microelectronics Reliability, 49*(2), 139—149.

Wong, E. H., Teo, Y. C., & Lim, T. B. (1998). Moisture diffusion and vapour pressure modeling for electronic packaging. In *Proc. 48th electron. compon. technol. conf., Seattle* (pp. 1372—1378).

Yaguchi, A., Yamada, M., & Yamamoto, K. (2003). Reliability evaluation of solder joints in ball-grid-array-type packages by impact bending test. *Journal of Japan Institute of Electronics Packaging, 6*(4), 314—321.

Yeh, C.-L., Lai, Y.-S., Chang, H.-C., & Chen, T.-H. (2005). Correlation between package-level ball impact test and board-level drop test. In *Proc. 7th electronic packaging technology conference* (pp. 270—275).

Yeh, C.-L., Lai, Y.-S., Chang, H.-C., & Chen, T.-H. (2007). Empirical correlation between package- level ball impact test and board-level drop reliability. *Microelectronics Reliability, 47*(7), 1127—1134.

Yoon, S. W., Hsiao, Y. K., Dzafir, S., Bum, A. Y. C., Choi, W. K., Kim, Y. C., et al. 3D TSV mid-end processes and assembly/packaging technology. *IMAPS-European Microelectronics and Packaging Conference (EMPC-2011), September 12—15, 2011, (Brighton, UK).*

Zhao, X. J., Caers, J. F. J. M., de Vries, J. W. C., Wong, E. H., & Rajoo, R. (2007). A component level test method for evaluating the resistance of Pb-free BGA solder joints to brittle fracture under shock impact. In *Proc. 57th electronic component technology conference* (pp. 1522—1529).

Zhu, L. (2001). Submodeling technique for BGA reliability analysis of CSP packaging subjected to an impact loading. In *Proc. InterPack 2001, IPACK2001—15873.*

Part One

Advances in robust design against temperature-induced failures

Robust design of microelectronic assemblies against mismatched thermal expansion

2

2.1 Introduction

Microelectronic assembly is made up of a multilayer structure. The integrated circuit (IC) component is made of a silicon chip that is interconnected to the substrate through either the die-attach adhesive for wire-bonded packaging or the solder joints reinforced with underfill for flip-chip packaging. The printed circuit board (PCB) assembly is made of IC components that are interconnected to the PCB by solder joints. The layers are made of different materials with different thermal expansion coefficients. As the microelectronic assembly undergoes thermal excursion during the assembly processes, or the accelerated test, or field use, individual layers undergo different thermal expansion. The mismatch in the expansion leads to internal stresses, and failure along the bonding interfaces is frequently reported. An IC component can be modelled most simply as a trilayer structure. The analytical solutions for thermal stress in a trilayer structure are presented herein.

The simplest method for estimating the interfacial stress in a trilayer structure is to assume that the layers are absolutely rigid such that each layer will experience expansion given by $\alpha \cdot \Delta T \cdot L$, where L is the half-length of the assembly. The differential expansion leads to interfacial shear stress. However, this approach is obviously too simplified. Reviews of various studies in thermoelasticity and applications to microelectronic packaging have been reported by Suhir (1988), Wen and Basaran (2004) and Jiang, Huang, and Chandra (1997). The problem of thermoelastic stresses was first addressed by Poisson in 1829. However, it was Duhamel who presented the equations of thermoelasticity in their modern forms in 1838. Thermodynamic theory was used to evaluate thermal stress by Kelvin in 1878. General methods for the solution of two- and three-dimensional problems of thermoelasticity were developed by Borchardy in 1873 and Rayleigh in 1901. The earliest treatment of bimetallic structure can be traced to Stoney (1909, p. 172). The problem of thermally induced deformation arising in bimetal thermostats subjected to uniform temperature loading was addressed by Timoshenko (1925). He used the elementary beam theory to obtain the curvature of a bimetallic beam due to a uniform temperature change. A more advanced treatment was provided by Aleck (1949), wherein the bi-material structure, bonded along the edges, was modelled using the two-dimensional theory of elasticity. The first analysis of the trilayer structure, in which the middle layer was an adhesive, was performed by Volkersen (1938) for a mechanically loaded single lap-shear joint. The trilayer structure was assumed to experience no bending and the adhesive was modelled as a distribution of shear

Robust Design of Microelectronics Assemblies Against Mechanical Shock, Temperature and Moisture
http://dx.doi.org/10.1016/B978-1-84569-528-6.00002-2

springs, experiencing only uniform through-thickness shear stress. Improved analysis of the single lap-shear structure incorporating bending effects was developed by Goland and Reissner (1944) who performed analysis for two cases of adhesives of two extreme compliances: flexible and inflexible. The adherends were modelled as beams. In the case of a flexible adhesive, the adhesive layer was modelled as a system of infinitesimal coil springs capable of transferring both shear and normal stresses, which were assumed to be uniform in the through-thickness of the adhesive; the condition of zero shear stress at the free edge was not satisfied. In the case of an inflexible adhesive, the adhesive layer was modelled using the elaborate theory of two-dimensional elasticity and the condition of nil shear stress at the free edge was satisfied.

The analysis of interfacial thermal stress in a tri-material structure in a semiconductor device subjected to temperature loading was first performed by Taylor and Yuan (1962), following the lap joint theory of Volkersen – the bonding layer was assumed as a distribution of shear springs and the structure was assumed to experience no bending. Chen and Nelson (1979) extended the thermal stress analysis, using Volkersen's theory, to more general sandwiched structures typical of microelectronics, including the penta-layer structure of three elastic layers with two bonding layers and circular trilayer structure with the middle layer as a bonding layer. Suhir (1986, 1988, 1989) improved upon Timoshenko's bi-thermostat beam theory with relatively simple calculations using longitudinal and transverse interfacial compliances. Pao and Eisele (1991) extended Suhir's model to multilayered thin stacks without imposing any additional assumptions on the interface. Jiang et al. (1997) improved the solution of the transverse interfacial stress by Suhir. Wong, Lim, and Mai (2009) incorporated the contribution of the adhesive thickness to in-plane compliance of the system and hence improved the accuracy of the shear stress solution. (See the Appendix at the end of this chapter for a summary of the major prior studies.)

The analysis of thermal stresses can be categorised into two approaches. In the first, pioneered by Timoshenko (1925), each layer is treated as a beam. Examples of this approach are Chen and Nelson (1979), Suhir (1986, 1988, 1989), Pao and Eisele (1991), Jiang et al. (1997), Heinrich (1997) and Wong et al. (2009), among others. This approach has the advantage of providing a simple solution. In the second approach, the layered structure is modelled as a continuum. Examples of this approach are Chen, Cheng, and Gerhardt (1982), Willams (1985), Kuo (1989) and Lee and Jasiuk (1991), among others. This method is more elaborate but does not provide a simple, easy-to-use analytical solution needed by engineers for design analysis.

The sandwiched (or intermediate or interconnecting) layer in the three-layer structure of a microelectronic assembly may be a continuous layer such as die-attached or underfilled flip-chip solder joints. The sandwiched layer may also be a discrete layer such as in non-underfilled board-level solder joints. In the former case, only shear and peeling stresses can be transferred across the interface. In the latter case, shear force, peeling force and moment are transferred across the interface. Analytical solutions for these two cases are presented.

2.2 Fundamentals

The theories of bending of a simple beam with and without a sectional force are revisited and analytical equations for simplified bilayer structure are developed in this section as foundations for the more advanced analysis in later sections.

2.2.1 Bending of a homogeneous beam

2.2.1.1 Pure bending (plane remains plane)

We have learnt from elementary strength of materials that the *conditions for compatibility* require that a beam which experiences pure bending will have its plane cross-sections remain plane and perpendicular to the longitudinal axis (Figure 2.1). The consequences of these conditions are:

- a beam that is deforming in the x–z plane must have its cross-section symmetric about the plane $y = 0$, where the y-axis passes through the neutral axis of the beam;
- the beam deforms into a perfect arc in the x–z plane with a radius of curvature ρ; and
- the strain of an element at distance z from the undeformed neutral axis is: $\varepsilon_z(z) = -z/\rho$.

The *constitutive equations* are

$$\begin{aligned} \varepsilon_x &= \frac{\sigma_x - \nu(\sigma_y + \sigma_z)}{E}, & \gamma_{xz} &= 0 \\ \varepsilon_y &= \frac{\sigma_y - \nu(\sigma_z + \sigma_x)}{E}, & \gamma_{yx} &= 0 \\ \varepsilon_z &= \frac{\sigma_z - \nu(\sigma_x + \sigma_y)}{E}, & \gamma_{zy} &= \frac{\tau_{zy}}{G} \end{aligned} \tag{2.1}$$

The condition $\gamma_{zy} = \tau_{zy}/G$ suggests that the beam could experience pure torsion about its x-axis. However, this is provided the condition 'plane remains plane' is not violated. Assuming σ_z is negligible, then

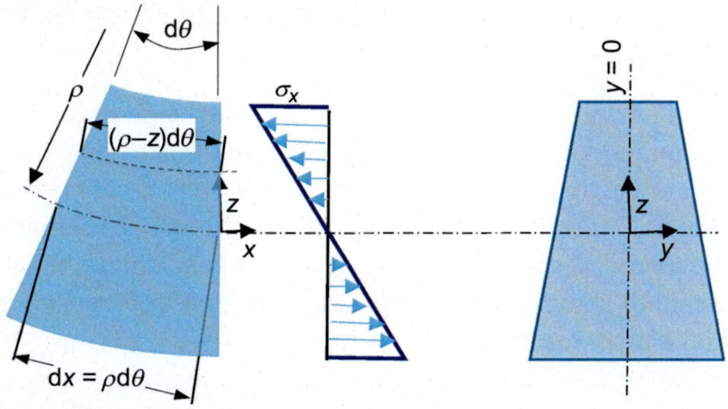

Figure 2.1 Plane remains plane and perpendicular to the longitudinal axis.

$$\sigma_x = E\varepsilon_x = -\frac{Ez}{\rho} \quad \text{for plane stress when } \sigma_y \text{ and } \sigma_{yz} = 0 \tag{2.2}$$

$$\sigma_x = \frac{E\varepsilon_x}{1-\nu^2} = -\frac{Ez}{(1-\nu^2)\rho} \quad \text{for plane strain when } \varepsilon_y \text{ and } \gamma_{yz} = 0 \tag{2.2a}$$

We shall henceforth write in the form of Eqn (2.2) for both plane stress and plane strain, but bearing in mind that:

$$E = \begin{cases} E & \text{for plane stress} \\ \dfrac{E}{1-\nu^2} & \text{for plane strain} \end{cases}.$$

The *equilibrium of forces* in the x-direction defines the position of the neutral axis:

$$\int_{\text{Area}} \sigma_x \mathrm{d}A = -\int_{\text{Area}} \frac{Ez}{\rho} \mathrm{d}A = -\frac{E}{\rho} \int_{\text{Area}} z\mathrm{d}A = 0 \tag{2.3}$$

The neutral axis, the longitudinal axis that experiences no deformation, passes through the centroid of the cross-section of the beam.

The *equilibrium of moment* for the differential element about the y-axis defines the moment–curvature relation:

$$M = -\int_{\text{Area}} z(\sigma_x \mathrm{d}A) = \frac{E}{\rho} \int_{\text{Area}} z^2 \mathrm{d}A = \frac{EI}{\rho} \tag{2.4}$$

Thus, the following relations are obtained:

$$\frac{M}{I} = \frac{E}{\rho} = -\frac{\sigma_x}{z} \tag{2.5}$$

2.2.1.2 Bending in the presence of sectional force (plane does not remain plane)

In the presence of a nonzero force in the z-direction, the beam experiences a sectional force Q and an accompanying shear stress τ_{xz}. The condition of pure bending is strictly invalid because plane section will not remain plane; however, the magnitude of out-of-plane deformation is generally very small, and the moment–curvature relation of Eqn (2.5) may be assumed valid. Referring to Figure 2.2, where the only unbalanced forces and moment are shown, the *equilibrium condition of moment* about the y-axis for the differential element $h \cdot \mathrm{d}x$ gives

$$\mathrm{d}M = -Q\mathrm{d}x \tag{2.6}$$

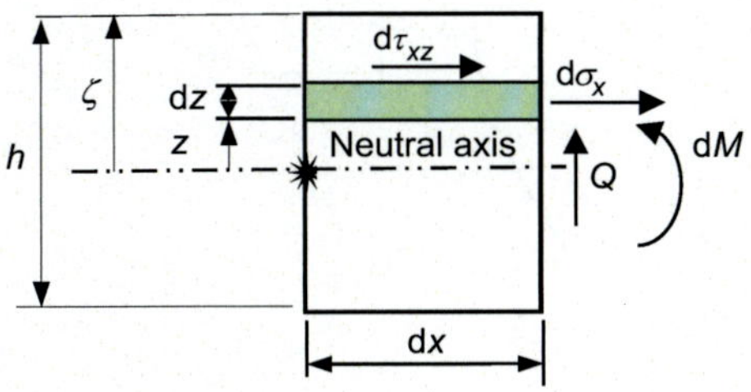

Figure 2.2 Equilibrium of a differential element with sectional force.

The *equilibrium of forces* in the x-direction for a differential element $\mathrm{d}x\cdot\mathrm{d}z$ gives

$$\mathrm{d}\tau_{xz} + \frac{\partial \sigma_x}{\partial x}\mathrm{d}z = 0 \tag{2.7}$$

Substituting Eqns (2.5) and (2.6) into Eqn (2.7) gives

$$\mathrm{d}\tau_{xz} + Q\frac{z\mathrm{d}z}{I} = 0 \text{ or } \tau_{xz} = -\frac{Q}{I}\int z\mathrm{d}z \tag{2.8}$$

Using the boundary condition $\tau_{xz}|_{z=\zeta} = 0$ for Eqn (2.8), the shear stress at distance z from the neutral axis is

$$\tau_{xz}(z) = \frac{Q}{2I}(\zeta^2 - z^2) \tag{2.9}$$

Thus, the beam experiences maximum shear stress in the cross-section along its neutral axis. The rotation of the neutral axis due to shearing is given by

$$\phi = \left.\frac{\tau_{xz}}{2G}\right|_{z=0} = \frac{Q\zeta^2}{4GI} \tag{2.10}$$

where G is the shear modulus. This is superimposed onto the rotation of the neutral axis of the beam due to bending, θ. Notice that (1) the cross-section of the beam is now at an angle 2ϕ from the normal to the longitudinal axis; and (2) $\mathrm{d}u_s/\mathrm{d}z = \mathrm{d}w/\mathrm{d}x = \varepsilon_{xz} = \tau_{xz}/2G$, and the x-direction displacement of the cross-section with respect to the neutral axis becomes

$$[u_s]_0^z = \frac{1}{2G}\int_0^z \tau_{xz}\mathrm{d}z = \frac{Q\zeta^3}{4GI}\left(Z - \frac{Z^3}{3}\right) \tag{2.11}$$

where $Z = z/\zeta$. This is illustrated in Figure 2.3. It is hence clear that the conditions that plane remains plane and normal to the neutral axis are strictly invalid. However, the magnitude of deviation is usually too small to exclude the applicability of Eqn (2.5).

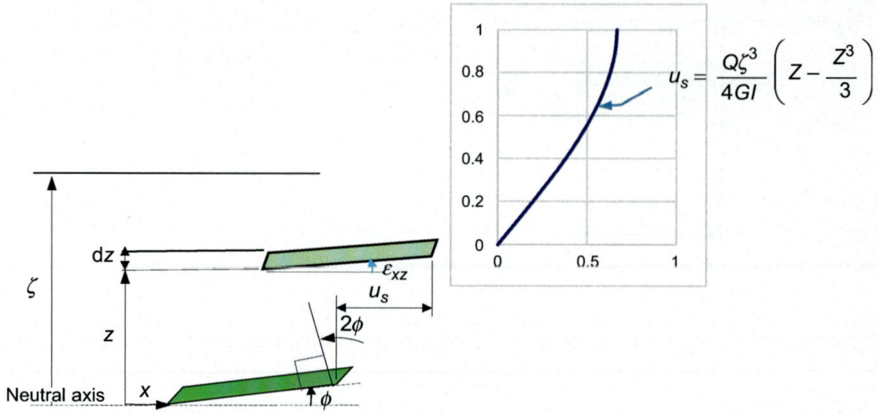

Figure 2.3 Shear-induced rotation and deformation.

2.2.2 Stretching of a bilayer (material) structure of absolute flexural rigidity and assuming no interfacial shear stress

The case of a bilayer structure experiencing mismatched thermal expansion is examined. It is assumed that the structure has absolute flexural rigidity and experiences no interfacial shear stress. The only mode of deformation is stretching. The free-body diagram of such a bi-material structure is shown in Figure 2.4.

The mismatch between the two members leads to traction F_i, which acts along the centroid (longitudinal) axis of the member. The force–strain relation is given by

$$F_i = \frac{1}{\lambda_{xi}}\left(\frac{\mathrm{d}u_i}{\mathrm{d}x} - \alpha_i \Delta T\right) \tag{2.12}$$

where u_i is the x-direction displacement of the longitudinal axis of member #i; α_i is the thermal expansion coefficient of member #i; λ_{xi} is the stretch compliance of the

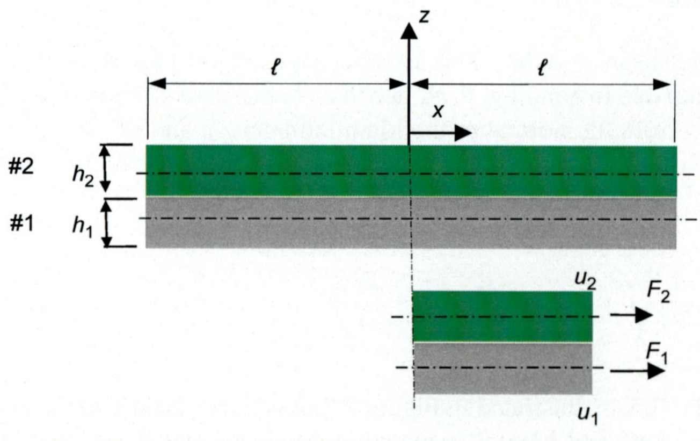

Figure 2.4 A bi-material structure with absolute flexural and shear rigidity.

member defined as the force (per unit width) required to cause unit mechanical strain in the longitudinal axis and

$$\lambda_{xi} = \begin{cases} \dfrac{1}{E_i h_1} & \text{for plane stress} \\[2ex] \dfrac{1-\nu^2}{E_i h_1} & \text{for plane strain} \end{cases} \tag{2.13}$$

In the absence of external forces, the condition of *equilibrium* gives

$$\frac{1}{\lambda_{x1}}\left(\frac{\mathrm{d}u_1}{\mathrm{d}x} - \alpha_1 \Delta T\right) + \frac{1}{\lambda_{x2}}\left(\frac{\mathrm{d}u_2}{\mathrm{d}x} - \alpha_2 \Delta T\right) = 0 \tag{2.14}$$

The condition of *compatibility* gives

$$u_1 = u_2 = u \tag{2.15}$$

Substituting Eqn (2.15) into Eqn (2.14) yields:

$$\frac{\mathrm{d}u}{\mathrm{d}x} = \frac{\lambda_{x2}\alpha_1 \Delta T + \lambda_{x1}\alpha_2 \Delta T}{\lambda_{x1} + \lambda_{x2}} \tag{2.16}$$

Inserting Eqn (2.16) into Eqn (2.12) gives the traction force as

$$F = \frac{\alpha_{21}\Delta T}{\lambda_x} \tag{2.17}$$

where $\alpha_{21} = \alpha_2 - \alpha_1$ and $\lambda_x = \lambda_{x1} + \lambda_{x2}$. Thus, the beams experience a uniform magnitude of traction along their length, except at $x = 0$ and $x = L$.

2.2.3 Bending of a bilayer (material) structure assuming no interfacial shear stress (planes remain planes)

If we assume that the plane cross-sections of a bilayer structure remains plane and perpendicular to the longitudinal axis of the structure, then the same compatibility conditions stated for the pure bending of a homogeneous beam apply to the composite beam, that is:

- the beam deforms into a perfect arc in the x–z plane with a radius of curvature ρ;
- the strain of an element at distance z from the undeformed neutral axis is $\varepsilon_z(z) = -z/\rho$.

This together with the constitutive equation gives the bending stress as

$$\sigma_x = E_i \varepsilon_x = -\frac{E_i z}{\rho}; \quad i = 1, 2, \ldots n \tag{2.18}$$

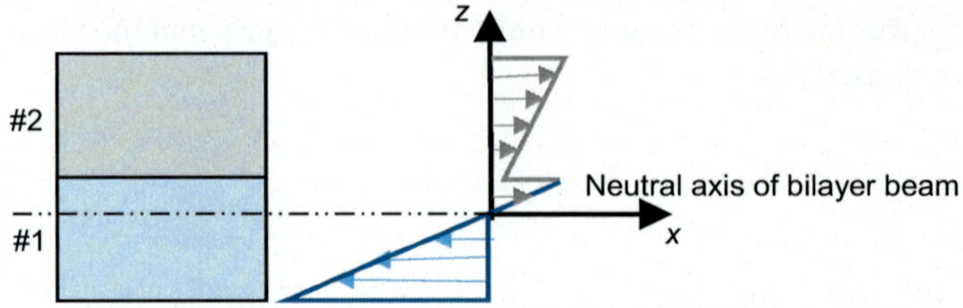

Figure 2.5 Bending stress distribution across the thickness of a bilayer composite beam.

The bending stress profile of a bilayer beam is illustrated in Figure 2.5. It should be noted that the stress is linearly distributed within each material but experiences discontinuity at the interface between two materials.

The *equilibrium of forces* in the *x*-direction defines the neutral axis of the beam in bending:

$$\sum_{i=1}^{n} \int_{\text{Area } i} \sigma_x \mathrm{d}A_i = \sum_{i=1}^{n} E_i \int_{\text{Area } i} z \mathrm{d}A_i = 0 \tag{2.19}$$

Thus, the neutral axis passes through not the axis of centroid but the elastic modulus-weighted centroid. The *equilibrium of moment* about the *y*-axis defines the moment–curvature relation:

$$M = -\sum_{i=1}^{n} \int_{\text{Area}} z(\sigma_x \mathrm{d}A) = \frac{1}{\rho} \sum_{i=1}^{n} E_i \int_{\text{Area}} z^2 \mathrm{d}A_i = \frac{1}{\rho} \sum E_i I_{iG} \tag{2.20}$$

where I_{iG} is the second moment of area of member #*i* about the neutral axis of the composite beam. Thus, the flexural rigidity of a composite beam is given by the sum of the flexural rigidity of the individual layer, with respect to the neutral axis of the composite beam. Designating D_s as $\sum E_i I_{iG}$, Eqn (2.20) is rearranged as

$$\frac{M}{D_s} = \frac{1}{\rho} = -\frac{\varepsilon_x}{z} = -\frac{\sigma_{xi}}{E_i z} \tag{2.21}$$

Figure 2.6 shows the bilayer beam deforming into a negative curvature with absolute shear rigidity for $\alpha_2 > \alpha_1$.
Note that in this case, $M/D_s = -\mathrm{d}\theta/\mathrm{d}x = -1/\rho$. The traction is related to the stretching of the member through the equation:

$$F_i \cos\theta = \frac{1}{\lambda_{xi}} \left(\frac{\mathrm{d}u_i}{\mathrm{d}x} - \alpha_i \Delta T \right) \tag{2.22}$$

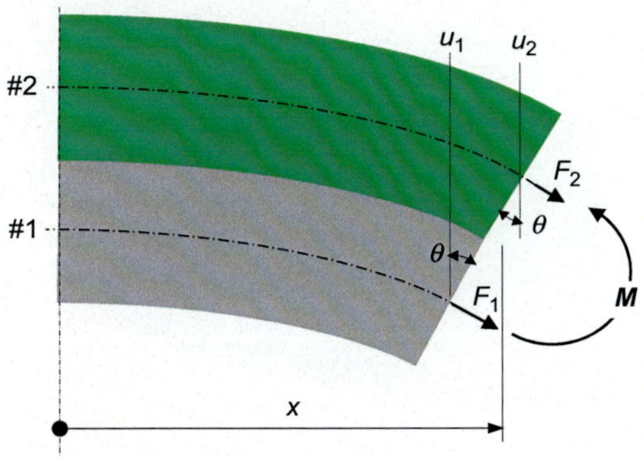

Figure 2.6 Pure bending of a bilayer beam — negative curvature.

The condition of equilibrium is given by Eqn (2.14) and is repeated below

$$\frac{1}{\lambda_{x1}}\left(\frac{\mathrm{d}u_1}{\mathrm{d}x} - \alpha_1 \Delta T\right) + \frac{1}{\lambda_{x2}}\left(\frac{\mathrm{d}u_2}{\mathrm{d}x} - \alpha_2 \Delta T\right) = 0 \tag{2.14bis}$$

The condition of compatibility gives

$$u_1 + \frac{h_1 \sin\theta}{2} = u_2 - \frac{h_2 \sin\theta}{2} \tag{2.23}$$

or, for small angle, we have

$$u_1 + \frac{h_1\theta}{2} = u_2 - \frac{h_2\theta}{2} \tag{2.23a}$$

Differentiating Eqn (2.23a) with respect to x and substituting into Eqn (2.14) gives

$$\frac{\mathrm{d}u_1}{\mathrm{d}x} = \frac{\Delta T(\lambda_{x1}\alpha_2 + \lambda_{x2}\alpha_1) - \lambda_{x1}X}{\lambda_{x1} + \lambda_{x2}} \tag{2.24}$$

where $X = (h_1 + h_2/2)(\mathrm{d}\theta/\mathrm{d}x)$. Substituting Eqn (2.24) into Eqn (2.22) and putting $\cos\theta \approx 1$ gives

$$F_1 = F = \frac{\alpha_{21}\Delta T - X}{\lambda_{x1} + \lambda_{x2}} \tag{2.25}$$

where $\alpha_{21} = \alpha_2 - \alpha_1$. The equilibrium of moment of the bilayer beam gives

$$M + \frac{F(h_1 + h_2)}{2} = 0 \tag{2.26}$$

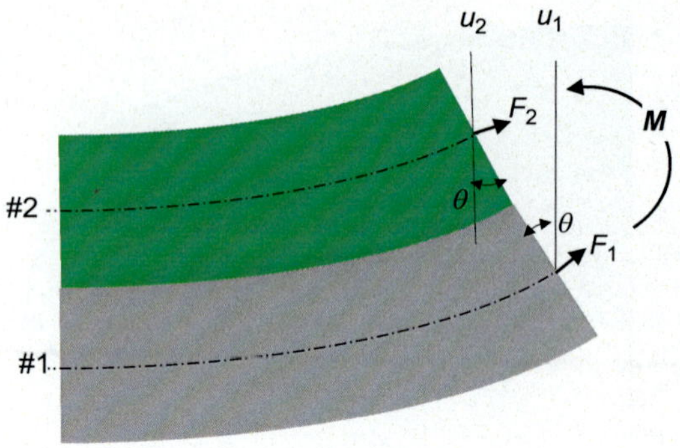

Figure 2.7 Pure bending of a bilayer beam – positive curvature.

and substituting into the moment–curvature relation gives

$$\frac{d\theta}{dx} = -\frac{M}{D_s} = \frac{F(h_1 + h_2)}{2D_s} \tag{2.27}$$

Note that $d\theta/dx$ decreases with increasing M. Substituting Eqn (2.27) into X gives $X = -F\lambda_{x\theta}$, where $\lambda_{x\theta} = (h_1 + h_2)^2/4D_s$, $D_s = \sum E_i I_{iG}$; and Eqn (2.25) becomes

$$F = \frac{\alpha_{21}\Delta T}{\lambda_x} \tag{2.28}$$

where $\lambda_x = \lambda_{x1} + \lambda_{x2} + \lambda_{x\theta}$, independent of x. Comparing with Eqn (2.17) suggests the ability of the bi-material to undergo flexural deformation has effectively increased its stretch compliance by $\lambda_{x\theta} = (h_1 + h_2)^2/4D_s$.

If we were to show the structure to deform into a positive curvature as shown in Figure 2.7, then, working through the same procedure as described above leads to

$$F_2 = \frac{\alpha_{21}\Delta T - X}{\lambda_{x1} + \lambda_{x2}} \tag{2.25a}$$

The equilibrium of moment of the bilayer beam together with the moment–curvature relation gives

$$\frac{d\theta}{dx} = \frac{M}{D_s} = \frac{F_2(h_1 + h_2)}{2D_s} \tag{2.27a}$$

which still leads to the same equation, that is Eqn (2.28).

2.2.4 Deformation of a bilayer (material) structure of absolute flexural rigidity

It is assumed that the structure has absolute flexural rigidity such that individual members could experience only stretching and shearing. The free-body diagram of a bi-material structure is shown in Figure 2.8.

Differential stretching between the two members leads to interfacial shear stress f_s giving rise to a resultant differential traction dF in the differential element $h \cdot \mathrm{d}x$, whose magnitude is given by

$$\mathrm{d}F_i = \mp f_s \mathrm{d}x \tag{2.29}$$

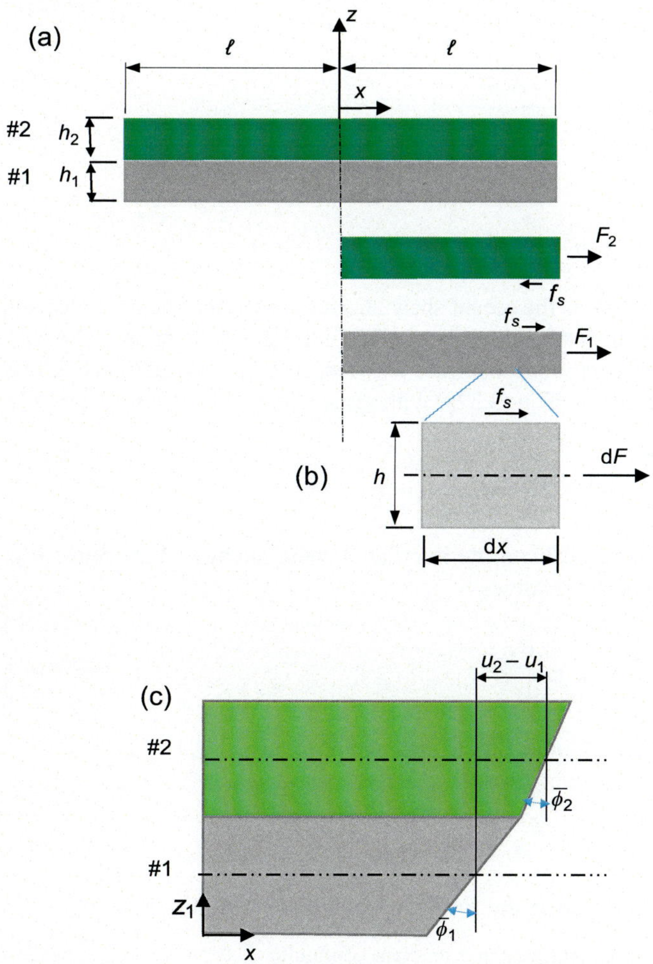

Figure 2.8 A bi-material structure with absolute flexural and rigidity: (a) free-body diagram, (b) a differential element, and (c) enlarged deformation.

where the top and bottom signs in $\mp$ are associated with member #1 and member #2, respectively. Applying Eqn (2.12) to the two members, respectively, and taking the difference,

$$\frac{d(u_2 - u_1)}{dx} = -(\lambda_{x1} + \lambda_{x2})F + \alpha_{21}\Delta T \tag{2.30}$$

where $F = F_1 = -F_2$. Differentiating Eqn (2.30) with respect to x and substituting with Eqn (2.29) gives

$$\frac{d^2(u_2 - u_1)}{dx^2} = (\lambda_{x1} + \lambda_{x2})f_s = \lambda_x f_s \tag{2.31}$$

The condition of compatibility dictates that

$$u_2 - u_1 = \frac{\overline{\phi}_1 h_1 + \overline{\phi}_2 h_2}{2} \tag{2.32}$$

or

$$u_2 - u_1 = \kappa_{s2} f_s + \kappa_{s1} f_s = \kappa_s f_s \tag{2.32a}$$

where $\overline{\phi}_i = cf_s/G_i$ is the secant shear strain between the shear surface and the longitudinal axis of member #i, and $\kappa_{si} = ch_i/2G_i$ is the shear compliance of the layer defined as the displacement of the shear surface with respect to its longitudinal axis due to unit shear stress. Differentiating Eqn (2.32a) with respect to x and substituting into Eqn (2.30):

$$\kappa_s \frac{df_s}{dx} = -\lambda_x F + \alpha_{21}\Delta T \tag{2.33}$$

We proceed to evaluate the interfacial shear stress f_s. Differentiating Eqn (2.32a) twice with respect to x gives

$$\frac{d^2(u_2 - u_1)}{dx^2} = \kappa_s \frac{d^2 f_s}{dx^2} \tag{2.34}$$

Equating Eqn (2.31) and Eqn (2.34), we have

$$\frac{d^2 f_s}{dx^2} - \beta^2 f_s = 0 \tag{2.35}$$

where $\beta^2 = \lambda_x/\kappa_s$. Using the following boundary conditions:

- $df_s(0)/dx = 0$, symmetry and gradient continuity
- $\kappa_s df_s(L)/dx = -\lambda_x F(L) + \alpha_{21}\Delta T = \alpha_{21}\Delta T$, from Eqn (2.33),

the solution to Eqn (2.35) is given by

$$f_s = A_s \frac{\cosh \beta x}{\sinh \beta L} \tag{2.36}$$

where

$$A_s = \frac{\alpha_{21}\Delta T}{\beta \kappa_s} = \frac{\alpha_{21}\Delta T}{\sqrt{\lambda_x \kappa_s}} \tag{2.37}$$

Thus, the magnitude of the shear stress decreases with increasing stretch and shear compliances of the structure.

At $x = 0$, $f_s(0) = (\alpha_{21}\Delta T/\beta\kappa_s)(1/\sinh \beta L)$. For a bilayer structure with large βL, say $\beta L > 3$, $f_s(0) \approx 0$. In such cases, the boundary conditions may be assumed to be:

- $f_s(0) = 0$, and
- $\kappa_s \mathrm{d}f_s(L)/\mathrm{d}x = \alpha_{21}\Delta T$,

which lead to the approximate solution:

$$\tilde{f}_s = A_s \frac{\sinh \beta x}{\cosh \beta L} \tag{2.38}$$

Equation (2.38) has been used in a number of articles (Jiang et al., 1997; Suhir, 1988; Wong et al., 2009). Note that while it satisfies the symmetry requirement about $x = 0$, it does not satisfy the gradient continuity at $x = 0$.

The constant of proportionality c in the expression, $\overline{\phi}_i = cf_s/G_i$, is evaluated as follows: Let there be no shear rotation of the longitudinal axis of the beam, then the shear strain at any distance z_i (referring to Figure 2.8(c)) is

$$\phi_i = \frac{\mathrm{d}u_i}{\mathrm{d}z_i} = \frac{\tau}{G_i} \tag{2.39}$$

Assume the shear stress in member #i varies linearly from the magnitude f_s on the shear surface to nil at the opposite surface, the displacement of the shear surface with respect to its longitudinal axis is

$$\Delta u_i|_{Z=1-0.5} = \int_{0.5}^{1} \frac{h_i f_s Z}{G_i} \mathrm{d}Z = \frac{3h_i f_s}{8G_i} \tag{2.40}$$

where $Z = z_i/h_i$. The ratio $\Delta u_i|_{Z=1-0.5}/f_s$ equates to κ_{si}. Thus, $\kappa_{si} = 3h_i/8G_i$ and $c = 3/4$.

2.3 Comprehensive analysis of a bilayer structure

The individual member in a bilayer structure is assumed to obey the moment–curvature relation of a simple beam in global deformation while behaving as a 2-D elastic body in local deformation. The flexural, shear, transverse and stretch compliances of the members are developed in this section, and the interfacial stresses in a bilayer structure due to mismatched thermal expansion are expressed as a function of these compliances.

2.3.1 The equilibrium and constitutive relations

A shear stress f_s accompanied by a normal stress f_a must necessarily develop along the interface of the bi-material so as to enforce the compatibility of displacement. Figure 2.9 shows the free-body diagram of the two members.

The global equilibrium of the bilayer structure gives the following conditions:

$$F_1 + F_2 = 0 \tag{2.41}$$

$$Q_{a1} + Q_{a2} = 0 \tag{2.42}$$

$$M = M_1 + M_2 \tag{2.43}$$

The interfacial shear stress f_s gives rise to a resultant differential force dF in the longitudinal direction whose magnitude is given by

$$\mathrm{d}F_i = \mp f_s \mathrm{d}x \tag{2.29bis}$$

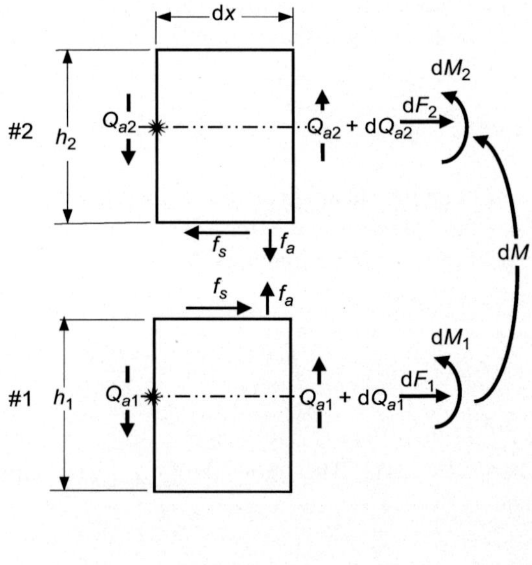

Figure 2.9 Free-body diagram of a differential element in member #1 and member #2.

Similarly, the interfacial normal force leads to a differential sectional force dQ_a whose magnitude is

$$dQ_{ai} = \mp f_a dx \tag{2.44}$$

Taking moment about the centroid axis of the differential element $h \cdot dx$ marked with ✵, the sectional force and the surface shear stress induce a moment in member #i:

$$dM_i = \left(\frac{f_s h_i}{2} - Q_{ai}\right) dx \tag{2.45}$$

which in turn causes changes in the curvature of the member – assuming the moment–curvature relation for pure bending remains applicable in this case:

$$\frac{1}{D_i}\frac{dM_i}{dx} = -\frac{d^2\theta_i}{dx^2} = \frac{d^3 w_{bi}}{dx^3} \tag{2.46}$$

where $D_i = E_i I_i$, and w_{bi} is associated with the deflection of the centroid axis of member #i due to bending. Combining Eqns (2.45) and (2.46) gives

$$\frac{d^2\theta_i}{dx^2} = \frac{1}{D_i}\left(Q_{ai} - \frac{f_s h_i}{2}\right) = \frac{1}{D_i}\left(\pm Q_a - \frac{f_s h_i}{2}\right) \tag{2.47}$$

and differentiating Eqn (2.47) with respect to x gives

$$\frac{d^4 w_{bi}}{dx^4} = -\frac{d^3\theta_i}{dx^3} = \frac{1}{D_i}\left(\frac{h_i}{2}\frac{df_s}{dx} \pm f_a\right) \tag{2.48}$$

2.3.2 *The governing equations*

Figure 2.10 shows the deformed state of a bilayer beam for $\alpha_2 > \alpha_1$. The angles θ and ϕ are associated with the rotation of the axis due to bending and shearing, respectively, and the angle $\overline{\phi}$ is associated with the shear-induced displacement of the surface with respect to its centroid axis.

2.3.2.1 *Enforcing u-compatibility*

Referring to Fig. 2.10, the condition of u-compatibility suggests that

$$u_2 - u_1 = \frac{h_1 \sin(\theta_1 + \overline{\phi}_1)}{2\cos\overline{\phi}_1} + \frac{h_1 \sin(\theta_2 + \overline{\phi}_2)}{2\cos\overline{\phi}_2} \tag{2.49}$$

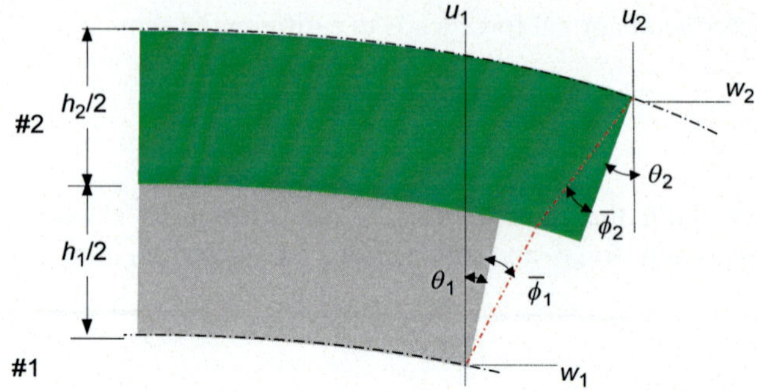

Figure 2.10 Bending and shearing of a bilayer beam: deformed state.

or for small values of θ_i and $\overline{\phi}_i$:

$$u_2 - u_1 = \kappa_s f_s + \frac{h_1\theta_1 + h_2\theta_2}{2} \tag{2.49a}$$

where $\kappa_s = \kappa_{s1} + \kappa_{s2}$ (the exact expression of κ_{si} will be given in Section 2.3.4.1). Differentiating Eqn (2.49a) twice with respect to x gives

$$\frac{d^2(u_2 - u_1)}{dx^2} = \kappa_s \frac{d^2 f_s}{dx^2} + \frac{1}{2}\left(h_1 \frac{d^2\theta_1}{dx^2} + h_2 \frac{d^2\theta_2}{dx^2}\right) \tag{2.50}$$

Substituting Eqn (2.47) into the last term of Eqn (2.50) gives

$$\frac{1}{2}\left(h_1 \frac{d^2\theta_1}{dx^2} + h_2 \frac{d^2\theta_2}{dx^2}\right) = -\lambda_{x\theta} f_s - \overline{D}_{21} Q_a \tag{2.51}$$

where $Q_a = Q_{a1} = -Q_{a2}$, $\lambda_{x\theta} = 1/4[(h_1^2/D_1) + (h_2^2/D_2)]$, and $\overline{D}_{21} = 1/2[(h_2/D_2) - (h_1/D_1)]$. Substituting Eqn (2.51) into Eqn (2.50):

$$\frac{d^2(u_2 - u_1)}{dx^2} = \kappa_s \frac{d^2 f_s}{dx^2} - \lambda_{x\theta} f_s - \overline{D}_{21} Q_a \tag{2.52}$$

Equation (2.31) is repeated:

$$\frac{d^2(u_2 - u_1)}{dx^2} = (\lambda_{x1} + \lambda_{x2}) f_s \tag{2.31bis}$$

Equating Eqn (2.31bis) and Eqn (2.52),

$$\frac{d^2 f_s}{dx^2} - \beta^2 f_s = \frac{\overline{D}_{21}}{\kappa_s} Q_a \tag{2.53}$$

where $\beta^2 = \lambda_x/\kappa_s$, $\lambda_x = \lambda_{x1} + \lambda_{x2} + \lambda_{x\theta}$

2.3.2.2 Enforcing w-compatibility

Referring to Figure 2.10, the condition of z-displacement compatibility suggests that

$$w_2 - w_1 = (\lambda_{z1} + \lambda_{z2}) f_a \tag{2.54}$$

where λ_{zi} is the transverse compliance of member #i defined as the z-displacement due to unit normal stress – the exact expression of λ_{zi} will be presented in Section 2.3.4.2. Differentiating Eqn (2.54) repeatedly with respect to x gives

$$\frac{d^4}{dx^4}(w_2 - w_1) = (\lambda_{z1} + \lambda_{z2}) \frac{d^4 f_a}{dx^4} \tag{2.55}$$

The w-displacement of the neutral axis of the members is made up of the bending component and the shearing component, that is:

$$\frac{d^4}{dx^4}(w_2 - w_1) = \frac{d^4}{dx^4}(w_{b2} - w_{b1}) + \frac{d^4}{dx^4}(w_{s2} - w_{s1}) \tag{2.56}$$

From Eqn (2.48) and taking the difference between the two members, we have

$$\frac{d^4}{dx^4}(w_{b2} - w_{b1}) = \overline{D}_{21} \frac{df_s}{dx} - \overline{D}_e f_a \tag{2.57}$$

where $\overline{D}_e = \frac{1}{D_2} + \frac{1}{D_1}$. It is given here, with proof presented in Section 2.3.4.1, that the rotation of the neutral axis of the member due to shearing is

$$\phi_i = \frac{dw_{si}}{dx} = \frac{3Q_i}{4G_i h_i} - \frac{f_s}{8G_i} \tag{2.58}$$

Substituting $dQ_i = \mp f_a dx$ from Eqn (2.44) into Eqn (2.58) and differentiating three times gives

$$\frac{d^4 w_{si}}{dx^4} = \mp \frac{3}{4G_i h_i} \frac{d^2 f_a}{dx^2} - \frac{1}{8G_i} \frac{d^3 f_s}{dx^3} \tag{2.59}$$

Taking the difference between the two members:

$$\frac{d^4}{dx^4}(w_{s2} - w_{s1}) = \kappa_\phi \frac{d^2 f_a}{dx^2} - \kappa_{21} \frac{d^3 f_s}{dx^3} \tag{2.60}$$

where $\kappa_{21} = \frac{1}{8}\left(\frac{1}{G_2} - \frac{1}{G_1}\right)$, $\kappa_\phi = \kappa_{\phi 1} + \kappa_{\phi 2}$ and $\kappa_{\phi i} = 3/4G_i h_i$. Thus,

$$\frac{d^4}{dx^4}(w_2 - w_1) = \overline{D}_{21} \frac{df_s}{dx} - \overline{D}_e f_a + \kappa_\phi \frac{d^2 f_a}{dx^2} - \kappa_{21} \frac{d^3 f_s}{dx^3} \tag{2.61}$$

Equating Eqn (2.55) with Eqn (2.61) gives

$$\frac{d^4 f_a}{dx^4} - \frac{\kappa_\phi}{\lambda_z}\frac{d^2 f_a}{dx^2} + \frac{\overline{D}_e}{\lambda_z} f_a = \frac{\overline{D}_{21}}{\lambda_z}\frac{df_s}{dx} - \frac{\kappa_{21}}{\lambda_z}\frac{d^3 f_s}{dx^3} \tag{2.62}$$

where $\lambda_z = \lambda_{z1} + \lambda_{z2}$.

2.3.3 *Evaluation of interfacial stresses*

2.3.3.1 *Interfacial shear stress, $f_s(x)$*

It shall be shown that the term containing Q_a on the right-hand side of Eqn (2.53) can be neglected. The differential equation (2.53) is reduced to

$$\frac{d^2 f_s}{dx^2} - \beta^2 f_s = 0 \tag{2.35bis}$$

The simplest function of f_s that satisfies symmetry about $x = 0$ and gradient continuity at $x = 0$ is

$$f_s = A_s \frac{\cosh \beta x}{\sinh \beta L} \tag{2.36bis}$$

where the constant A_s may be evaluated as follows: From Eqn (2.50),

$$\frac{d(u_2 - u_1)}{dx} = \kappa_s \frac{df_s}{dx} + \frac{1}{2}\left(h_1 \frac{d\theta_1}{dx} + h_2 \frac{d\theta_2}{dx}\right) \tag{2.63}$$

From Eqn (2.30), we have

$$\frac{d(u_2 - u_1)}{dx} = -(\lambda_{x1} + \lambda_{x2})F + \alpha_{21}\Delta T \tag{2.30bis}$$

Equating Eqn (2.30) with Eqn (2.63) gives

$$\kappa_s \frac{df_s}{dx} = \alpha_{21}\Delta T - (\lambda_{x1} + \lambda_{x2})F - \frac{1}{2}\left(h_1 \frac{d\theta_1}{dx} + h_2 \frac{d\theta_2}{dx}\right) \tag{2.64}$$

Using the boundary conditions, $F(L) = d\theta_i(L)/dx = 0$, Eqn (2.64) is reduced to

$$\frac{df_s(L)}{dx} = \frac{\alpha_{21}\Delta T}{\kappa_s} \tag{2.65}$$

Differentiating Eqn (2.36) with respect to x and setting $x = L$,

$$\frac{df_s(L)}{dx} = A_s \beta \tag{2.66}$$

Comparing Eqn (2.65) and Eqn (2.66) gives

$$A_s = \frac{\alpha_{21}\Delta T}{\beta\kappa_s} \tag{2.37bis}$$

2.3.3.2 *Interfacial transverse stress,* $f_a(x)$

Ignoring the shear-induced deflection of the neutral axis for members #1 and #2, which is small compared to the bending-induced deflection, Eqn (2.62) becomes

$$\frac{d^4 f_a}{dx^4} + \frac{\overline{D}_e}{\lambda_z} f_a = \frac{\overline{D}_{21}}{\lambda_z}\frac{df_s}{dx} \tag{2.67}$$

Differentiating Eqn (2.53) with respect to x, followed by substitution with Eqn (2.36) gives

$$\frac{df_s}{dx} = \frac{1}{\beta^2}\frac{d^3 f_s}{dx^3} + \frac{\overline{D}_{21}}{\beta^2\kappa_s} f_a = A_s\beta\frac{\sinh\beta x}{\sinh\beta L} + \frac{\overline{D}_{21}}{\lambda_x} f_a \tag{2.68}$$

Substituting Eqn (2.68) into Eqn (2.67) yields

$$\frac{d^4 f_a}{dx^4} + 4\alpha_z^4 f_a = \frac{A_s\beta\overline{D}_{21}}{\lambda_z}\frac{\sinh\beta x}{\sinh\beta L} \tag{2.69}$$

where $4\alpha_z^4 = \frac{\overline{D}_e - \overline{D}_{21}^2/\lambda_x}{\lambda_z}$. In most cases, the second term is insignificantly small compared to the first term. Thus, $4\alpha_z^4 \approx \overline{D}_e/\lambda_z$.

Using the boundary conditions:

- $\lambda_z\frac{df_a(0)}{dx} = \frac{dw_2(0)}{dx} - \frac{dw_1(0)}{dx} = 0$, from Eqn (2.54) and gradient continuity,
- $\lambda_z\frac{d^2 f_a(L)}{dx^2} = \frac{d^2 w_2(L)}{dx^2} - \frac{d^2 w_1(L)}{dx^2} = 0$, nil curvature at free edge,
- $\lambda_z\frac{d^3 f_a(L)}{dx^3} = \frac{d^3[w_2(L) - w_1(L)]}{dx^3} = \overline{D}_{21} f_s(L) + \overline{D}_e Q_a(L) = \overline{D}_{21} A_s \coth\beta L$; from Eqn (2.57), and
- $\int_0^L f_a(x)dx = 0$, from z-directional force equilibrium,

the solution to Eqn (2.69) has been obtained as

$$f_a(x) = C_S\left(B_1\xi_{1x} + B_2\xi_{2x} + \frac{\sinh\beta x}{\sinh\beta L}\right) \tag{2.70}$$

where

$$C_S = \frac{\overline{D}_{21}\alpha_{21}\Delta T}{\lambda_z\kappa_s\left(4\alpha_z^4 + \beta^4\right)} \tag{2.71}$$

$$B_1 = \frac{1}{\alpha_z^2\beta}\frac{\beta^3(\xi_{4L}-\xi_{3L})\cosh\beta L - 4\alpha_z^3\xi_{1L}\sinh\beta L}{\xi_{5L}}$$
$$B_2 = -\frac{1}{\alpha_z^2\beta}\frac{\beta^3(\xi_{4L}+\xi_{3L})\cosh\beta L + 4\alpha_z^3\xi_{2L}\cosh\beta L}{\xi_{5L}} \qquad (2.72)$$

$$\xi_{1x} = \cos(\alpha_z x)\cosh(\alpha_z x),\ \ \xi_{2x} = \sin(\alpha_z x)\sinh(\alpha_z x),$$
$$\xi_{3L} = \cos(\alpha_z L)\sinh(\alpha_z L),\ \ \xi_{4L} = \sin(\alpha_z L)\cosh(\alpha_z xL),$$
$$\xi_{5L} = \sin(2\alpha_z L) + \sinh(2\alpha_z L)$$

For $\alpha_z L > 2$, $\sinh\alpha_z L \approx \cosh\alpha_z L$, $\xi_{5L} \to \sinh(2\alpha_z L)$, Eqn (2.72) is reduced to

$$B_1 \approx \frac{\beta^3\sin(\alpha_z L) - \left(4\alpha_z^3+\beta^3\right)\cos(\alpha_z L)}{2\alpha_z^2\beta\sinh(\alpha_z L)}$$
$$B_2 \approx -\frac{\beta^3\cos(\alpha_z L) + \left(4\alpha_z^3+\beta^3\right)\sin(\alpha_z L)}{2\alpha_z^2\beta\sinh(\alpha_z L)} \qquad (2.73)$$

For $\alpha_z L$ and $\beta L > 2$, $e^{-\beta(x+L)} \approx e^{-\alpha_z(x+L)} \approx 0$, Eqn (2.73) is reduced to (Jiang et al., 1997)

$$f_a(x) = C_s\left\{e^{\alpha_z(x-L)}[C_1\sin\alpha_z(x-L) + C_2\cos\alpha_z(x-L)] + e^{\beta(x-L)}\right\} \qquad (2.74)$$

where

$$C_1 = -\frac{\beta^2}{2\alpha_z^2}$$
$$C_2 = C_1 - \frac{2\alpha_z}{\beta} \qquad (2.75)$$

2.3.4 Evaluation of compliances

2.3.4.1 Shear compliance, κ_{si}

Figure 2.11 shows a segment of the differential beam element of member #1 and member #2 in equilibrium in the x-direction. Note that only the relevant out-of-balance force and moment are shown. The global forces over the differentiable element $h\cdot dx$ are presented in dotted arrows, while the local stresses around the segment of interest are presented in solid arrows. Note the new Cartesian coordinate orientation of member #2; the solution developed in this section will be applicable to both members.

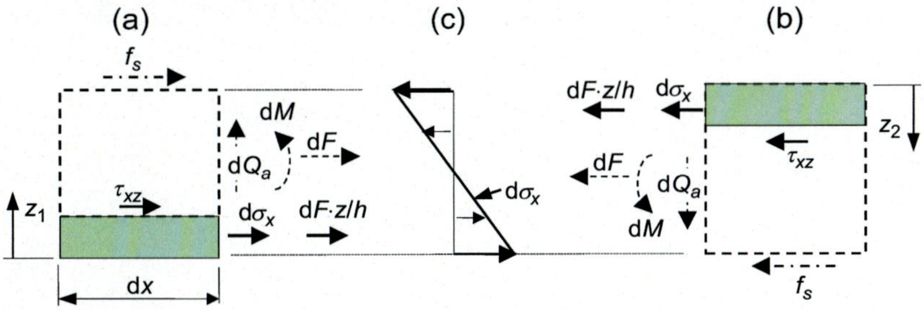

Figure 2.11 x-Direction equilibrium of a segment in a differential element of (a) member #1 and (b) member #2; profile of differential bending stress over the thickness of the differential element.

The figure also shows the profile of the differential bending stress $d\sigma_x$ due to the differential moment dM acting over the thickness of the element. This is given by

$$d\sigma_x = \frac{h \cdot dM}{2I}\left(1 - \frac{2z}{h}\right) \tag{2.76}$$

where $h \cdot dM/2I$ is the differential bending stress at $z = 0$. The out-of-balance bending force over the segment is

$$\int_0^z d\sigma_x d\xi = \frac{dM}{2I}\left(hz - z^2\right) \tag{2.77}$$

The differential traction dF gives rise to a traction force of magnitude, $dF \cdot z/h$, assuming uniform distribution over the thickness of the segment. The equilibrium of the segment in the x-direction gives the shear stress τ_{xz} on the top face of the segment as

$$\tau_{xz} dx + \frac{dM}{2I}\left(hz - z^2\right) + dF\frac{z}{h} = 0 \tag{2.78}$$

Note that Eqn (2.78) applies to both member #1 and member #2. Substituting $dM/dx = (h/2)f_s - Q_a$ from Eqn (2.45) and $f_s = -dF/dx$ from Eqn (2.29) into Eqn (2.78) gives

$$\tau(Z) = f_s\left(3Z^2 - 2Z\right) + \frac{6Q_a}{h}\left(Z - Z^2\right) \tag{2.79}$$

where $Z = z/h$. It can be shown that integrating τ_{xz} over the thickness of the member yields the sectional shear force Q_a.

The internal shear stress $\tau(Z)$ is induced by the surface shear stress f_s and the sectional shear force Q_a. This is depicted in Figure 2.12. The shear stress τ due to

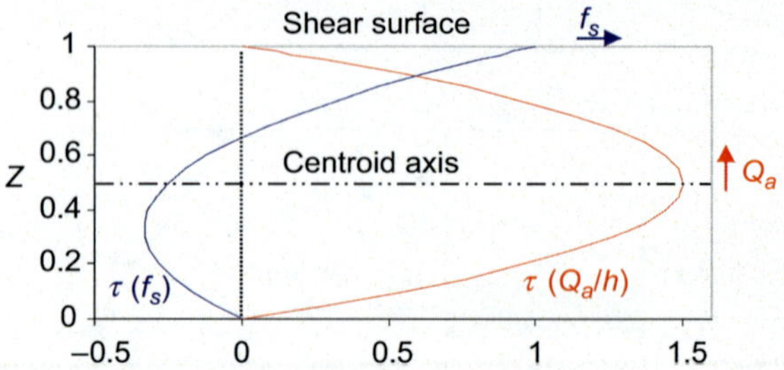

Figure 2.12 Shear stress distribution in the cross-section.

the sectional shear force Q_a has a symmetric parabolic profile over the thickness, while that due to the surface shear stress f_s has a minimum (negative) value at $Z = 1/3$ and a maximum magnitude at the surface.

The corresponding shear strain is given by

$$\gamma = \gamma_{xz} + \gamma_{zx} = \frac{\mathrm{d}w_s}{\mathrm{d}x} + \frac{\mathrm{d}u_s}{\mathrm{d}z} = \frac{\tau_{xz}}{G} \tag{2.80}$$

Noting that $\gamma_{xz} = \gamma_{zx}$, the shear-induced rotation of the neutral axis is simply

$$\phi = \gamma_{xz}|_{Z=0.5} = \frac{\tau}{2G}\Big|_{Z=0.5} = \frac{1}{4G}\left(\frac{3Q_a}{h} - \frac{f_s}{2}\right) \tag{2.81}$$

Assuming shearing causes no rotation of the centroid axis then

$$\frac{\mathrm{d}u_s}{\mathrm{d}z} = \frac{\tau}{G} \tag{2.82}$$

The shear-induced x-directional displacement of the shear surface, Δu_s, with respect to the centroid axis is given by

$$\begin{aligned}\Delta u_s|_{Z-0.5} &= \int_{0.5}^{z} \gamma_{zx}\mathrm{d}z = \frac{h}{G}\int_{0.5}^{z} \tau(Z)\mathrm{d}Z \\ &= \frac{hf_s}{G}\left[(Z^3 - Z^2) + \frac{1}{8}\right] + \frac{Q_a}{G}\left[(3Z^2 - 2Z^3) - \frac{1}{2}\right]\end{aligned} \tag{2.83}$$

This is displayed in Figure 2.13. The displacement of the shear surface with respect to the neutral axis is

$$\Delta u_s|_{Z=1-0.5} = \frac{f_s h}{8G} + \frac{Q_a}{2G} \tag{2.84}$$

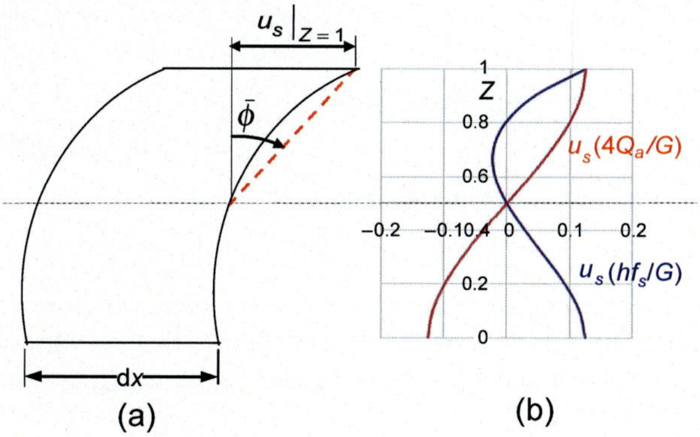

Figure 2.13 Shear deformation (a) of a differential element, (b) as a function of Z.

While it is clear that the cross-section of the element does not remain plane, it is nevertheless convenient to define a secant shear angle between the shear surface and the neutral axis, which is given by

$$\bar{\phi} = \frac{2\Delta u_s|_{Z=1-0.5}}{h} = \frac{f_s}{4G} + \frac{Q_a}{Gh} \tag{2.85}$$

It is also convenient to define a shear compliance κ_{si} as the differential x-displacement between the shear surface and the neutral axis per unit shear stress f_s:

$$\kappa_{si} = \frac{\Delta u_{si}|_{Z=1-0.5}}{f_s} = \frac{h_i}{8G_i} + \frac{Q_{ai}}{2G_i f_s} \tag{2.86}$$

If $Q_{ai} \ll f_s h_i$, then

$$\kappa_{si} \approx \frac{h_i}{8G_i} \tag{2.87}$$

However, if $Q_{ai}/2G_i f_s \approx h_i/8G_i$, then

$$\kappa_{si} \approx \frac{h_i}{4G_i} \tag{2.88}$$

2.3.4.2 Stretch compliance, λ_{xi}, and transverse compliance, λ_{zi}

We define the transverse compliance of the beam element as the differential z-displacement between the shear surface and the neutral axis per unit stress f_a. This can be expressed as

$$\lambda_z = \frac{\Delta w_s|_{Z=1-0.5}}{f_a} = \frac{h}{f_a} \int_{0.5}^{1} \varepsilon_z \mathrm{d}Z \tag{2.89}$$

The transverse strain ε_z is given by

$$\varepsilon_z = \begin{cases} \dfrac{1}{E}(\sigma_z - \nu\sigma_x) & \text{for plane stress} \\ \\ \dfrac{1}{E}\left[(1-\nu^2)\sigma_z - \nu(1+\nu)\sigma_x\right] & \text{for plane strain} \end{cases} \tag{2.90}$$

If $\sigma_z = f_a$ and $\sigma_x = 0$ then the compliance for plane stress is $\lambda_z = h/2E$, which was the relation used by Suhir (1988). The fact that $\sigma_z|_{z=0} = 0$ suggests that σ_z varies across the thickness of the beam. The transverse stress in the beam, σ_z, may be obtained using the equilibrium relationship:

$$\frac{\partial\sigma_z}{\partial z} + \frac{\partial\tau_{xz}}{\partial x} = 0 \tag{2.91}$$

Substituting Eqn (2.79) into Eqn (2.91), followed by integration from $Z = 0.5$ to $Z = 1$, gives

$$\sigma_z = h\frac{\mathrm{d}f_s}{\mathrm{d}x}\left(Z^2 - Z^3\right) + f_a\left(3Z^2 - 2Z^3\right) \tag{2.92}$$

This is shown in Figure 2.14. Similarly, the x-direction stress, σ_x, may be obtained using the equilibrium relationship:

$$\frac{\partial\sigma_x}{\partial x} + \frac{\partial\tau_{xz}}{\partial z} = 0 \tag{2.93}$$

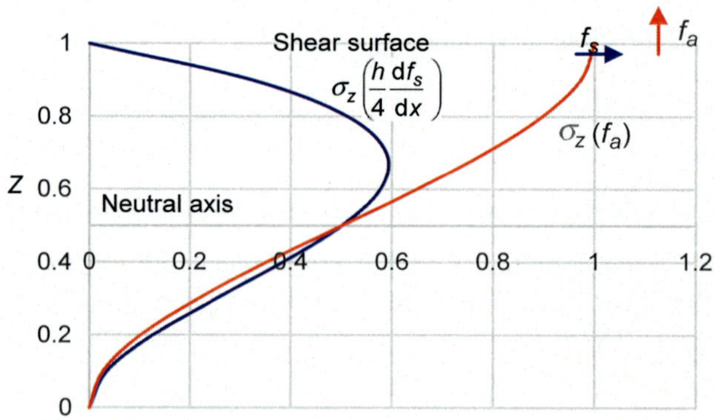

Figure 2.14 Variation of σ_z across thickness.

Substituting Eqn (2.79) into Eqn (2.93) and integrating from $\xi = x$ to L gives

$$[\sigma_x]_L^x = \frac{(2-6Z)}{h}\int_L^x f_s \mathrm{d}x + \frac{6(2Z-1)}{h^2}\int_L^x Q_a \mathrm{d}x \tag{2.94}$$

Substituting $f_s = -\mathrm{d}F/\mathrm{d}x$ into Eqn (2.94) together with the boundary conditions $\sigma_x|_{x=L} = F|_{x=L} = 0$ gives

$$\sigma_x = \frac{2F(3Z-1)}{h} + \frac{6(2Z-1)}{h^2}\int_L^x Q_a \mathrm{d}x \tag{2.95}$$

Substituting Eqn (2.92) and Eqn (2.95) into Eqn (2.90) and then into Eqn (2.89) gives

$$\lambda_z = \begin{cases} \dfrac{h}{Ef_a}\left[\left(\dfrac{11h}{192}\dfrac{\mathrm{d}f_s}{\mathrm{d}x} + \dfrac{13}{32}f_a\right) + \nu\left(\dfrac{5F}{4h} + \dfrac{3}{2h^2}\displaystyle\int_L^x Q_a \mathrm{d}x\right)\right] & \text{for plane stress} \\ \dfrac{h}{Ef_a}\left[(1-\nu^2)\left(\dfrac{11h}{192}\dfrac{\mathrm{d}f_s}{\mathrm{d}x} + \dfrac{13}{32}f_a\right) + \nu(1+\nu)\left(\dfrac{5F}{4h} + \dfrac{3}{2h^2}\displaystyle\int_L^x Q_a \mathrm{d}x\right)\right] & \text{for plane strain} \end{cases} \tag{2.96}$$

The transverse compliance of the beam therefore varies over the length of the beam, which would have rendered the subsequent solution excessively complicated. Assuming λ_{zi} to be independent of x, then for plane stress:

$$\lambda_{zi} \approx \frac{13}{32}\frac{h_i}{E_i} \approx \frac{3h_i}{8E_i} \tag{2.97}$$

The stretch compliance of the beam element has been defined as the strain of the neutral axis per unit force F:

$$\lambda_x = \frac{\varepsilon_{x,\mathrm{NA}}}{F} \tag{2.98}$$

The x-direction strain ε_x is given by

$$\varepsilon_x = \begin{cases} \dfrac{1}{E}(\sigma_x - \nu\sigma_z) & \text{for plane stress} \\ \dfrac{1}{E}\left[(1-\nu^2)\sigma_x - \nu(1+\nu)\sigma_z\right] & \text{for plane strain} \end{cases} \tag{2.99}$$

Substituting Eqn (2.92) and Eqn (2.95) into Eqn (2.99) and then Eqn (2.98) shows

$$\lambda_x = \begin{cases} \dfrac{1}{EF}\left[\dfrac{F}{h} - \nu\left(\dfrac{h}{8}\dfrac{\mathrm{d}f_s}{\mathrm{d}x} + \dfrac{1}{2}f_a\right)\right] & \text{for plane stress} \\ \dfrac{1}{EF}\left[\left(1-\nu^2\right)\dfrac{F}{h} + \nu(1+\nu)\left(\dfrac{h}{8}\dfrac{\mathrm{d}f_s}{\mathrm{d}x} + \dfrac{1}{2}f_a\right)\right] & \text{for plane strain} \end{cases} \tag{2.100}$$

Assuming λ_x is independent of x, then for plane stress,

$$\lambda_{xi} \approx \frac{1}{E_i h_i} \tag{2.101}$$

For plane strain, simply substitute E_i with $E_i' = 1 - \nu_i^2/E_i$ in Eqn (2.97) and Eqn (2.101).

2.4 Microelectronic assembly as a sandwich structure with a continuous bonding layer

Many trilayer structures are made of two relatively rigid layers and a relatively compliant sandwiched layer. A microelectronic assembly that is made up of an IC chip that is adhesively bonded to a substrate is one such example. In this section, we develop the analytical solutions for robust design of microelectronic assembly against interfacial delamination, fracturing of IC chip, and excessive out-of-plane deformation of the assembly.

2.4.1 The model and the linear component of normal stress

The free-body diagram of the microelectronic assembly is illustrated in Figure 2.15. The IC chip and the substrate are modelled as beams that are assumed to obey the moment–curvature relation of a simple beam in global deformation while behaving as a 2-D elastic body in local deformation. The adhesive layer is modelled as a 2-D elastic body and is assumed to have indefinite stretch compliance. The elemental representations of the IC device, the substrate and the adhesive are illustrated in Figure 2.16.

It is worth highlighting that an addition term of normal stress f_b has been introduced to the members, and the value of this stress varies from $-f_b$ on the bottom interface of member #3 to $+f_b$ on the top interface of member #3. There is a reason for introducing f_b: the adhesive layer experiences a uniform shear stress f_s across its thickness. The equilibrium equation below

$$\frac{\partial \sigma_z}{\partial z} + \frac{\partial \tau_{xz}}{\partial x} = 0 \tag{2.91bis}$$

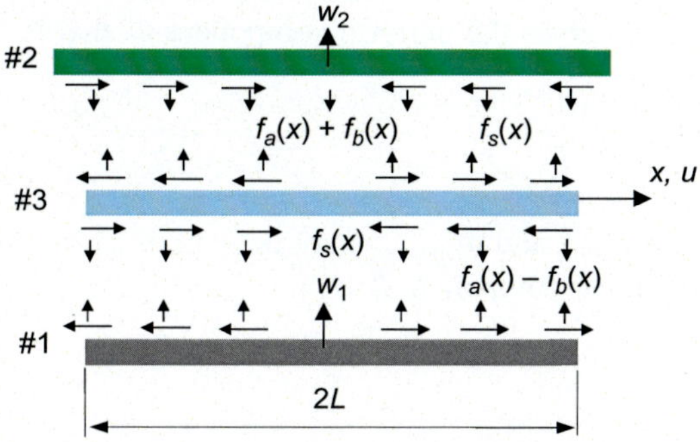

Figure 2.15 Free-body diagram of printed circuit board (PCB) assembly subjected to mismatched thermal expansion – interconnects modelled as axial, shear and flexural spring elements.

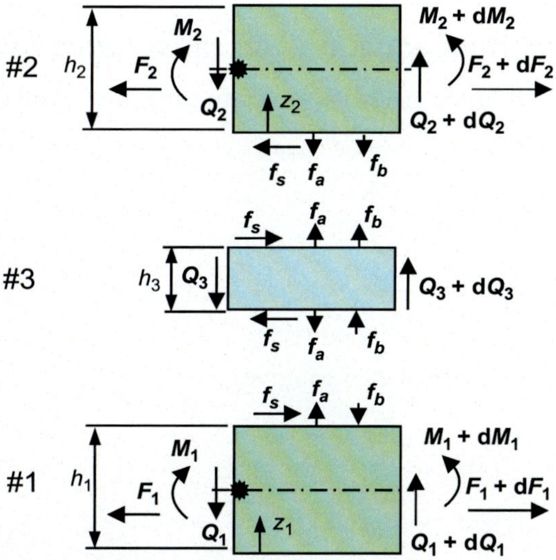

Figure 2.16 Elemental representation of the integrated circuit (IC) component, printed circuit board (PCB) and the interconnect.

suggests that a linear stress gradient $\Delta\sigma_z/\Delta z = -\mathrm{d}f_s/\mathrm{d}x$ must have been induced within the adhesive layer. Setting $\Delta z = h_3/2$, then we have

$$\frac{\sigma_{z,\mathrm{top}} - \sigma_{z,\mathrm{bottom}}}{2} = -\frac{h_3}{2}\frac{\mathrm{d}f_s}{\mathrm{d}x} = f_b \tag{2.102}$$

When the thickness of the sandwiched layer is small, as in the case of a bilayer structure, the magnitude of f_b is negligible for most of the x-value except when $x \to L$. However, the magnitude of f_b becomes significant as the thickness of the sandwiched layer becomes significant. The average of the normal stresses at the top and the bottom interfaces gives the uniform component f_a, that is,

$$\frac{\sigma_{z,\text{top}} + \sigma_{z,\text{bottom}}}{2} = f_a \tag{2.103}$$

The normal stresses $\sigma_{z,\text{top}}$ and $\sigma_{z,\text{bottom}}$ may be expressed in terms of the uniform component f_a and the linear component f_b as

$$\begin{aligned} \sigma_{z,\text{top}} &= f_a + f_b \\ \sigma_{z,\text{bottom}} &= f_a - f_b \end{aligned} \tag{2.104}$$

This is as shown in the free-body diagram in Figure 2.16. The presence of this 'unbalanced' stress sets up a sectional shear force Q_3.

2.4.2 The equilibrium equation and constitutive relations

The global equilibrium of the sandwich structure gives the following conditions:

$$F_1 + F_2 = 0 \tag{2.41bis}$$

$$Q_1 + Q_2 + Q_3 = 0 \tag{2.105}$$

$$M = M_1 + M_2 \tag{2.43bis}$$

The equilibrium of a differential element gives

$$\mathrm{d}F_i = \mp f_s \mathrm{d}x \tag{2.29bis}$$

$$\begin{aligned} \mathrm{d}Q_i &= (\mp f_a + f_b)\mathrm{d}x; \quad i = 1,2 \\ \mathrm{d}Q_3 &= -2 f_b \mathrm{d}x \end{aligned} \tag{2.106}$$

$$\mathrm{d}M_i = \left(\frac{f_s h_i}{2} - Q_i\right)\mathrm{d}x \tag{2.45bis}$$

Substituting $f_b = -(h_3/2)(\mathrm{d}f_s/\mathrm{d}x)$ from Eqn (2.102) into Eqn (2.106) and performing integration, the condition for transverse force-equilibrium may also be expressed as

$$\begin{aligned} Q_i &= \pm Q_a - Q_b; \quad i = 1,2 \\ Q_3 &= 2Q_b \end{aligned} \tag{2.106a}$$

where

$$Q_a = -\int_L^x f_a \mathrm{d}x$$
$$Q_b = -\int_L^x f_b \mathrm{d}x = \frac{h_3 f_s}{2} \tag{2.106b}$$

Assuming the moment–curvature relation for pure bending remains applicable in this case, then

$$\frac{1}{D_i}\frac{\mathrm{d}M_i}{\mathrm{d}x} = \frac{\mathrm{d}^2\theta_i}{\mathrm{d}x^2} = \frac{\mathrm{d}^3 w_{bi}}{\mathrm{d}x^3} \tag{2.107}$$

Combining Eqn (2.46) and Eqn (2.107) and the relation $Q_i = \pm Q_a - \frac{h_3 f_s}{2}$, we obtain

$$\frac{\mathrm{d}^2\theta_i}{\mathrm{d}x^2} = \frac{1}{D_i}\left(\frac{f_s h_i}{2} - Q_i\right) = \frac{1}{D_i}\left(\frac{h_i + h_3}{2} f_s \mp Q_a\right) \tag{2.108}$$

$$\frac{\mathrm{d}^4 w_{bi}}{\mathrm{d}x^4} = \frac{\mathrm{d}^3\theta_i}{\mathrm{d}x^3} = \frac{1}{D_i}\left(\frac{h_i + h_3}{2}\frac{\mathrm{d}f_s}{\mathrm{d}x} \pm f_a\right) \tag{2.109}$$

2.4.3 Compatibility of displacements

Figure 2.17 shows the deformed state of the sandwich structure. The condition of u-compatibilities gives

$$u_2 - u_1 = \frac{1}{2}(h_1\overline{\phi}_1 + h_2\overline{\phi}_2 + 2h_3\phi_3) - \frac{1}{2}(h_1\theta_1 + h_2\theta_2) \tag{2.110}$$

or

$$u_2 - u_1 = \kappa_s f_s - \frac{1}{2}(h_1\theta_1 + h_2\theta_2) \tag{2.111}$$

where $\kappa_s = \kappa_{s1}^* + \kappa_{s2}^* + \kappa_{s3}$, whose expressions will be given in Section 2.4.5.1. Note that the bilayer example assumes a negative curvature, while a positive curvature is assumed in this derivation.

The condition of w-compatibilities requires that

$$w_2 - w_1 = \lambda_z f_a \tag{2.112}$$

where $\lambda_z = \lambda_{z1} + \lambda_{z2} + \lambda_{z3}$.

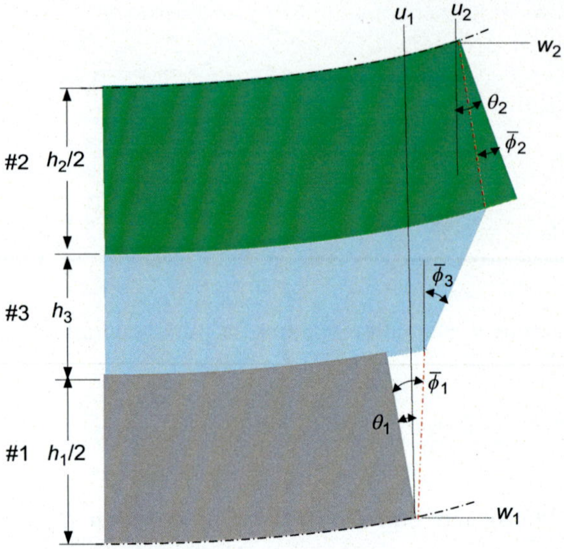

Figure 2.17 Deformed state of a sandwich structure.

2.4.4 Interfacial stresses

2.4.4.1 Interfacial shear stress, $f_s(x)$

From Eqn (2.111), we have

$$\frac{d^2(u_2 - u_1)}{dx^2} = \kappa_s \frac{d^2 f_s}{dx^2} - \frac{1}{2}\left(h_1 \frac{d^2\theta_1}{dx^2} + h_2 \frac{d^2\theta_2}{dx^2}\right) \tag{2.113}$$

Substituting Eqn (2.108) into the last term gives

$$\begin{aligned}\frac{1}{2}\left(h_1 \frac{d^2\theta_1}{dx^2} + h_2 \frac{d^2\theta_2}{dx^2}\right) &= \frac{1}{2}\left(\frac{h_2}{D_2} - \frac{h_1}{D_1}\right)Q_a + \frac{1}{4}\left(\frac{h_1^2 + h_1h_3}{D_1} + \frac{h_2^2 + h_2h_3}{D_2}\right)f_s \\ &= \overline{D}_{21}Q_a + \lambda_{x\theta}^* f_s\end{aligned} \tag{2.114}$$

where $\lambda_{x\theta}^* = \frac{1}{4}\left(\frac{h_1^2 + h_1h_3}{D_1} + \frac{h_2^2 + h_2h_3}{D_2}\right)$; thus, the thickness of member #3 has contributed to additional stretch compliance of the structure. Substituting Eqn (2.114) into Eqn (2.113) gives

$$\frac{d^2(u_2 - u_1)}{dx^2} = \kappa_s \frac{d^2 f_s}{dx^2} - \lambda_{x\theta}^* f_s - \overline{D}_{21}Q_a \tag{2.115}$$

Equating Eqn (2.115) with Eqn (2.31) leads to the differential equation for the interfacial shear stress f_s:

$$\frac{d^2 f_s}{dx^2} - \beta^2 f_s = \frac{\overline{D}_{21}}{\kappa_s} Q_a \tag{2.116}$$

where $\beta^2 = \lambda_x/\kappa_s$, $\lambda_x = \lambda_{x1} + \lambda_{x2} + \lambda^*_{x\theta}$. It shall be shown that the magnitude of Q_a is negligibly small, thus Eqn (2.116) may be reduced to

$$\frac{d^2 f_s}{dx^2} - \beta^2 f_s = 0 \tag{2.35bis}$$

and the solution is

$$f_s(x) = A_s \frac{\cosh \beta x}{\sinh \beta L} \tag{2.36bis}$$

where

$$A_s = \frac{\alpha_{21} \Delta T}{\beta \kappa_s} \tag{2.37bis}$$

However, the suggested expression of shear stress implies that $f_s(L) \approx A_s \neq 0$ and violates the boundary condition that a free edge does not support shearing. This can be addressed by introducing a second term to the expression of f_s, that is,

$$f_s = A_s \left(\frac{\cosh \beta x}{\sinh \beta L} - \coth \beta L \frac{\cosh n\beta x}{\cosh n\beta L} \right) \tag{2.117}$$

The second term in the equation compels the shear stress to dip to a nil value as $x \to L$, wherein n is a positive real number, whose value dictates the rate of dipping. Hence, the shear stress $f_s(x)$ rises monotonically from a minimum value at $x = 0$ until it reaches a peak value (a stationary point) before dipping toward nil at $x = 0$. The stationary point is defined by the condition $df_s/dx = 0$, which gives

$$\frac{df_s}{dx} = A_s \beta \left(\frac{\sinh \beta x_{\text{stationary}}}{\sinh \beta L} - n \coth \beta L \frac{\sinh n\beta x_{\text{stationary}}}{\cosh n\beta L} \right) = 0 \tag{2.118}$$

It is postulated that the stationary point of df_s/dx occurs at approximately half the thickness of the adhesive layer away from the edge. Putting $x_{\text{stationary}} = L - h_3/2$, we obtain

$$\cosh n\beta L \sinh \beta(L - h_3/2) - n \cosh \beta L \sinh n\beta(L - h_3/2) = 0 \tag{2.119}$$

The values of n for the range of $0.05 \leq \beta h_3 \leq 0.9$ have been evaluated and found to be approximated reasonably well by the equation:

$$n = 4(\beta h_3)^{-1.37} \tag{2.120}$$

Differentiating Eqn (2.117) with respect to x:

$$\frac{\mathrm{d}f_s(x)}{\mathrm{d}x} = \frac{\alpha_{21}\Delta T}{\kappa_s}\left(\frac{\sinh \beta x}{\sinh \beta L} - n \coth \beta L \frac{\sinh n\beta x}{\cosh n\beta L}\right) \tag{2.121}$$

and putting $x = L$ gives

$$\frac{\mathrm{d}f_s(L)}{\mathrm{d}x} = A_s\beta(1 - n \coth \beta L \tanh n\beta L) \tag{2.122}$$

For $\beta L > 2$, $\coth \beta L \tanh n\beta L \approx 1$ and Eqn (2.122) is reduced to

$$\frac{\mathrm{d}f_s(L)}{\mathrm{d}x} = A_s\beta(1 - n). \tag{2.123}$$

Thus, the magnitude of the shear stress gradient at the free edge increases with n.

2.4.4.2 The uniformed component of the transverse stress, $f_a(x)$

From Eqn (2.112), we have

$$\frac{\mathrm{d}^4(w_2 - w_1)}{\mathrm{d}x^4} = \lambda_z \frac{\mathrm{d}^4 f_a}{\mathrm{d}x^4} \tag{2.124}$$

Ignoring the deflection of the centroid axis of the members due to shearing, then upon substituting with Eqn (2.109), the LHS of Eqn (2.124) becomes

$$\frac{\mathrm{d}^4(w_2 - w_1)}{\mathrm{d}x^4} = \overline{D}_{21}^* \frac{\mathrm{d}f_s}{\mathrm{d}x} - \overline{D}_e f_a \tag{2.125}$$

where $\overline{D}_{21}^* = \frac{1}{2}\left(\frac{h_2 + h_3}{D_2} - \frac{h_1 + h_3}{D_1}\right)$. Substituting Eqn (2.125) into Eqn (2.124) gives the differential equation for the uniform component of the transverse stress:

$$\frac{\mathrm{d}^4 f_a}{\mathrm{d}x^4} + \frac{\overline{D}_e}{\lambda_z} = \frac{\overline{D}_{21}^*}{\lambda_z} \frac{\mathrm{d}f_s}{\mathrm{d}x} \tag{2.126}$$

For $(\alpha_z L,\ \beta L) > 3$, substituting $\tilde{f}_s = A_s \frac{\sinh \beta x}{\cosh \beta L} f_s(x)$ into Eqn (2.116) and then substituting df_s/dx into Eqn (2.126) gives

$$\frac{d^4 f_a}{dx^4} + 4\alpha_2^{*4} = \frac{A_s \beta \overline{D}_{21}^*}{\lambda_z} \frac{\sinh \beta x}{\sinh \beta L} \tag{2.127}$$

where $\alpha_z^* = \sqrt[4]{\frac{\overline{D}_e - \overline{D}_{21}^* \overline{D}_{21}/\lambda_x}{4\lambda_z}} \approx \alpha_z$. For a trilayer structure with $\beta L > 3$, the solution is given approximately by

$$f_a(x) \approx C_s^* \left\{ e^{\alpha_z (x-L)} [C_1 \sin \alpha_z (x - L) + C_2 \cos \alpha_z (x - L)] + e^{\beta (x-L)} \right\} \tag{2.128}$$

where

$$C_s^* = \frac{\overline{D}_{21}^* \alpha_{21} \Delta T}{\lambda_z \kappa_s (4\alpha_z^4 + \beta^4)} \tag{2.129}$$

$$\begin{aligned} C_1 &= -\frac{\beta^2}{2\alpha_z^2} \\ C_2 &\approx -\left(\frac{\beta^2}{2\alpha_z^2} + \frac{2\alpha_z}{\beta}\right) \end{aligned} \tag{2.130}$$

The maximum magnitude of f_a occurs at $x = L$, which is

$$f_a(L) \approx C_s^*(1 + C_2) = \frac{A_s \overline{D}_{21}^*}{\lambda_z} \left[-\frac{2\alpha_z + \beta}{2\alpha_z^2 (2\alpha_z^2 + \beta^2 + 2\alpha_z \beta)} \right] \tag{2.131}$$

2.4.4.3 *The linear component of the transverse stress, $f_b(x)$*

From Eqn (2.102), the linear component of the transverse normal stress f_b is given by

$$f_b = -\frac{h_3}{2} \frac{df_s}{dx} = -\frac{h_3 \alpha_{21} \Delta T}{2\kappa_s} \left(\frac{\sinh \beta x}{\sinh \beta L} - n \coth \beta L \frac{\sinh n\beta x}{\cosh n\beta L} \right) \tag{2.132}$$

The maximum magnitude of f_b occurs at $x = L$:

$$f_b(L) = -\frac{h_3}{2} \frac{df_s(L)}{dx} = \frac{A_s \beta h_3}{2} (n - 1) \tag{2.133}$$

Substituting $n = 4(\beta h_3)^{-1.37}$ into the above and for $n \gg 1$, then

$$f_b(L) \approx \frac{2A_s}{(\beta h_3)^{0.37}} \tag{2.134}$$

For a bilayer structure, $h_3 \to 0$, and $f_b(L)$ becomes singular.

The ratios of $|f_s(L)|:|f_a(L)|:|f_b(L)|$ are approximately:

$$1 : \frac{\overline{D}_{21}^*}{\lambda_z}\left[\frac{2\alpha_z + \beta}{2\alpha_z^2(2\alpha_z^2 + \beta^2 + 2\alpha_z\beta)}\right] : \frac{2}{(\beta h_3)^{0.37}} \tag{2.135}$$

The ratio $\overline{D}_{21}^*/\lambda_z[2\alpha_z + \beta/2\alpha_z^2(2\alpha_z^2 + \beta^2 + 2\alpha_z\beta)]$ is generally less than 1 and much less in many cases, which justifies the exclusion of the term containing Q_a in the differential equation f_s when solving for f_s. By contrast, the ratio $2/(\beta h_3)^{0.37}$ is invariably larger than 1. A more sophisticated comparison of the magnitudes of $f_s(L)$ and $f_b(L)$ will be given later in this chapter.

2.4.5 The compliances

2.4.5.1 Shear compliance, κ_{si}

Equation (2.86) is repeated below

$$\kappa_{si} = \frac{h_i}{8G_i} + \frac{Q_{ai}}{2G_i f_s} \tag{2.86bis}$$

Referring to Figure 2.18, it is clear that the introduction of f_b has reduced the effective sectional force in member #1 and increased it in member #2. Substituting Q_{ai} in Eqn (2.86) with $Q_a \mp \frac{h_3 f_s}{2}$, we have

$$\begin{aligned} \kappa_{s1}^* &= \frac{1}{8G_1}(h_1 - 2h_3) + \frac{Q_a}{2G_1 f_s} \\ \kappa_{s2}^* &= \frac{1}{8G_2}(h_2 + 2h_3) + \frac{Q_a}{2G_2 f_s} \end{aligned} \tag{2.136}$$

Therefore, the presence of the adhesive layer has reduced the shear compliance of member #1 and increased that of member #2. This is not intuitive. For $Q_a \ll f_s$,

$$\begin{aligned} \kappa_{s1}^* &\approx \frac{1}{8G_1}(h_1 - 2h_3) \\ \kappa_{s2}^* &\approx \frac{1}{8G_2}(h_2 + 2h_3) \end{aligned} \tag{2.137}$$

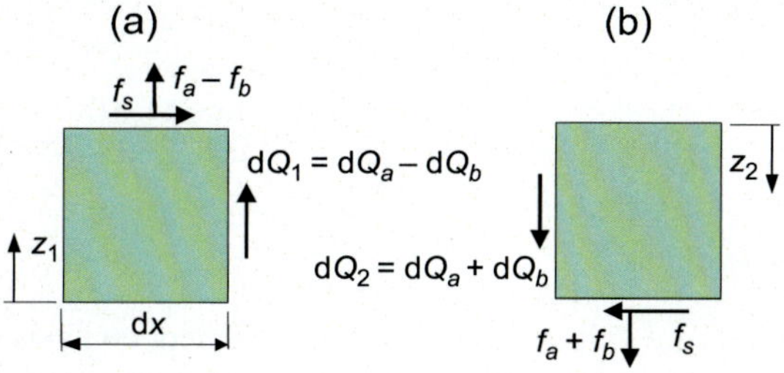

Figure 2.18 x-Direction equilibrium of a segment in a differential element of (a) member #1 and (b) member #2; profile of differential bending stress over the thickness of the differential element.

For member #3, which experiences uniform shear stress f_s over its thickness, the shear compliance is

$$\kappa_{s3} = \frac{h_3}{G_3} \tag{2.138}$$

2.4.5.2 Stretch compliance, λ_{xi}, and transverse compliance, λ_{zi}

From Eqn (2.96),

$$\lambda_{zi} = \frac{h_i}{E_i f_a}\left[\left(\frac{11h_1}{192}\frac{df_s}{dx} + \frac{13}{32}f_a\right) + \nu\left(\frac{5F_i}{4h_i} + \frac{3}{2h_i^2}\int_L^x Q_{ai}dx\right)\right] \tag{2.96bis}$$

Substituting f_a in the bracket with $f_a \mp f_b = f_a \pm [(h_3/2)(df_s/dx)]$ and eliminating those terms that are functions of x gives

$$\lambda_{zi} \approx \frac{3h_i}{8E_i} \tag{2.97bis}$$

From Eqn (2.91) and Eqn (2.102), the transverse stress in member #3 is given by

$$\sigma_z(Z) = f_a + f_b(2Z - 1) \tag{2.139}$$

The transverse compliance of member #3 becomes

$$\lambda_{z3} = \frac{\Delta w_3|_{Z=1-0}}{f_a} = \frac{h_3}{E_3 f_a}\int_0^1 \sigma_z dZ = \frac{h_3}{E_3} \tag{2.140}$$

It is clear from Eqn (2.100) that

$$\lambda_{xi} \approx \frac{1}{E_i h_i} \tag{2.101bis}$$

2.4.6 Validations

The analytical solutions are validated against finite element analysis (FEA) in this section. Comparisons are also made with the solutions of Suhir (1986) and Jiang et al. (1997). The basic dimensions and material properties for the models used in the validations are shown in Table 2.1. The compliances and characteristic constants, including those in references Suhir (1986) and Jiang et al. (1997), are given in Table 2.2. The units for individual parameters are listed underneath the table. It is worth highlighting that the stretch compliance λ_x used by Suhir (1986) and Jiang et al. (1997) is identical to that for a bilayer. By contrast, the contribution of the adhesive thickness to the stretch compliance is captured in the derivation shown in this work, which explains the significantly higher magnitude of λ_x compared to that of Suhir (1986) and Jiang et al. (1997).

Two FE models were used for validation. In one model, the IC device, laminated substrate and adhesive were all modelled using quadratic plane elements (Figure 2.19(a)) – this shall be referred to as the FE plane model. In another model, the adhesive was modelled using quadratic plane elements, while the IC device and the substrate were modelled using Timoshenko beam elements with appropriate offsets for their respective thicknesses (Figure 2.19(b)) – this shall be referred to as an FE-beam model. To simulate the adhesive as incapable of transferring lateral forces, the plane elements in the adhesive layer were defined with orthotropic material properties wherein the lateral elastic modulus (E_x and E_y) and the out-of-plane shear modulus (G_{xy} and G_{yz}) were assigned negligible values compared to the transverse modulus (E_z) and in-plane shear modulus (G_{xz}). The interfacial shear stress $f_s(x)$ for the two FE models is shown in Figure 2.19(c). It is noted that there is a slight difference between the results of the two FE models – the model in which the IC device and the

Table 2.1 Basic parameters used for validation analyses

	Dimensions		Material properties			Diff. Temp.
Component	L	h	E	G	α	ΔT
#1, IC device	5	1	24	9.2	5	100
#2, substrate		1	24	9.2	15	
#3, adhesive		0.5	4.9	1.9	0	

Note: L, h (mm); E, G (GPa); α (10^{-6}/°C); ΔT (°C).The elastic modulus of integrated circuit (IC) package provided is an estimation of the composite property for silicon and molding compound.

Table 2.2 Derived parameters used for validation

Source	Compliances			Characteristic constants	
	κ_s	λ_x	λ_z	β	α_z
Wong et al. (2009)	2.91×10^{-4}	4.49×10^{-4}	1.32×10^{-4}	1.24	1.17
Suhir (1986)	2.47×10^{-4}	3.27×10^{-4}	–	1.15	–
Jiang et al. (1997)	2.65×10^{-4}	3.27×10^{-4}	1.02×10^{-4}	1.11	1.25

Note: λ_x (N^{-1} mm); λ_z, κ_s (N^{-1} mm^3); α_z, β (mm^{-1}).

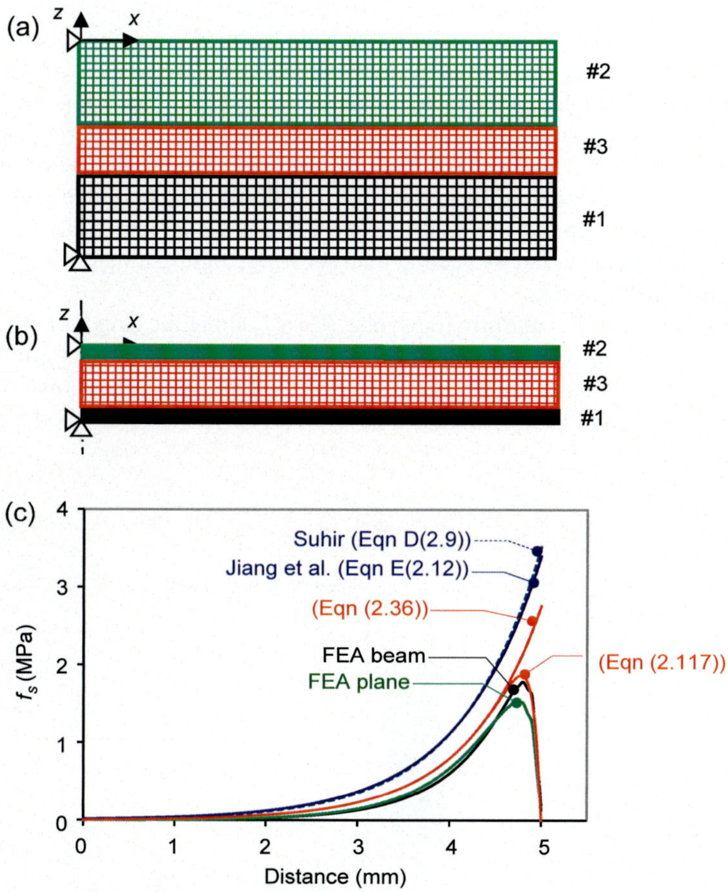

Figure 2.19 (a) FE model where all members are modelled using plane elements; (b) FE model where members #1 and #2 are modelled as beams; and (c) shear stress distribution.

substrate were modelled as beams (FE-beam model) gave rise to a higher peak shear stress. The FE-beam model shall be used as a benchmark for the analytical solutions.

Substituting $h_3 = 0.5$ mm and $\beta = 1.24$ mm^{-1} into Eqn (2.119) gives $n = 7.5$. The interfacial shear stresses $f_s(x)$ are obtained from Eqn (2.36) and Eqn (2.117) and compared against those evaluated using Eqn D(2.9) of Suhir (1986) and Eqn E(2.12) of Jiang et al. (1997) and against the solution from the FE-beam model in Figure 2.19(c). The simpler equation (2.36) gives a similar trend as that of Suhir and Jiang, but much better agreement with the FE-beam results. Conversely, the more complex equation, Eqn (2.117), gives a solution that has good agreement with the FE-beam solution for almost the entire length of the assembly, including satisfying the boundary condition $f_s(L) = 0$.

The deformation and the z-direction stress contours for the outer column of the finite elements for the adhesive layer are shown in Figure 2.20, which shows that the z-normal stress indeed varies linearly along its thickness. The magnitude of the uniform component f_a and the linear component f_b are evaluated using Eqn (2.103) and Eqn (2.102), respectively; these are $f_a = -0.05$ MPa and $f_b = 6.32$ MPa − the linear component is two orders of magnitude larger than the uniform component.

The interfacial normal stress σ_z at the top and bottom surfaces of the adhesive layer from the middle to the edge of the assembly is shown in Figure 2.21, together with the shear stress f_s. It is noted that (1) the stationary point of $f_s(x)$ occurs at $x = 4.75$ mm, which is indeed half the thickness of the adhesive layer away from the edge; (2) the stationary point coincides with the instance of $f_b = 0$; (3) the magnitude of the normal stress is a few times higher than that of the shear stress, which implies that the assembly is much more likely to fail by peeling than shearing, highlighting the importance of modelling the peeling stress.

The distributions of the uniform transverse stress f_a along the length of the IC package evaluated using the analytical equations Eqn (2.128), Eqn D(2.16) of Suhir (1986), and Eqn E(2.13) of Jiang et al. (1997) are shown in Figure 2.22. Also shown is the $f_a(x)$ from FE-beam, which is evaluated using Eqn (2.103). There is good agreement

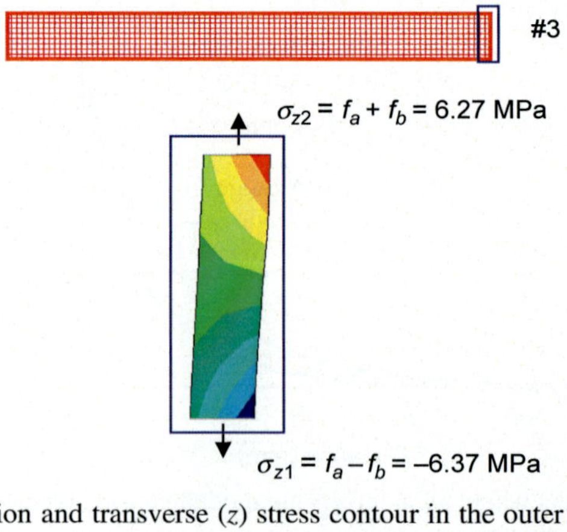

Figure 2.20 Deformation and transverse (z) stress contour in the outer FE layer.

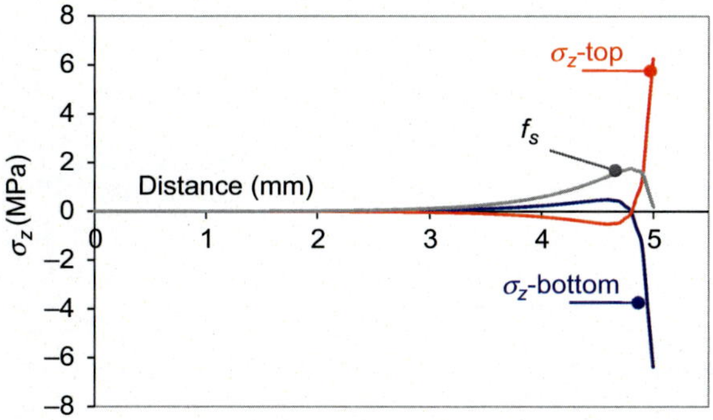

Figure 2.21 σ_z and f_s along the top and bottom surfaces of the adhesive layer.

between the FE results and the analytical results, except for the solution of Suhir (1986), Eqn C(2.16), which has violated global equilibrium; that is $\int_0^L f_a dx \neq 0$. It is clear by comparing Figures 2.19 and 2.22 that the magnitude of f_a is significantly smaller than that of the shear stress f_s. This has justified the exclusion of the sectional force Q from the differential equation of shear stress in Eqn (2.53).

The fact that the linear component of the peeling stress is a few orders of magnitude larger than the uniform component and a few times higher than the interfacial shear stress highlights the importance of modelling this stress. In Figure 2.23, the distributions of the linear transverse stress f_b along the length of the IC component evaluated using the analytical equation (2.132) are compared against that evaluated using the FE-beam model followed by manipulation with Eqn (2.102). No other analytical solution methods are available for comparison. The agreement between the analytical equation (2.132) and the FE-beam results is remarkable – the data practically overlap.

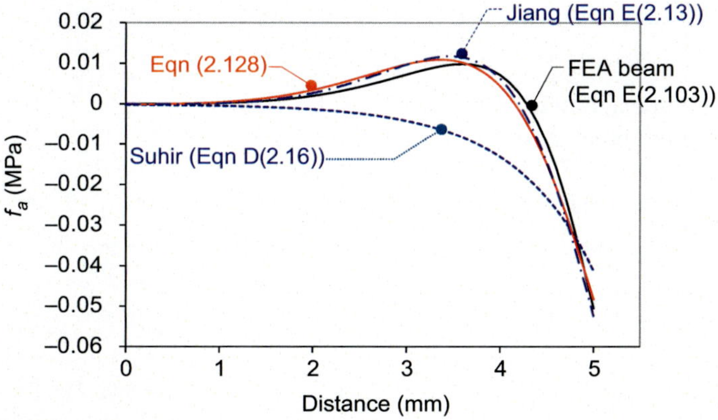

Figure 2.22 Comparisons of f_a.

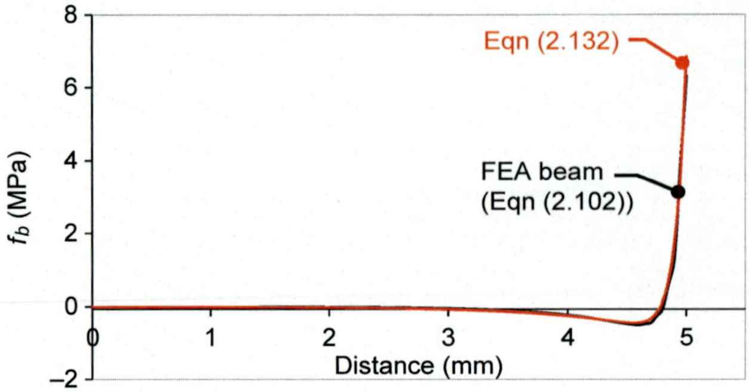

Figure 2.23 Comparisons of f_b.

2.4.7 Robust design analysis

2.4.7.1 Interfacial delamination

The delamination of the interfaces is a major concern for trilayer structure in service. There are two driving forces:
(1) shearing stress:

$$f_s = A_s\left(\frac{\cosh \beta x}{\sinh \beta L} - \coth \beta L \frac{\cosh n\beta x}{\cosh n\beta L}\right) \tag{2.115bis}$$

and (2) peeling stress:

$$\sigma_{\text{peeling}} \approx f_b = -\frac{h_3 A_s \beta}{2}\left(\frac{\sinh \beta x}{\sinh \beta L} - n \coth \beta L \frac{\sinh n\beta x}{\cosh n\beta L}\right) \tag{2.141}$$

where

$$A_s = \frac{\alpha_{21}\Delta T}{\beta \kappa_s} \tag{2.37bis}$$

The maximum magnitude of the shearing stress occurs at $x = L - h_3/2$; ignoring the second term in the bracket, we have

$$f_{s,\max} \approx A_s \frac{\cosh \beta(L - h_3/2)}{\sinh \beta L} \tag{2.142}$$

For $x > 2$, $\frac{\cosh(x-y)}{\sinh x} \approx \exp(-y)$, thus

$$f_{s,\max} \approx A_s \exp\left(-\frac{\beta h_3}{2}\right) \tag{2.142a}$$

In contrast, the maximum magnitude of the peeling stress occurs at $x = L$; from Eqn (2.133),

$$\sigma_{\text{peeling,max}} \approx f_b(L) \approx \frac{A_s\beta h_3}{2}(n-1) \approx \left[4(\beta h_3)^{-1.37} - 1\right]\frac{\beta h_3}{2} \quad (2.143)$$

Figure 2.24 shows the magnitude of the interfacial stresses versus βh_3, which suggests that the magnitude of the peeling stress is always higher than that of the interfacial shear stress. The difference is accentuated with decreasing magnitude of βh_3, which could be brought about by decreasing thickness of the adhesive layer or by increasing thicknesses of the outer layers.

Probably the most frequently asked question about managing interfacial delamination is: Will increasing the thickness of the adhesive layer reduce the risk of interfacial delamination? A glance at Eqn (2.143) suggests that the magnitude of the peeling stress can be decreased by reducing the adhesive layer thickness; however, a closer look at the equation indicates that this is not that obvious since decreasing the adhesive thickness will increase the shear and the stretch compliances of the structure, which will increase the peeling stress. A more detailed analysis is given below.

Expanding Eqn (2.143) and focusing on the solutions for $n \gg 1$,

$$\sigma_{\text{peeling,max}} \approx \frac{nh_3\alpha_{21}\Delta T}{2\kappa_s} = \frac{2(\beta h_3)^{-1.37}h_3\alpha_{21}\Delta T}{\kappa_s} = \frac{2\alpha_{21}\Delta T}{\lambda_x^a\kappa_s^b h_3^c} \quad (2.144)$$

where $a = 0.68$, $b = 1 - a$ and $c = 2a - 1$. It is clear that the maximum magnitude of peeling stress can be reduced by increasing the stretch and the shear compliances of the system. This can be accomplished by

- decreasing the moduli of the constituent materials or
- increasing the adhesive layer thickness, which not only increases the factor h_3^c in the denominator but also the stretch and the shear compliances.

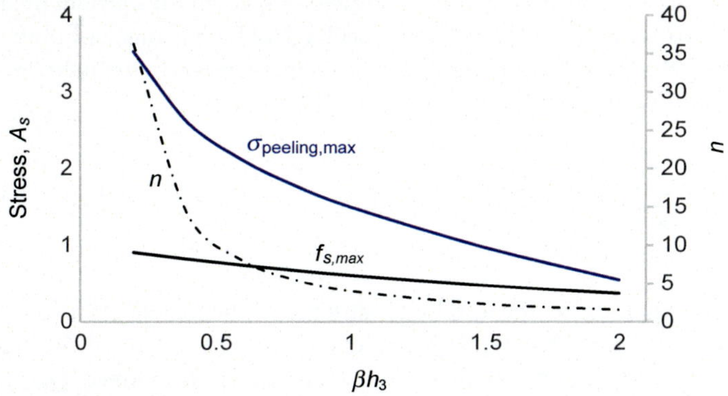

Figure 2.24 Comparisons of $\sigma_{\text{peeling,max}}$ and $f_{s,\text{max}}$.

The thickness effect of the outer member is not entirely clear and requires further analysis. Differentiating $\sigma_{\text{peeling,max}}$ with respect to parameter Y of the member gives, after some manipulation,

$$\frac{d\sigma_{\text{peeling,max}}}{dY} \approx -\sigma_{\text{peeling,max}}\left(\frac{a}{\lambda_x}\frac{d\lambda_x}{dY}+\frac{b}{\kappa_s}\frac{d\kappa_s}{dY}+\frac{c}{h_3}\frac{dh_3}{dY}\right) \tag{2.145}$$

where Y may represent h, E or G. Table 2.3 lists the rate of change of λ_x and κ_s with Y. Note that the analysis for member #2 is given by swapping the subscripts 1 and 2 in the table. Equation (2.145), supported by Table 2.3, allows the sensitivity study of changing individual parameters.

The maximum peeling stress and its gradient at the extreme ends of the thicknesses and the moduli of the members have been computed and are shown in Figure 2.4. Also given are the sketches of the postulated variations of $\sigma_{\text{peeling,max}}$. Thus, the magnitude of peeling stress can be reduced by increasing the thickness of the bonding layer while reducing its shear modulus and by reducing the modulus of the outer members. It is noted that the maximum magnitude of the peeling stress increases initially with increasing thickness of the outer member before falling off. There is a particular thickness of the outer member that leads to a peak magnitude of $\sigma_{\text{peeling,max}}$. The condition of peak $\sigma_{\text{peeling,max}}$ is given by

$$3bE_ih_i^3\sum_{i=1}^{2}\frac{1}{E_ih_i}\left(4+\frac{3h_3}{h_i}\right) = 16aG_i(2h_i+3h_3)\left(\sum_{i=1}^{2}\frac{3h_i}{8G_i}+\frac{h_3}{G_3}\right), \quad i=1,2 \tag{2.146}$$

which should be avoided.

2.4.7.2 Fracture of IC chip

In-plane stresses are induced in the outer members by the traction force F that could potentially lead to tensile fracture of the members, especially if they are made of brittle materials, such as for an IC device in an IC package, which is made of pure silicon that is highly susceptible to tensile fracture. The in-plane stresses in the member have been analysed in the bilayer structure and are given below

$$\sigma_{xi} = \frac{2F(3Z-1)}{h_i}+\frac{6(2Z-1)}{h_i^2}\int_L^x Q_a dx, \quad i=1,2 \tag{2.95bis}$$

To extend this equation to the sandwich structure, and considering the worst-case scenario, Q_a is replaced by $Q_a \mp Q_b$ (refers to Figure 2.16, where the top sign refers to member #1 and the bottom sign to member #2). Since $Q_b \gg Q_a$, then

Table 2.3 The basic operations

Parameters	λ_x	$d\lambda_x/dY$	κ_s	$d\kappa_s/dG_i$
h_3	$\sum_{i=1}^{2}\frac{1}{E_ih_i}\left(4+\frac{3h_3}{h_i}\right)$		$\sum_{i=1}^{2}\frac{h_i}{8G_i}+\frac{h_3}{G_3}$	
$h_3 \to 0$	$\sum_{i=1}^{2}\frac{4}{E_ih_i}$	$\sum_{i=1}^{2}\frac{3}{E_ih_i^2}$	$\sum_{i=1}^{2}\frac{h_i}{8G_i}$	$\frac{1}{G_3}$
$h_3 \to \infty$	$\sum_{i=1}^{2}\frac{3(\infty)}{E_ih_i^2}$		$\frac{(\infty)}{G_3}$	
h_1	$\sum_{i=1}^{2}\frac{1}{E_ih_i}\left(4+\frac{3h_3}{h_i}\right)$	$\frac{-2}{E_1h_1^2}\left(2+\frac{3h_3}{h_1}\right)$	$\sum_{i=1}^{2}\frac{h_i}{8G_i}+\frac{h_3}{G_3}$	
$h_1 \to 0$	$\frac{3h_3}{E_1(0)^2}$	$\frac{-6h_3}{E_1(0)^3}$	$\frac{h_2}{8G_2}+\frac{h_3}{G_3}$	$\frac{1}{8G_1}$
$h_1 \to \infty$	$\frac{1}{E_2h_2}\left(4+\frac{3h_3}{h_2}\right)$	$\frac{-4}{E_1(\infty)^2}$	$\frac{(\infty)}{8G_1}$	
E_3			$\sum_{i=1}^{2}\frac{h_i}{8G_i}+\frac{h_3}{G_3}$	$-\frac{h_3}{G_3{}^2}$
$E_3 \to 0$	$\sum_{i=1}^{2}\frac{1}{E_ih_i}\left(4+\frac{3h_3}{h_i}\right)$	0	$\frac{h_3}{(0)}$	$-\frac{h_3}{(0)^2}$
$E_3 \to \infty$			$\sum_{i=1}^{2}\frac{h_i}{8G_i}$	$-\frac{h_3}{(\infty)^2}$
E_1	$\sum_{i=1}^{2}\frac{1}{E_ih_i}\left(4+\frac{3h_3}{h_i}\right)$	$-\frac{1}{E_1^2h_1}\left(4+\frac{3h_3}{h_1}\right)$	$\sum_{i=1}^{2}\frac{h_i}{8G_i}+\frac{h_3}{G_3}$	$-\frac{h_1}{8G_1{}^2}$
$E_1 \to 0$	$\frac{1}{(0)h_1}\left(4+\frac{3h_3}{h_1}\right)$	$-\frac{1}{(0)^2h_1}\left(4+\frac{3h_3}{h_1}\right)$	$\frac{h_1}{8(0)}$	$-\frac{h_1}{8(0)^2}$
$E_1 \to \infty$	$\frac{1}{E_2h_2}\left(4+\frac{3h_3}{h_2}\right)$	$-\frac{1}{(\infty)^2h_1}\left(4+\frac{3h_3}{h_1}\right)$	$\frac{h_2}{8G_2}+\frac{h_3}{G_3}$	$-\frac{h_1}{8(\infty)^2}$

$Q_a \mp Q_b \approx \mp Q_b$. That is, $\int_L^x Q_a \mathrm{d}x \rightarrow \mp \int_L^x Q_b \mathrm{d}x$. Noting that $Q_b = h_3 f_s/2$ from Eqn (2.105) and $F = -\int_L^x f_s \mathrm{d}x$, this is further transformed to $\int_L^x Q_a \mathrm{d}x \rightarrow \mp \int_L^x Q_b \mathrm{d}x \rightarrow \pm F(x)$. Substituting this into (2.95bis) gives

$$\sigma_{xi} = \left[\frac{2(3Z-1)}{h_i} \pm \frac{3h_3(2Z-1)}{h_i^2}\right] F(x), \quad i = 1,2 \tag{2.147}$$

The maximum in-plane stress occurs in the middle of the member ($x = 0$); and for a positive $\sigma_{21}\Delta T$, the maximum in-plane stress occurs in member #1 at $Z = 1$, along its interface with member #3, that is,

$$\sigma_{x1,\max} = \left(\frac{4}{h_1} + \frac{3h_3}{h_1^2}\right) F(0) \tag{2.148}$$

Using the simple expression of $f_s(x) = \frac{A_s \cosh \beta x}{\sinh \beta L}$, the surface traction at $x = 0$ becomes

$$\begin{aligned} F(0) &= -\int_L^0 f_x \mathrm{d}x = \frac{\alpha_{21}\Delta T}{\lambda_x}\left(1 - \frac{1}{\sinh \beta L}\right) \\ &\approx \frac{\alpha_{21}\Delta T}{\lambda_x} \quad \text{for } \beta L > 3 \end{aligned} \tag{2.149}$$

whereupon substitution of the above into Eqn (2.148) gives

$$\sigma_{x1,\max} = \frac{\alpha_{21}\Delta T}{\lambda_x}\left(\frac{4}{h_1} + \frac{3h_3}{h_1^2}\right) \tag{2.150}$$

where $\lambda_x = \lambda_{x1} + \lambda_{x2} + \lambda_{x\theta}^*$. Putting $\lambda_x = X_1 + X_2$, where $X_i = \frac{4}{E_i h_i} + \frac{3h_3}{E_i h_i^2}$, and substituting into Eqn (2.150) gives

$$\sigma_{x1,\,\max} = \alpha_{21}\Delta T \frac{E_1 X_1}{X_1 + X_2} \tag{2.150a}$$

Differentiating Eqn (2.150a) with respect to h_1 gives

$$\frac{\mathrm{d}\sigma_{x1,\max}}{\mathrm{d}h_1} = \frac{E_1\alpha_{21}\Delta T}{(X_1 + X_2)^2}\left(X_2 \frac{\mathrm{d}X_1}{\mathrm{d}h_1}\right), \quad \frac{\mathrm{d}\sigma_{x1,\max}}{\mathrm{d}h_2} = \frac{E_1\alpha_{21}\Delta T}{(X_1 + X_2)^2}\left(-X_1 \frac{\mathrm{d}X_2}{\mathrm{d}h_2}\right) \tag{2.150b}$$

Since X_i is always positive and dX_i/dh_i always negative, then $d\sigma_{x1,max}/dh_1 = -ve$ and $d\sigma_{x1,max}/dh_2 = +ve$. Thus, the risk of IC chip fracture may be lowered by increasing its thickness while reducing the thickness of the substrate.

Differentiating Eqn (2.150a) with respect to E_i, and after some manipulation, leads to

$$\frac{d\sigma_{x1,max}}{dE_1} = \frac{\alpha_{21}\Delta T}{(X_1+X_2)^2}X_1^2, \quad \frac{d\sigma_{x1,max}}{dE_2} = \frac{\alpha_{21}\Delta T}{(X_1+X_2)^2}\frac{E_1 X_1 X_2}{E_2} \tag{2.150c}$$

Thus, the risk of IC chip fracture may be lowered by reducing the modulus of the substrate and (if possible) the IC chip. Since $d\sigma_{x1,max}/dE_3 = 0$, the modulus of the adhesive layer has no impact on the magnitude of the fracture stress in the IC chip.

Let $a_i + b_i h_3 = \frac{4}{E_i h_i} + \frac{3h_3}{E_i h_i^2}$, and substituting into Eqn (2.150) gives

$$\sigma_{x1,max} = E_1\alpha_{21}\Delta T\frac{a_1 + b_1 h_3}{(a_1+a_2)+(b_1+b_2)h_3} \tag{2.150d}$$

Differentiating Eqn (2.150d) with respect to h_3 leads to

$$\begin{aligned}\frac{d\sigma_{x1,max}}{dh_3} &= \frac{E_1\alpha_{21}\Delta T}{[(a_1+a_2)+(b_1+b_2)h_3]^2}(a_2 b_1 - a_1 b_2)\\ &= \frac{12\alpha_{21}\Delta T}{E_2 h_1 h_2[(a_1+a_2)+(b_1+b_2)h_3]^2}\left(\frac{1}{h_1}-\frac{1}{h_2}\right)\end{aligned} \tag{2.150e}$$

In general, the thickness of the IC chip is smaller than that of the substrate; that is, $(1/h_1 - 1/h_2) > 0$; thus, increasing the thickness of the bonding adhesive increases the risk of IC chip fracture.

2.4.7.3 *Warping of microelectronic assembly*

The microelectronic assembly will be attached onto a PCB through arrays of solder balls. This requires the solder balls (and hence the substrate) to be reasonably planer. However, the mismatched thermal expansion inevitably leads to out-of-plane deformation of microelectronic assembly.

The z-deflection of the neutral axis of the members consists of two components, one attributed to bending and another attributed to shearing, that is, $w_i = w_{bi} + w_{si}$. From Eqn (2.48) and boundary conditions: $w_{bi}(0) = dw_{bi}(0)/dx = 0$ and $d^2w_{bi}(L)/dx^2 = d^3w_{bi}(L)/dx^3 = 0$, the relative z-deflection of the neutral axis of the member due to bending between $x = 0$ and $x = L$ is given by

$$\Delta w_{bi} = \frac{h_i}{2D_i}\int_0^L\int_0^\xi\int_L^\varsigma f_s(\gamma)d\gamma d\varsigma d\xi \pm \frac{1}{D_i}\int_0^L\int_0^\xi\int_L^\upsilon\int_L^\zeta f_a(\gamma)d\gamma d\zeta d\upsilon d\xi \tag{2.151}$$

Ignoring the contribution from f_a and for $\alpha_Z L$, $\beta L > 3$ Eqn (2.151) is reduced to

$$\Delta w_{bi} \approx \frac{h_i \kappa_s \alpha_{21} \Delta T (\beta^2 L^2 - 2)}{2D_i \lambda_x^2} \tag{2.152}$$

Substituting Eqn (2.44) into Eqn (2.58) and using the boundary conditions, $w_{si}(0) = \mathrm{d}w_{si}(0)/\mathrm{d}x = 0$, the z-deflection of the neutral axis of the member due to shearing is

$$\Delta w_{si} = \frac{1}{8G_i} \int_0^L f_s(\varsigma)\mathrm{d}\varsigma + \frac{3}{4G_i h_i} \int_0^L \int_0^\xi f_a(\zeta)\mathrm{d}\zeta \mathrm{d}\xi \tag{2.153}$$

Neglecting the contribution from f_a and for $\alpha_z L$, $\beta L > 3$, Eqn (2.153) is reduced to

$$\Delta w_{si} \approx \frac{\alpha_{21} \Delta T}{8G_i \lambda_x} \tag{2.154}$$

Combining Eqn (2.152) and Eqn (2.154), we have

$$\Delta w_i \approx \alpha_{21} \Delta T \left[\frac{h_i \kappa_s (\beta^2 L^2 - 2)}{2D_i \lambda_x^2} + \frac{1}{8G_i \lambda_x} \right] \tag{2.155}$$

Thus, the out-of-plane deformation of the arrays of solder balls on the substrate of a microelectronic assembly may be reduced by increasing the flexural rigidity of the substrate while increasing the in-plane compliance of the microelectronic assembly.

2.5 PCB assembly as a sandwich structure with a layer of solder joints

Many trilayer structures are made of two relatively rigid layers that are bonded by discrete joints. An electronic assembly that is made up of an IC component interconnected to a PCB using solder joints is one such example. This assembly is referred to as a PCB assembly. The solder joints in a PCB assembly are susceptible to creep fatigue due to excessive stresses. A two-step approach will be used to model the stresses in the solder joints. In the first step, the PCB assembly will be modelled as a sandwich structure with a continuous bonding layer for which the interfacial stresses are evaluated as discussed in Section 2.4. The solder joints are modelled as short beams, which are acted upon by an axial force F_a, a shear force F_s, and a moment M_s. The axial and shear forces are evaluated from integrating the interfacial axial stress f_a and shear stress f_s, respectively, over an effective area around an individual solder joint. The moment M_s is obtained from the moment equilibrium of the short beam.

The interfacial shear stress f_s and the interfacial normal stress are repeated below:

$$f_s = A_s\left(\frac{\cosh \beta x}{\sinh \beta L} - \coth \beta L \frac{\cosh n\beta x}{\cosh n\beta L}\right) \tag{2.117bis}$$

$$f_a(x) \approx C_s^*\left\{e^{\alpha_z(x-L)}[C_1 \sin \alpha_z(x-L) + C_2 \cos \alpha_z(x-L)] + e^{\beta(x-L)}\right\} \tag{2.128bis}$$

where

$$A_s = \frac{\alpha_{21}\Delta T}{\beta\kappa_s} \tag{2.37bis}$$

$$C_s^* = \frac{\overline{D}_{21}^*\alpha_{21}\Delta T}{\lambda_z\kappa_s(4\alpha_z^4 + \beta^4)} \tag{2.129bis}$$

$$C_1 = -\frac{\beta^2}{2\alpha_z^2} \qquad C_2 \approx C_1 - \frac{2\alpha_z}{\beta} \tag{2.130bis}$$

Note that the compliances of the sandwiched layer are now a smeared property of the discrete joints:

$$\kappa_{s3} = \frac{h_3}{n \cdot G_3 A_3} \tag{2.156}$$

$$\lambda_{z3} = \frac{h_3}{n \cdot E_3 A_3} \tag{2.157}$$

where A_3 is the cross-sectional area of a single discrete member and G_3 is the shear modulus of the discrete member; n is the number of discrete members per unit area. This evaluates to $n = 1/(p \cdot q)$ for discrete members that are distributed uniformly with a pitch p in the x-direction and a pitch q in the y-direction. In this case,

$$\kappa_{s3} = \frac{pqh_3}{G_3 A_s} \tag{2.158}$$

$$\lambda_{z3} = \frac{pqh_3}{E_3 A_s} \tag{2.159}$$

2.5.1 Average stresses in the discrete solder joints

Modelling the solder joints as an area array with a spacing pitch of $p \cdot q$, the axial force F_a and shear force F_s, in the solder joint that is at distance Γ from the middle of a PCB assembly can be obtained through definite integration of $f_a(x)$ and $f_s(x)$ over a pitch, p, between the limit $(\Gamma - p/2)$ to $(\Gamma + p/2)$ and for a width of q. This is shown in Figure 2.25.

The magnitudes of the forces are

$$F_s(\Gamma) = q \int_{\Gamma - p/2}^{\Gamma + p/2} f_s(x)\mathrm{d}x \tag{2.160}$$

$$F_a(\Gamma) = q \int_{\Gamma - p/2}^{\Gamma + p/2} f_a(x)\mathrm{d}x \tag{2.161}$$

A moment of magnitude

$$M_s(\Gamma) = \frac{h_3 F_s(\Gamma)}{2} \tag{2.162}$$

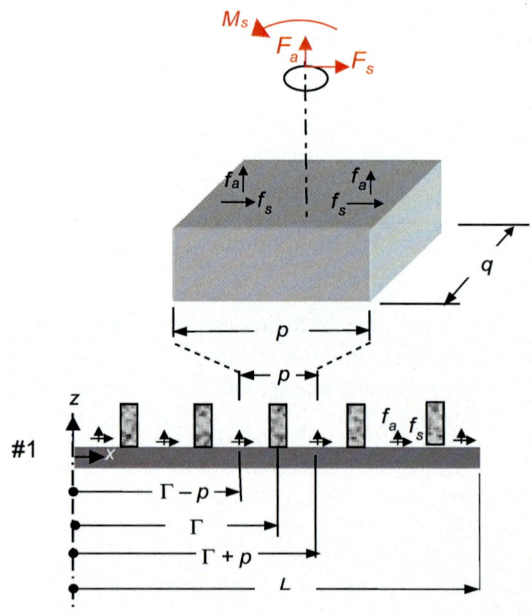

Figure 2.25 Integration of interfacial stresses into forces over an area equivalent to that covered by a single discrete member.

acts at each end of the discrete member in order to maintain equilibrium. It is noted that the moment M_s is equivalent to: $q\int_{\Gamma-p/2}^{\Gamma+p/2} Q_b(x)\mathrm{d}x$, where Q_b is the linear component of the sectional force that could exist only in a continuous sandwiched layer. Thus, the linear component of the sectional force is replaced by a moment $M_s(\Gamma)$ for a discrete sandwiched layer.

Comparing with a continuous layer, a discrete member has additional shear compliance in that it has the new freedom of undergoing flexural deformation. Figure 2.26 shows the u-direction deformation of a discrete member due to bending.
The moment along the member at distance z away from the top end is given by

$$m_s(z) = M_s - F_s z = F_s\left(\frac{h_3}{2} - z\right) \tag{2.163}$$

The differential u-displacement between the two ends of the member is

$$\Delta u_{b3} = \frac{1}{E_3 I_3}\int_0^{h_3}\int_0^{z} m_s(\zeta)\mathrm{d}\zeta\mathrm{d}z = \frac{F_s h_3^3}{12E_3I_3} \tag{2.164}$$

The shear compliance due to the flexing of the discrete member is defined as $\Delta u_{b3}/\bar{f}_s$, where $\bar{f}_s$ is the average shear stress over the area of integration. Putting $F_s = \bar{f}_s pq$, the shear compliance becomes

$$\kappa_{s3\vartheta} = \frac{\Delta u_{b3}}{\bar{f}_s} = \frac{h_3^3 pq}{12E_3I_3} \tag{2.165}$$

The total shear compliance of a sandwich structure with a discrete sandwiched layer is therefore

$$\kappa_s = \kappa_{s1}^* + \kappa_{s2}^* + \kappa_{s3} + \kappa_{s3\vartheta} \tag{2.166}$$

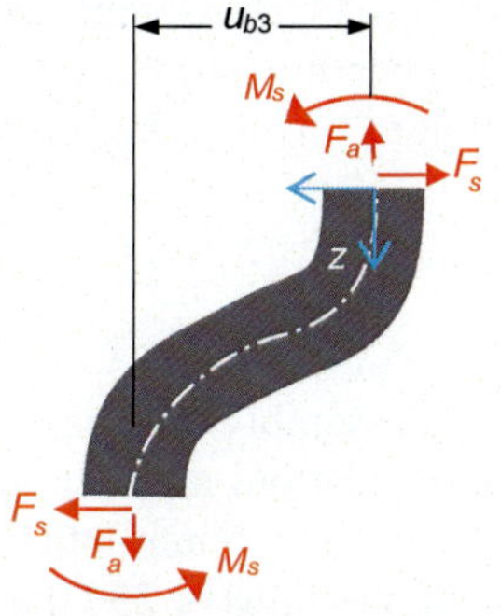

Figure 2.26 The deformation of a discrete member under shearing and bending.

By assuming the discrete member to be uniform in its cross-section with area A_3, the average shear and average axial stresses in the member are given, respectively, by

$$\bar{\tau}(\Gamma) = \frac{q}{A_3} \int_{\Gamma-p/2}^{\Gamma+p/2} f_s(x)\mathrm{d}x \tag{2.167}$$

$$\bar{\sigma}_a(\Gamma) = \frac{q}{A_3} \int_{\Gamma-p/2}^{\Gamma+p/2} f_a(x)\mathrm{d}x \tag{2.168}$$

Substituting Eqn (2.117) and Eqn (2.128), respectively, into the above equations gives

$$\bar{\tau}(\Gamma) = \frac{qA_s}{\beta A_3}\left[\frac{\sinh \beta x}{\sinh \beta L} - \frac{\coth \beta L}{n}\,\frac{\sinh n\beta x}{\cosh n\beta L}\right]_{\Gamma-p/2}^{\Gamma+p/2} \tag{2.169}$$

$$\begin{aligned}\bar{\sigma}_a(\Gamma) = \frac{qC_s^*}{A_3}\Bigg[\frac{\mathrm{e}^{\beta(x-L)}}{2\alpha_z}&[(C_1 + C_2)\sin \alpha_z(x-L) \\ &-(C_1 - C_2)\cos \alpha_z(x-L)] + \frac{\mathrm{e}^{\beta(x-L)}}{\beta}\Bigg]_{\Gamma+p/2}^{\Gamma-p/2}\end{aligned} \tag{2.170}$$

where $\beta = \sqrt{\lambda_x/\kappa_s}$, $\lambda_x = \lambda_{x1} + \lambda_{x2} + \lambda_{x\theta}^*$, $\kappa_s = \kappa_{s1}^* + \kappa_{s2}^* + \kappa_{s3} + \kappa_{s3\vartheta}$.

Assuming further that the discrete member is made of a cylindrical beam and behaves as a perfect beam – plane section remains plane – under the bending moment M_s, the bending stress on the outer fibre of the beam becomes

$$\bar{\sigma}_b(\Gamma) = \frac{d_3 M_s(\Gamma)}{2I_3} = \frac{4h_3}{d_3}\left(\frac{F_s(\Gamma)}{A_3}\right) = \frac{4h_3}{d_3}\bar{\tau}(\Gamma) \tag{2.171}$$

The ratio $4h_3/d_3$ of a solder joint in a PCB assembly is typically larger than 1, which suggests that the magnitude of the bending stress in the solder joints of a PCB assembly will always be larger than that of the shearing stress.

2.5.2 Validations

The FE model used in the validation analysis is depicted in Figure 2.27(a), which shows a symmetric half of the PCB assembly. The basic dimensions and material properties for the models used in the validations are identical to those listed in Table 2.1. The diameter of the cylindrical joints is 0.5 mm and the joints are spaced uniformly with a pitch of 1 × 1 mm. A stretch modulus of 25 GPa is assigned to the solder joints.

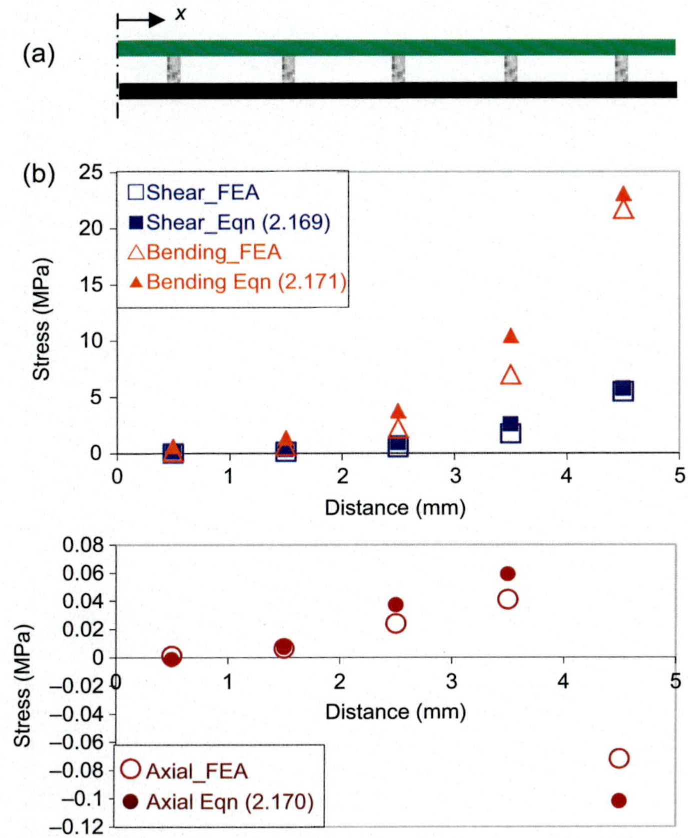

Figure 2.27 (a) FE model and (b) comparisons of analytical solutions and finite element analysis (FEA) for discrete interconnects.

The IC package and the PCB are modelled using 10 Timoshenko beam elements with appropriate thickness offset, while each interconnect is modelled using a single Timoshenko beam element. The FE results are compared with the corresponding analytical solutions and are displayed in Figure 2.27(b), which shows good agreement between the two solutions.

2.5.3 Robust design analysis

The discrete member that is near the edge experiences the highest magnitude of stress. From Eqn (2.171) and Eqn (2.169), the maximum bending stresses on the fibre of this cylindrical discrete element are given by

$$\overline{\sigma}_{b,\max} = \frac{16h_3qA_s}{\pi d_3^3\beta}\left[\frac{\sinh \beta x}{\sinh \beta L} - \frac{\coth \beta L}{n}\,\frac{\sinh n\beta x}{\cosh n\beta L}\right]_{L-p}^{L} \tag{2.172}$$

Table 2.4 Analysis of $\sigma_{\text{peeling,max}}$

Parameters		$\sigma_{\text{peeling,max}}$	$\frac{d\sigma_{\text{peeling,max}}}{dY}$	Illustration
h_3	$h_3 \to 0$	∞	$-\infty$	
	$h_3 \to \infty$	0	-0	
h_1	$h_1 \to 0$	0	$\frac{16\alpha_{21}\Delta T b G_2 G_3}{G_1 G_3 h_2 + G_2(8G_1 h_3)}$	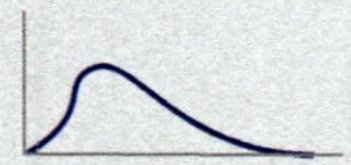
	$h_1 \to \infty$	0	-0	
G_3	$E_3 \to 0$	$+0$	$+\infty$	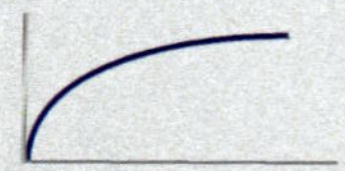
	$E_3 \to \infty$	$\frac{2\alpha_{21}\Delta T}{\left(\sum_{i=1}^{2}\frac{4+\frac{3h_3}{h_2}}{E_2 h_2}\right)^a \left(\sum_{i=1}^{2}\frac{h_2}{8G_2}\right)^b h_3^c}$	0	
E_1	$E_1 \to 0$	$+0$	$2\alpha_{21}\Delta T \cdot 8^b h_1^{-2-b} h_3^{-c}\left(\frac{4h_1+3h_3}{h_1^2}\right)^{-1-a} \times \left(4b + a\left[4+\frac{3h_3}{h_1}\right]h_1 + 3bh_3\right)$	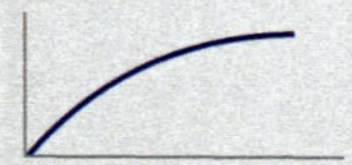
	$E_1 \to \infty$	$\frac{2\alpha_{21}\Delta T}{\left(\frac{4+\frac{3h_3}{h_2}}{E_2 h_2}\right)^a \left(\frac{h_2}{8G_2}+\frac{h_3}{G_3}\right)^b h_3^c}$	0	

Note: $a = 0.68$, $b = 1 - a$ and $c = 2a - 1$.

Ignoring the second term in the bracket for reason of simplicity, but with little loss of generosity

$$\overline{\sigma}_{b,\max} = \frac{16 h_3 q \alpha_{21} \Delta T}{\pi {d_3}^3 \lambda_x} f(\beta x, \beta L) \tag{2.173}$$

where

$f(\beta p, \beta L) = [\sinh \beta x]_{L-p/2}^{L+p/2} / \sinh \beta L = 2 \coth \beta L \sinh(\beta p/2) \approx 2 \sinh(\beta p/2)$.

Substituting this back into Eqn (2.173) gives

$$\overline{\sigma}_{b,\max} = \frac{32 q \alpha_{21} \Delta T}{\pi {d_3}^3} \frac{h_3}{\lambda_x} \sinh \frac{\beta p}{2} \tag{2.174}$$

Table 2.5 **Analysis of $\overline{\sigma}_{b,\max}$**

Parameters		$\overline{\sigma}_{b,\max}$	$\frac{\mathrm{d}\overline{\sigma}_{b,\max}}{\mathrm{d}Y}$	Illustration
h_3	$h_3 \to 0$	0	$\frac{\sinh\left[\sqrt[1/2p]{\frac{\sum_{i=2}^{2}\frac{4}{E_i h_i}}{\sum_{i=2}^{2}\frac{h_i}{8G_i}}}\right]}{\sum_{i=2}^{2}\frac{4}{E_i h_i}}$	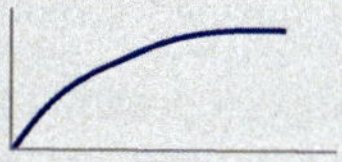
	$h_3 \to \infty$	$\frac{\sinh\left[\sqrt[1/2p]{\frac{\frac{1}{E_1 h_1^2}+\frac{1}{E_2 h_2^2}}{\frac{1}{G_3}}}\right]}{\frac{1}{E_1 h_1^2}+\frac{1}{E_2 h_2^2}}$	$+0$	
h_1	$h_1 \to 0$	∞	$-\infty$	
	$h_1 \to \infty$	0	-0	
E_3	$E_3 \to 0$	0	$+\infty$	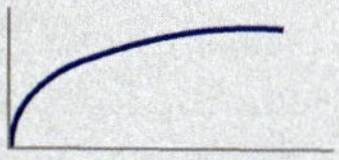
	$E_3 \to \infty$	$\frac{\sinh\left[\sqrt[1/2p]{\frac{\sum_{i=2}^{2}\frac{4+\frac{3h_3}{h_i}}{E_i h_i}}{\sum_{i=2}^{2}\frac{h_i}{8G_i}}}\right] h_3}{\sum_{i=2}^{2}\frac{4+\frac{3h_3}{h_i}}{E_i h_i}}$	0	
E_1	$E_1 \to 0$	0	$\frac{\mathrm{e}\sqrt[p]{\frac{8h_1+6h_3}{h_1^3}}\, h_1^2 h_3}{4h_1+3h_3}$	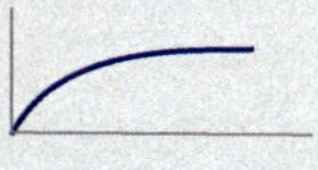
	$E_1 \to \infty$	$\frac{\mathrm{e}\sqrt[\frac{1}{2p}]{\frac{\frac{4+\frac{3h_3}{h_2}}{e_2 h_2}}{\frac{h_2}{8G_2}+\frac{h_3}{G_3}}}\, h_3}{\frac{4+\frac{3h_3}{h_2}}{e_2 h_2}}$	0	

It is obvious that the maximum magnitude of the bending stress on the discrete member can be reduced by

- increasing its diameter;
- reducing the pitch of the neighboring members,
- increasing the stretch compliance λ_x of the structure, and
- decreasing the characteristic constant $\beta = \sqrt{\lambda_x/\kappa_s}$ of the structure.

The last two conditions are obviously contradictory. Nevertheless, it can be deduced that reducing the moduli of the constituent materials or increasing the thickness of the adhesive layer will increase the stretch and the shear compliances, but will alter little of β and, hence, will be beneficial. By contrast, increasing the thickness of the outer layers will reduce the stretch compliance (negative) while also reducing the characteristic constant β (positive). The exact effect is not clear.

A detailed parametric analysis has been performed and these are summarised in Table 2.5. Thus, maximum magnitude of the bending stress on the discrete member can be reduced by

- reducing the length of the discrete member,
- increasing the thickness of the outer members, and
- reducing the moduli of the discrete and the outer members.

Comparing with Table 2.4, it is noted that while it is beneficial to have a thick adhesive layer, a short discrete element is preferred.

References

Aleck, B. (1949). Thermal stresses in a rectangular plate clamped along an edge. *ASME Journal of Applied Mechanics, 16*, 118–122.

Chen, D., Cheng, S., & Gerhardt, T. D. (1982). Thermal stresses in laminated beams. *Journal of Thermal Stresses, 5*, 67–84.

Chen, W. T., & Nelson, C. W. (1979). Thermal stresses in bolted joints. *IBM Journal of Research and Development, 23*, 178–188.

Goland, M., & Reissner, E. (1944). Stresses in cemented joints. *ASME Journal of Applied Mechanics, 11*, A17–A27.

Heinrich, S. S. (1997). Improved analytical estimate of global CTE mismatch displacement in area-array solder joints. *ASME Transactions, Journal of Electronic Packaging, 119*, 218–227.

Jiang, Z. Q., Huang, Y., & Chandra, A. (1997). Thermal stresses in layered electronic assemblies. *ASME Transactions, Journal of Electronic Packaging, 119*, 1127–1132.

Kuo, A. (1989). Thermal stress at the edge of a bi-metallic thermostat. *ASME Journal of Applied Mechanics, 56*, 585–589.

Lee, M., & Jasiuk, I. (1991). Asymptotic expansions for the thermal stresses in bonded semi-infinite biomaterial strips. *ASME Journal of Electronic Packaging, 113*, 173–177.

Pao, Y. H., & Eisele, E. (1991). Interfacial shear and peel stresses in multi-layered thin stacks subjected to uniform thermal loading. *ASME Journal of Electronic Packaging, 113*, 164–172.

Stoney, G. G. (1909). The tension of metallic films deposited by electrolysis. *Proceedings of the Royal Society London* (Vol. 82). London: Royal Society London.

Suhir, E. (1986). Stresses in bi-metal thermostats. *ASME Journal of Applied Mechanics, 53*, 657–660.
Suhir, E. (1988a). An approximate analysis of stresses in multilayered elastic thin films. *ASME Journal of Applied Mechanics, 55*, 143–148.
Suhir, E. (1988b). In A. B.-C. Kraus (Ed.), *Thermal stress failures in microelectronic components – Review and extension, advances in thermal modeling of electronic components and systems* (Vol. 1). New York: Hemisphere.
Suhir, E. (1989). Interfacial stresses in bimetal thermostats. *ASME Journal of Applied Mechanics, 56*, 595–600.
Taylor, T. C., & Yuan, F. L. (1962). Thermal stress and fracture in shear-constrained semiconductor device structures. *IRE Tansactions on Electronic Devices, 9*, 303–308.
Timoshenko, S. (1925). Analysis of bi-metal thermostats. *Journal of the Optical Society of America, 11*, 233–255.
Volkersen, O. (1938). Die nietkraftverteilung in zugbeanspruchten mit konstanten laschenquerschritten. *Luftfahrtforschung, 15*, 41–47.
Wen, Y., & Basaran, C. (2004). An analytical model for thermal stress analysis of multi-layered microelectronic packaging. *Mechanics of Materials, 36*, 369–385.
Willams, H. (1985). Asymptotic analysis of the thermal stresses in a two-layer composite with an adhesive layer. *Journal of Thermal Stresses, 8*, 183–203.
Wong, E. H., Lim, K.-M., & Mai, Y.-W. (2009). Analytical solutions for PCB assembly subjected to mismatched Thermal expansion. *IEEE Transactions on Advanced Packaging, 32*, 602–611.

Appendix: prior published works in thermoelasticity

This section details prior articles that contributed significantly to the development of the analytical work in thermoelasticity and hence to its applications for electronic packaging. Instead of reviewing the articles in chronological order, which does present a good review of historical development, this may not be ideal for developing understanding. Instead, the articles are reviewed in the order of logical development, which we believe shall facilitate reading of the subject.

A. Stoney (1909, p. 172) *The Tension of Metallic Films Deposited by Electrolysis*

This paper provides an estimation of the uniform in-plane stress in a thin film deposited on a substrate when the bi-material layers experience differential expansion/contraction such that the structure deforms into a shape with spherical radius of curvature (equi-biaxial was assumed). This condition may be valid in the inner region of a plane structure, away from the boundary.

Assumptions: (1) the individual members are linear elastic and isotropic; (2) the members are modelled as plates such that plane sections remain plane and the members experience equi-bending about the x and y axes; (3) the individual member experiences no transverse and shear stresses; and (4) the members have identical curvatures.

The structure is depicted in Figure 2.A.

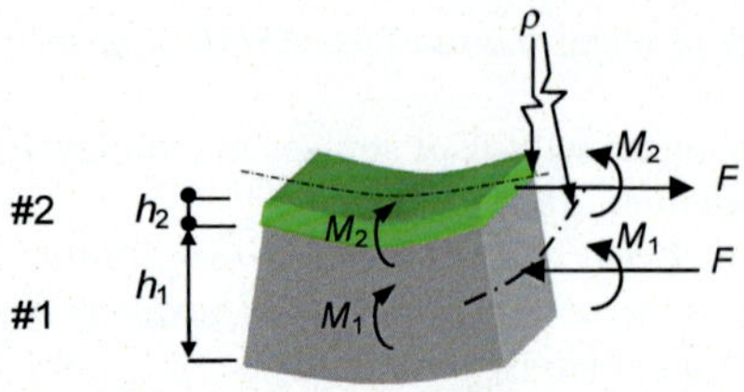

Figure 2.A Free-body diagram of the structure analysed by Stoney.

The equilibrium of the bilayer structure gives

$$M_1 + M_2 - \frac{h_1 + h_2}{2} F = 0 \qquad \text{A(2.1)}$$

The curvature of a plate subjected to equi-biaxial bending moments per length of moment is given by

$$\frac{1}{\rho} = \frac{M}{E'I} \qquad \text{A(2.2)}$$

where $E' = E/(1 - \nu^2)$ is the bi-axial modulus. Substituting A(2.1) into A(2.2) gives the axial force per unit length as

$$F = \frac{2\left(E_1' I_1 + E_2' I_2\right)}{\rho(h_1 + h_2)} \qquad \text{A(2.3)}$$

For the thickness of member #2 that is much smaller than member #1; that is, $h_2 \ll h_1$, A(2.3) is reduced to

$$F \approx \frac{E_1' h_1^2}{6\rho} \qquad \text{A(2.4)}$$

The average axial stress in member #2, assuming uniform distribution, is simply

$$\sigma_2 \approx \frac{E_1' h_1^2}{6\rho h_2} \qquad \text{A(2.5)}$$

B. Timoshenko (1925), *Analysis of Bi-Metal Thermostat*

This paper provides estimates of the radius of curvature, deflection, and the maximum in-plane stress in a bi-metal structure, when the structure experiences uniform temperature loading. The structure is assumed to deform into a uniform radius of curvature along its length of interest.

Assumptions: (1) the individual members are linear elastic and isotropic; (2) the members are modelled as simple beams such that plane sections in the members remain plane; (3) the individual member experiences no transverse and shear stresses; (4) the members have identical curvatures; and (5) the magnitude of moment, curvature, stress and strain in the individual member is uniform along its length.

The structure is depicted in Figure 2.B.

The condition of equilibrium gives

$$F = \frac{2(D_1 + D_2)}{\rho h} \quad \text{B(2.1)}$$

The condition of compatibility of strain along the interface is

$$F = \frac{\alpha_{12}\Delta T - \frac{h}{2\rho}}{\lambda_{x1} + \lambda_{x2}} \quad \text{B(2.2)}$$

where $D_i = E_1 h_i^3/12$ and $\lambda_{xi} = 1/E_i h_i$. Equating the two expressions gives the curvature

$$\frac{1}{\rho} = \frac{\alpha_{12}\Delta T}{\frac{h}{2} + \frac{2(D_1+D_2)(\lambda_{x1}+\lambda_{x2})}{h}} \quad \text{B(2.3)}$$

The deflection of the bi-metal over a predeformed length of ℓ is given by

$$\delta = \frac{\ell^2}{8\rho} \quad \text{B(2.4)}$$

The maximum normal stresses in the member in the direction of the length of the structure are the total of the axial stress and the bending stress at the surface fibre given by

$$\sigma_{i,\max} = \frac{F}{h_i} + \frac{E_i h_i}{2\rho}; \quad i = 1, 2 \quad \text{B(2.5)}$$

C. Chen and Nelson (1979), *Thermal Stress in Bonded Joints*

This paper provides a comprehensive treatment of thermal stress in multilayer structure covering multilayer plate and circular disc. The solutions are developed

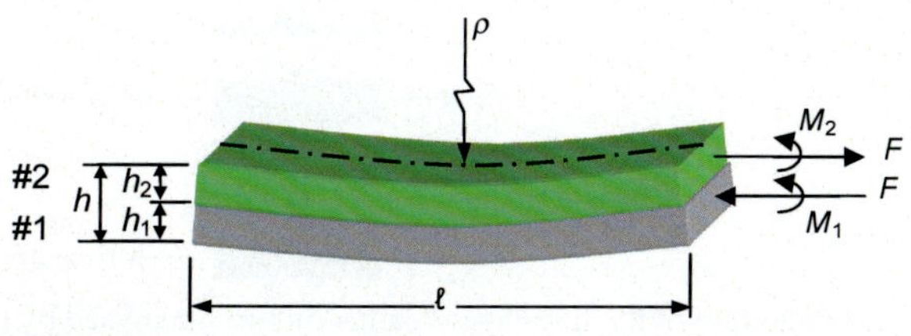

Figure 2.B Free-body diagram of structure analysed by Timoshenko.

from first principle using infinitesimal elements. The elastic layers are modelled as plates, and the bonding layer is modelled as a distribution of shear and normal springs. Shear deformation of the elastic members is ignored. Several cases are analysed: (1) three plate elastic layers with two bonding layers in which the elastic layers are assumed to be absolutely rigid to bending; (2) two plate elastic layers with one bonding layer in which the elastic layers are assumed to be absolutely rigid to bending; (3) two circular elastic layers with one bonding layer in which the elastic layers are assumed to be absolutely rigid to bending; and (4) two plate elastic layers with one bonding layer in which the elastic layers are assumed to be compliant in bending. The bonding layer was assumed to be a distributed spring layer. Only cases (1) and (4) are discussed below.

C1. Three elastic layers with two bond layers – no bending

Assumptions: (1) the adherents are linear elastic and isotropic; (2) the adherents are modelled as truss such that the members are capable of only uniform tension or compression stress along its length; (3) the bond layers are modelled as a distribution of shear springs capable of transferring only shear loads, but this offers no stiffness in normal directions and in bending.

The sandwich beam structure is shown in Figure 2.C1.

Force equilibrium equations of the elastic layers:

$$\begin{aligned} \frac{dF_1}{dx} + \tau_4 &= 0 \\ \frac{dF_2}{dx} + \tau_5 - \tau_4 &= 0 \qquad \text{C(2.1)} \\ \frac{dF_3}{dx} - \tau_5 &= 0 \end{aligned}$$

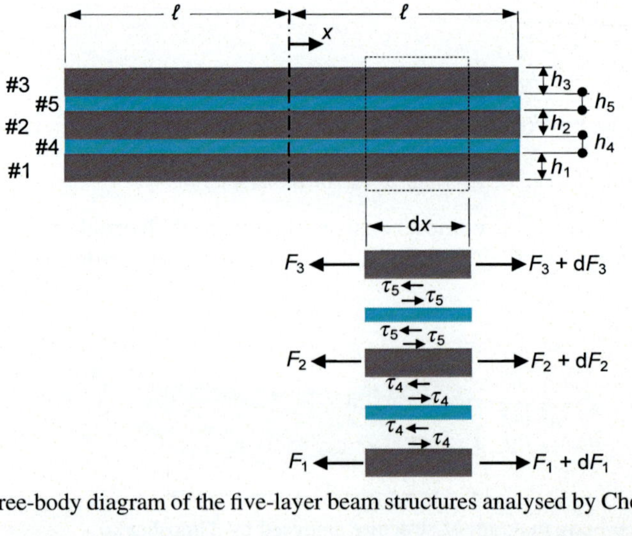

Figure 2.C1 Free-body diagram of the five-layer beam structures analysed by Chen and Nelson.

The constitutive relations of the elastic layers:

$$F_i = \frac{1}{\lambda_{xi}}\left(\frac{\mathrm{d}u_i}{\mathrm{d}x} - \alpha_i T\right), \quad i = 1, 2, 3 \qquad \text{C(2.2)}$$

where $\lambda_{xi} = 1/E_i h_i$, and the constitutive relations of the bonding layers:

$$\begin{aligned} \frac{\tau_4}{G_4} &= \frac{u_2 - u_1}{h_4} \\ \frac{\tau_5}{G_5} &= \frac{u_3 - u_2}{h_5} \end{aligned}, \quad i = 1, 2, 3 \qquad \text{C(2.3)}$$

Combining Eqns C(2.1)–C(2.3) gives the two differential equations:

$$\begin{aligned} \frac{\mathrm{d}^2\tau_4}{\mathrm{d}x^2} &= \frac{G_4}{h_4}[(\lambda_{x1} + \lambda_{x2})\tau_4 - \lambda_{x2}\tau_5] \\ \frac{\mathrm{d}^2\tau_5}{\mathrm{d}x^2} &= \frac{G_5}{h_5}[(\lambda_{x2} + \lambda_{x3})\tau_5 - \lambda_{x2}\tau_4] \end{aligned} \qquad \text{C(2.4)}$$

which can be combined to give

$$\begin{aligned} &\frac{\mathrm{d}^4\tau_4}{\mathrm{d}x^4} - \left[\frac{G_4}{h_4}(\lambda_{x1} + \lambda_{x2}) + \frac{G_5}{h_5}(\lambda_{x2} + \lambda_{x3})\right]\frac{\mathrm{d}^2\tau_4}{\mathrm{d}x^2} \\ &\quad + \frac{G_4 G_5}{h_4 h_5}[\lambda_{x1}(\lambda_{x2} + \lambda_{x3}) + \lambda_{x2}\lambda_{x3}]\tau_4 = 0 \end{aligned} \qquad \text{C(2.5)}$$

and a similar expression for τ_5. The general solutions for τ_4 and τ_5 are

$$\begin{aligned} \tau_4 &= A_1 \sinh \beta_1 x + A_2 \sinh \beta_2 x + A_3 \cosh \beta_1 x + A_4 \cosh \beta_2 x \\ \tau_5 &= A_1 k_1 \sinh \beta_1 x + A_2 k_2 \sinh \beta_2 x + A_3 k_3 \cosh \beta_1 x + A_4 k_4 \cosh \beta_2 x \end{aligned} \qquad \text{C(2.6)}$$

where $\pm\beta_1$ and $\pm\beta_2$ are the roots of the equation

$$\begin{aligned} &\beta^4 - \left[\frac{G_4}{h_4}(\lambda_{x1} + \lambda_{x2}) + \frac{G_5}{h_5}(\lambda_{x2} + \lambda_{x3})\right]\beta^2 \\ &\quad + \frac{G_4 G_5}{h_4 h_5}[\lambda_{x1}(\lambda_{x2} + \lambda_{x3}) + \lambda_{x2}\lambda_{x3}]\beta = 0 \end{aligned} \qquad \text{C(2.7)}$$

Applying symmetric boundary conditions yields:

$$\begin{aligned}\tau_4 &= C_1 \frac{\sinh \beta_1 x}{\cosh \beta_1 \ell} + C_2 \frac{\sinh \beta_2 x}{\cosh \beta_2 \ell} \\ \tau_5 &= C_1 k_1 \frac{\sinh \beta_1 x}{\cosh \beta_1 \ell} + C_2 k_2 \frac{\sinh \beta_2 x}{\cosh \beta_2 \ell}\end{aligned} \qquad \text{C(2.8)}$$

where

$$k_i = \frac{1}{\lambda_{x2}}\left(\lambda_{x1} + \lambda_{x2} - \frac{\beta_i^2 h_4}{G_4}\right), \quad i = 1, 2.$$

$$C_1 = \frac{G_4 \lambda_{x2} \beta_1 (\overline{\alpha}_1 k_2 + \overline{\alpha}_3)}{h_4(\beta_1^2 - \beta_2^2)}, \quad C_2 = \frac{G_4 \lambda_{x2} \beta_2 (\overline{\alpha}_2 k_1 + \overline{\alpha}_3)}{h_4(\beta_1^2 - \beta_2^2)}$$

$$\overline{\alpha}_1 = \frac{(\lambda_{x1}\alpha_{32} - \lambda_{x3}\alpha_{21})\Delta T}{\lambda_{x1}\lambda_{x2} + \lambda_{x2}\lambda_{x3} + \lambda_{x3}\lambda_{x1}}, \quad \overline{\alpha}_2 = \frac{(\lambda_{x2}\alpha_{31} + \lambda_{x3}\alpha_{21})\Delta T}{\lambda_{x1}\lambda_{x2} + \lambda_{x2}\lambda_{x3} + \lambda_{x3}\lambda_{x1}},$$

$$\overline{\alpha}_3 = -\frac{(\lambda_{x1}\alpha_{32} + \lambda_{x2}\alpha_{31})\Delta T}{\lambda_{x1}\lambda_{x2} + \lambda_{x2}\lambda_{x3} + \lambda_{x3}\lambda_{x1}}$$

The maximum shear stress occurs at the free edge and, for very large value of length-to-joint thickness ratio, is given by

$$\begin{aligned}\tau_{4,\max} &= C_1 + C_2 \\ \tau_{5,\max} &= C_1 k_1 + C_2 k_2\end{aligned} \qquad \text{C(2.9)}$$

The solution can be reduced to that of two elastic layers and one bond layer by eliminating those terms containing the subscripts 3 and 5.

C2. Two elastic layers with one bond layer – bending

Assumptions: (1) the adherents are linear elastic and isotropic; (2) the adherents are modelled as plates such that plane sections in the members remain plane and the members experience bending about the *y*-axis and nil strain in the *y*-direction; (3) the adherents experience negligible transverse stress and strain; (4) the bond layer is modelled as a distribution of coil springs capable of transferring only transverse normal and shear loads, but offers no stiffness in the *x*- and *y*-normal directions and in bending.

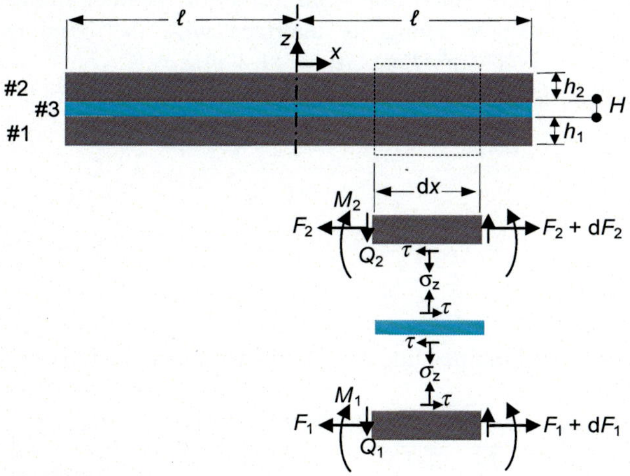

Figure 2.C2 Free-body diagram of the three-layer beam structures analysed by Chen and Nelson.

The structure is shown in Figure 2.C2; note the presence of the transverse stress. Moment equilibrium equations of the elastic members:

$$\frac{\mathrm{d}M_i}{\mathrm{d}x} + Q_i - \frac{\tau h_i}{2} = 0; \quad i = 1, 2 \qquad \text{C(2.10)}$$

Equilibrium of horizontal forces for the elastic members:

$$\frac{\mathrm{d}F_i}{\mathrm{d}x} \pm \tau = 0; \quad i = 1, 2; \ \text{+ve for } i = 1 \text{ and } -\text{ve for } i = 2 \qquad \text{C(2.11)}$$

Equilibrium of vertical forces for the elastic members:

$$\frac{\mathrm{d}Q_i}{\mathrm{d}x} \pm \sigma_z = 0; \quad i = 1, 2; \ \text{+ve for } i = 1 \text{ and } -\text{ve for } i = 2 \qquad \text{C(2.12)}$$

The moment–force–deflection relations of the elastic members give

$$\frac{\mathrm{d}^2 w_i}{\mathrm{d}x^2} = \frac{M_i}{D_i}; \quad i = 1, 2 \qquad \text{C(2.13)}$$

and

$$\frac{\mathrm{d}u_i}{\mathrm{d}x} = F_i \lambda_{xi} \mp \frac{h_i M_i}{2D_i} + \alpha_i' T; \ i = 1, 2; \ -\text{ve for } i = 1 \text{ and } +\text{ve for } i = 2 \qquad \text{C(2.14)}$$

where w_i is the z-displacement of the neutral axis of member i, while u_i is the x-displacement of member $\#i$ along its interface with the bonding layer: $D_i = E_i h_i^3/12(1-\nu_i^2)$, $\lambda_{xi} = 1-\nu_i^2/E_i h_i$, $\alpha_i' = (1+\nu_i)\alpha_i$.

The bonding layer is assumed to experience transversely uniform normal and shear strains:

$$\frac{\tau_3}{G_3} = \frac{u_2-u_1}{h_3}$$
$$\frac{\sigma_z}{E_3} = \frac{w_2-w_1}{h_3} \qquad \text{C(2.15)}$$

These equations are combined into a single differential equation of sixth order in σ_z:

$$\frac{\partial^6\sigma_z}{\partial x^6} - \frac{G_3 c}{h_3}\frac{\partial^4\sigma_z}{\partial x^4} + \frac{E_3 b}{h_3}\frac{\partial^2\sigma_z}{\partial x^2} - \frac{E_3 G_3(bc-a^2)}{h_3^2}\sigma_z = 0 \qquad \text{C(2.16)}$$

where the constants a, b, and c are defined as $a = 1/2(D_2h_2 - D_1h_1)$, $b = D_2 + D_1$, $c = 4(\lambda_{x1} + \lambda_{x2})$.

For symmetric boundary condition, the normal stress σ_z and shear stress τ are obtained from

$$\sigma_z = A_1 \cosh\beta_1 x + A_2\cosh\beta_2 x\cos\beta_3 x + A_3\sinh\beta_2 x\sin\beta_3 x$$
$$\tau = C_1\sinh\beta_1 x + C_2\sinh\beta_2 x\cos\beta_3 x + C_3\cosh\beta_2 x\sin\beta_3 x \qquad \text{C(2.17)}$$

where $\pm\beta_1$ and $\pm\beta_2 \pm i\beta_3$ are the roots of the algebraic equation:

$$\beta^3 - \frac{G_3\mathrm{c}}{h_3}\beta^2 + \frac{E_3\mathrm{b}}{h_3}\beta - \frac{E_3G_3(\mathrm{bc}-\mathrm{a}^2)}{h_3^2} = 0. \qquad \text{C(2.18)}$$

The constants A_1, A_2 and A_3 are the roots of the simultaneous equations:

$$A_1\beta_1^2\cosh\beta_1\ell + A_2\left[(\beta_2^2-\beta_3^2)\cosh\beta_2\ell\cos\beta_3\ell - 2\beta_2\beta_3\sinh\beta_2\ell\sin\beta_3\ell\right]$$
$$+A_3\left[(\beta_2^2-\beta_3^2)\sinh\beta_2\ell\sin\beta_3\ell + 2\beta_2\beta_3\cosh\beta_2\ell\cos\beta_3\ell\right] = 0 \qquad \text{C(2.19)}$$

$$A_1\frac{\sinh\beta_1\ell}{\beta_1} + A_2\left[\frac{\beta_3}{\beta_2^2-\beta_3^2}\cosh\beta_2\ell\sin\beta_3\ell + \frac{\beta_2}{\beta_2^2-\beta_3^2}\sinh\beta_2\ell\cos\beta_3\ell\right]$$
$$+A_3\left[\frac{\beta_2}{\beta_2^2+\beta_3^2}\cosh\beta_2\ell\sin\beta_3\ell - \frac{\beta_3}{\beta_2^2+\beta_3^2}\sinh\beta_2\ell\cos\beta_3\ell\right] = 0 \qquad \text{C(2.20)}$$

$$A_1\left[\beta_1^4\frac{E_3\mathrm{b}}{h_3}\cosh\beta_1\ell\right]+A_2\left\{\left[\left(\beta_2^2-\beta_3^2\right)^2-4\beta_2^2\beta_3^2+\frac{E_3\mathrm{b}}{h_3}\right]\cosh\beta_2\ell\sin\beta_3\ell\right.$$
$$\left.-4\beta_2\beta_3\left(\beta_2^2-\beta_3^2\right)\sinh\beta_2\ell\cos\beta_3\ell\right\}+A_3\left\{\left[\left(\beta_2^2-\beta_3^2\right)^2-4\beta_2^2\beta_3^2+\frac{E_3\mathrm{b}}{h_3}\right]\right.$$
$$\left.\sinh\beta_2\ell\sin\beta_3\ell+4\beta_2\beta_3\left(\beta_2^2-\beta_3^2\right)\cosh\beta_2\ell\cos\beta_3\ell\right\}=0 \quad \mathrm{C}(2.21)$$

The constants C_1, C_2 and C_3 are given by

$$C_1=\frac{1}{\beta_1}\left(\frac{h_3}{E_3\mathrm{b}}\right)\left(\frac{\mathrm{b}}{\mathrm{a}}\right)\left(\beta_1^4+\frac{E_3\mathrm{b}}{h_3}\right)A_1$$
$$C_2=\left(\frac{h_3}{E_3\mathrm{b}}\right)\left(\frac{\mathrm{b}}{\mathrm{a}}\right)(\gamma_2\mathrm{A}_2-\gamma_3\mathrm{A}_3) \quad \mathrm{C}(2.22)$$
$$C_5=\left(\frac{h_3}{E_3\mathrm{b}}\right)\left(\frac{\mathrm{b}}{\mathrm{a}}\right)(\gamma_2\mathrm{A}_3+\gamma_3\mathrm{A}_2)$$

where

$$\gamma_2=\beta_2\left[\frac{E_3\mathrm{b}}{h_3\left(\beta_2^2+\beta_3^2\right)}+\beta_2^2-3\beta_3^2\right]$$
$$\gamma_3=\beta_3\left[\frac{E_3\mathrm{b}}{h_3\left(\beta_2^2+\beta_3^2\right)}+\beta_3^2-3\beta_2^2\right] \quad \mathrm{C}(2.23)$$

D. Suhir (1988b), *Thermal Stress Failures in Microelectronic Components—Review and Extension*

This paper provides an approximate solution for the distributions of shear and normal stresses in the bonding layer of a sandwich structure. The assumption of equal radius of curvature for the three layers has greatly simplified the solutions while not sacrificing much in accuracy. The greatest shortfall of this solution is that the normal stress does not satisfy the condition of global equilibrium. The outer members are modelled as plates with shear deformation. Two cases are considered, one in which the bonding layer is assumed to be of equivalent stiffness as the adherent and is treated as an elastic body and another in which the in-plane and bending stiffnesses of the

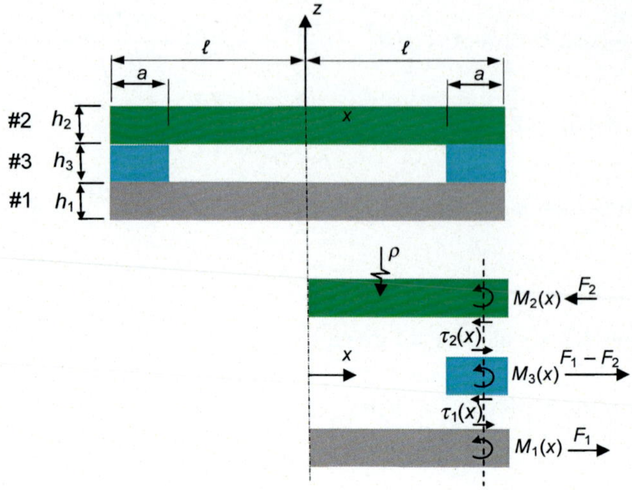

Figure 2.D Free-body diagram of the structure analysed by Suhir.

bonding layer are assumed to be much smaller than the adherent and assumed to be a distributed spring layer.

The free-body diagram of a tri-material layer used by Suhir is shown in Figure 2.D.

D1. Three elastic members

Assumptions: (1) the individual members are linear elastic and isotropic; (2) the members are modelled as plates such that plane sections in the members remain plane and the members experience bending about the *y*-axis and nil strain in the *y*-direction; (3) the individual member experiences no transverse stress and strain; and (4) the members have identical curvatures.

The moment equilibrium of the entire assembly gives

$$\frac{h_1 + h_3}{2} F_1(x) + \frac{h_2 + h_3}{2} F_2(x) + M_1(x) + M_2(x) + M_3(x) = 0 \qquad \text{D(2.1)}$$

Substituting $\rho = M_i/D_i$ into Eqn D(2.1) gives

$$\frac{1}{\rho(x)} = -\sqrt{\frac{\lambda'_{13}}{D_T}} F_1(x) - \sqrt{\frac{\lambda'_{23}}{D_T}} F_2(x) \qquad \text{D(2.2)}$$

where $\lambda'_{i3} = (h_i + h_3)^2/4D_T$, $D_T = \sum_{i=1}^{3} D_i$, and $D_i = E_i h_i^3/12(1 - \nu_i^2)$.

Compatibility of the members along the interfaces gives

$$u_1^+ = u_3^- \text{ and } u_3^+ = u_2^- \qquad \text{D(2.3)}$$

where the displacements are due to the balancing of the thermal expansion with the longitudinal force, shear stress and bending as given in Eqn D(2.4)

$$
\begin{aligned}
u_1^+(x) &= \alpha_1 \Delta T \cdot x + \lambda_{x1} \int_0^x F_1(\xi)\mathrm{d}\xi + \kappa_1 \tau_1(x) - \frac{h_1}{2} \int_0^x \frac{\mathrm{d}\xi}{\rho(\xi)} \\
u_2^-(x) &= \alpha_2 \Delta T \cdot x - \lambda_{x2} \int_0^x F_2(\xi)\mathrm{d}\xi - \kappa_2 \tau_2(x) + \frac{h_2}{2} \int_0^x \frac{\mathrm{d}\xi}{\rho(\xi)} \\
u_3^+(x) &= \alpha_3 \Delta T \cdot x - \lambda_{x3} \int_0^x [F_1(\xi) - F_2(\xi)]\mathrm{d}\xi + \kappa_3 \tau_2(x) - \frac{h_3}{2} \int_0^x \frac{\mathrm{d}\xi}{\rho(\xi)} \\
u_3^-(x) &= \alpha_3 \Delta T \cdot x - \lambda_{x3} \int_0^x [F_1(\xi) - F_2(\xi)]\mathrm{d}\xi - \kappa_3 \tau_1(x) + \frac{h_3}{2} \int_0^x \frac{\mathrm{d}\xi}{\rho(\xi)}
\end{aligned}
\qquad \text{D(2.4)}
$$

where $\lambda_{xi} = 1 - \nu_i^2 / E_i h_i$ and $\kappa_{si} = h_i / 3G_i$. Substituting D(2.4) into D(2.3) gives rise to two compatibility equations expressed in terms of force:

$$
\begin{aligned}
\alpha_{31} \Delta t \cdot x &= \lambda_{13}^{\mathrm{o}} \int_0^x F_1(\xi)\mathrm{d}\xi + \lambda_3^{\mathrm{o}} \int_0^x F_2(\xi)\mathrm{d}\xi + \kappa_{13} \tau_1(x) \\
\alpha_{23} \Delta t \cdot x &= \lambda_{23}^{\mathrm{o}} \int_0^x F_2(\xi)\mathrm{d}\xi + \lambda_3^{\mathrm{o}} \int_0^x F_1(\xi)\mathrm{d}\xi + \kappa_{23} \tau_2(x)
\end{aligned}
\qquad \text{D(2.5)}
$$

where $\alpha_{ij} = \alpha_i - \alpha_j$, $\lambda_{i3}^o = \lambda_{xi} + \lambda_{x3} + \lambda_{i3}'$, $\lambda_3^o = -\lambda_3 + \frac{(h_1+h_3)(h_2+h_3)}{4D_T}$, and $\kappa_{\mathrm{ij}} = \kappa_i + \kappa_j$.

The shear stresses that satisfy D(2.5) are given by

$$\tau_i(x) = C_i \sinh \beta_s x, \quad i = 1, 2 \qquad \text{D(2.6)}$$

where β_s are the solutions to the algebraic equation:

$$\kappa_{13} \kappa_{23} \beta_s^4 - \left(\kappa_{13} \lambda_{23}^o + \kappa_{23} \lambda_{13}^o\right) \beta_s^2 + \zeta_s = 0 \qquad \text{D(2.7)}$$

where $\zeta_s = \lambda_{x1}\lambda_{23}^{o} + \lambda_{x3}\left[\lambda_{x2} + \left(\sqrt{\lambda_{13}'} + \sqrt{\lambda_{23}'}\right)^2\right]$ and the constants C_1 and C_2 are given by

$$\begin{aligned} C_1 &= \frac{\beta_s \Delta T}{\zeta_s \cosh \beta_s \ell}\left[\lambda_{x3}\alpha_{21} + (\lambda_{x1} + \lambda_{13}')\alpha_{23} - \sqrt{\lambda_{13}'\lambda_{23}'}\alpha_{31}\right] \\ C_2 &= \frac{\beta_S \Delta T}{\zeta_s \cosh \beta_s \ell}\left[\lambda_{x3}\alpha_{21} + (\lambda_{x2} + \lambda_{13}')\alpha_{31} - \sqrt{\lambda_{13}'\lambda_{23}'}\alpha_{21}\right] \end{aligned} \qquad \text{D(2.8)}$$

D2. Two elastic members and one bonding layer

Assumptions: (1) the adherents are linear elastic and isotropic; (2) the adherents are modelled as plates such that plane sections in the members remain plane and the members experience bending about the y-axis and nil strain in the y-direction; (3) the adherents experience negligible transverse stress and strain; (4) the bond layer is modelled as a distribution of coil springs capable of transferring only transverse normal and shear loads, but offers no stiffness in the x- and y-normal directions and in bending; and (5) the members have identical curvatures.

From assumption (4), $F_1 = F_2 = F$ and $\tau_1 = \tau_2 = \tau$. The shear stress along the interface is given by

$$\tau(x) = \frac{\beta_S \alpha_{21} \Delta T}{\lambda_S} \frac{\sinh \beta_S x}{\cosh \beta_S \ell} \qquad \text{D(2.9)}$$

or

$$\tau(x) \approx \frac{\beta_S \alpha_{21} \Delta T}{\lambda_S} e^{-\beta_S(\ell - x)} \qquad \text{D(2.10)}$$

for large value of $\beta_s \ell$, where $\beta_S = \sqrt{\lambda_S/\kappa_S}$, $\lambda_S = \lambda_{x1} + \lambda_{x2} + \frac{(h_1+h_2)^2}{4(D_1+D_2)}$, $\kappa_S = \kappa_1 + \kappa_2 + 2\kappa_3$.

The uniform in-plane force in the members is given by

$$F(x) = \pm \frac{\alpha_{21}\Delta T}{\lambda_S}\left[1 - \frac{\cosh \beta_S(\ell - x)}{\cosh \beta_S \ell}\right]; \quad i = 1, 2; \ \text{+ve for } i = 1 \qquad \text{D(2.11)}$$

The maximum x-directional stress is the resultant of the uniform and bending stresses, which occurs at $x = a$ and is given by

$$\sigma_{xi,\max} = F_{i,\max}\left[1 + 3\frac{(h_1 + h_2)D_i}{(D_1 + D_2)h_i}\right]; \quad i = 1, 2; \ \text{+ve for } i = 1 \qquad \text{D(2.12)}$$

The equilibrium of moment of member #2 gives

$$Q + \frac{h_2}{2}\frac{\mathrm{d}F}{\mathrm{d}x} + \frac{\mathrm{d}M_2}{\mathrm{d}x} = 0 \qquad \text{D(2.13)}$$

Substituting

$$\frac{1}{\rho(x)} = \frac{M_b}{D_T} - \frac{h}{2D_T}F(x) \qquad \text{D(2.14)}$$

where $D_T = D_1 + D_2 + D_3$ into D(2.14) gives

$$Q(x) = -\frac{D_2 h - D_T h_2}{2D_T} f_s(x) \qquad \text{D(2.15)}$$

Differentiating Eqn D(2.15) gives

$$f_a(x) = \frac{h_2 D_1 - h_1 D_2}{2(D_1 + D_2)\kappa_S} \frac{\alpha_{21}\Delta T \cosh(\beta_S x)}{\cosh(\beta_S L)} \qquad \text{D(2.16)}$$

E. Jiang et al. (1997), *Thermal Stresses in Layered Electronic Assemblies*

This paper advances the treatment of tri-material layers of Suhir (1988), focusing on developing an accurate solution for the transverse normal stress in the bonding layer. The approach is similar to that of Chen and Nelson (1979) with improved definition of the transverse normal strain. The use of singular perturbation method has greatly simplified the solutions.

Shear deformation of the elastic members has been ignored.

Assumptions: (1) the adherents are linear elastic and isotropic; (2) the adherents are modelled as plates such that plane sections in the members remain plane and the members experience bending about the y-axis and nil strain in the y-direction; (3) the adherents experience negligible transverse stress and strain; and (4) the bond layer is modelled as a simplified plane strain body. It offers no stiffness in the x-direction and in bending; and (5) the magnitudes of the transverse normal and shear strains do not vary through the thickness.

The structure is depicted in Figure 2.E.

The equilibrium and constitutive equations for the adherents are similar to those of Chen and Nelson (1979) except for the force–displacement relation:

$$\frac{\mathrm{d}u_i}{\mathrm{d}x} = F_i\lambda_{xi} + \alpha_i' T \qquad \text{E(2.1)}$$

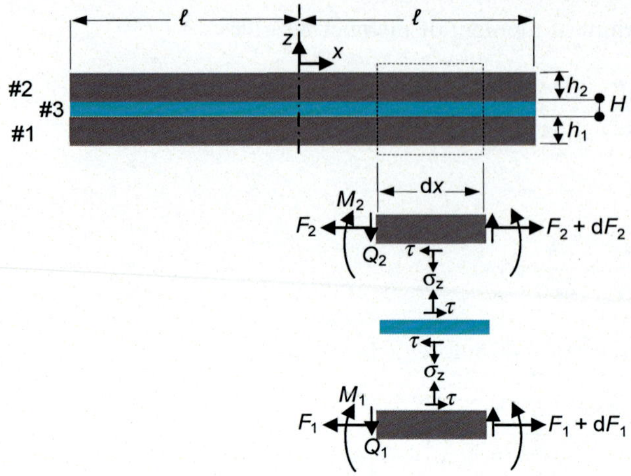

Figure 2.E Free-body diagram of the structures analysed by Jiang et al.

The bonding layer is assumed to experience transversely uniform normal and shear strains given by

$$\varepsilon_z = \frac{w_2 - w_1}{\mathrm{d}x}$$
$$\gamma = \frac{u_3^+ - u_3^-}{h_3} = \frac{\tau}{G_3} \qquad \text{E(2.2)}$$

where the displacements of the bonding layer are

$$u_3^+ = u_2 + \frac{h_2}{2}\frac{\mathrm{d}w_2}{\mathrm{d}x}$$
$$u_3^- = u_1 - \frac{h_1}{2}\frac{\mathrm{d}w_1}{\mathrm{d}x} \qquad \text{E(2.3)}$$

Assuming negligible normal stress in the x-direction, the transverse strain in the bonding layer becomes

$$\varepsilon_z = \frac{\sigma_z\left(1 - \nu_3{}^2\right)}{E_3} + (1 + \nu_3)\alpha_3\Delta T \qquad \text{E(2.4)}$$

Combining E(2.1) to E(2.4) gives the differential equations for shear and transverse normal stresses in the bonding layer as

$$\frac{h_3}{G_3}\frac{d^3\tau}{dx^3} - 4(\lambda_{x1} + \lambda_{x2})\frac{d\tau}{dx} + \overline{D}_{21}\sigma_z = 0 \qquad \text{E(2.5)}$$

$$\frac{h_3}{E_3'}\frac{d^4\sigma_z}{dx^4} + \overline{D}_e\sigma_z - \overline{D}_{21}\frac{d\tau}{dx} = 0 \qquad \text{E(2.6)}$$

where $E_3' = E_3/(1 - \nu_3{}^2)$, $\overline{D}_e = \frac{1}{D_1} + \frac{1}{D_2}$, $\overline{D}_{21} = \frac{1}{2}\frac{h_2}{D_2} - \frac{h_1}{D_1}$

Eliminating σ_z from the two differential equations gives

$$\frac{d^7\tau}{dx^7} - 4\frac{G_3}{h_3}(\lambda_{x1} + \lambda_{x2})\frac{d^5\tau}{dx^5} + \frac{E_3'\overline{D}_e}{h_3}\frac{d^3\tau}{dx^3} + \frac{E_3'G_3}{h_3^2}\left[\overline{D}_{21}^2 - 4(\lambda_{x1} + \lambda_{x2})\overline{D}_e\right]\frac{d\tau}{dx} = 0$$

E(2.7)

The symmetric solution gives

$$\tau = K_1 \sinh \beta_2 x + K_2 \sinh \beta_3 x \cos \beta_4 x + K_3 \cosh \beta_3 x \sin \beta_4 x \qquad \text{E(2.8)}$$

where $\pm\beta_2$, and $\pm\beta_3 \pm i\beta_4$ are the roots of the algebraic equation:

$$\beta^7 - 4\frac{G_3}{h_3}(\lambda_{x1} + \lambda_{x2})\beta^5 + \frac{E_3'\overline{D}_e}{h_3}\beta^3 + \frac{E_3'G_3}{h_3^2}\left[\overline{D}_{21}^2 - 4(\lambda_{x1} + \lambda_{x2})\overline{D}_e\right]\beta = 0$$

E(2.9)

For $h_3 \ll h_1$ and h_2, an approximate solution based on singular perturbation method is proposed. Introducing a new variable:

$$\xi = \frac{\ell - x}{\sqrt{h_3(h_1 + h_2)}} \qquad \text{E(2.10)}$$

Equation E(2.10) is reduced to

$$\frac{d^7\tau}{d\xi^7} - 4G_3(h_1 + h_2)(\lambda_{x1} + \lambda_{x2})\frac{d^5\tau}{dx^5} + O\left(\frac{h_3}{h_1 + h_2}\right) = 0 \qquad \text{E(2.11)}$$

where the last term reflects the order of the neglected term. The solution to E(2.11) using the boundary conditions: $\int_0^{\ell} \sigma_z \mathrm{d}x = 0$, $\left.\frac{\mathrm{d}\tau}{\mathrm{d}x}\right|_{x=\ell} = \frac{G_3 \alpha'_{21} \Delta T}{h_3}$ and $\left.\frac{\mathrm{d}^2\sigma_z}{\mathrm{d}x^2}\right|_{x=\ell} = 0$ gives

$$\tau = \frac{G_3 \alpha'_{21} \Delta T}{\beta_J h_3} \mathrm{e}^{\beta_J (x-\ell)} \quad \text{E(2.12)}$$

and the peeling stress is given by

$$\sigma_z(x) = \sigma_o \mathrm{e}^{\beta_J(x-\ell)} - \sigma_o \mathrm{e}^{\alpha_J(x-\ell)} \left\{ \frac{\beta_J^2}{2\alpha_J^2} \sin[\alpha_J(x-\ell)] + \left(\frac{\beta_J^2}{2\alpha_J^2} + \frac{2\alpha_J}{\beta_J} \right) \cos[\alpha_J(x-\ell)] \right\} \quad \text{E(2.13)}$$

where $\sigma_o = \frac{\overline{D}_{21} \alpha'_{21} \Delta T}{8(1-\nu_3)(\lambda_{x1}+\lambda_{x2})^2 + \left(\frac{h_3 \overline{D}_e}{G_3}\right)}$, $\alpha_J = \sqrt[4]{\frac{E'_3 \overline{D}_e}{4h_3}}$, and $\beta_J = 2\sqrt{\frac{G_3(\lambda_{x1}+\lambda_{x2})}{h_3}}$.

Nomenclature

X_i	Subscript $i = 1,2,3$ for IC package, PCB, and interconnect, respectively.
A_3, d_3, I_3	Cross-sectional area, diameter, second moment of area of a single discrete interconnect.
L, h_i	Half length and thickness of the section of PCB assembly of interest.
D_i, E_i, G_i, I_i	Flexural rigidity, elastic modulus, shear modulus, second moment of inertia.
$\overline{D}_e$, $\overline{D}_{21}$	Effective flexural compliance of members #1 and #2; differential flexural compliance between members #2 and #1.
f_a, f_b, f_s	Interfacial stresses: uniform transverse stress, linear transverse stress, shear stress.
F_a, F_s	Forces in a discrete element: axial force, shear force.
σ_x, σ_b, σ_z, τ_{xz}	Stresses along the thickness of the outer members: x-directional stress, x-directional bending stress, z-directional stress, shear stress.
F_i	x-Directional sectional traction (per unit y-width) acting along the neutral axis of member i.
M_i	Sectional moment (per unit y-width).
m_s	Shear-couple moment in a discrete element.
P	Pitch between neighbouring interconnects.
Q_i	Sectional shear force (per unit y-width).
u	x-Directional displacement of the neutral axis due to stretching.
Δu	Differential x-directional displacement.
w_b, w_s	z-Directional displacement of the neutral axis due to bending, shearing.

α_z, β	Characteristic constants for z-directional deformation, shear deformation.
α_{ij}	Differential coefficient of thermal expansion between members #i and #j.
$\kappa_{si}, \kappa_{s3\vartheta}, \kappa_s$	Shear compliances (shear displacement between the neutral axis and the interface per unit f_s for $i = 1{,}2$ and shear displacement between the two interfaces for member #3): of member i, of discrete member #3 due to flexing, of the structure between the neutral axis of member #1 and member #2.
$\kappa_{\phi i}, \kappa_\phi$	Shear compliance (shear-induced rotation per unit Q_a): of member #i, of the system.
κ_{21}	Differential shear compliance (shear-induced rotation per unit f_s) between member #2 and #1.
$\lambda_{xi}, \lambda_{x\theta}, \lambda_x$	Stretch compliances (displacement per unit traction force): of the neutral axis of member #i, due to rotation of members, of the system.
λ_{zi}, λ_z	Transverse compliance (displacement between the neutral axis and the interface per unit f_a): of member #i; of the system (between the neutral axis of member #1 and member #2).
$\theta, \phi, \overline{\phi}$	Rotation of neutral axis: due to bending, due to shearing; mean shear strain between shear surface and neutral axis.
$\overline{\sigma}_a, \overline{\tau}, \overline{\sigma}_b$	Average stresses in a discrete interconnect: axial stress, shear stress, bending stress.
ΔT	Temperature change.
Γ	Distance of a discrete interconnect from the mid-plane of the structure.

Mathematical symbols

$$A_s = \frac{\alpha_{21}\Delta T}{\beta\kappa_s}$$

$$\alpha_z = \sqrt[4]{\frac{\overline{D}_e}{4\lambda_z}} \approx \begin{cases} \sqrt[4]{\dfrac{\overline{D}_e - \overline{D}_{21}^2/\lambda_x}{4\lambda_z}} & \text{for bilayer structure} \\ \sqrt[4]{\dfrac{\overline{D}_e - \overline{D}_{21}^*\overline{D}_{21}/\lambda_x}{4\lambda_z}} & \text{for trilayer structure} \end{cases}$$

$$\beta^2 = \lambda_x/\kappa_s$$

$$C_s = \begin{cases} \dfrac{\overline{D}_{21}\alpha_{21}\Delta T}{\lambda_z\kappa_s(4\alpha_z^4 + \beta^4)} & \text{for bilayer structure} \\ \dfrac{\overline{D}_{21}^*\alpha_{21}\Delta T}{\lambda_z\kappa_s(4\alpha_z^4 + \beta^4)} & \text{for trilayer structure} \end{cases}$$

$$\overline{D}_e = \frac{1}{D_2} + \frac{1}{D_1}$$

$$D_i = \frac{E_i h_i^3}{12}, \quad i = 1, 2, \text{ (plane stress; substitutes } E_i \text{ with } E_i/(1 - \nu_i^2) \text{ for plane strain)}$$

$$\begin{matrix}\overline{D}_{21} \\ \overline{D}_{21}^*\end{matrix} = \begin{cases} \dfrac{1}{2}\left(\dfrac{h_2}{D_2} - \dfrac{h_1}{D_1}\right) \\ \dfrac{1}{2}\left(\dfrac{h_2 + h_3}{D_2} - \dfrac{h_1 + h_3}{D_1}\right) \end{cases}$$

$$\kappa_s \approx \begin{cases} \kappa_{s1} + \kappa_{s2} & \text{for bilayer structure} \\ \kappa_{s1}^* + \kappa_{s2}^* + \kappa_{s3} & \text{for trilayer structure with continuous sandwiched layer} \\ \kappa_{s1}^* + \kappa_{s2}^* + \kappa_{s3} + \kappa_{s3\vartheta} & \text{for trilayer structure with discrete sandwiched layer} \end{cases}$$

$$\begin{matrix} \kappa_{si} \\ \kappa_{si}^* \end{matrix} = \begin{cases} \dfrac{h_i}{8G_i} + \dfrac{Q_{ai}}{2G_if_s} \approx \dfrac{h_i}{8G_i} \\ \dfrac{1}{8G_i}(h_i \mp 2h_3) + \dfrac{Q_a}{2G_if_s} \approx \dfrac{1}{8G_i}(h_i \mp 2h_3) \end{cases}, \quad i = 1,2$$

$$\kappa_{s3} = \begin{cases} \dfrac{h_3}{G_3} & \text{for continuous sandwiched layer} \\ \dfrac{pqh_3}{G_3A_3} & \text{smeared property for discrete sandwiched layer} \end{cases}$$

$$\kappa_{s3\vartheta} = \frac{{h_3}^3 pq}{12E_3I_3}$$

$$\kappa_\phi = \kappa_{\phi 1} + \kappa_{\phi 2}, \quad \kappa_{\phi i} = \frac{3}{4G_ih_i}$$

$$\kappa_{21} = \frac{1}{2}\left(\frac{1}{G_2} - \frac{1}{G_1}\right)$$

$$\lambda_x = \begin{cases} \lambda_{x1} + \lambda_{x2} + \lambda_{x\theta} \approx \displaystyle\sum_{i=1}^{2} \frac{4}{E_ih_i} & \text{for bilayer structure} \\ \lambda_{x1} + \lambda_{x2} + \lambda_{x\theta}^* \approx \displaystyle\sum_{i=1}^{2} \frac{1}{E_ih_i}\left(4 + \frac{3h_3}{h_i}\right) & \text{for trilayer structure} \end{cases}$$

$$\lambda_{xi} = \frac{1}{E_iF}\left[\frac{F}{h_i} - \nu_i\left(\frac{h_i}{8}\frac{\mathrm{d}f_s}{\mathrm{d}x} + \frac{1}{2}f_a\right)\right] \approx \frac{1}{E_ih_i},$$ $i = 1,2$ (plane stress; substitutes E_i with $E_i/(1-\nu_i^2)$ for plane strain)

$$\begin{matrix} \lambda_{x\theta} \\ \lambda_{x\theta}^* \end{matrix} = \begin{cases} \dfrac{1}{4}\left(\dfrac{h_1^2}{D_1} + \dfrac{h_2^2}{D_2}\right) \\ \dfrac{1}{4}\left(\dfrac{h_1^2 + h_1h_3}{D_1} + \dfrac{h_2^2 + h_2h_3}{D_2}\right) \end{cases}, \quad i = 1,2$$

$$\lambda_z = \lambda_{z1} + \lambda_{z2} + \lambda_3$$

$$\lambda_{zi} = \frac{h}{Ef_a}\left[\left(\frac{11h}{192}\frac{\mathrm{d}f_s}{\mathrm{d}x} + \frac{13}{32}f_a\right) + \nu\left(\frac{5F}{4h} + \frac{3}{2h^2}\int_L^x Q_a\mathrm{d}x\right)\right] \approx \frac{13h_i}{32E_i},$$ $i = 1,2$ (plane stress; substitutes E_i with $E_i/(1-\nu_i^2)$ for plane strain)

$$\lambda_{z3} = \begin{cases} \dfrac{h_3}{E_3} & \text{for continuous sandwiched layer} \\ \dfrac{pqh_3}{E_3 A_3} & \text{smeared property for discrete sandwiched layer} \end{cases}$$

$$\chi_i = \tfrac{\xi_{3L} \mp \xi_{4L}}{\xi_{5L}}, \ \xi_{1x} = \cos(\alpha_z x)\cosh(\alpha_z x), \ \xi_{2x} = \sin(\alpha_z x)\sinh(\alpha_z x),$$
$$\xi_{3x} = \cos(\alpha_z x)\sinh(\alpha_z x),$$

$$\xi_{4x} = \sin(\alpha_z x)\cosh(\alpha_z x), \ \xi_{5L} = \sin(2\alpha_z L) + \sinh(2\alpha_z L)$$

Advances in creep-fatigue modelling of solder joints

3

3.1 Introduction

In the ball grid array packaging, the integrated circuit (IC) component, solder joints and printed circuit board (PCB) form a three-layer construction. As the ICs are powered on and off, the IC chip within the IC component experiences heating and cooling leading to thermal expansion and contraction of the IC component. The mismatch in thermal expansions between the IC component and PCB is accommodated by the interconnecting solder joints, resulting in creep-fatigue.

The systematic investigation of fatigue leading to the stress-life (S-N) curve for high-cycle fatigue can be dated to Wöhler (1955). Basquin (1910) described the stress-life data in the form of a power law. The extension of stress life to strain life for low-cycle fatigue was reported by Coffin (1954) and Manson (1954) independently. This led to the Coffin–Manson equation:

$$\Delta\varepsilon_{\mathrm{p}} = C_{\mathrm{o}} N^{-\beta_{\mathrm{o}}} \tag{3.1}$$

where C_o and β_o are constants, referred to as the fatigue ductility coefficient and the fatigue ductility exponent, respectively. The studies of creep-fatigue began in the 1950s pioneered by the power and the aviation industries (Penny & Marriott, 1995). Despite the tremendous efforts, creep-fatigue remains an unsolved puzzle (Schutz, 1996). Interested readers should review Schutz (1996) for a detailed narration of the history of fatigue.

The main stumbling block for creep-fatigue is indeed the indefiniteness of creep strain. The general expression of creep strain rate takes the form

$$\dot{\varepsilon}_{\mathrm{c}} = f(M, \sigma, T, t, \varepsilon) \tag{3.2}$$

or assuming the various effects are separable, we have

$$\dot{\varepsilon}_{\mathrm{c}} = f_1(M) f_2(\sigma) f_3(T) g_1(t) g_2(\varepsilon) \tag{3.3}$$

where $f_1(M)$ represents the microstructure of the test specimen, which can take the form (Bird, 1969)

$$f_1 = \frac{Gb}{kT}\left(\frac{b}{d}\right)^p \tag{3.4}$$

where G, d and b are shear modulus, grain size and Burgers vector, respectively, and p is a fitting constant. In practice, however, the role of microstructures and their

Robust Design of Microelectronics Assemblies Against Mechanical Shock, Temperature and Moisture
http://dx.doi.org/10.1016/B978-1-84569-528-6.00003-4

evolution during the process of creep are typically ignored due to the difficulty of characterisation in a test specimen and implementation in an analysis. The functions $g_1(t)$ and $g_2(\varepsilon)$ embody the hypotheses of time hardening and strain hardening, respectively, during the primary phase of creep. In practice, neither time hardening nor strain hardening is satisfactory, and a reliable hardening model for creep strain remains elusive (Penny & Marriott, 1995). Most researchers have evaded this challenge by assuming only steady-state creep, which implies $g_1(t) = g_2(\varepsilon) = 1$; however, the accuracy of the evaluated creep strain must necessarily suffer.

3.2 Life-prediction models for creep-fatigue

The creep-fatigue issues of solder materials and solder joints in microelectronic assembly have been extensively studied over the last three decades, and several life-prediction models have been proposed (Frear, 1991; Lau, 1991; Lee, 2000; Ross, 1993). These models can be traced back to prior studies in the power and aviation industries.

Since both plastic strain and creep strain are inelastic strains, it appears that the Coffin–Manson equation can be extended to creep-fatigue by simply replacing the plastic strain range $\Delta\varepsilon_p$ in Eqn (3.1) with the inelastic strain range $\Delta\varepsilon_{in}$, which is the sum of the plastic strain range $\Delta\varepsilon_p$ and the creep strain range $\Delta\varepsilon_c$; that is,

$$\Delta\varepsilon_p \rightarrow \Delta\varepsilon_{in} = \Delta\varepsilon_p + \Delta\varepsilon_c \tag{3.5}$$

However, the assumed additions of $\Delta\varepsilon_p$ and $\Delta\varepsilon_c$ have not been supported by experimental evidence.

3.2.1 Review of models

A number of life-prediction models for creep-fatigue have been proposed (Penny & Marriott, 1995), and some of these have been adopted and extended by the electronic packaging community:

1. *The linear damage summation rule* (Taira, 1962), which assumes that damage, regardless of whether it comes from creep or fatigue, is cumulative in a linear fashion. That is, failure occurs when:

$$D_f + D_c = 1 \tag{3.6}$$

where $D_f = \sum_i \frac{n_i}{N_i}$ is the damage caused by fatigue based on the linear damage hypothesis (Miner, 1945; Palmgren, 1924), $D_c = \sum_i \frac{t_i}{t_{Ri}}$ is the damage due to creep based on the life-fraction hypothesis (Robison, 1938) or $D_c = \sum_i \frac{\varepsilon_{ci}}{\varepsilon_{Ri}}$ based on the strain-fraction hypothesis (Goldhoff, 1965); where n_i is the number of cycles at a given strain range $\Delta\varepsilon_p$; N_i is the fatigue life at the same strain range at a temperature at which the mechanism of creep is inactive; t_i and ε_{ci} are time and creep strain, respectively, at a given stress and temperature;

t_R and ε_R are time-to-rupture and rupture strain, respectively, at the same stress and temperature.

However, the linear addition of damages is inconsistent with the microstructural characteristics of fatigue damage and creep damage (Dowling, 2012; Hertzbery, 1996; Penny & Marriott, 1995; Zhuang, 1998). Cyclic strain induces damage through the formation of slips in the lattice of the material leading to the formation of persistent slip bands as exemplified by the formation of intrusions and extrusions on the surface of the structure; alternations of these intrusions and extrusions lead to the nucleation of microcracks. In contrast, creep induces damage through diffusion of dislocations along the grain boundaries ($0.4\ T_m < T < 0.6\ T_m$) and within the lattice ($T > 0.6\ T_m$) that accumulate into microvoids, which do not necessarily occur on the surface of the structure. Microstructurally, fatigue cracking is typically transgranular with damaged confined to the slip planes, while creep damage is intergranular and dispersed.

Despite its inconsistence and inaccuracy (Beres, 2000; Penny & Marriott, 1995), this method is popularly used in the power and aviation industries owing to its simplicity. However, this method is not used in the microelectronic assembly community for the simple reason that it is impractical to characterise the strain life of microelectronics assembly materials at the temperature at which the mechanism of creep is dormant; for example, the melting temperature, T_m, of the eutectic SnPb solder is 456 K and $0.4\ T_m$ is 110 °C below room temperature.

2. *The strain range partitioning method* (Manson, 1976), which apportions the damage within a cycle using the following rule:

$$\frac{1}{N} = \frac{\Delta\varepsilon_{pp}}{\Delta\varepsilon_{in}N_{pp}} + \frac{\Delta\varepsilon_{cc}}{\Delta\varepsilon_{in}N_{cc}} + \frac{\Delta\varepsilon_{cp}}{\Delta\varepsilon_{in}N_{cp}} + \frac{\Delta\varepsilon_{pc}}{\Delta\varepsilon_{in}N_{pc}} \tag{3.7}$$

where $\Delta\varepsilon_{pp}$ is plastic strain reversed by plastic strain, $\Delta\varepsilon_{cc}$ is tensile creep strain reversed by compressive creep, $\Delta\varepsilon_{cp}$ is tensile creep reversed by compressive plasticity and $\Delta\varepsilon_{pc}$ is tensile plasticity reversed by compressive creep. Individual strain components are assumed to obey their respective Coffin–Manson relation; that is $N^{\beta_{jj}} = C_{jj}^{-1}\Delta\varepsilon_{jj}$.

By partitioning the four mechanisms of creep-fatigue interaction, this model has been shown to be capable of modelling a wide range of creep-fatigue phenomena with reasonable accuracy (Penny & Marriott, 1995) and has been adopted by the power and aviation industries; however, this method has not found favour with the microelectronic packaging industry in view of its complexity.

3. *The hysteresis energy method* (Ostergren, 1976), based on experimental observations that there is little creep damage during compression, suggests that the inelastic damage accumulated in a cycle is given by the tensile portion of the stress–strain hysteresis energy:

$$w = \int_{\text{tension cycle}} \sigma\varepsilon_{in}\,d\varepsilon \tag{3.8}$$

where w may be used in place of $\Delta\varepsilon_{in}$ in Eqn (3.2); that is,

$$w = C_o N^{-\beta_o} \tag{3.9}$$

A similar method that uses the entire stress–strain hysteresis energy is often adopted by the microelectronics assembly community for modelling the creep-fatigue of solder joints (Liang, 1997). Inherent in this method is the linear summation of plastic and creep damages, which, as discussed, is not supported by microstructural evidence.

4. *The fracture mechanics-based method* (Ainsworth, 1992), resting on the observation that low-cycle fatigue is dominated by crack propagation, suggests that the rate of crack growth is driven collectively by cyclic fatigue and creep; that is,

$$\left.\frac{da}{dN}\right|_{cycle} = \left.\frac{da}{dN}\right|_{fatigue} + \left.\frac{da}{dt}\right|_{cycle} \tag{3.10}$$

where

$$\left.\frac{da}{dN}\right|_{fatigue} = C_1 \Delta J_{eff}^{n/2}, \left.\frac{da}{dt}\right|_{cycle} = \int_{\text{tension cycle}} C_2 C^{*n'} dt; \tag{3.10a}$$

ΔJ_{eff} is the effective range of J-integral; and C^* is the time-dependent fracture parameter. The Darveaux's model (Darveaux, 2002) that is used extensively for modelling solder joints assumes crack initiation and propagation are power-law functions of the total stress–strain hysteresis energy:

$$N = N_o + \frac{a_{allow}}{da/dN}, \tag{3.11}$$

where

$$N_o = C_1 \Delta w^{C_2}, \quad \frac{da}{dN} = C_3 \Delta w^{C_4}; \tag{3.11a}$$

and a_{allow} is the allowable crack length and is commonly taken as the smallest diameter of a solder joint.

5. *The mechanism-based method* (Majumder, 1976) assumes that fatigue damage is characterised by crack size, a, and creep damage by cavity size, c, each governed by their respective equations:

$$\begin{aligned} \frac{1}{a}\frac{da}{dt} &= \begin{Bmatrix} T \\ C \end{Bmatrix} \left(1 + \alpha \ln \frac{c}{c_o}\right) \left|\varepsilon_{in}^{m}\right| \left|\dot{\varepsilon}_{in}^{k}\right| \\ \frac{1}{c}\frac{dc}{dt} &= \begin{Bmatrix} G_T \\ -G_C \end{Bmatrix} \left|\varepsilon_{in}^{m}\right| \left|\dot{\varepsilon}_{in}^{k}\right| \end{aligned} \tag{3.12}$$

where T and G_T are for tension, C and $-G_C$ for compression and c_o is a threshold cavity size below which cavities will be sintered away. While academically interesting, this method is too complicated to be adopted for industry practices.

6. *The frequency modified method* (Coffin, 1976) accounts for the effects of creep through the introduction of a frequency term into the Coffin–Manson equation:

$$\Delta \varepsilon_p = C_c \left(N f^{k-1}\right)^{-\beta_o} \tag{3.13}$$

where C_c, k and β_o are constants. Comparison of Eqn (3.13) with Eqn (3.1) suggests that the fatigue ductility coefficient C_o is a function of frequency given by $C_c f^{\beta_o(1-k)}$; thus the effects of cyclic frequency are to alter the magnitude of the fatigue coefficient.

Solomon (1991) performed lap shear test on eutectic SnPb solder at 35 °C and reported values of the frequency exponent in Eqn (3.14) to be $k = -0.42$ for the frequency range 5×10^{-5} Hz $< f < 3 \times 10^{-4}$ Hz and $k = 0.84$ for the frequency range 3×10^{-4} Hz $< f < 3 \times 10^{-1}$ Hz.

7. *The frequency–temperature modified method* accounts for the effects of creep by introducing a frequency and a temperature term into the Coffin–Manson equation, which is a natural extension of the frequency-modified method, and the inclusion of temperature completes the description of creep. In analysing the creep-fatigue of eutectic SnPb solder joints in leaded IC packaging, Engelmaier (1983) came to the following frequency–temperature modified strain-life equation:

$$\Delta\varepsilon_p = C_o N^{-\beta(T,f)} \tag{3.14}$$

where

$$\beta = 0.442 + 6 \times 10^{-4}\overline{T} - 1.74 \times 10^{-2}\ln(1 + 43200f). \tag{3.15}$$

Here, $\overline{T}$ is the mean cyclic solder joint temperature in °C. Unlike Coffin (1976), Engelmaier (1983) has assumed that the effects of frequency–temperature is in altering the magnitude of the fatigue ductility.

Shi (2000) performed an elaborate characterisation of the eutectic SnPb solder under uniaxial tension-compression loading over the temperature range from −40 to 150 °C and frequency range from 10^{-4} to 1 Hz. They proposed the following frequency–temperature modified strain-life equation:

$$\Delta\varepsilon_p = C_c(T) f^{\beta(T)[1-k(T)]} N^{-\beta(T)} \tag{3.16}$$

where

$$\begin{aligned} C_c &= 2.122 - 3.57 \times 10^{-3}T + 1.329 \times 10^{-5}T^2 - 2.502 \times 10^{-7}T^3 \\ \beta &= 0.731 - 1.63 \times 10^{-4}T + 1.392 \times 10^{-6}T^2 - 1.151 \times 10^{-8}T^3 \\ k_1 &= 0.919 - 1.765 \times 10^{-4}T - 8.634 \times 10^{-7}T^2 \\ k_2 &= 0.437 - 3.753 \times 10^{-4}T - 8.04 \times 10^{-7}T^2 \end{aligned} \tag{3.17}$$

k_1 and k_2 represent the frequency exponent for 10^{-3} Hz $< f < 1$ Hz and 10^{-4} Hz $< f < 10^{-3}$ Hz, respectively.

3.2.2 The ideal model

Many of the above life-prediction models are highly empirical with fitting constants that are valid only for a limited window. Except for the frequency-modified and the frequency–temperature-modified methods, the rest also suffer from the indefiniteness of creep strain. An ideal life-prediction model is hence one that is capable of modelling the entire range of creep-fatigue from pure fatigue to pure creep rupture and does not require the evaluation of creep strain. This leads to the unified equation in Section 3.3.

3.3 The unified equation

3.3.1 Rationalisation

At conditions when creep is dormant, a material would fail at a definite plastic strain capacity, ε_p, under monotonic loading. At conditions when creep becomes active, the plastic strain capacity would decrease with increasing temperature/time; and this may be expressed as $\varepsilon_p \cdot c(T, t)$. This is illustrated in Figure 3.1. As the loading condition changes from monotonic to cyclic, the plastic strain capacity decreases with increasing number of cycle given by: $\Delta\varepsilon_p = C_o N^{-\beta_o}$. Assuming the cyclic plastic strain capacity, $\Delta\varepsilon_p$, shares identical sensitivity as the monotonic plastic strain capacity, ε_p, then this gives

$$\Delta\varepsilon_p = C_o c(T, t) N^{-\beta_o} \tag{3.18}$$

where t is now the cycle time.

It is known that creep strain increases with increasing applied stress (Dorn, 1955); that is, there is more creep strain associated with higher cyclic plastic strain. Increasing the creep strain implies depleting the cyclic plastic strain capacity. Effectively, this implies that the fatigue ductility β_o decreases with increasing creep as shown in Figure 3.2. This gives the unified creep-fatigue equation:

$$\Delta\varepsilon_p = C_o c(T, t) N^{-\beta_o s(T, f)} \tag{3.19}$$

It should be noted that $0 \le c(T, t) \le 1$ and $0 \le s(T, t) \le 1$. At $c(T, t) = 1$ and $s(T, t) = 1$, the Coffin–Manson equation is recovered. At $c(T, t) = 0$, creep rupture occurs within a cycle of fatigue. The unified equation covers the full range of creep-fatigue from 'pure fatigue' to 'pure creep rupture'.

3.3.2 The reference frame and the transformation equations

At the reference condition when the mechanism of creep becomes dormant, a state of 'pure fatigue' prevails; the plastic strain range, $\Delta\varepsilon_{p,ref}$, has the maximum capacity (for

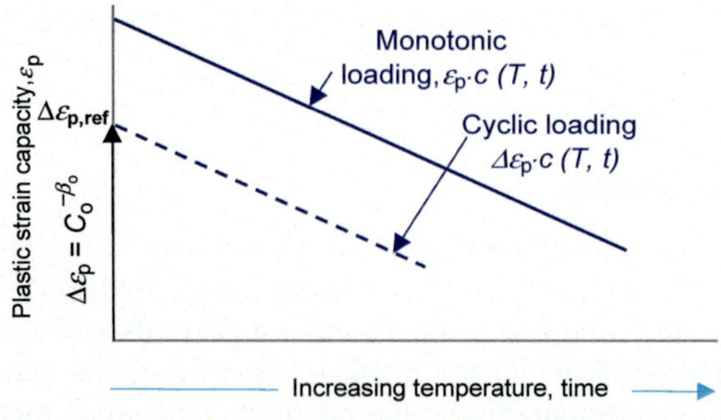

Figure 3.1 Reduction of plastic and cyclic plastic strain capacities due to creep.

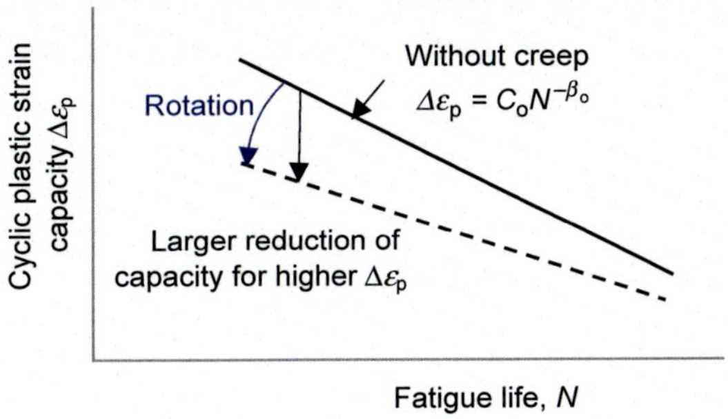

Figure 3.2 Rotation of strain life due to creep.

the same fatigue life) as shown in Figure 3.1; and the strain life is described by the Coffin–Manson equation:

$$\Delta\varepsilon_{p,ref} = C_o N^{-\beta_o} \tag{3.20}$$

Comparing Eqn (3.20) with Eqn (3.19) gives

$$\Delta\varepsilon_p = C_o^{1-s(T,t)} c(T,t) \Delta\varepsilon_{p,ref}^{s(T,t)} \tag{3.21}$$

Performing the logarithmic operation on Eqn (3.21),

$$\log\Delta\varepsilon_p = s(T,t)\log\Delta\varepsilon_{p,ref} + \log\left[C_o^{1-s(T,t)} c(T,t)\right]. \tag{3.22}$$

It becomes clear that the effect of creep is to reduce the allowable plastic strain range through two transformations in the logarithmic scale: shrinking $\log\Delta\varepsilon_{p,ref}$ by a factor $s(T, t)$ followed by translating the transformed product along the strain-axis by a magnitude $\log\left[C_o^{1-s(T,t)} c(T,t)\right]$. This is illustrated in Figure 3.3, in which the reduction in the plastic strain-range capacity is designated as the effective creep strain, $\Delta\varepsilon_{c,eff}$.

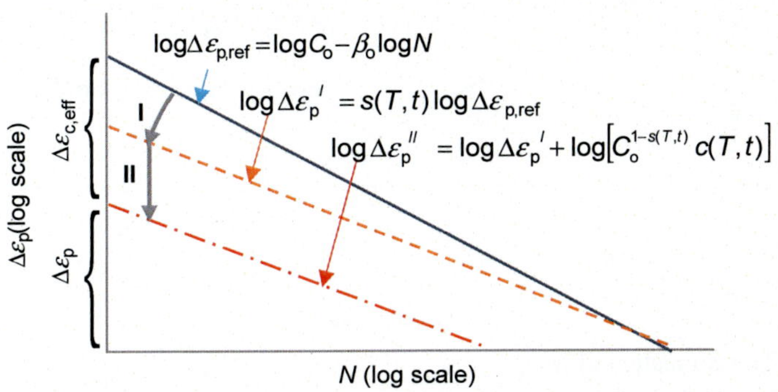

Figure 3.3 Transformation of $\Delta\varepsilon_p$.

A similar analogy exists for fatigue life. At the reference condition, the fatigue life N_{ref} has the maximum magnitude (for the same $\Delta\varepsilon_p$), and the strain life is described by the Coffin–Manson equation:

$$\Delta\varepsilon_p = C_o N_{ref}^{-\beta_o} \tag{3.23}$$

Comparing Eqn (3.23) with Eqn (3.19) we obtain

$$N = c(T,t)^{1/\beta_o s(T,t)} N_{ref}^{1/s(T,t)} \tag{3.24}$$

Performing the logarithmic operation on Eqn (3.24) gives

$$\log N = \frac{\log N_{ref}}{s(T,t)} + \frac{\log c(T,t)}{\beta_o s(T,t)} \tag{3.25}$$

The effect of creep reduces the fatigue life via two transformations in the logarithmic scale: stretching $\log N_{ref}$ with a factor $1/s(T, t)$ followed by translation of a magnitude $\frac{\log[c(T,t)]}{\beta_o s(T,t)}$. This is illustrated in Figure 3.4, in which the reduction in the fatigue-life capacity is taken to be consumed by creep and is designated as N_{creep}.

With creep and stress functions defined, creep-fatigue data generated at arbitrary cycle time and temperature, $(\Delta\varepsilon_p, T, t, N)$, can be transformed to the reference condition, thus reducing the number of variables from four to two. This may be expressed either in the form of $\Delta\varepsilon_{p,ref} - N$ or $\Delta\varepsilon_p - N_{ref}$. The transformation equations are

$$\Delta\varepsilon_{p,ref} = C_o \left[\frac{\Delta\varepsilon_p}{C_o c(T,t)}\right]^{1/s(T,t)} \tag{3.21a}$$

$$N_{ref} = N^{s(T,t)} [c(T,t)]^{-1/\beta_o} \tag{3.24a}$$

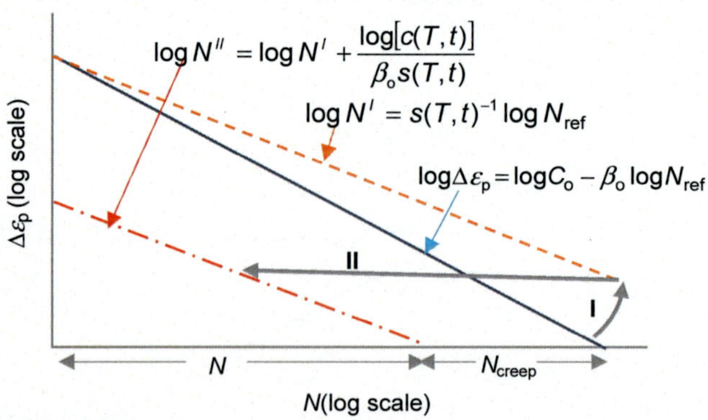

Figure 3.4 Transformation of N.

3.3.3 The forms of creep and stress functions

We shall refer to the functions $c(T, t)$ and $s(T, t)$ as creep function and stress function, respectively. A number of time–temperature relations have been proposed for the creep-rupture time, t_R, and two among these are particularly popular. The Larson–Miller time–temperature relation (Larson, 1952):

$$P_{LM}(\sigma) = T(a + \log t_R) \tag{3.26}$$

where T is in Kelvin; it is derived from the popular exponential constitutive relation, $\dot{\varepsilon}_c \propto \exp(1/RT)$. The Manson–Haferd time–temperature relation (Manson, 1953),

$$P_{MH}(\sigma) = \frac{T - T_a}{\log(t_R/t_a)} \tag{3.27}$$

where ($\log t_a$, T_a) is the point of convergence of the $T - \log t_R$ lines; it is empirically established through analysis of numerous experimental data of more than 40 steel alloys.

The L–M relation suggests a linear relation between $\log t$ and $1/T$, and the creep function becomes

$$c(T,t) = 1 + c_1\left(\frac{1}{T - T_{ref}}\right) + c_2\left(\log\frac{t}{t_{ref}}\right) \tag{3.28}$$

where T_{ref} and t_{ref} are reference temperature and reference cycle time, respectively, below which creep is assumed to be dormant. The M–H relation suggests a linear relation between $\log t$ and T, which was first proposed by Bailey (1935). The creep function is

$$c(T,t) = 1 + c_1(T - T_{ref}) + c_2\left(\log\frac{t}{t_{ref}}\right) \tag{3.29}$$

It is postulated that the stress function follows similar time–temperature relations as the creep function; that is,

$$s(T,t) = 1 + s_1(T - T_{ref}) + s_2\left(\log\frac{t}{t_{ref}}\right) \tag{3.30}$$

3.4 Self-validations and benchmarking

The experimental data of Shi (2000) for the eutectic SnPb solder are used to self-validate the unified equation, which are also benchmarked against the equations of Shi and Engelmaier (1983).

3.4.1 Preliminary analysis of the experimental data

3.4.1.1 Extractions of fatigue ductility coefficient C_o and fatigue ductility exponent β_o

The coefficient C_o and exponent β_o at five test temperatures (−40, 25, 75, 125 and 150 °C) at the cyclic frequency of 1 Hz have been extracted from Figure 3.6 of Ref. Shi (2000) and summarised in Table 3.1. Similarly, the fatigue ductility coefficient C_o and fatigue ductility exponent β_o at five cyclic frequencies (10^{-4}, 10^{-3}, 10^{-2}, 10^{-1} and 1 Hz) at the test temperature of 25 °C have been extracted from Figure 3.12 of Ref. Shi (2000) and summarised in Table 3.2.

3.4.1.2 Preliminary analysis of the creep and the stress functions

Figure 3.5(a) and (b) shows the plots of C_o versus $T - T_{ref}$ and C_o versus $1/(T - T_{ref})$, respectively, at five temperatures at 1 Hz. The reference temperature is taken as 298 K for convenience. The plot of C_o versus $T - T_{ref}$ gives a much better fit and overwhelmingly supports the Manson–Haferd relation over the Larson–Miller relation. The coefficients C_o and c_1 of Eqn (3.29) are: $C_o = 2.19$ and $c_1 = -0.0088$.

Figure 3.6 shows the plot of C_o versus f/f_{ref} at five frequencies at 25 °C for $f_{ref} = 1$ Hz, which suggests that a logarithmic function gives a much better fit than a power law function – in contrast to the assumption of Coffin (1976), Solomon (1991) and Shi (2000). The coefficients C_o and c_2 of Eqn (3.29) are: $C_o = 2.32$ and $c_2 = -0.167$.

Table 3.1 Coefficient C_o and exponent β_o at five temperatures at $f = 1$ Hz

	Temperature (°C)				
Coef.	**−40**	**25**	**75**	**125**	**150**
C_o	2.74	2.22	1.71	1.45	0.995
β_o	0.769	0.753	0.734	0.724	0.709

Table 3.2 Coefficient C_o and exponent β_o at five frequencies at $T = 25$ °C

	Cyclic frequency (Hz)				
Coef.	10^{-4}	10^{-3}	10^{-2}	10^{-1}	**1**
C_o	0.629	1.26	1.68	1.93	2.23
β_o	0.674	0.714	0.724	0.732	0.753

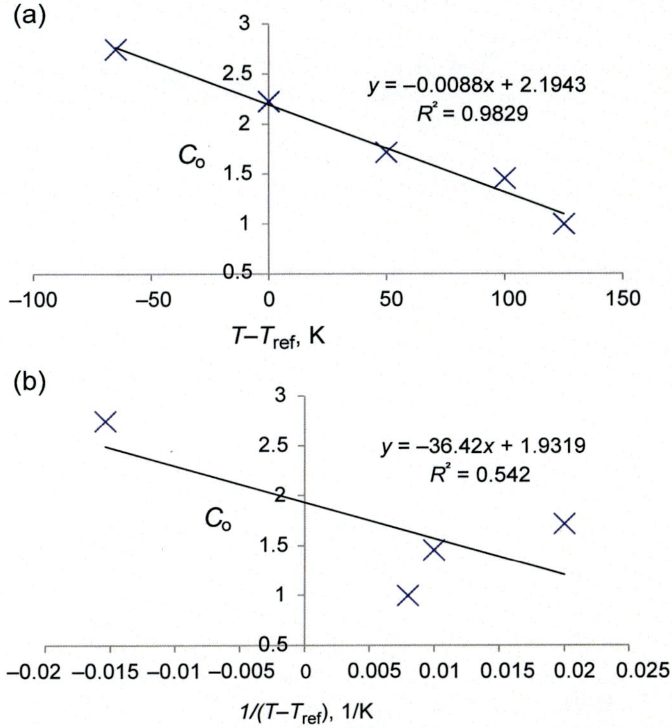

Figure 3.5 C_o versus (a) $(T - T_{ref})$ and (b) $1/(T - T_{ref})$ at 1 Hz, $T_{ref} = 298$ K.

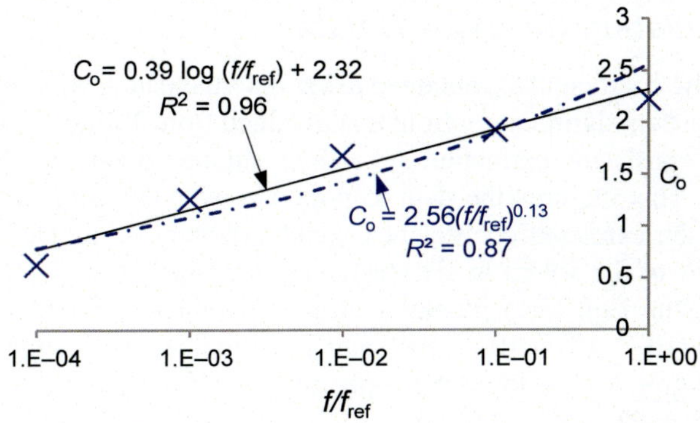

Figure 3.6 C_o versus f/f_{ref} at 25 °C, $f_{ref} = 1$ Hz.

Performing similar analysis for β_o, Figure 3.7 shows the plots of β_o versus $(T - T_{ref})$ at five temperatures at 1 Hz for $T_{ref} = 298$ K and Figure 3.8 β_o versus f/f_{ref} at five frequencies at 25 °C for $f_{ref} = 1$ Hz. The reasonably good quality of fit supports the postulated form of Eqn (3.30).

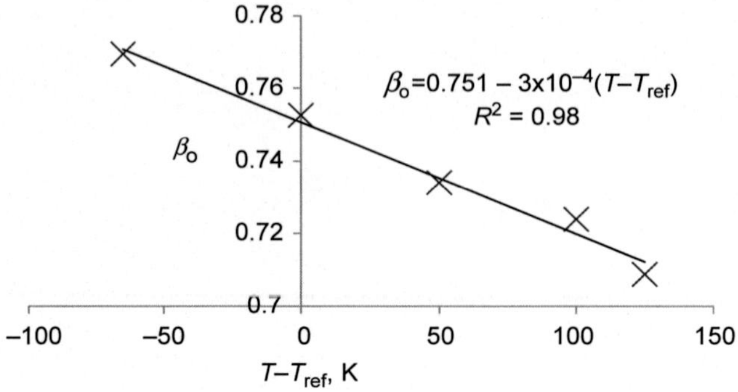

Figure 3.7 β_o versus $(T - T_{ref})$ at 1 Hz, $T_{ref} = 298$ K.

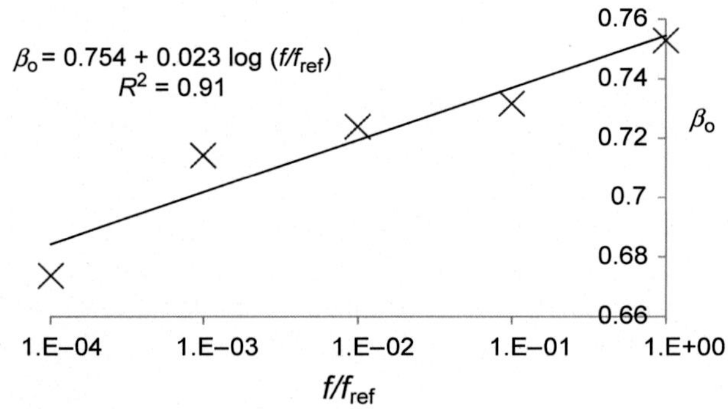

Figure 3.8 β_0 versus f/f_{ref} at 25 °C, $f_{ref} = 1$ Hz.

3.4.1.3 Populating the $\Delta\varepsilon_p - N$ data

It is noted that the coefficient C_o obtained using only the data from Table 3.1 (temperature function) differs significantly from that obtained from Table 3.2 (frequency function). A more consistent extraction procedure should extract all the coefficients simultaneously. This requires the data to populate over the temperature–frequency domain. Using the extracted coefficient C_o and exponent β_o in Tables 3.1 and 3.2, the fatigue life N of the solder at six plastic strain ranges (1, 2, 5, 10, 25 and 50%) are populated using Eqn (3.1). Figure 3.9(a) shows the populated $\Delta\varepsilon_p - N$ data at five frequencies (10^{-4}, 10^{-3}, 10^{-2}, 10^{-1} and 1 Hz) and at 25 °C. Figure 3.9(b) shows the populated $\Delta\varepsilon_p - N$ data for five temperatures (−40, 25, 75, 125 and 150 °C) at cyclic frequency 1 Hz.

3.4.2 The unified creep-fatigue equation for eutectic SnPb solder

The creep-dormant temperature of the eutectic SnPb solder is taken as two-thirds the melting temperature of the solder (456 K), which gives $T = 180$ K. The reference

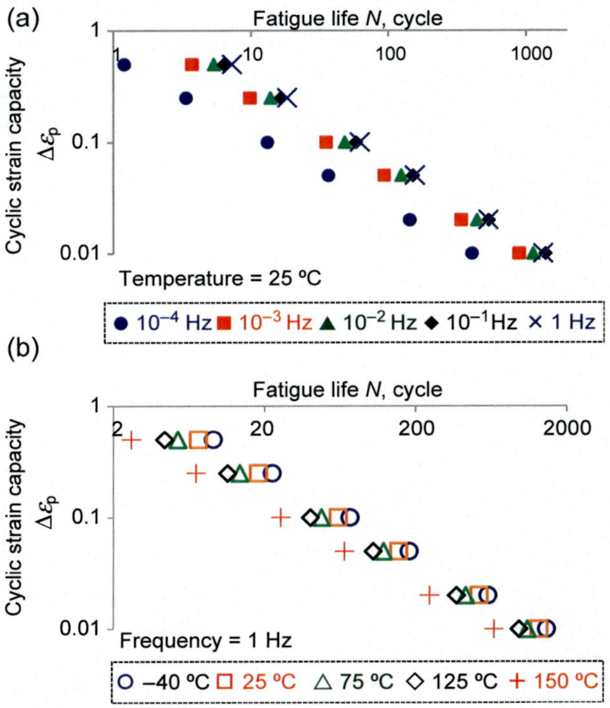

Figure 3.9 Populated $\Delta\varepsilon_p - N$ data (a) at five cyclic frequencies at 25 °C and (b) at five temperatures at 1 Hz.

condition for the eutectic SnPb solder is assumed as $T_{ref} = 180$ K and $f_{ref} = 1$ Hz. Substituting Eqn (3.29), Eqn (3.30) and the reference conditions into Eqn (3.19) gives the unified creep-fatigue equation for the eutectic SnPb solder as:

$$\Delta\varepsilon_p = C_o[1 + c_1(T - 180) + c_2 \log f]N^{-\beta_o[1+s_1(T-180)+s_2 \log f]} \tag{3.31}$$

The ratios c_2/c_1 and s_2/s_1 define the time-temperature creep relations of the eutectic solder and should be identical. By defining the fatigue life of the populated ($\Delta\varepsilon_p$, T, f, N) data of Figure 3.9 as N_{pop} and that computed by Eqn (3.31) as N_{eq}, and minimising the error squares sum$\sum_{i=1}^{n}\left(\ln(N_{eq,i}) - \ln(N_{pop,i})\right)^2$, where i includes the data at five frequencies, five temperatures and six plastic strain ranges for each frequency and each temperature, a total of 60 sets of data, together with the constraint $c_2/c_1 = s_2/s_1$, the coefficients have been evaluated as $C_o = 3.89$, $c_1 = -3.10 \times 10^{-3}$ and $c_2 = 0.117$, $\beta_o = 0.839$, $s_1 = -7.25 \times 10^{-4}$, $s_2 = 0.0273$. The average fitting error per data point defined by $\xi = \sqrt{\sum_{i=1}^{n}\left(\ln(N_{eq,i}) - \ln(N_{reg,i})\right)^2/n}$ has been evaluated as 0.010.

The 60 sets of populated ($\Delta\varepsilon_p$, T, f, N) data of Figure 3.9 have been transformed to the reference frames ($\Delta\varepsilon_{p,ref} - N$) and ($\Delta\varepsilon_p - N_{ref}$) using Eqns (3.21a) and (3.24a), respectively. These are plotted in Figure 3.10(a) and (b), respectively. Remarkably, the 60 sets of populated ($\Delta\varepsilon_p$, T, f, N) data collapsed almost perfectly ($R^2 = 0.998$) into a straight line (on a log–log scale) in the reference frames, which has self-validated the unified equation, Eqn (3.31). It is worth highlighting that only the frequency–temperature modified models are capable of self-validation.

3.4.3 Benchmarking

The unified equation, Eqn (3.31), for the eutectic SnPb solder is benchmarked against the equations of Engelmaier (1983), Eqn (3.14) and Shi (2000), Eqn (3.16), for the same solder alloy.

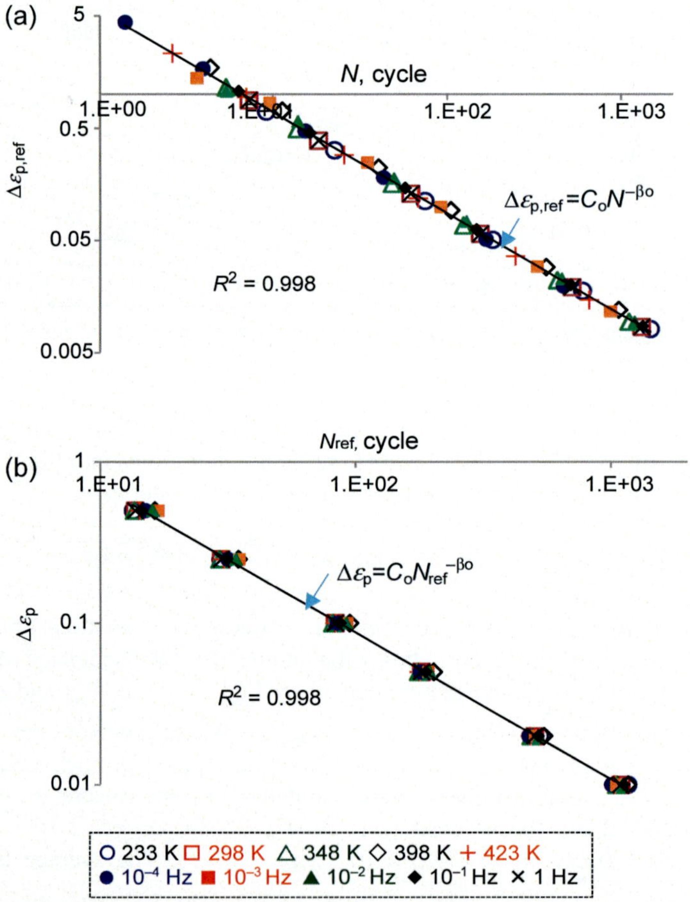

Figure 3.10 (a) $\Delta\varepsilon_{p,ref}$ versus N and (b) $\Delta\varepsilon_p$ versus N_{ref} for eutectic SnPb solder.

3.4.3.1 The average fitting error

Fitting the 60 sets of populated ($\Delta\varepsilon_p$, T, f, N) data with Eqns (3.16) and (3.17) for Shi (2000), the average fitting error has been evaluated as $\xi = 0.160$, which is alarmingly large. It is noted that the strain-life data at the frequency of 10^{-4} Hz is mainly responsible for the error. Excluding the six sets of data at this frequency, the average fitting error for the remaining 54 sets of data for the frequency range $10^{-3}\text{Hz} < f < 1$ Hz is reduced significantly to $\xi = 0.014$.

The equation of Engelmaier (1983) is given in Eqn (3.14). The exponential function is modelled as $\beta = b_o + b_1 T + b_2 \ln(1 + 43200f)$, wherein the coefficients b_o, b_1 and b_2 are established by regression with the 60 sets of populated ($\Delta\varepsilon_p$, T, f, N) data; the average fitting errors, including and excluding the six sets of data at 10^{-4} Hz, have been evaluated as 0.039 and 0.042, respectively.

The average fitting error ξ for Shi (2000), Engelmaier (1983), and the unified equation, Eqn (3.31) for the eutectic SnPb solder are summarised in Table 3.3. The superior fitting of the unified equation is remarkable, especially considering the fact that up to third order in temperature was used in the equations of Shi (2000).

3.4.3.2 Discontinuity of frequency exponent

Both Solomon (1991) and Shi (2000) have reported a discontinuity of the frequency exponent k at $\sim 10^{-3}$ Hz. Such discontinuity is physically unintuitive. It is worth noting that both Solomon and Shi have assumed a power law relation between fatigue life and frequency, which has been shown in Figure 3.6 to give rise to a poor fit for the creep function. When a power law creep function, $c(T, f) = T^p f^q$, is used to fit the 60 sets of populated ($\Delta\varepsilon_p$, T, f, N) data, and then transformed to the reference frames ($\Delta\varepsilon_{p,ref} - N$) using Eqn (3.21a), the data points at the frequency of 10^{-4} Hz do appear as outliers in Figure 3.11. The reported discontinuity in the frequency exponent k is thus an artefact of enforcing a power law relation over what appears to be a logarithmic relation between fatigue life and frequency.

3.4.3.3 Check for ideal characteristics

An ideal creep-fatigue equation should be capable of modelling the entire range of creep-fatigue from pure fatigue to pure creep rupture. That is, it should have the

Table 3.3 Comparisons of the average fitting error

Creep-fatigue equations	Average fitting error ξ: Include data at 10^{-4} Hz	Exclude data at 10^{-4} Hz
Shi	0.160	0.014
Engelmaier	0.039	0.042
The unified equation	0.010	0.008

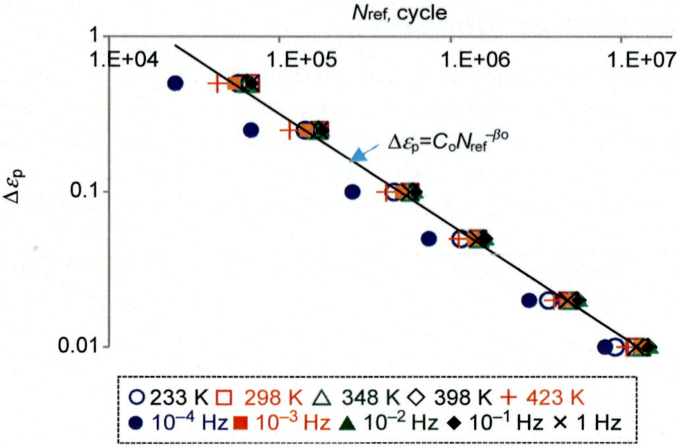

Figure 3.11 $\Delta\varepsilon_p$ versus N_{ref} for $c(T, f) = T^p f^q$, $s(T, f) = 1$, $\beta_0 = 0.733$, $p = -1.18$ and $q = 0.049$.

following characteristics: (1) The Coffin–Manson equation is recovered at the reference condition and (2) the condition $C(T, f) = 0$ corresponds to a viable state of creep rupture. Moreover, (3) the creep and stress functions should have consistent time–temperature characteristics.

Characteristics (1) and (3) have been inherently built into the unified equation, Eqn (3.31). The condition of 'creep rupture' for the eutectic SnPb solder is given by: $c(T,\ f) = 1 - 3.10 \times 10^{-3}(T - 180) + 0.117 \log f = 0$. Substituting $T = (T_m - 10)$ $K = 446$ K into the above function gives $f = 0.03$ Hz, which corresponds to a cycle time of 32 s. The condition $c(T, f) = 0$ is associated with $\Delta\varepsilon_p = 0$ and $N < 1$. Most materials exhibit cyclic strain hardening, that is,

$$\Delta\sigma = \Delta\sigma_{yield}\left(1 + \Delta\varepsilon_p^{\gamma}\right) \tag{3.32}$$

The condition $\Delta\varepsilon_p = 0$ gives $\Delta\sigma = \Delta\sigma_{yield}$. Hence, it is postulated that creep rupture of the solder will occur in less than 32 s at $T = 446$ K when subjected to a harmonically varying stress of amplitude $\Delta\sigma = \Delta\sigma_{yield}$ at the frequency of 0.03 Hz. This is a viable condition, and, hence, the unified equation, Eqn (3.31), satisfies characteristic (2)

By replacing T in Eqn (3.17) with $(T - T_{ref})$ and f with f/f_{ref} and letting $T_{ref} = 0$ °C and $f_{ref} = 1$ Hz, Eqn (3.16) of Shi (2000) is reduced to the Coffin–Manson equation at the reference condition, satisfying characteristic (1). The condition $C_c(T)$ $f^{\beta(T)[1-k(T)]} = 0$ implies either $f = 0$ or $C_c = 2.122 - 3.57 \times 10^{-3}T + 1.329 \times 10^{-5}T^2 - 2.502 \times 10^{-7}T^3 = 0$; the former indicates indefinite creep life, which is unreal; the latter leads to $T = 198$ °C, which is above the melting temperature of the eutectic SnPb solder (183 °C); therefore, characteristic (2) is violated. Finally, the exponent function $\beta(T)$ does not embody a time–temperature relation, violating characteristic (3).

By replacing $\overline{T}$ in Eqn (3.15) with $(\overline{T} - T_{ref})$ and letting $T_{ref} = 0$ °C and $f_{ref} = 0$ Hz, Eqn (3.14) of Engelmaier is reduced to the Coffin–Manson equation at the reference condition, satisfying characteristic (1), although the reference condition of $f_{ref} = 0$ Hz is unrealistic. The coefficient $C = C_o$ also does not allow modelling of creep rupture, violating characteristic (2). The coefficient C_o does not embody a time–temperature relation, violating characteristic (3).

3.5 Applications

3.5.1 Extension to thermomechanical fatigue of solder joints

The creep and the stress functions $c(T, f)$ and $s(T, f)$ for the eutectic SnPb solder in Section 3.4 are derived from standard test specimens that experience uniform cross-sectional strain under isothermal conditions. Question naturally arises as to the applicability of these functions to the thermomechanical fatigue of engineering assemblies like solder joints in a microelectronics assembly. This will be addressed by examining the impacts of strain singularity and non-isothermal conditions on the creep functions.

3.5.1.1 Strain singularity

Solder joints, as with many practical structures, have geometrical discontinuity where the plastic strain becomes singular. When evaluated using finite element method, the magnitude of the plastic strain becomes finite, and is dependent on the fineness of the volume discretisation – a phenomenon referred to as mesh dependency. To circumvent the difficulty of mesh dependency, some researchers have resorted to averaging the plastic strain over a selected volume, assuming the total plastic strain within the selected volume of solder is relatively independent of mesh. However, the computed average strain must necessarily depend on the size and location of the selected volume, which gives rise to inconsistency in the evaluation of $\Delta\varepsilon_p$. It is clear from the unified equation, $\Delta\varepsilon_p = C_o c(T, f) N^{-\beta_o s(T,f)}$, that any variation in $\Delta\varepsilon_p$ will lead to only proportionate change in the coefficient C_o, whose value needs to be calibrated with the volume-averaging practice used in the evaluation of $\Delta\varepsilon_p$. The presence of strain singularity has no bearing on the creep and the functions, $c(T, f)$ and $s(T, f)$.

3.5.1.2 Temperature cycling

Many electronic and mechanical assemblies experience temperature fluctuations when in service. The non-isothermal condition is typically simulated in the laboratory using constant amplitude temperature cycling with possible dwell time at extreme temperatures. Plastic strain is induced by the mismatched thermal expansions between the components constituting the assembly, while creep strain is accumulated over the temperature cycle. The thermomechanical fatigue life can be modelled with the unified equation, $\Delta\varepsilon_p = C_o c(T, f) N^{-\beta_o s(T,f)}$, but the isothermal creep functions are substituted with non-isothermal creep functions. For first approximation, the non-isothermal creep functions are simply time-averaging of the isothermal creep functions over the duration of a temperature cycle $1/f$, that is,

$$\overline{c}(T,f) = \frac{\int_0^{1/f} c(T,f)\mathrm{d}t}{1/f} \quad \text{and} \quad \overline{s}(T,f) = \frac{\int_0^{1/f} s(T,f)\mathrm{d}t}{1/f} \tag{3.33}$$

Using the function $c(T,f) = 1 + c_1(T - T_{\mathrm{ref}}) + c_2 \log(f/f_{\mathrm{ref}})$ for illustration, the non-isothermal function $\overline{c}(T,f)$ is given by

$$\overline{c}(T,f) = 1 + c_1\left(\overline{T} - T_{\mathrm{ref}}\right) + c_2 \log(f/f_{\mathrm{ref}}) \tag{3.34}$$

where $\overline{T} = \frac{\int_0^{1/f} T\mathrm{d}t}{1/f}$.

It is acknowledged that the above approach may not account for dwell sensitivity (Bache, 2003), whose physics is still not well understood.

3.5.2 Evaluation of acceleration factor

Engineering components are typically designed for many years of creep-fatigue life in service. Controlled tests for design evaluation and quality assurance are typically performed at overstressed conditions so as to reduce the test duration. The total duration of a fatigue test is given by the product Nf^{-1}. The quotient $N_1f_1^{-1}/N_2f_2^{-1}$ gives the acceleration factor for test condition 2 to test condition 1. From the unified equation, $\Delta\varepsilon_{\mathrm{p}} = C_{\mathrm{o}}c(T,f)N^{-\beta_{\mathrm{o}}s(T,f)}$, expressing N in terms of $\Delta\varepsilon_{\mathrm{p}}$, $c(T,f)$ and $\beta_{\mathrm{o}}s(T,f)$, the acceleration factor is given by

$$\mathrm{AF} = \frac{N_1f_1^{-1}}{N_2f_2^{-1}} = \frac{\Delta\varepsilon_{p1}^{-1/\beta_{\mathrm{o}}s(T_1,f_1)}}{\Delta\varepsilon_{p2}^{-1/\beta_{\mathrm{o}}s(T_2,f_2)}} \frac{c(T_1,f_1)^{1/\beta_{\mathrm{o}}s(T_1,f_1)}}{c(T_2,f_2)^{1/\beta_{\mathrm{o}}s(T_2,f_2)}} \frac{f_1^{-1}}{f_2^{-1}} \tag{3.35}$$

As an example, substituting $c(T,f) = 1 + c_1(T - 180) + c_2 \log f$, $s(T,f) = 1 + s_1(T - 180) + s_2 \log f$ and $c_1 = -3.10 \times 10^{-3}$, $c_2 = 0.117$, $\beta_{\mathrm{o}} = 0.839$, $s_1 = -7.25 \times 10^{-4}$ and $s_2 = 0.0273$ obtained for the eutectic SnPb solder into Eqn (3.35), and for $\{\Delta\varepsilon_2 = \Delta\varepsilon_1 = 0.1,\ T_2 = 423\ \mathrm{K},\ T_1 = 233\ \mathrm{K},\ f_2 = f_1 = 1\ \mathrm{Hz}\}$, we obtain $\mathrm{AF} = 3.8$.

It is indeed much easier to express the creep and the stress functions in terms of simple power laws, which though less accurate are often good enough for the purpose of estimating the acceleration factor. Thus, substituting $c(T,f) = T^pf^q$, $s(T,f) = 1$ into Eqn (3.35), we obtain

$$\mathrm{AF} = \left(\frac{\Delta\varepsilon_{\mathrm{p1}}}{\Delta\varepsilon_{\mathrm{p2}}}\right)^{-1/\beta_{\mathrm{o}}} \left(\frac{T_1}{T_2}\right)^{p/\beta_{\mathrm{o}}} \left(\frac{f_1}{f_2}\right)^{q/\beta_{\mathrm{o}}-1}. \tag{3.36}$$

For $\beta_o = 0.733$, $p = -1.18$ and $q = 0.049$ evaluated through best fitting, the conditions, $\{\Delta\varepsilon_2 = 1.66\Delta\varepsilon_1,\ T_2 = T_1,\ f_2 = f_1\}$ or $\{\Delta\varepsilon_2 = \Delta\varepsilon_1,\ T_2 = 1.54T_1,\ f_2 = f_1\}$ or $\{\Delta\varepsilon_2 = \Delta\varepsilon_1,\ T_2 = T_1,\ f_2 = 2.1f_1\}$, yield an acceleration factor AF = 2.

3.5.3 Evaluation of equivalent test parameter

Temperature cycling is used routinely as a test to assure the quality of a design or a manufacturing process. However, there is strong motivation to substitute a temperature cycling test, which is expensive to perform, with a mechanical fatigue test (Shi, 2000). The simplest equivalent mechanical fatigue test is one that applies a plastic strain range of magnitude identical to that experienced by the test object in the temperature cycling test, at identical cyclic frequency, while holding the object in a temperature chamber set at the time-average temperature, $\overline{T} = \frac{\int_0^{1/f} T\mathrm{d}t}{1/f}$. In the event that it is only possible to subject the test object at a different condition from the above, then it is possible to substitute the damage contribution of one test parameter with another. As an example, a mechanical cyclic test is to be conducted at a room temperature of 293 K instead of the time-average temperature of $\overline{T} = 323$ K. The '+30 K' of temperature-induced damage has to be provided by equivalent damage in frequency or strain range. The equivalence relation may be established as follows: from the unified equation, $\Delta\varepsilon_\mathrm{p} = C(T, f)N^{-\beta(T,f)}$, the condition for identical fatigue life is given by

$$N = C(T_1, f_1)^{1/\beta(T_1,f_1)}\Delta\varepsilon_{\mathrm{p}1}^{-1/\beta(T_1,f_1)} = C(T_2, f_2)^{1/\beta(T_2,f_2)}\Delta\varepsilon_{\mathrm{p}2}^{-1/\beta(T_2,f_2)} \tag{3.37}$$

The condition for the equivalent test parameters that return the same fatigue life is simply

$$\begin{aligned} &\left.\frac{C(T_1, f_1)^{1/\beta(T_1,f_1)}\Delta\varepsilon_{\mathrm{p}1}^{-1/\beta(T_1,f_1)}}{C(T_2, f_2)^{1/\beta(T_2,f_2)}\Delta\varepsilon_{\mathrm{p}2}^{-1/\beta(T_2,f_2)}}\right|_{T_1=T_2,f_1=f_2} \\ &= \left.\frac{C(T_1, f_1)^{1/\beta(T_1,f_1)}\Delta\varepsilon_{\mathrm{p}1}^{-1/\beta(T_1,f_1)}}{C(T_2, f_2)^{1/\beta(T_2,f_2)}\Delta\varepsilon_{\mathrm{p}2}^{-1/\beta(T_2,f_2)}}\right|_{\Delta\varepsilon_{\mathrm{p}1}=\Delta\varepsilon_{\mathrm{p}2},T_1=T_2} \\ &= \left.\frac{C(T_1, f_1)^{1/\beta(T_1,f_1)}\Delta\varepsilon_{\mathrm{p}1}^{-1/\beta(T_1,f_1)}}{C(T_2, f_2)^{1/\beta(T_2,f_2)}\Delta\varepsilon_{\mathrm{p}2}^{-1/\beta(T_2,f_2)}}\right|_{f_1=f_2,\Delta\varepsilon_{\mathrm{p}1}=\Delta\varepsilon_{\mathrm{p}2}} \end{aligned} \tag{3.38}$$

Substituting $c(T, f) = 1 + c_1T + c_2 \log f$ and $s(T, f) = 1$ into Eqn (3.38) gives

$$\frac{\Delta\varepsilon_{\mathrm{p}2}}{\Delta\varepsilon_{\mathrm{p}1}} = \frac{1 + c_1T + c_2 \log f_1}{1 + c_1T + c_2 \log f_2} = \frac{1 + c_1T_1 + c_2 \log f}{1 + c_1T_2 + c_2 \log f} \tag{3.39}$$

For $c_1 = -1.72 \times 10^{-3}$ and $c_2 = 0.0719$ (obtained through best data fitting) and $f = f_1 = 1$ Hz, $T = T_1 = 293$ K, a 30 K increase in temperature may be substituted with a 12% increase in plastic strain or an 80% reduction in frequency.

References

Ainsworth, R. A. (1992). Flaw assessment procedure for high temperature reactor components. *Journal of Pressure Vessel Technology, 114*, 166–170.

Bache, M. R. (2003). A review of dwell sensitivity fatigue in titanium alloys: the role of microstructure, texture and operation conditions, Int. *J. Fatigue, 25*, 1079–1087.

Bailey, R. (1935). *The utilistation of creep test data in engineering design.* I Mech E. p. 131.

Basquin, O. (1910). The exponential law of endurance tests. In *ASTM Annual Meeting* (Vol. 10, pp. 625–630). ASTM.

Beres, W. (2000). *Mechanics of materials failure, RTO-TR-28-Appendix 3.* NATO Research and Technology Organisation.

Bird, J. M. (1969). *Correlations between high-temperature creep behaviour and structure.* Haifa University Press.

Coffin, L. F., Jr. (1954). A study of the effects of cyclic thermal stresses on a ductile metal. *Transactions of the ASME, 76*, 923–930.

Coffin, L. F. (1976). The concept of frequency separation in life prediction for time-dependent failure. In *ASME-MPC symposium on creep/fatigue interaction, MPC-3. New York.*

Darveaux, R. (2002). Effects of simulation methodology on solder joint crack growth correlation and faituge life prediction. *Transactions of the ASME, Journal of Electronic Packaging, 124*, 147.

Dorn, J. (1955). Some fundamental experiments on high temperature creep. *Journal of the Mechanics and Physics of Solids, 3*, 85–116.

Dowling, N. (2012). *Mechanical behaviour of materials* (4th ed.). Prentice Hall.

Engelmaier, W. (1983). Fatigue life of leadless chip carrier solder joint during power cycling. *IEEE Transactions on Components, Hybrids and Manufacturing Technology, 6*, 232–237.

Frear, D. J. (1991). *Solder mechanics-a state of the art assessment.* TMS Publication.

Goldhoff, R. (1965). Uniaxial creep – rupture behaviour of low alloy steel under variable loading conditions. *Journal of Basic Engineering, 86.*

Hertzbery, R. (1996). *Deformation and fracture mechanics of engineering materials* (4th ed.). New York: John Wiley & Sons.

Larson, F. M. (1952). A time-temperature relationship for rupture and creep stresses. *Transactions of the ASME, 74*, 765–771.

Lau, J. H. (1991). *Solder joint reliability: Theory and application.* New York: Van Nostrand Reinhold.

Lee, W. N. (2000). Solder joint fatigue models: review and applicability to chip scale packages. *Microelectronic Reliability, 40*, 231–244.

Liang, J. G. (1997). *A integrated fatigue life prediction methodology for optimum design and reliability assemssment of solder integ-connections.* International Intersociety Electronic and Photonic Packaging. pp. 1583–1592.

Majumder, S. M. (1976). A damage equation for creep-fatigue interaction. In *ASME-MPC symposium on creep/fatigue interaction, MPC-3, New York.*

Manson, S. H. (1953). *A linear time-temperature relation for extrapolation of creep and stress-rupture data, NACA TN 2890.* National Advisory Committee for Aeronautics.

Manson, S. (1954). *Behavior of materials under conditions of thermal stress*. National Advisory Committee for Aeronautics.

Manson, S. H. (1976). Creep-fatigue analysis by strain range partitioning. In *ASME-MPC symposium on creep/fatigue interaction, MPC-3, New York.*

Miner, M. A. (1945). Cumulative damage in fatigue. *Journal of Applied Mechanics, 12,* A159–A164.

Ostergren, W. (1976). Correlation of hold time effects in elevated temperature fatigue using a frequency modified damage function. In *ASME-MPC symposium on creep/fatigue interaction, MPC-3, New York.*

Palmgren, A. (1924). Die Lebensdauer von Kugellagern. *Zeitschrift des Vereines Deutscher Ingenieure, 68*, 339–341.

Penny, R., & Marriott, D. (1995). *Design for creep* (2nd ed.). Chapman & Hall.

Robison, R. (1938). Effect of temperature variations on the creep strength of steels. *Transactions of the ASME, 60*, 253–259.

Ross, J. R.-C. (1993). Solder creep-fatigue interactions with flexible leaded surface mount components. In J. H. Lau (Ed.), *Thermal stress and strain in microelectronics packaging*. Van Nostrand Reinhold.

Schutz, W. (1996). A history of fatigue. *Engineering fracture mechanics, 54*, 263–300.

Shi, X. P. (2000). Low cycle fatigue analysis of temperature and frequency effects in eutectic solder alloy. *International Journal of Fatigue, 22*, 217–228.

Solomon, H. (1991). Predicting thermal and mechanical fatigue life from isothermal low cycle data. In J. H. Lau (Ed.), *Solder joint reliability: Theory and application*. New York: Van Nostrand Reinhold.

Taira, S. (1962). In *Creep in structures* (pp. 96–124). Academic Press.

Wöhler, A. (1955). Theorie rechteckiger eiserner Brückenbalken mit Gitterwänden und mit Blechwänden. *Zeitschrift für Bauwesen, 5*, 121–166.

Zhuang, W. S. (1998). *Thermo-mechanical fatigue life prediction – A critical review, DSTO-TR-0609*. Australian Defence Science Technology Organisation.

Part Two

Advances in robust design against moisture-induced failures

Moisture properties and their characterisations

4

4.1 Introduction

Microelectronic packaging has migrated from ceramic body to plastic encapsulation. The change is motivated entirely by cost consideration – the raw material cost of plastics, even including the silica fillers, is significantly less than that of ceramics; more importantly, plastic encapsulation facilitates the use of high-production-rate (transfer) moulding as the production technique that has led to significant reduction in the production cost.

The bulk of the microelectronic packaging materials today are made of polymeric materials; for example die-attached materials found in almost all electronic packaging; moulding compound in over-mould packaging; underfill materials in flip-chip packaging; adhesives and waveguides in optoelectronics packaging; conductive/nonconductive adhesives in fine-pitch flip-chip interconnection; and laminated substrates in plastic ball grid array (PBGA) packaging; etc. However, these polymers are capable of absorbing, retaining and transporting moisture. Microscopically, the molecules of water can be present within the molecular structure of the host polymer in two basic forms: physically residing in the 'free volume' within the molecular structure of the polymer or chemically attached to the hydrogen bonding sites along the polymer chains. The physical and chemical interactions of moisture with polymers lead to physical, mechanical and, in many instances, chemical changes in the polymer. The extent of these changes varies amongst the polymers.

Volumetric expansion, typically referred to as hygroscopic swelling, is probably the most well-known physical change associated with absorption of moisture by polymers. Dielectric constant is another physical property of polymers that is associated with moisture absorption. The absorbed moisture may also react chemically with the molecules of the host polymer leading to 'plasticisation' of the polymer. The result is a decrease in the glass-transition temperature of the polymer accompanied by reductions in strength and modulus of the polymer but increases in its ductility and fracture toughness. The water molecules may also react chemically with the molecular bonds between the interfaces of two polymeric surfaces (e.g., formation of hydrogen bond by the oxygen ion in water with the hydrogen ion in the polymeric interface) leading to reduction in the interfacial bond strength. Hence, the absorption of moisture by the polymeric members of microelectronic packaging can adversely reduce its mechanical/structure integrity. Modern microelectronic packaging may be simplistically visualised as being made up of layers of organic (polymers) and non-organic (silicon and copper) materials that are bonded together. Volumetric expansion of polymers upon absorption of moisture can lead to mismatched hygroscopic expansion between layers of the constituent materials, not unlike the mismatched thermal expansion, leading to hygroscopic stresses in the microelectronic packaging. The risk to structural

Robust Design of Microelectronics Assemblies Against Mechanical Shock, Temperature and Moisture
http://dx.doi.org/10.1016/B978-1-84569-528-6.00004-6

integrity is further aggregated by the reduction of interfacial bond strength due to the water molecules.

The absorption of moisture by the polymeric members of electronic packaging can also adversely reduce the electrical performance of the packaging. With a dielectric constant (permittivity) that is more than 10 times that of a typical polymer, the presence of water molecules in a polymer can significantly increase the dielectric constant of the polymer, thus, decreasing the propagation speed of electrical signal through the circuits. The characteristic impedance of the micro-strip waveguide within the electrical circuits will also be altered by the change in the permittivity of the dielectric, and the resulting mismatched impedance will increase the transmission loss of the electric signal.

The negative moisture effects on the constituent polymers of electronic packaging may be accentuated by elevated temperature as is experienced by microelectronic packaging during its processing and field application. For example, the reduction in strength in the presence of both moisture and temperature is typically more pronounced than the sum of the individual effect. Of particular interest is the phase transformation of moisture by temperature. The retained moisture in the constituent polymers could turn into pressurised vapour at elevated temperature, which in turn could act in conjunction with the mismatched thermal and moisture expansions, and aided by the reduction in chemical bond strength of the constituting materials and interfaces, lead to premature failure of microelectronic packaging. The well-known 'popcorn' cracking of electronic packaging during reflow of solder is a result of such potent combinations.

The less common, volumetric expansion could also happen at cryogenic temperature when the retained moisture is transformed into ice from liquid water. Repeated cyclic loading at cryogenic temperature could drive crack formation. The negative effects of moisture to microelectronic packaging are categorically referred to as moisture-induced damage in this book.

Figure 4.1 summarises the effects of moisture on a PBGA package – a quarter of which is shown without the silicon chip. The PBGA package is assumed to have some residue thermomechanical stresses prior to moisture sorption (top left). The absorption of moisture by the constituent polymeric materials alters the elastic modulus and the glass transient temperature of the materials, leading to redistribution of the residue thermomechanical stresses. The absorption of moisture by the constituent polymers leads to hygroscopic swelling of these materials (top right). The differential swelling between the moisture absorbants and non-absorbants, between polymers of different swelling characteristics, and between different parts of a polymer, which have different moisture content, result in mismatched deformation and hence stresses within the PBGA package. The retained moisture in electronic packages will turn into high-pressure vapour when heated. The magnitude of the vapour pressure depends on the ambient temperature and on the specific volume of the vapour as defined by its equation of state.

The characterisation of moisture properties – sorption, diffusion and swelling – will be discussed in various sections of this chapter. But first, the fundamentals of the thermodynamics of water are introduced in Section 4.2 as a prelude to the following sections.

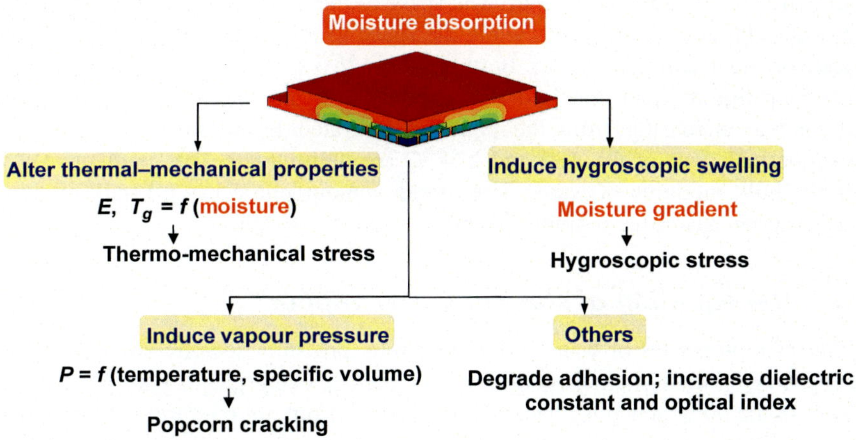

Figure 4.1 An overview of moisture-induced damage in microelectronic packaging.

4.2 Thermodynamics of water

4.2.1 The p–v *diagram*

Under normal conditions, water exists in one of three states: solid, liquid and vapour. The transition between two states is referred to as phase transformation. Figure 4.2 shows the graph of pressure versus specific volume of water, v, under isothermal conditions. The blue line defines the line of saturated liquid and the

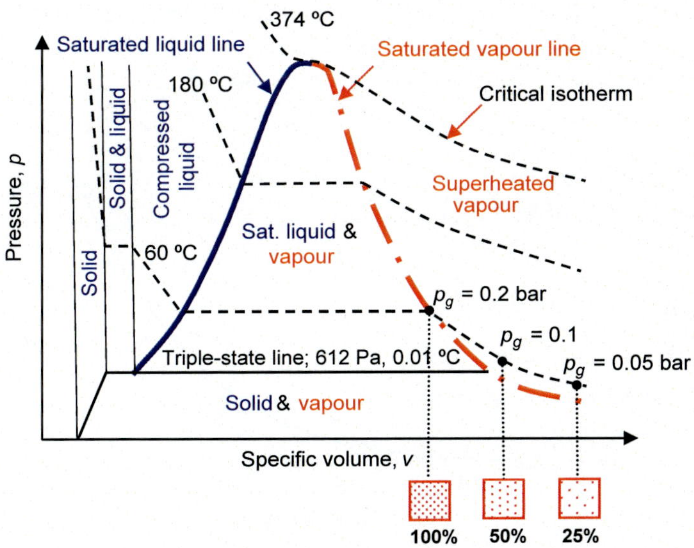

Figure 4.2 Pressure-specific volume of water (not to scale).

red line the line of saturated vapour. The space to the left of the blue line defines the state of unsaturated liquid water, while the space to the right of the red line defines the state of supersaturated vapour. Between these two saturated lines lies the mixture of saturated liquid water and saturated vapour, with the fractional weight of saturated vapour increasing from the point of saturated liquid to saturated vapour. The isothermal lines of 60, 180 and 274 °C are highlighted. The gradient, $\Delta p/\Delta v$, gives the bulk modulus of water. The liquid–vapour phase has nil bulk modulus, but the vapour has a decreasing bulk modulus.

4.2.2 The equation of state of water vapour

The intrinsic properties of water are temperature, pressure, specific volume, specific heat capacity, specific internal energy, specific enthalpy and specific entropy. The state of water in the homogeneous phase (liquid or vapour but not mixture) is fully defined by any two intrinsic properties. At liquid or vapour saturation, the state of water is fully defined by a single property. The subscript g is generally used to define the saturation vapour condition; for example, the saturated vapour pressure is written as $p_g(T)$.

4.2.2.1 Saturation vapour pressure

The saturation pressures, $p_g(T)$, are freely available in many Web-based calculators, for example TLV (n.d.); and may be expressed rather well using the Arrhenius equation (Wong, 2003):

$$p_g = p_o \exp\left(\frac{-E_v}{RT}\right) \tag{4.1}$$

where $R = 8.314$ J/K mol, $p_o = 18.87$ GPa, $E_v = 36.98$ kJ/mol for temperature between 25 and 100 °C and $p_o = 30.05$ GPa, $E_v = 38.83$ kJ/mol for temperature between 100 and 250 °C. This is illustrated in Figure 4.3.

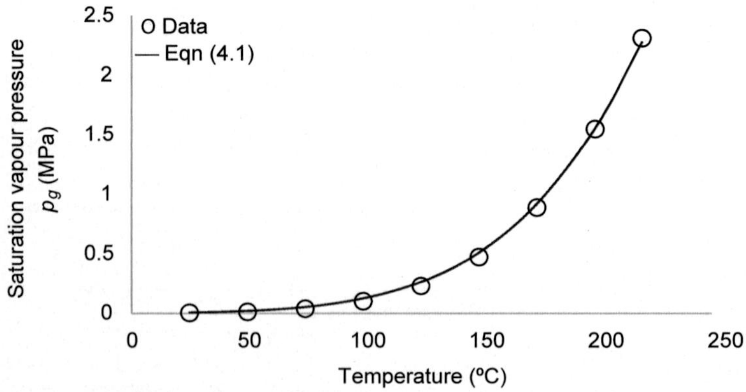

Figure 4.3 The saturation vapour pressure of water, $p_g(T)$.

4.2.2.2 Supersaturated vapour

The supersaturated vapour at low pressure and high temperature may be approximated as an ideal gas:

$$\begin{aligned} pv_m &= RT \\ pv &= R_g T \end{aligned} \tag{4.2}$$

where v_m is the volume per unit mole and $R_g = 461$ J/K/kg is the gas constant of water vapour. Better approximation is given by the van der Waals equation:

$$\left(P + \frac{\mathrm{a}}{v_m^2}\right)(v_m - \mathrm{b}) = RT \tag{4.3}$$

which includes the contribution of the intermolecular forces (the constant a) and excludes the volume taken up by the gas molecules (the constant b). The constants, a and b, for water vapour are 0.547 Pa m^3 and 30.52×10^{-6} m^3/mol, respectively. The states of water vapour may be more conveniently observed using the Web-based calculators such as 'the steam table online' (Steamtable, 2014).

4.2.3 The relative humidity

Atmospheric air is a mixture of air and water vapour, which exists in the supersaturated state. Relative humidity (RH) describes the 'degree of supersaturation' of the water vapour. It is defined as the partial pressure of the vapour in the air–vapour mixture, p_v, to the saturation pressure of the vapour, p_g, at the same temperature:

$$\%\mathrm{RH} = \frac{p_v(T)}{p_g(T)} \times 100\% \tag{4.4}$$

The RH is 100% at the saturation pressure; the lower the partial pressure, the higher the degree of superheating; this is illustrated in Figure 4.2. As an approximate ideal gas, the RH may also be written as

$$\%\mathrm{RH} \approx \frac{\rho_v(T)}{\rho_g(T)} \times 100\% = \frac{C_v(T)}{C_g(T)} \times 100\% \tag{4.5}$$

where ρ (kg/m^3) is the density of water vapour and C (kg/m^3) is the concentration of water vapour in the air.

4.3 Sorption and its characterisation

Adsorption is the enrichment of the guest substance on the surface of the host substance, while *absorption* describes the solution of the guest substance into the body of the host substance through diffusion. The term *sorption* was introduced by McBain (1909) to describe the processes of surface adsorption and bulk absorption. Adsorption must precede absorption in all sorption processes.

4.3.1 Adsorption and isotherms

The driving force for adsorption is the reduction in the interfacial potential energy between the fluid and the solid. These forces are of two main kinds – physical and chemical – and they give rise to physisorption and chemisorption, respectively. Chemisorption involves a chemical reaction between the adsorbate and the adsorbant leading to the formation of chemical bonds; the process is irreversible and only a monolayer of molecules is bonded onto the adsorbant. Physisorption, by contrast, is associated with intermolecular forces (van der Waal's forces) between the adsorbate and the adsorbant. See Table 4.1 for a comparison of physisorption and chemisorption.

4.3.1.1 Adsorption isotherms

Modelling water vapour as an ideal gas and that it spreads as a monolayer on the surface of the adsorbant (Stadie, 2013), then

$$p_a A_a = n_a RT \tag{4.6}$$

where p_a is the spreading pressure, A_a the surface area of the coverage, n_a the number of moles of the water vapour within A_a. Assuming the spreading pressure to be proportionate to the partial pressure of the water vapour, p_v, then Eqn (4.6) may be rearranged to give

$$\frac{n_a}{A_a} = \frac{c_O p_v}{RT} = k_H p_v \tag{4.7}$$

Table 4.1 Comparisons between physisorption and chemisorption

Physisorption	Chemisorption
Reversible	Irreversible
Not very specific	Highly specific
Force of attraction are van der Waal's forces	Forces of attraction are chemical bond forces
Forms multimolecular layers	Forms monomolecular layers
Occurs at low temperature and decreases with increasing temperature	Occurs at high temperature
Does not require any activation energy	Requires activation energy

where k_{H} is Henry's constant, which is a function of the intermolecular forces between adsorbate and adsorbant (Henry, 1803). The intermolecular forces are a function of temperature and Henry's constant decreases with temperature and can be expressed in the form of the Arrhenius equation:

$$k_{\mathrm{H}} = A_{\mathrm{H}}\mathrm{e}^{-E_{\mathrm{H}}/RT} \tag{4.8}$$

Equation (4.7) is generally referred to as Henry's law, and in its original form it indicates that: 'at a constant temperature, the amount of a given gas that dissolves in a given type and volume of liquid is directly proportional to the partial pressure of that gas in equilibrium with that liquid'. It also implies a linear relationship between the quantity of water vapour being adsorbed and the partial pressure of the vapour in the surround, which is generally valid for systems with relatively low concentration of water molecules. At high concentration of water molecules, the Hill–de Boer equation

$$\frac{\theta}{1-\theta}\mathrm{e}^{\frac{\theta}{1-\theta}-k_2\theta} = k_{\mathrm{H}}p_v \tag{4.9}$$

is more satisfactory, where $\theta = \frac{n_a}{n_{\mathrm{site}}}$ is the fractional site on the surface of the adsorbant that has been occupied.

A more general analysis based on rate equilibrium was proposed by Langmuir (1918). The adsorption process is assumed to be kinetic: both adsorption and desorption take place simultaneously and adsorption rate equals desorption rate at equilibrium. The adsorption is assumed to be asymptotic; that is, the adsorption rate is assumed to be proportional not only to the partial pressure but also the residual site, $(1-\theta)$, while the desorption rate is assumed to be proportionate to θ. Thus,

$$\begin{aligned} \left.\frac{\mathrm{d}\theta}{\mathrm{d}t}\right|_{\mathrm{adsorption}} &= k_{\mathrm{ads}}p_v(1-\theta) \\ \left.\frac{\mathrm{d}\theta}{\mathrm{d}t}\right|_{\mathrm{desorption}} &= k_{\mathrm{des}}\theta \end{aligned} \tag{4.10}$$

At equilibrium,

$$k_{\mathrm{ads}}p_v(1-\theta) = k_{\mathrm{des}}\theta \tag{4.11}$$

Rearranging Eqn (4.11) and designating the Langmuir constant k_{L} as $k_{\mathrm{L}} = \frac{k_{\mathrm{ads}}}{k_{\mathrm{des}}}$ gives the Langmuir isotherm equation:

$$\theta = \frac{k_{\mathrm{L}}p_v}{1+k_{\mathrm{L}}p_v} \tag{4.12}$$

The rates of adsorption and desorption are a function of temperature; k_{ads} and k_{des} can be expressed in the form of the Arrhenius equation such that

$$k_L = \frac{A_{abs} e^{-E_{abs}/RT}}{A_{des} e^{-E_{des}/RT}} = A_L e^{\Delta E/RT} \tag{4.13}$$

where $\Delta E = E_{des} - E_{abs}$. For very low partial pressure of the vapour, Eqn (4.12) is reduced to

$$\theta \approx k_L p_v \tag{4.14}$$

and Henry's law is recovered. For high pressures, Eqn (4.12) is reduced to $\theta \approx 1$, implying the occupation of all possible sites by the adsorbate.

One of the basic assumptions of the Langmuir adsorption isotherm is that adsorption is monolayer in nature. Under the conditions of low pressure and high temperature, gaseous molecules possess high thermal energy and high escape velocity; as a result, fewer gaseous molecules would be available near the surface of the adsorbant, promoting the formation of only a monolayer of adsorbate. Under the condition of high pressure and low temperature, thermal energy of gaseous molecules decreases and more and more gaseous molecules would be available per unit surface area, encouraging the formation of multilayer adsorbate on adsorbant. The Brunauer–Emmett–Teller (BET) isotherm is an extension of the Langmuir theory to multilayer formation of adsorbate on adsorbant and takes the form

$$\theta = \frac{c p_v/p_g}{\left(1 - p_v/p_g\right)\left(1 - p_v/p_g + c p_v/p_g\right)} \tag{4.15}$$

where c is the BET constant given by

$$c = e^{\frac{\Delta E_1 - \Delta E_L}{RT}} \tag{4.16}$$

where ΔE_1 is the heat of adsorption for the first layer, and ΔE_L for the second and the higher layers. The BET isotherm is reduced to the Langmuir isotherm when $p_v/p_g \ll 1$ and $c \gg 1$. For very low fractional pressure,

$$\theta \approx c p_v/p_g \tag{4.17}$$

and the BET isotherm is again reduced to Henry's isotherm. At very high fractional pressure, as $p_v/p_g \rightarrow 1$, then $\theta \rightarrow \infty$, implying adsorption is not limited by the available site on the surface of the adsorbant.

Figure 4.4 shows the six types of adsorption isotherm first presented by Brunauer (1940), in which Type I is the Langmuir isotherm and Type II and Type III are the BET isotherms.

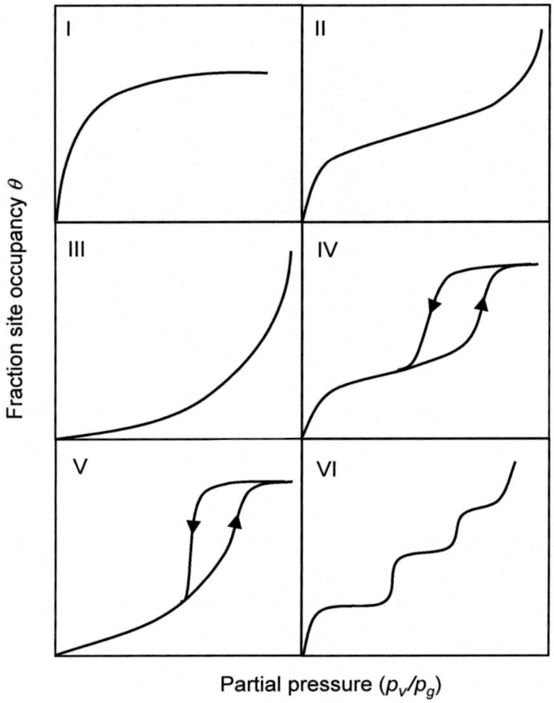

Figure 4.4 The adsorption isotherm (Brunauer, 1940).

4.3.1.2 The pores

The absorbate must interact with the internal surfaces and the internal volume in the adsorbant. These internal surfaces and volumes come in the form of pores as shown in Figure 4.5 (Akshaya, 2014). These pores may be through, blind, or closed. Highly specialised instruments are available that are capable of characterising the structure, the total volume of through and blind pores, the total surface area of pores and the fractional distribution of pore diameter in a solid absorbant (Akshaya, 2014). Among these techniques, the vapour adsorption and vapour condensation are of special interest to the adsorption and absorption of moisture in polymers. The former is capable of characterising the total surface area of pores, both through and blind pores, within an absorbant, while the latter is capable of characterising the total volume and the fractional distribution of the pore size in an absorbant.

The size of the pores may be subdivided into three classes: micropores (less than 2 nm), mesopores (between 2 and 50 nm) and macropores (larger than 50 nm) (IUPAC, 1972). The size of the pores together with the affinity of the gas to the solid determines the shape of the adsorption isotherm. In general, increasing sorption in the initial isotherm curve suggests strong vapour–solid affinity; for example Types I, II, IV and VI isotherms. Saturation characteristics as seen in Type I isotherm suggest the dominance of micropores. Increasing adsorption at high pressure as seen in Types II

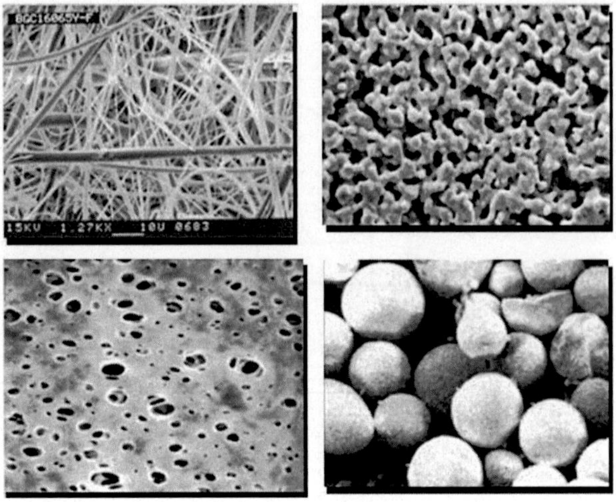

Figure 4.5 Typical pores in the adsorbant.
Porous Materials, Inc.

and III isotherms indicates dominance of macropores. The presence of hysteresis as seen in Types IV and V shows capillary condensation that is usually associated with mesopores or macropores. Table 4.2, characteristics of isotherms (IUPAC, 1972), summarises the characteristics of the first five isotherms.

4.3.2 Capillary condensation

It was observed that the volume of the gas that is adsorbed into a piece of charcoal was a few times more than the volume of the charcoal. This could only be possible if the adsorbed gas condensed into liquid in the micropores of the charcoal. To understand this phenomenon, we shall visit the theory of capillary condensation first proposed by Kelvin. Referring to Figure 4.6, capillary action leads to the formation of a meniscus in a capillary tube of radius r at a height h above the water level; the condition of vertical force-equilibrium gives

$$h = \frac{2\gamma}{\rho_w g r} \tag{4.18}$$

where γ (Nm^{-1}) is surface tension between air and water, ρ_w water density and g gravitational acceleration. However, the pressure differential due to a differential height of water vapour is given by

$$\mathrm{d}p_v = -\rho_v g \mathrm{d}h \tag{4.19}$$

Assuming ideal gas for the water vapour, then

$$\rho_v = \frac{p_v}{R_w T} \tag{4.20}$$

Table 4.2 Characteristics of isotherms (IUPAC, 1972)

Isotherm type	Equation	Characteristics				
		Nature of adsorption	**Strength of interaction**	**Molecular layer**	**Pore size**	**Capillary condensation**
I	Langmuir	Chemisorption or physisorption	Strong	Monolayer	Micropores	No
II	BET, $c \gg 1$	Physisorption	Strong	Monolayer followed by multilayer	Nonporous or a wide range of pore sizes	
III	BET, $c \ll 1$	Physisorption	Weak	Multilayer	Nonporous or a wide range of pore sizes	
IV		Physisorption	Strong	Monolayer followed by multilayer	Mesopores	Yes
V		Physisorption	Weak		Mesopores	Yes

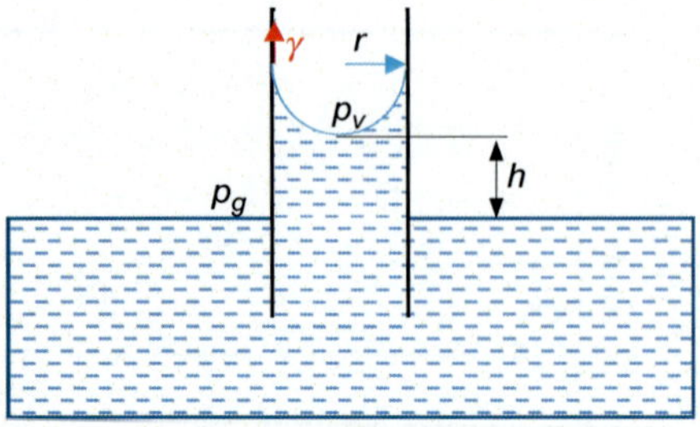

Figure 4.6 Capillary action.

where R_w is the gas constant of water vapour. Substituting Eqn (4.20) into Eqn (4.19) and performing integration gives

$$\ln\frac{p_v}{p_g} = -\frac{gh}{R_w T} \tag{4.21}$$

where p_g is saturated vapour pressure at the water level. Equating Eqn (4.21) with Eqn (4.18) gives the Kelvin's equation:

$$\ln\frac{p_v}{p_g} = -\frac{2\gamma}{\rho_w r R_w T} = -\frac{2\gamma v_m}{rRT} \tag{4.22}$$

where v_m is the specific molar volume and R the universal gas constant; or

$$\frac{p_v}{p_g} = e^{-\frac{2\gamma v_m}{rRT}} \tag{4.23}$$

Kelvin's equation suggests that in the presence of a manicure, which is promoted by a small geometrical dimension of a pore, condensation will occur at a partial pressure that is lower than the saturated pressure measured in the absence of a manicure.

4.3.3 Absorption and saturated concentration

4.3.3.1 Characteristics

In most cases, the solute that is adsorbed onto the surface of the adsorbant would migrate into the interior of the adsorbant – the phenomenon of absorption. The term *solubility* is used to describe the solution of solute (usually gas or vapour) into the absorbant (usually liquid or solid). The most common definition of solubility used in the microelectronic packaging industry is the saturated concentration C_{sat}, which is defined as the saturated mass of solute absorbed in a unit volume of the absorbant under the condition of constant temperature and environmental pressure; that is,

$$C_{\text{sat}} = \frac{m_{\text{solute, saturated}}}{V_{\text{absorbant}}} \left(\text{kg/m}^3\right) \tag{4.24}$$

The absorbed solute may exist in the form of vapour or condensed liquid. In view of the significantly higher density of the condensed liquid over vapour, we shall assume the mass, $m_{\text{solute, saturated}}$ to be made up of only condensed liquid and may be expressed as

$$m_{\text{solute, saturated}} = \rho_w V_{\text{sites}} \theta \tag{4.25}$$

where ρ_w is the density of condensed solute and V_{sites} is the total volume of the available condensation sites within the bulk of an absorbant. Substituting Eqn (4.24) into Eqn (4.25) gives

$$C_{\text{sat}} = \rho_w f_v \theta \tag{4.26}$$

where f_v is the volume fraction of the available condensed sites to the total volume of the absorbant. The concentration of water in a polymeric microelectronic packaging material is insignificantly low and can be assumed to obey Henry's law. Substituting Eqn (4.14) into Eqn (4.26) gives

$$C_{\text{sat}} = \rho_w f_v k_{\text{L}} p_v \tag{4.27}$$

and substituting k_{L} with $k_{\text{L}} = A_{\text{L}} e^{\Delta E/RT}$ yields

$$C_{\text{sat}} = \rho_w p_v f_v A_{\text{L}} e^{\Delta E/RT} \tag{4.28}$$

Expressing Eqn (4.28) in the form of Henry's equation:

$$C_{\text{sat}} = k_{\text{H}} p_v \tag{4.29}$$

suggests that

$$k_{\text{H}} = \rho_w f_v A_{\text{L}} e^{\Delta E/RT} \tag{4.30}$$

It has been shown in Section 4.2.2 that the saturation pressure, $p_g(T)$, of water vapour could be expressed in the form of the Arrhenius equation, Eqn (4.1), $p_g = p_o e^{-E_v/RT}$, where $E_v = 36.98$ kJ/mol for temperature between 25 and 100 °C. For many polymer matrix composites, as are most microelectronic packaging materials, it has been reported (ASTM-D5229M, 1998; CMH-17, 2011) that the saturated concentration of water decreases very mildly with increasing temperature, which implies that: $0 < \Delta E - E_v \leq 10R$.

4.3.3.2 Characterisation

A number of test standards have been developed for measuring the saturated concentration of water in polymers (ASTM-D570, 2010; BS-2782, 1983; ISO-62, 2008; JESD22-A120A, 2008; JIS-K7209, 1984; SEMI-G66, 1996) or polymer matrix composite materials (ASTM-D5229M, 1998). The experimental procedures typically involve drying a specimen, estimating its volume, measuring its weight, putting the specimen in a controlled temperature–humidity environment and measuring its weight at saturation. C_{sat} of the material for the particular temperature–humidity is then calculated using Eqn (4.24), which is

$$C_{sat} = \frac{m_{sat}}{V_{absorbant}} \left(\text{kg/m}^3\right). \tag{4.31}$$

The specimen shall have adequate volume such that the absorbed mass of the water can be measured accurately with the available instrument. To accelerate the water uptake by the specimen, the specimen typically has a small transverse dimension. The single largest source of error in such measurement is the condensation of water on the surface of the specimen. It is therefore advisable to have a specimen that has a large ratio of volume-to-surface area.

From Eqn (4.28), the pre-exponential constant A_L, void fraction f_v and differential activation energy ΔE are characteristics of the solute–absorbant combination. These three material constants can be extracted by characterising C_{sat} of the absorbant at multiple temperature and humidity. This is illustrated using the experimental C_{sat} of water in the moulding compound reported by Galloway and Miles (1996) as shown in Table 4.3, C_{sat} of moulding compound (Galloway & Miles, 1996).

By regression analysis of the six readings of C_{sat} obtained from experiments and computed by Eqn (4.28), the three material constants of the moulding compound are evaluated as: $A_L = 3.65 \times 10^{-5}$ mm^2/N, $\Delta E = 36$ kJ/mol and $f_v = 3.6 \times 10^{-4}$; that is, approximately 0.036% of the volume of the moulding compound are potential sites for moisture condensation.

Table 4.3 C_{sat} of moulding compound (Galloway & Miles, 1996)

		C_{sat} (kg/m^3)
Temperature (°C)	**% (RH)**	**Exp**
30	30	3.1
	60	6.9
	85	10.5
85	30	5.8
	60	8.9
	85	11.8

4.4 Diffusivity and its characterisation

The term *diffusivity* is used in mass transportation, heat transportation and charge transportation. Diffusivity, or diffusion coefficient, D, in mass transportation is the proportional constant between the mass concentration (kg/m^3) gradient and the flux (kg/s m^2) (Crank, 1956); that is

$$J_x = -D_x \frac{\partial C}{\partial x} \tag{4.32}$$

Diffusion is the macroscopic result of random thermal motion on a microscopic scale. Therefore, it increases with temperature and is typically described using the Arrhenius equation:

$$D = D_o \exp\left(\frac{-E_d}{RT}\right) \tag{4.33}$$

The pre-exponent constant D_o and the activation energy E_d can be extracted by characterising diffusivity at two or more temperatures. The diffusivity of water in an absorbant also depends on the structure and size of the pores, which, for polymers, undergoes large transformation across the glass-transition temperature. The constants D_o and E_d are expected to be different for polymers before and after the glass-transition temperature.

4.4.1 Characterisation by absorption

A number of test standards have been developed for characterising the diffusivity of water in polymers or polymer matrix composites. The experimental procedures typically involve drying a specimen, estimating its volume, measuring its weight, putting the specimen in a controlled temperature–humidity environment and measuring its weight periodically until saturation. Many standards, particularly (ASTM-D5229M, 1998), provide details on the requirements of test equipment and measuring instrument, as well as detailed description of the test procedures. Most test standards recommend the use of planar specimen that promotes one-dimensional (1-D) diffusion in the transverse direction. Assuming the bulk of the specimen to be completely dry ($C = 0$) at $t = 0$, and the adsorption of the water molecules ($C = C_{sat}$) on the outer surface of the specimen occurs instantaneous at $t = 0$, the moisture uptake in a planar specimen may be estimated using the simple 1-D moisture uptake equation (Crank, 1956):

$$\frac{M_t}{M_{sat}} = 1 - \frac{8}{\pi^2} \sum_{n=0}^{\infty} \frac{1}{(2n+1)^2} \exp\left[-\pi^2 (2n+1)^2 \frac{D_z t}{z_o^2}\right] \tag{4.34}$$

where M_t and M_{sat} are the instantaneous and the saturated mass of moisture in the specimen; D_z is the diffusivity of moisture in the transverse direction; and z_o is the specimen thickness.

4.4.1.1 Extractions of D and M_{sat} using numerical regression

The diffusivity, D_z, and the saturation moisture uptake, M_{sat}, of the specimen may be extracted simultaneously through regression by minimising $\sum_i \left(\left.\frac{M_{t,i}}{M_{sat}}\right|_{exp} - \left.\frac{M_{t,i}}{M_{sat}}\right|_{Eqn\ (4.34)} \right)^2$. Equation (4.34) is an infinite series; for reason of efficient regression iterations, it is essential that only the necessary number of terms is used. The rate of convergence of the infinite series is a function of the normalised time, $D_z t / z_o^2$. Also, the number of terms required may be estimated by examining the ratio of the nth term to the first term:

$$R_n = \frac{1}{(2n^2+1)^2} \exp\left\{ -\pi^2 \frac{D_z t}{z_o^2} \left[1 + (2n^2+1)^2 \right] \right\} \tag{4.35}$$

The nth term is required to assure $R_n \leq 1 \times 10^{-6}$ has been evaluated and may be described by a power law relation:

$$n \geq 0.4537 \left(\frac{D_z t}{z_o^2} \right)^{-0.481} \tag{4.36}$$

which suggests that more terms are required during the early phase of sorption.

It is a good practice in regression analysis to spread the data uniformly over the range of interest. The rate of moisture uptake is given by

$$\begin{aligned} \frac{\mathrm{d}}{\mathrm{d}t} \frac{M_t}{M_{sat}} &= \frac{8D_z}{z_o^2} \sum_{n=0}^{\infty} \exp\left[-\pi^2 (2n+1)^2 \frac{D_z t}{z_o^2} \right] \\ &\approx \frac{8D_z}{z_o^2} \exp\left[-\pi^2 \frac{D_z t}{z_o^2} \right] \end{aligned} \tag{4.37}$$

Thus, the rate of moisture uptake decreases exponentially from $8D_z/z_o^2$ at $t = 0$ and tends toward 0 as $M_t/M_{sat} \rightarrow 1$. The exponential decay rate suggests that taking the moisture uptake at a regular interval in an absorption experiment will give rise to a set of data that is skewed heavily toward M_{sat}, which is not ideal. The ideal measurement intervals may be estimated as follows: setting $n = 0$ in Eqn (4.34), the time for fractional moisture uptake of $x = M_t/M_{sat}$ is given approximately by

$$\frac{D_z t_x}{z_o^2} \approx -\frac{1}{\pi^2} \ln\left[\frac{\pi^2}{8} \left(1 - \frac{M_t}{M_{sat}} \right) \right] \tag{4.38}$$

Figure 4.7 shows the graph of Eqn (4.38). Also shown are the data (in red circles) generated using Eqn (4.34) and a power point fit through the data. It appears that

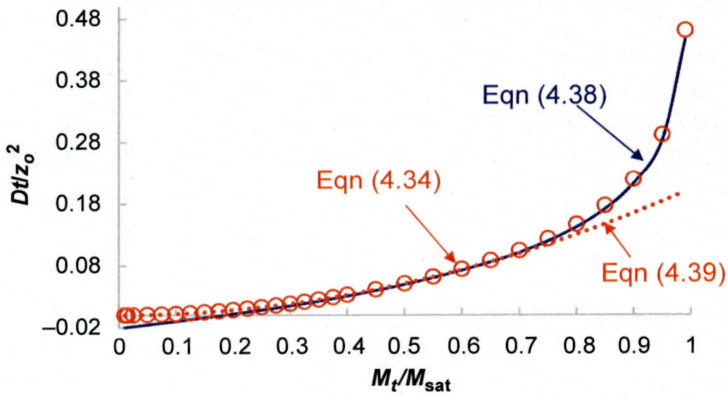

Figure 4.7 Predicting the sorption interval.

Eqn (4.38) predicts the sorption interval rather well for $M_t/M_{\text{sat}} > 0.35$, but underpredicts the sorption duration, even giving negative values, for smaller M_t/M_{sat}. The sorption duration for the range $0 \leq M_t/M_{\text{sat}} \leq 0.5$ can indeed be more accurately modelled using the simple power law relation:

$$\frac{D_z t_x}{z_o^2} \approx 0.1965 \left(\frac{M_t}{M_{\text{sat}}} \right)^2 \tag{4.39}$$

By taking p-number of measurements, the interval between measurements is simply $1/p$; and the time $t_{i/p}$ at which data number i is taken may be estimated:

$$\frac{D_z t_{i/p}}{z_o^2} \approx \begin{cases} 0.1965 \left(\dfrac{i}{p} \right)^2 & \text{for} \quad i/p \leq 0.5 \\ -\dfrac{1}{\pi^2} \ln \left[\dfrac{\pi^2}{8} \left(1 - \dfrac{i}{p} \right) \right] & \text{for} \quad i/p \geq 0.35 \end{cases} \; ; \quad i = 1, 2, 3 \ldots p \tag{4.40}$$

It is noted that at $i = p$, $M_t = M_{\text{sat}}$ and $t_p = \infty$. In practice, a time corresponding to M_t that is marginally smaller than M_{sat} shall be used in place of $i = p$. Assuming $M_t = 99\%\ M_{\text{sat}}$, the corresponding time is given by

$$\frac{D_z t_{99\%}}{z_o^2} \approx 0.45 \tag{4.41}$$

The computation of $t_{i/p}$ does require a ballpark estimate of D_z, which does not differ too much for most polymeric materials.

4.4.1.2 *Extractions of* D *using graphical method*

Regression techniques are typically not recommended in the test standards for fear of being intimating. Instead, user-friendly equations are used in the test standards for extracting D_z. Invariably these methods assume prior knowledge of m_{sat}, which can be tricky. Three methods that do not require regression are presented:

1. Referring to Eqn (4.38), by plotting $\ln\left[\frac{\pi^2}{8}\left(1-\frac{M_t}{M_{sat}}\right)\right]$ versus t for $M_t/M_{sat} \geq 0.35$, the gradient, G_t, may be determined graphically, and the diffusivity D_z is estimated as

$$D_z \approx \frac{-G_t z_o^2}{\pi^2} \tag{4.42}$$

2. From Eqn (4.39), by plotting $(M_t/M_{sat})^2$ versus t for $M_t/M_{sat} \leq 0.5$, the gradient, G_t, may be determined graphically, and the diffusivity D_z is estimated as

$$D_z \approx 0.1965 G_t z_o^2 \tag{4.43}$$

3. In view of the exponential nature of moisture uptake, the sorption duration, $t_{x\%}$, corresponding to $x\%$ of moisture uptake is more reliably determined experimentally during the initial stage. The normalised times $D_z t/z_o^2$, which correspond to 10%, 20%, 30%, 40% and 50% of moisture uptake have been estimated using Eqn (4.34) as 0.001963, 0.007854, 0.01767, 0.03142 and 0.04918. By estimating the actual time $t_{x\%}$ corresponds to the x-fractional moisture uptake, the diffusivity $D_{z,x\%}$ may then be estimated using the following equations:

$$\begin{aligned} D_{z,10\%} &\approx 0.001963\frac{z_o^2}{t_{10\%}} \\ D_{z,20\%} &\approx 0.007854\frac{z_o^2}{t_{20\%}} \\ D_{z,30\%} &\approx 0.01767\frac{z_o^2}{t_{30\%}} \\ D_{z,40\%} &\approx 0.03142\frac{z_o^2}{t_{40\%}} \\ D_{z,50\%} &\approx 0.04918\frac{z_o^2}{t_{50\%}} \end{aligned} \tag{4.44}$$

Thus, the diffusivity D_z may now be estimated as the average of the five $D_{z,x}$.

As an illustration, Table 4.4 gives the actual readings obtained for a laminated microelectronic packaging material of dimensions $50 \times 50 \times 0.4\ \mathrm{mm}^3$ in an absorption experiment.

Applying the regression technique that minimises $\sum_i \left(\left.\frac{M_{t,i}}{M_{\mathrm{sat}}}\right|_{\mathrm{exp}} - \left.\frac{M_{t,i}}{M_{\mathrm{sat}}}\right|_{\mathrm{Eqn\ (4.34)}} \right)^2$, M_{sat} and D_z have been evaluated as 46.6 mg and $5.80 \times 10^{-4}\ \mathrm{mm}^2/\mathrm{h}$, respectively. By contrast, assuming M_{sat} to be 45.3 mg, Eqns (4.42) and (4.43) yield D_z as $6.27 \times 10^{-4}\ \mathrm{mm}^2/\mathrm{h}$ and $5.91 \times 10^{-4}\ \mathrm{mm}^2/\mathrm{h}$, respectively. By underestimating the true magnitude of M_{sat}, both Eqns (4.42) and (4.43) have overestimated the magnitude of D_z. Between these two equations, Eqn (4.43) that uses the data range $0 \leq M_t/M_{\mathrm{sat}} \leq 0.4$ gives a better estimate.

4.4.1.3 Sources of experimental errors

Measuring moisture diffusivity of water in plastics using the absorption method described above is prone to a host of errors:

1. In view of the minute mass of the absorbed moisture, especially during the initial uptake of moisture, it is essential that the weighting instrument has adequate resolution. If multiple specimens are used to improve the quality of measurement, it is essential that the specimens are made of identical shape and dimensions; and the moisture uptake in individual specimen is used in Eqn (4.34).
2. Condensation of water on the surface of the specimen can contribute to 'false mass' causing significant measurement error, especially during the initial moisture uptake. A number of test standards have given methods to remove the condensation prior to measurements. An obvious approach in minimising the contribution of condensation is to use specimens with large volume-to-surface area; however, this has to be balanced against the prolonged test duration.
3. The boundary condition around the test specimen is typically different from that of the test environment when the test specimen is taken out of the test chamber for measurement. This difference in the boundary condition introduces a disturbance to the distribution of water within the test specimen, resulting in irreversible changes. This disturbance must be kept as small as possible by minimising the duration of the specimen outside the test chamber or by minimising the difference in the environment between the test chamber and the holding chamber.
4. Probably the single largest source of error to the absorption experiment is the disturbance to the test environment when loading and unloading the test specimen (or other test specimens that are sharing the test chamber) during the duration of the measurements. Depending on the dimensions of the test specimen and the moisture diffusivity of the material, test interval as short as half an hour is not uncommon at the initial stage of the absorption experiment. By contrast, depending on the size of the test chamber and the power of the test equipment, it is common to take up to half an hour for the humidity in the test chamber to equilibrate back to the test condition. This has serious implication on the accuracy of the measurement during the initial stage of the absorption. More worrying is that this error will be propagated to the later stage of the absorption.

Table 4.4 Experimental moisture uptake of a planar specimen

Time, h	M_t, mg
0	0
0.75	8.15
1.75	10.35
2.75	11.45
4.25	13.85
5.75	15.6
7.75	17.85
23.75	29.95
28.25	32.1
32.9	34.35
47.75	39.45
55.75	40.9
79.75	45.3

4.4.1.4 Advanced instruments

Eliminating the disturbance mentioned above is not only desirable but essential in some cases. In practice, it is impossible to prepare specimens to have adequate thicknesses; examples of such test specimens include flexible substrate used in advanced microelectronic packaging, polyimide and benzocyclobutene (BCB) preservation on the wafer, and the polymeric dielectrics between the metallisation layers in the wafer. Specialised instruments with integrated temperature–humidity and microbalance capabilities (DVS, 2014; IGASorp, 2014; VSA, 2014) are available and capable of characterising these sophisticated specimens. These are generally referred to as 'dynamic vapour sorption analysers'. A schematic of the IGASorp instrument is presented in Figure 4.8. This has been demonstrated on a low-dielectric material, which has been specially prepared into a planar specimen with a thickness of 10 μm. Figure 4.9 shows the moisture uptake in the low-dielectric film at a constant temperature of 60 °C and with humidity stepping from 10% to 90% RH. The saturated concentration (C_{sat}) corresponding to %RH has been evaluated. It is noted that C_{sat} increases linearly with %RH up to 60% RH before rising at increasing rate. This appears to match the Type III isotherm and the rapid increase in C_{sat} at above 60% RH is postulated to be due to capillary condensation.

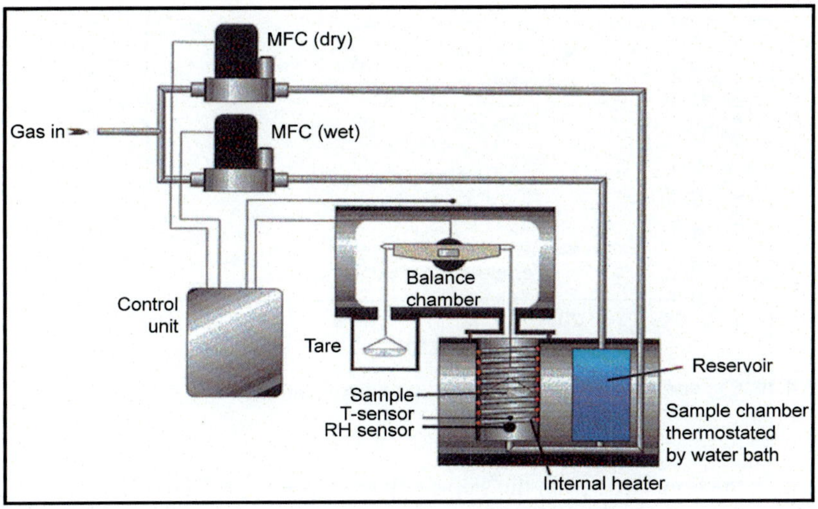

Figure 4.8 A schematic of IGASorp dynamic vapour sorption analyser.

Note the asymptotic rise of the absorbed moisture after each step-jump in %RH. Assuming moisture saturation at the end of each %RH, the asymptotic rise can be described by the equation:

$$\frac{\Delta M_t}{\Delta M_{\text{sat}}} = 1 - \frac{8}{\pi^2} \sum_{n=0}^{\infty} \frac{1}{(2n+1)^2} \exp\left[-\pi^2 (2n+1)^2 \frac{D_z \Delta t}{z_o^2} \right] \tag{4.45}$$

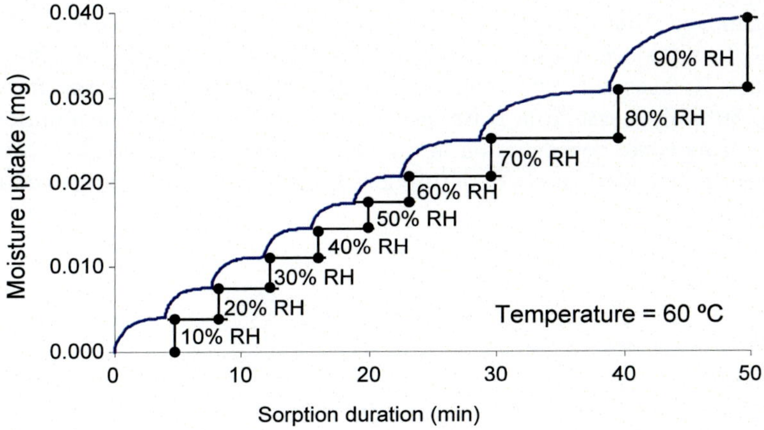

Figure 4.9 Water absorption isotherm of a low-*k* polymeric material of 10 μm in thickness.

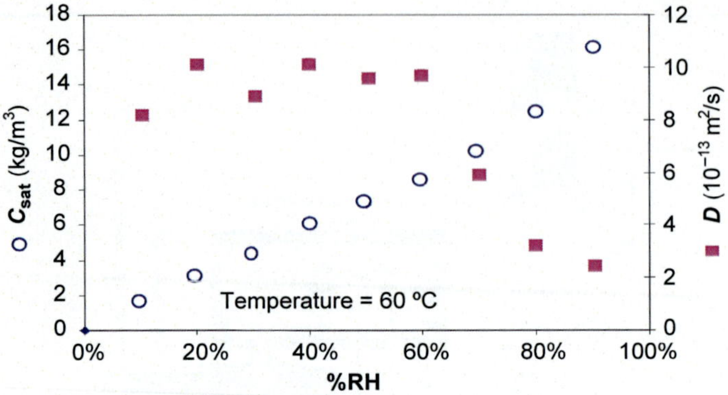

Figure 4.10 C_{sat} and D_z of the low-dielectric polymeric material.

where Δt is the time lapse from the step-jump of respective %RH; ΔM_t; and ΔM_{sat} are the instantaneous and saturated differential moisture uptake with respect to $\Delta t = 0$. Using Eqn (4.45), the diffusivity at the respective %RH has been evaluated and these values are shown in Figure 4.10. It is noted that the diffusivity stays pretty much constant up to 60% RH, before decreasing monotonically with increasing %RH. The fact that the reduction in the diffusivity coincides with the rapid rise in C_{sat} lends support to the postulation of condensation, which reduces the mean-free path of the water molecules.

4.4.2 Characterisation by desorption

Assuming absorption of moisture in the microelectronic packaging materials to be physisorption in nature, the moisture that has been introduced into a specimen can be removed from the specimen in a perfectly reverse order. That is, if it takes time $t_{20\%}$ to introduce moisture into a specimen from 0% to 20% of saturation in mass, then it takes the same time to remove moisture from a specimen from 100% to 80% of saturation in mass. For a planar specimen that has a uniform distribution of moisture concentration in the bulk volume and nil moisture concentration along the boundary at $t = 0$, desorption of moisture from the specimen may be described by

$$\frac{M_t}{M_{ini}} = \frac{8}{\pi^2} \sum_{n=0}^{\infty} \frac{1}{(2n+1)^2} \exp\left[-\pi^2 (2n+1)^2 \frac{D_z t}{z_o^2} \right] \tag{4.46}$$

where M_t and M_{ini} are the instantaneous and initial mass of moisture in the planar specimen or

$$\frac{M_{t,\text{loss}}}{M_{\infty,\text{loss}}} = 1 - \frac{8}{\pi^2}\sum_{n=0}^{\infty}\frac{1}{(2n+1)^2}\exp\left[-\pi^2(2n+1)^2\frac{D_z t}{z_o^2}\right] \tag{4.47}$$

where $M_{t,\text{loss}}$ and $M_{\infty,\text{loss}}$ are the instantaneous and eventual loss of moisture in the planar specimen.

Thus, the extraction techniques using numerical regression and graphical methods applied to 'absorption' are equally applicable for 'desorption'.

4.4.2.1 Advantages of characterisation by desorption

The use of desorption for characterising diffusivity offers the following advantages:

1. Compared to maintaining a specific temperature—humidity condition, which is required for absorption testing, it is relatively easy to achieve a temperature-nil humidity environment for desorption testing. Thermogravimetric analysis (TGA) that integrates a microbalance with a temperature-nil humidity environment is readily available in most microelectronic organisations.
2. Opposite to absorption, desorption tests can be conducted over a broad temperature range that will allow the characterisation of the Arrhenius constants across the glass-transition temperature. Indeed, desorption is probably the only feasible experimental technique for characterising diffusivity at elevated temperatures, for example, 95 °C.

4.4.2.2 Challenges of characterisation by desorption

Performing a desorption experiment at elevated temperature has some unique challenges:

1. Evaporation of solvents in the test specimen is promoted at elevated temperature. The mass of the evaporated solvents could contribute significantly to the total desorbed mass recorded by TGA. Figure 4.11 displays the TGA desorption characteristics of a typical die-attach material at temperatures 120, 170 and 220 °C. The vertical axis shows the fractional loss of moisture $M_{t,\text{loss}}/M_{\infty,\text{loss},120\ °\text{C}}$ where $M_{\infty,\text{loss},120\ °\text{C}}$ is the eventual loss of moisture measured by TGA for the specimen tested at 120 °C. Notice that the eventual loss of moisture recorded by TGA increases significantly with increasing temperature, showing the substantial contribution due to the out-gassing of solvents. In view of the relatively low rate of out-gassing compared to moisture diffusion, the presence of the pseudomass can lead to underestimation of moisture diffusivity.
2. Using TGA, temperature in the test chamber has to be ramped up from ambient to the test temperature at a finite rate. It is noted that the 1D-moisture-loss equation, Eqn (4.46), is strictly valid for a planar specimen that is uniformly saturated at $t = 0$. It is, hence, essential

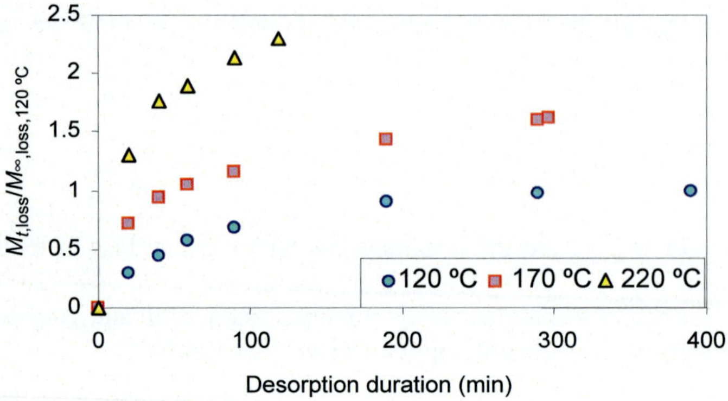

Figure 4.11 Desorption of a die-attach material using a thermogravimetric analysis (TGA).

to minimise the loss of moisture during the temperature ramp up. An aggressive temperature ramp-up rate, while reducing the total ramp-up time, will be followed by a relatively long stabilisation duration during which the temperature in the test chamber fluctuates wildly, rendering the 1-D moisture-loss equation inapplicable. Conversely, a conservative temperature ramp-up rate will lead to excessive loss of moisture during the ramp up. Figure 4.12 shows the desorption characteristics of the specimen whose moisture diffusivity at 210 °C has been measured. Because of the constraint of the instrument, the specimen experiences a temperature ramp up from 30 to 210 °C in the normalised time of $0.1D_z t/z_o^2$, then followed by isothermal desorption at 210 °C. Different from the standard moisture-loss characteristic as described by Eqn (4.47), the moisture-loss characteristics exhibit positive curvature during the initial stage of the moisture loss attributable to the increasing moisture diffusivity with increasing temperature. Up to 47% of moisture in the specimen has been lost before the test temperature of 210 °C is reached. Forced curve fitting the experimental data with the isothermal desorption equation Eqn (4.47) only leads to a grossly underestimated moisture diffusivity.

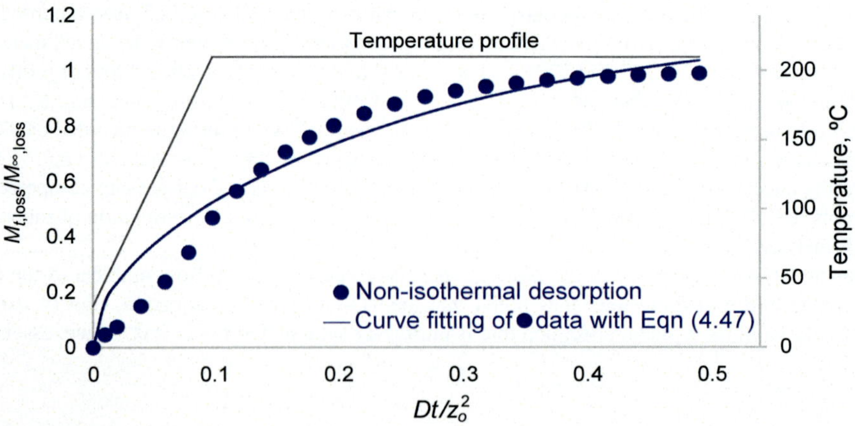

Figure 4.12 Non-isothermal desorption in thermogravimetric analysis (TGA).

4.4.2.3 The time-shift technique

Absorption or desorption is a path-independent process. That is, there is a unique concentration profile, $C(x,t)$, for every $M(t)$, independent of the history of moisture diffusion. This allows us to discard the data obtained during the ramp up. Using the data of Figure 4.12 as an example, we shall attempt to extract moisture diffusivity using the M_t data for $t \geq 0.1$ when desorption occurs isothermally at 210 °C. The initial diffusion at a lower diffusivity for $0 \leq t \leq 0.1$ has effectively time-shifted the M_t data for $t \geq 0.1$ to the right. By describing the experimental data using the desorption equation, we have

$$\frac{M_{t,\text{loss}}}{M_{\infty,\text{loss}}} = 1 - \frac{8}{\pi^2} \sum_{n=0}^{\infty} \frac{1}{(2n+1)^2} \exp\left[-\pi^2 (2n+1)^2 \frac{D_z\left(t_{\text{exp}} - \Delta t_{\text{shift}}\right)}{z_o^2}\right] \tag{4.48}$$

where t_{exp} is the experimental time, and Δt_{shift} is time-shift, D_z, M_{sat} and Δt_{shift} can be extracted simultaneously using the regression technique $\sum_i \left(\left.\frac{M_{t,\text{loss},i}}{M_{\infty,\text{loss}}}\right|_{\text{exp}} - \left.\frac{M_{t,\text{loss},i}}{M_{\infty,\text{loss}}}\right|_{\text{Eqn (4.48)}} \right)^2$. Figure 4.13 shows fitting of the experimental data for $t_{\text{exp}} \geq 0.1$ using Eqn (4.48).

4.4.2.4 The titration technique

The challenges of solvent evaporation and temperature ramp up may be mitigated using the titration technique (Shi, 2002). The titrator meters a precision quantity of chemical that corresponds to the amount of moisture desorbed. Being sensitive only to moisture and able to preheat the chamber to the required temperature, the titration technique does suffer from the challenges experienced by using TGA. Figure 4.14 depicts

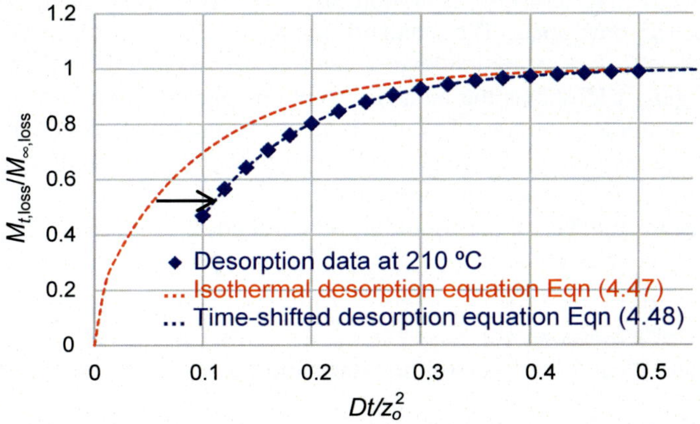

Figure 4.13 Extraction of D_z using the time-shift method.

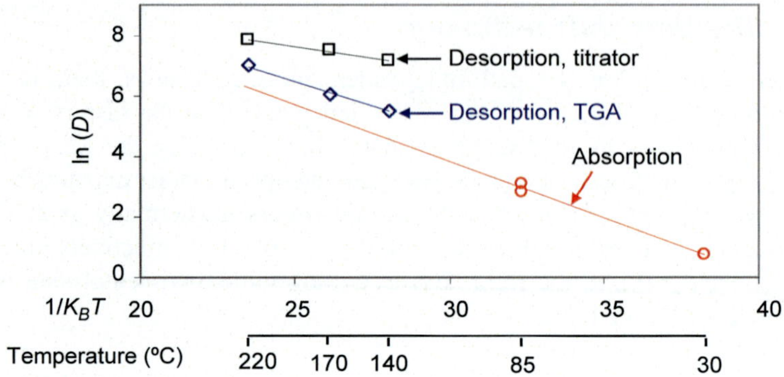

Figure 4.14 Comparisons of D_z characterised using absorption and desorption.

the Arrhenius plot of diffusivity for a typical moulding compound (T_g at 140 °C). The horizontal axis is expressed as $1/k_BT$, where k_B is Boltzmann's constant. The diffusivities from 30 to 85 °C are obtained by absorption, while those from 140 to 220 °C are by desorption using TGA and the titrator, individually. The dimensions of the specimen used in absorption, TGA, and titration tests are ∅50 × 1 mm, 4 × 4 × 3 mm and 10 × 10 × 1 mm, respectively. It is noted that (1) TGA technique tends to report much lower moisture diffusivity than the titration technique; and (2) the Arrhenius relationship projected from the TGA and the titration techniques at temperatures above the glass-transition temperature of the material deviate positively from that obtained using absorption method at the lower temperatures.

4.4.3 Non-asymptotic sorption

Equation (4.34) suggests an asymptotic saturation behaviour – that is, it will take infinite time for the moisture level in the specimen to reach saturation. ASTM standard D5229 differentiates *absolute* from *effective* moisture equilibrium (saturation) for practical reasons. While absolute moisture saturation requires no measurable uptake in moisture, effective moisture saturation allows a specified small moisture uptake over a specified time span. The standard applies to effective moisture saturation. The use of the effective saturation level is even more crucial in the presence of chemisorption. Figure 4.15 shows the moisture sorption characteristics of a typical silica-filled epoxy moulding compound (1 mm in thickness) at two temperature–humidity conditions. The asymptotic characteristics, suggested by Eqn (4.34), have disappeared and the sorption of moisture continues unabated.

The moisture uptake of the moulding compound at 85 °C/85% RH shown in Figure 4.15 is replotted in Figure 4.16. Using the regression technique, $\sum_i \left(\left.\frac{M_{t,i}}{M_{sat}}\right|_{exp} - \left.\frac{M_{t,i}}{M_{sat}}\right|_{Eqn\ (4.40)} \right)^2$, at sorption times of 100, 200, 400 and 750 h, respectively, C_{sat} and D_z at these four sorption durations have been obtained, which shows

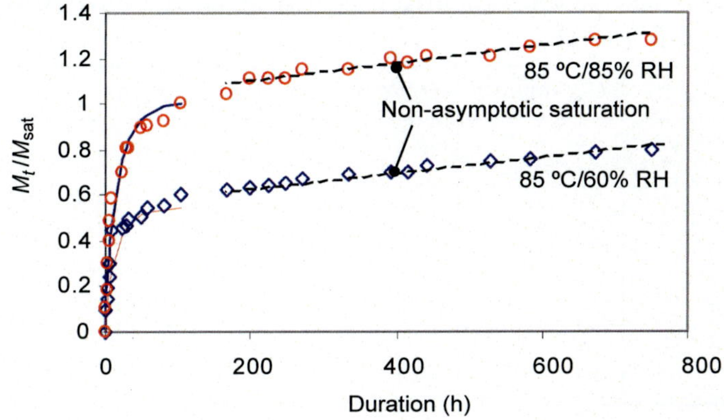

Figure 4.15 Non-asymptotic saturation characteristic of a typical moulding compound.

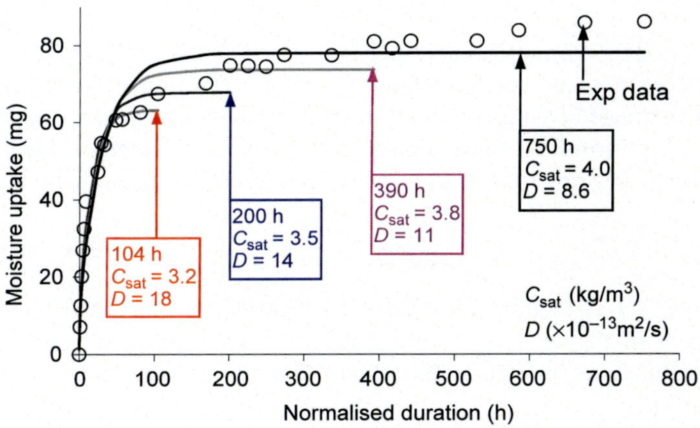

Figure 4.16 Strong dependence of C_{sat} and D on sorption duration for non-asymptotic sorption.

that C_{sat} is increased by 25% with increasing sorption duration, but there is a rapid decline of D_z (by more than 50%) with increasing sorption time.

4.4.3.1 The nonlinear sorption method

The chemisorption of moisture may be incorporated into the moisture uptake equation. Assuming the rate of chemisorption follows a power law relation, then this may be added to the moisture uptake equation:

$$M_t = M_{sat}\left\{1 - \frac{8}{\pi^2}\sum_{n=0}^{\infty}\frac{1}{(2n+1)^2}\exp\left[-\pi^2(2n+1)^2\frac{D_z t}{z_o^2}\right]\right\} + at^b \qquad (4.49)$$

Table 4.5 Extracted C_{sat} and D_z for moulding compound at 85 °C/85% RH

$t_{duration}$ (h)	C_{sat} (kg/m^3)	D_z ($\times 10^{-13}$ m^2/s)	a (mg/h)	b
30	1.8	50	1.3	0.79
55	2.0	41	1.1	0.75
104	2.3	33	0.9	0.69
200	2.4	29	1.1	0.61
392	2.4	25	1.9	0.48
750	2.4	20	3.5	0.37

By performing the regression $\sum_i \left(\frac{M_{t,i}}{M_{sat}}\Big|_{exp} - \frac{M_{t,i}}{M_{sat}}\Big|_{\text{Eqn (4.49)}} \right)^2$ at the sorption durations of 30, 55, 104, 200, 392 and 750 h, respectively, the values of C_{sat}, D_z, a and b at these six sorption durations have been tabulated in Table 4.5. Unfortunately, it is noted that the values of the extracted D_z again depend strongly on the duration of the experiment.

Instead of performing regressions at the six test duration individually, a regression was performed for the six sorption durations collectively; that is, $\sum_{\substack{\text{sorption}\\\text{duration}=1}}^{6} \left[\sum_i \left(\frac{M_{t,i}}{M_{sat}}\Big|_{exp} - \frac{M_{t,i}}{M_{sat}}\Big|_{\text{Eqn (4.49)}} \right)^2 \right]$, the C_{sat}, D_z, a and b have been evaluated as $C_{sat} = 2.4$ kg/m^3, $D_z = 23 \times 10^{-9}$ cm^2/s, $a = 2.44$ mg/h^b and $b = 0.43$. This is shown in Figure 4.17. Unfortunately, even this method is not immune to the sorption durations, though at less extent than the previous methods.

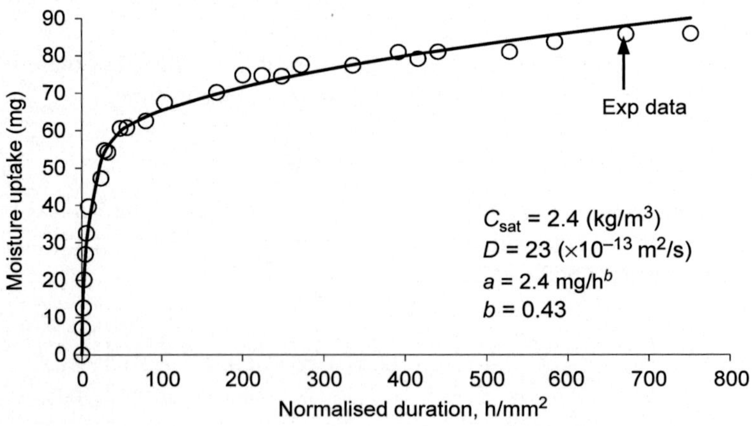

Figure 4.17 Moisture uptake due to physisorption and chemisorption.

Table 4.6 $t_{99\%}$ approximation for moulding compound at 85 °C and 85% RH

t_{guess} (h)	$C_{sat,\ guess}$ (kg/m^3)	D_{guess} ($\times 10^{-13}$ m^2/s)	$t_{99\%} = 0.45\, z_o^2 / D_{guess}$(h)
24	2.6	34	37
30	2.8	28	44
55	3.1	22	57
104	3.2	18	68
200	3.5	14	87
392	3.8	11	118
750	4.0	8.6	144

4.4.3.2 The 99% saturation method

The rate of chemisorption is typically an order lower than that of physisorption, thus reducing the duration of sorption shall reduce the error caused by chemisorption. A duration of 99% saturation is proposed (Wong, 2003). The duration required for 99% saturation of a planar specimen, assuming only physisorption, has been estimated from Eqn (4.34) as

$$t_{99\%} \approx \frac{0.45 z_o^2}{D_z} \tag{4.50}$$

The determination of the 99% saturation duration using Eqn (4.50) requires extracting multiple diffusivities $D_{extracted}$ at a few initial 'guessed' durations t_{guess}. The particular pair of (t_{guess}, D_{guess}) that satisfies Eqn (4.50) corresponds to ($t_{99\%}$, D_z).

The $C_{sat,\ guess}$ and D_{guess} for the same moulding compound at seven guess durations are tabulated in Table 4.6, which shows clearly the increasing $C_{sat,\ guess}$ and decreasing D_{guess} with increasing duration of t_{guess}. Also shown in the same table is the 99% saturation duration ($t_{99\%}$). It is noted that $t_{99\%}$ of 57 h is near the t_{guess} duration of 55 h; hence, it is suggested that the saturated concentration and diffusivity of the moulding compound at 85 °C/85% RH are $C_{sat} = 3.1$ kg/m^3 and $D_z = 22 \times 10^{-9}$ cm^2/s, respectively. It is noted that the so-extracted diffusivity is remarkably close to that using the nonlinear sorption method.

4.4.4 Corrections for specimen aspect ratio

The assumption of 1-D diffusion is strictly valid for a specimen of large planar-to-thickness aspect ratio when moisture uptake due to in-plane diffusion can be ignored. In practice, the aspect ratio of the specimen is frequently dictated by the fabrication process, by the cost of preparing the specimen, and at times simply by convenience. Thus, a specimen of large planar-to-thickness ratio is not always feasible.

Equation (4.34) can be extended to 2-D diffusion through superposition. Assuming diffusion occurs only in the y- and the z-directions of a specimen, then the commutation of moisture uptake is given by

$$\frac{M_t}{M_{\text{sat}}} = 1 - \left(\frac{8}{\pi^2}\right)^2 \sum_{m=0}^{\infty} \sum_{n=0}^{\infty} \frac{1}{M \cdot N} \exp\left[-\pi^2 t\left(\frac{D_y M}{y_o^2} + \frac{D_x N}{z_o^2}\right)\right] \tag{4.51}$$

where y_o and z_o are the dimensions of the test specimen in the y- and the z-directions, respectively, and $M = (2m + 1)^2$, $N = (2n + 1)^2$. An example of such a specimen is a rod of rectangular cross-section. Equation (4.51) can be further extended to 3-D diffusion through superposition. When diffusion occurs in the x-, y- and z-directions of a specimen, then the commutation of moisture uptake is given by

$$\frac{M_t}{M_{\text{sat}}} = 1 - \left(\frac{8}{\pi^2}\right)^3 \sum_{l=0}^{\infty} \sum_{m=0}^{\infty} \sum_{n=0}^{\infty} \frac{1}{L \cdot M \cdot N} \exp\left[-\pi^2 t\left(\frac{D_x L}{x_o^2} + \frac{D_y M}{y_o^2} + \frac{D_z N}{z_o^2}\right)\right] \tag{4.52}$$

where x_o, y_o and z_o are the dimensions of the test specimen in the x-, y- and z-directions, respectively, and $L = (2l + 1)^2$, $M = (2m + 1)^2$, $N = (2n + 1)^2$. Eqn (4.52) is applicable to cuboid specimen.

Mathematically, diffusion in a test specimen occurs in all three directions. The recommended aspect ratio in the standards ranges from 20 (JESD22-A120A, 2008), to 50 (SEMI-G66, 1996) and to 100 (ASTM-D5229M, 1998). What should be the aspect ratio that will allow one to use the 1-D diffusion equation Eqn (4.34)? This question is best answered by comparing the apparent diffusivity, $D_{z,1D}$, extracted using the 1-D equation, Eqn (4.34) and the true diffusivity, D_z, extracted using the 3-D equation, Eqn (4.52). Figure 4.18 shows the plot of the ratio $D_{z,1D}/D_z$ as a function of the

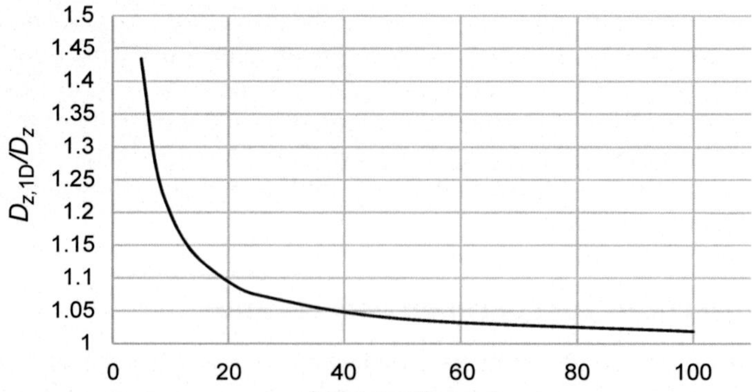

Figure 4.18 $D_{z,1D}/D_z$ versus x_o/z_o of test specimens that have identical planar dimensions ($x_o = y_o$).

aspect ratio, x_o/z_o, for a specimen of equal planar dimensions; that is $x_o = y_o$. In view of the contribution of moisture inflow from the in-plane directions, it is no surprise that the apparent diffusivity $D_{z,1D}$ has consistently overestimated the true diffusivity D_z. The ratio $D_{z,1D}/D_z$, which represents the magnitude of overestimation, decreases rapidly with increasing aspect ratio x_o/z_o and approaches the asymptotic value of 1 with increasing x_o/z_o. The errors for the aspect ratio of 20, 50 and 100 are 9.5%, 3.8% and 1.9%, respectively.

4.4.4.1 Correction factor for $x_o = y_o$

Armed with the graph of Figure 4.18, one could now apply a correction factor to the extracted apparent diffusivity to restore the true diffusivity; that is,

$$D_z = C_f D_{z,1D} \tag{4.53}$$

The following expression for the correction factor has been proposed (Wong, 2003):

$$C_f = \frac{1}{\left(1 + \frac{z_o}{x_o}\right)^2} \tag{4.54}$$

Figure 4.19 compares the difference between the true correction factor, $C_t = \frac{D_z|_{\text{Eqn (4.52)}}}{D_{z,1D}|_{\text{Eqn (4.34)}}}$, which is obtained from the ratio of Eqn (4.52) to Eqn (4.34), and the correction factor estimated using Eqn (4.54). The vertical axis in the figure, $(C_f - C_t)/C_t$, represents the percentage deviation of the estimated correction factor C_f from the true correction factor C_t. It is noted that Eqn (4.54) is capable of estimating the correction factor to within the accuracy of 3% for aspect ratio $x_o/z_o \geq 3$. In other

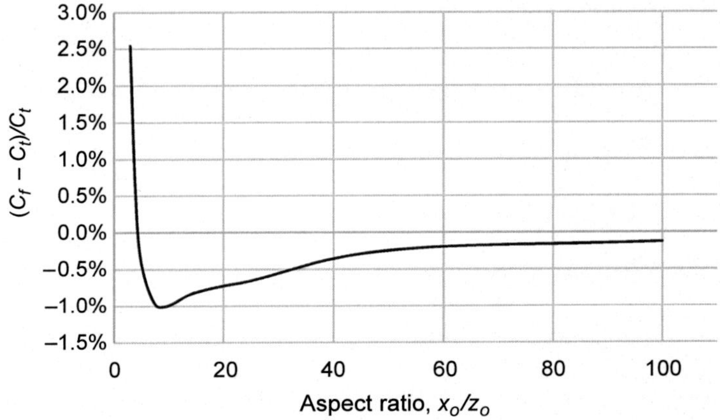

Figure 4.19 Validating Eqn (4.54) ($y_o = x_o$).

words, one could indeed use a planar specimen with an aspect ratio as low as three and achieve an accuracy of within 3% after applying a correction using Eqn (4.54). It is worth mentioning that the ASTM standard D5229M (ASTM-D5229M, 1998) has given the correction factor as $C_f|_{\text{ASME D5229}} = \frac{1}{\left(1+\frac{2z_o}{x_o}\right)^2}$, which will indeed result in an overcorrected diffusivity.

4.4.4.2 Correction factor for $x_o \neq y_o$

The above analysis assumes a planar specimen of equal planar dimensions; that is $x_o = y_o$. Figure 4.20 shows the graph of the ratio $D_{z,1D}/D_z$ as a function of the planar aspect ratio, y_o/x_o, for specimens with aspect ratio $x_o/z_o = 10$. The ratio $x_o/z_o = 10$ was chosen as this corresponds to a significant deviation of 1% between the computed correction factor C_f, using Eqn (4.54), and the true correction factor C_t. It is noted that the ratio $D_{z,1D}/D_z$, which represents the magnitude of overestimation, decreases with increasing y_o/x_o, attributable to reducing inflow from the y-direction. The magnitude of overestimation plateaus at 1.1% as $y_o \rightarrow \infty$. The overestimation is attributed to the small x-direction inflow of moisture due to the relatively small aspect ratio of $x_o = 10z_o$.

Equation (4.54) can be extended to estimate the correction factor for specimen with nonequal planar dimensions:

$$C_f = \frac{1}{\left[1 + \frac{z_o}{2}\left(\frac{1}{x_o} + \frac{1}{y_o}\right)\right]^2} \tag{4.55}$$

Figure 4.21 compares the difference between the true correction factor C_t and the estimated correction factor C_f given in Eqn (4.55), which shows that Eqn (4.55) is capable of estimating the true correction factor within 1% of margin for a wide range of y_o/x_o.

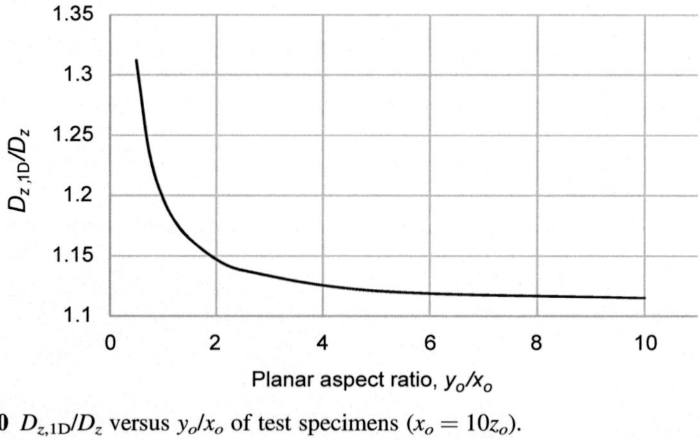

Figure 4.20 $D_{z,1D}/D_z$ versus y_o/x_o of test specimens ($x_o = 10z_o$).

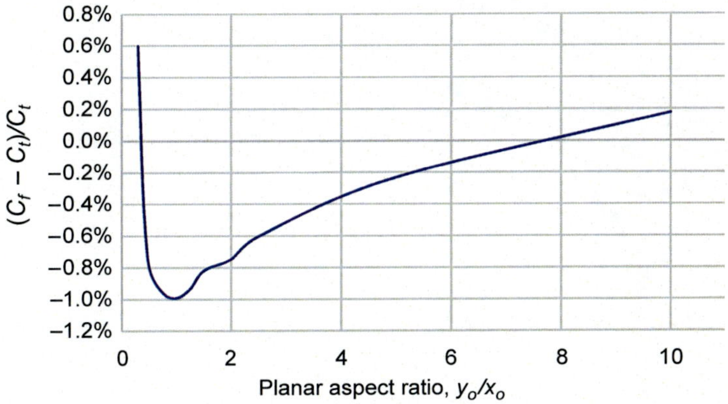

Figure 4.21 Validating Eqn (4.55) ($x_o = 10z_o$).

4.4.5 Characterising orthotropic diffusivity

4.4.5.1 The restrictive diffusion technique

The glass-reinforced layers of the core of the organic laminates give rise to transversely orthotropic properties, including diffusivity. The transverse and the in-plane diffusivities may be characterised using two separate experiments: the transverse diffusivity D_z may be characterised using a specimen with large x_o to z_o aspect ratio that promotes 1-D diffusion in the z-direction, while the in-plane diffusivity may be characterised using a specimen that restrict transverse flow. This is illustrated using a BT organic core laminate as the test vehicle, whose transverse diffusivity D_z was characterised using a standard membrane specimen of $50 \times 50 \times 0.4\ mm^3$. In order to facilitate characterisation of in-plane diffusivity, the laminate was applied with copper cladding on both its plane surfaces to restrict moisture diffusion only along the in-plane direction. In order to accelerate the rate of moisture uptake, the copper-cladded laminate was cut into multiple parallelepiped specimens of $3 \times 3 \times 0.4\ mm^3$. The moisture uptake of the two test specimens at 30 °C/60% RH is depicted in Figure 4.22.

The transverse diffusivity of the planar specimen was extracted using the 1-D equation, Eqn (4.34), which gives $D_z = 1.5 \times 10^{-13}\ m^2/s$; while the in-plane diffusivity of the Cu-cladded specimen was characterised using the 2-D equation, Eqn (4.51), which gives $D_x = D_y = 7.7 \times 10^{-13}\ m^2/s$. The BT laminate therefore has a diffusivity ratio of $D_x/D_z = 5.1$. The diffusivity ratio will most likely be dependent on the volume fraction and the ply structure of the glass fibre.

4.4.5.2 The aspect ratio technique

The transversely orthotropic diffusivity of materials may be more conveniently characterised using the equation of 'correction factor for the aspect ratio', as shall

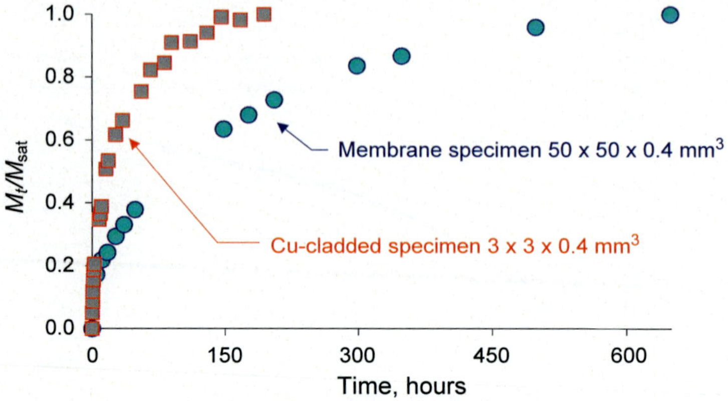

Figure 4.22 M_t/M_{sat} versus time of two dimensions of BT laminated core.

be elaborated: Figure 4.23 shows the plot of the ratio $D_{z,1D}/D_z$ as a function of the aspect ratio, x_o/z_o, for a specimen with equal planar dimensions ($y_o = x_o$) and with transversely orthotropic diffusivity $D_x = D_y = 5D_z$. The graph is similar to that of isotropic diffusivity as shown in Figure 4.18, but with significant higher magnitude of $D_{z,1D}/D_z$, which represents the magnitude of overestimation, because of the higher inflow of moisture along the planar direction. Figure 4.24 shows the plot of the ratio $D_{z,1D}/D_z$ as a function of the diffusivity ratio, D_x/D_z (assuming $D_y = D_x$), for a specimen with aspect ratio $x_o = y_o = 10z_o$. It is noted that the ratio $D_{z,1D}/D_z$ increases with increasing diffusivity ratio D_x/D_z, attributable to the increased inflow of moisture along the in-plane direction. Combining the data of Figures 4.23 and 4.24 and plotting $D_{z,1D}/D_z$ against X_o/Z_o, wherein $X_o = x_o/\sqrt{D_x}$ and $Z_o = z_o/\sqrt{D_z}$, the data are merged into a single coherent set as shown in Figure 4.25, which resembles that of Figure 4.18. The ratio X_o/Z_o is the effective aspect ratio for orthotropic diffusion.

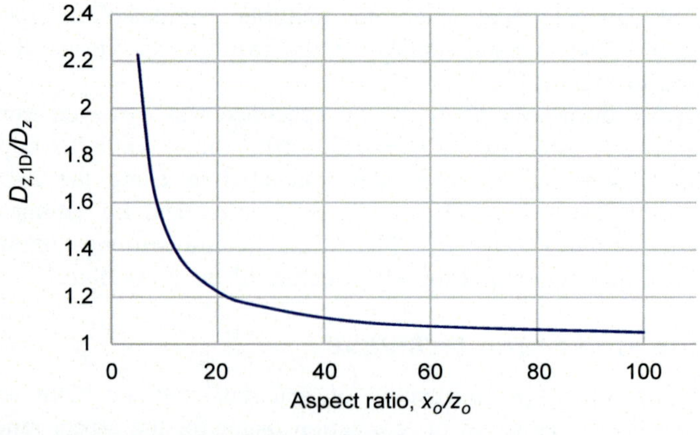

Figure 4.23 $D_{z,1D}/D_z$ versus x_o/z_o of test specimens ($D_x = D_y = 5D_z$, $x_o = y_o$).

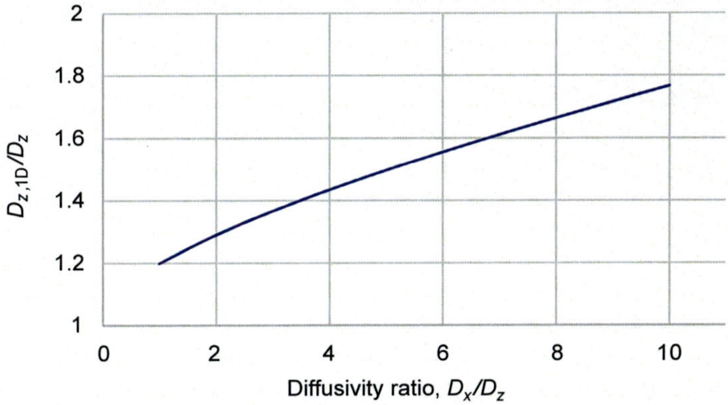

Figure 4.24 $D_{z,1\mathrm{D}}/D_z$ versus D_x/D_z of test specimens ($x_o = y_o = 10z_o$).

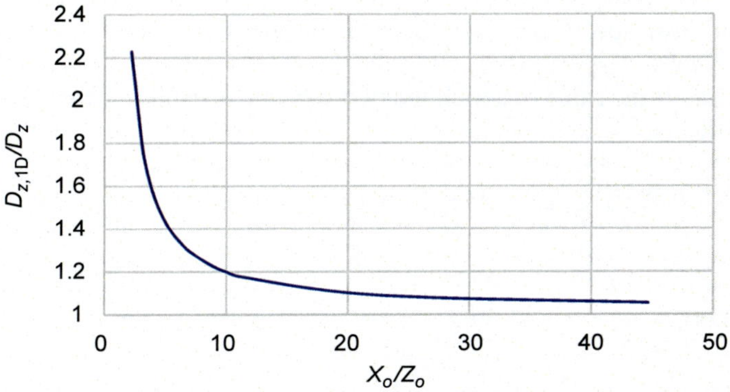

Figure 4.25 $D_{z,1\mathrm{D}}/D_z$ versus x_o/z_o ($X_o = Y_o$).

The correction factor for material with orthotropic diffusivity is therefore:

$$C_f = \frac{1}{\left[1 + \frac{z_o}{x_o}\sqrt{\frac{D_x}{D_z}}\right]^2} \tag{4.56}$$

and the correction factor for a specimen with nonequal planar dimensions and orthotropic diffusivity is

$$C_f = \frac{1}{\left[1 + \frac{z_o}{2\sqrt{D_z}}\left(\frac{\sqrt{D_x}}{x_o} + \frac{\sqrt{D_y}}{y_o}\right)\right]^2} \tag{4.57}$$

The true z-direction diffusivity of a specimen with orthotropic diffusivity is given by Eqn (4.53) as

$$D_z = C_f D_{z,1\mathrm{D}} \tag{4.53}$$

where

$$C_f = \frac{1}{\left[1 + \frac{z_o}{x_o}\sqrt{\frac{D_x}{D_z}}\right]^2} \tag{4.56}$$

This relation may be used to evaluate the orthotropic moisture diffusivities of a material by extracting the apparent z-diffusivity of the material using specimens of different dimensions. Figure 4.26 shows the moisture uptake of two specimens of an FR-4 core material under the condition 85 °C/85% RH. The two specimens have dimensions $100 \times 100 \times 1\ \mathrm{mm}^3$ and $30 \times 30 \times 1\ \mathrm{mm}^3$, respectively. Using Eqn (4.34), the apparent diffusivity ($D_{z,1\mathrm{D}}$) of the $30 \times 30 \times 1$ mm and $100 \times 100 \times 1\ \mathrm{mm}^3$ test specimens have been evaluated to be $4.00 \times 10^{-12}\ \mathrm{m}^2/\mathrm{s}$ and $3.62 \times 10^{-12}\ \mathrm{m}^2/\mathrm{s}$, respectively.

Being made of the same material, the two specimens exhibit identical true z-diffusivity (D_z); that is,

$$D_z = \left(C_f \cdot D_{z,1\mathrm{D}}\right)_{\frac{x_o}{z_o}=30} = \left(C_f \cdot D_{z,1\mathrm{D}}\right)_{\frac{x_o}{z_o}=100} \tag{4.58}$$

or

$$\frac{4.00}{\left(1 + \frac{1}{30}\sqrt{\frac{D_x}{D_z}}\right)^2} = \frac{3.62}{\left(1 + \frac{1}{100}\sqrt{\frac{D_x}{D_z}}\right)^2} \tag{4.59}$$

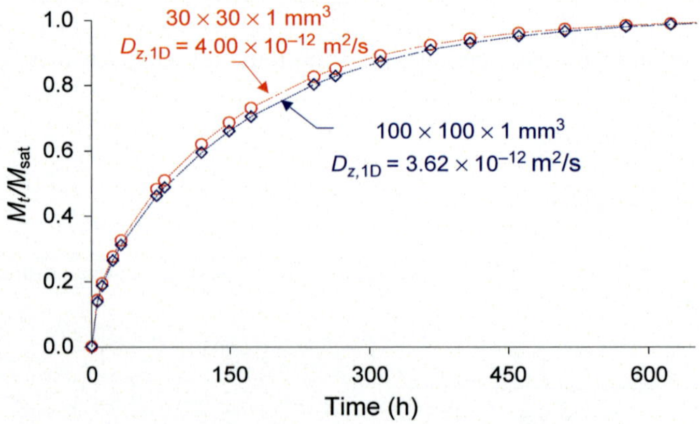

Figure 4.26 M_t/M_{sat} versus time of two specimens of FR-4 laminated core.

Solving Eqn (4.59) gives $D_x/D_z = 5$, which upon substituting into Eqn (4.58) gives $D_z = 3.5 \times 10^{-12}$ m²/s. The in-plane diffusivity is given by $D_x = 5D_z = 1.7 \times 10^{-11}$ m²/s.

The factor $\frac{x_o}{z_o}\sqrt{\frac{D_z}{D_x}}$ dictates the relative contribution of the inflow of moisture along the in-plane direction. The repeatability and accuracy of the experiment can be improved by using specimens of diverse $\frac{x_o}{z_o}\sqrt{\frac{D_z}{D_x}}$. Ideally, one of the specimens shall have a factor of 50 that allows negligible in-plane inflow of water (referring to Figure 4.18, this leads to a mere 3.8% increase in the apparent diffusivity); and one of the specimens shall have a factor of four that allows significant in-plane inflow of water (referring to Figure 4.18, this leads to 57% increase in the apparent diffusivity). Given $D_x/D_z = 5$, the factor $\frac{x_o}{z_o}\sqrt{\frac{D_z}{D_x}}$ for the $100 \times 100 \times 1$ and the $30 \times 30 \times 1$ specimens evaluates to 45 and 13, respectively. It would be ideal to repeat the test using a specimen with planar dimensions smaller than 30×30 mm.

4.5 Hygroscopic swelling and its characterisation

4.5.1 Introduction

Polymers swell upon absorbing moisture, which could be attributed to a reduction in the intermolecular forces between the molecules of the host polymer – a phenomenon known as plasticising – and an increase in the intermolecular forces between the molecules of the absorbed water and that of the host. The differential thermal expansion of components within a microelectronic assembly leads to thermal stress, which has been responsible for the bulk of the failure of microelectronic assemblies. Similarly, differential swelling occurs between the polymeric and the non-polymeric materials as well as between the different polymeric materials constituting the microelectronic packaging. This differential swelling induces hygroscopic stress, which acts on top of the thermal stress to cause cracking.

The studies of hygroscopic swelling by the first author were triggered by the experimental observations of colleagues in the Institute of Microelectronics, Singapore in 1989. A distinct difference in the life of a flip-chip PBGA packaging was observed for three under bump metallisation (UBM) with and without underfilling material when subjected to autoclave conditioning (131 °C/100% RH) (Teo, 2000). These are tabulated in Table 4.7. The failure mode was predominantly cracking of solder joints at the UBM with clear evidence of ‘opening’ with little evidence of ‘shearing’ as shown in Figure 4.27. Thermal stress was the only known mechanism then. However, the underfilling materials were cured at a temperature that would have induced compressive stress on the solder joint. Hygroscopic swelling of the underfilling material was postulated. The verification of the postulation requires the evaluation of the hygroscopic strain of the underfilling material.

Table 4.7 Normalised autoclave performance of different under bump metallisation (UBM), with and without underfill

UBM structure	Autoclave life	
	Without underfill	With underfill
UBM-A	1	1.3
UBM-B	1	0.3
UBM-C	1	0.5

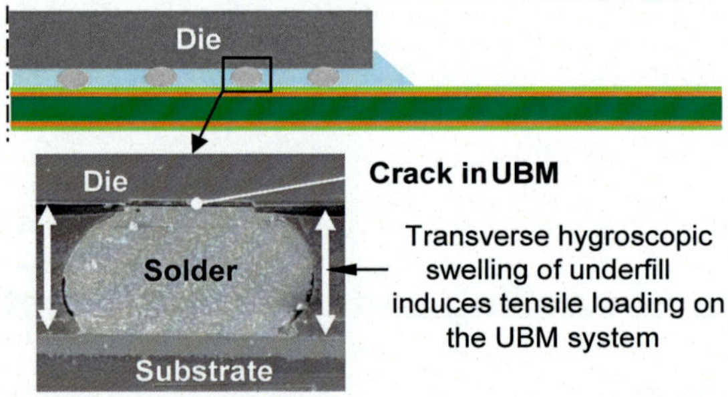

Figure 4.27 Hygroscopic swelling induced tensile failure in autoclave testing of flip-chip plastic ball grid array (PBGA) package.

4.5.2 Characterising hygroscopic swelling property

Polymers expand with increasing temperature attributed to an increase in internal energy. The coefficient of thermal expansion is traditionally measured used thermo-mechanical analysis (TMA), which measures a change of a dimension of a sample while it is subjected to a temperature regime and in a controlled atmosphere. A typical TMA instrument has submicrometer resolution and excellent temperature and environment control with nitrogen that keeps the environment dry. Polymers decompose at elevated temperature; this could be measured with TGA in which the mass of a substance is monitored as a function of temperature or time as the sample specimen is subjected to a controlled temperature program in a controlled atmosphere. A typical TGA instrument has microgram resolution and excellent temperature and environment control.

The hygroscopic swelling characteristic of polymers can be measured by combining the measurements from TMA and TGA. By taking measurements of the loss of moisture mass of a specimen and also the reduction in the dimension of an identical specimen, the change in dimension of the specimen can be related to the change in the mass of moisture. The exact procedures involved (Wong, 2002) are:

- saturate two specimens of identical material, shape and dimension with moisture to identical concentration;
- desorb moisture from the specimens with identical temperature and environmental history in TMA and TGA, respectively; it is essential to ensure that two apparatus are set to identical

ramp rate and are purged with adequate inert gas to assure that the environment is indeed 'dry'; it is also essential to ensure that the specimens are rested on identical surface in the apparatus.

- extract dimensional change and weight loss of the specimens at the same desorption duration;
- plot dimensional strain versus moisture concentration.

This is depicted in Figure 4.28. The curve defines the relation between hygroscopic swelling strain and moisture concentration. If a linear relation exists between these two characteristics, then the constant of linearity is the coefficient of moisture expansion (β), which is defined mathematically as

$$\beta = \frac{\varepsilon}{C} \tag{4.60}$$

The hygroscopic swelling characteristic of a typical moulding compound is presented in Figure 4.29, which shows a linear relation between hygroscopic swelling and moisture concentration. The observation of a linear relation supports the definition of the coefficient of moisture expansion as suggested by Eqn (4.60). It seems that the

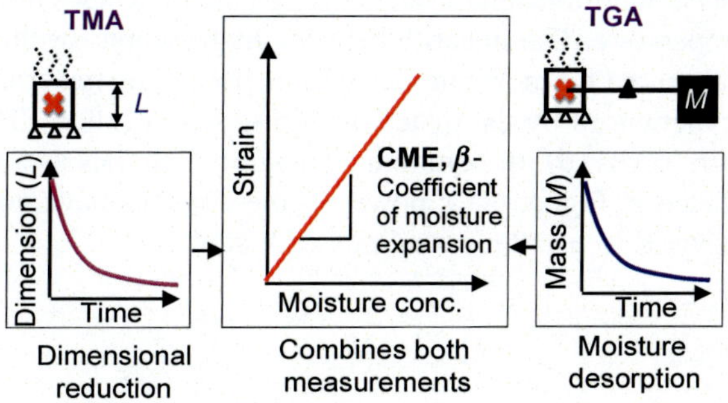

Figure 4.28 Set-up and technique for characterisation of hygroscopic swelling property of material.

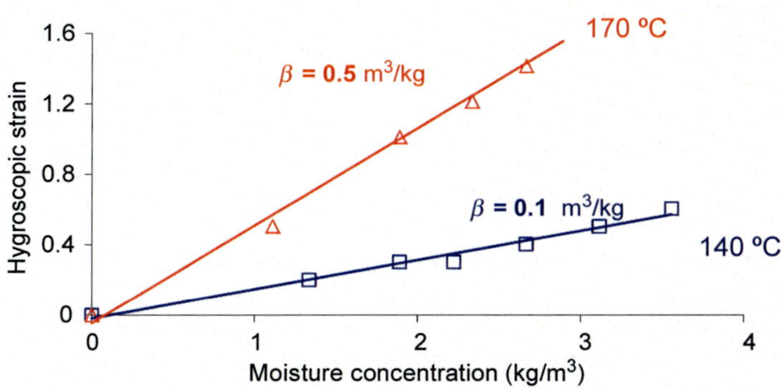

Figure 4.29 Set-up and technique for characterisation of hygroscopic swelling property of material.

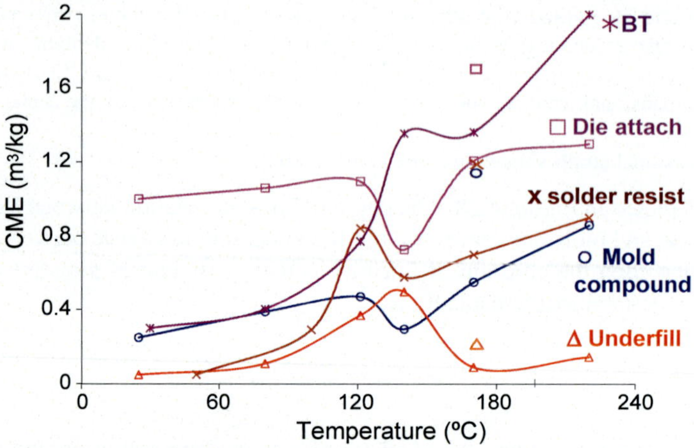

Figure 4.30 Coefficient of moisture expansion (CME) of packaging materials as a function of temperature.

coefficient of moisture expansion increases with increasing temperature, which could be due to the reduced intermolecular forces between the molecules of the host polymer with elevating temperature. The linearity between hygroscopic swelling and moisture content has been observed for most microelectronic packaging materials over the range of temperatures experienced by the materials during solder reflow. The coefficient of moisture expansion, CME (β), of common packaging materials as a function of temperature are presented in Figure 4.30; however, these results have not been validated and readers are advised to use these readings with care.

References

Akshaya, J. (2014). *Characterization of pore structure: Foundation.ppt*. Porous Materials, Inc. Retrieved from http://www.pmiapp.com/publications/docs/characterization_pore_structure_foundation.ppt.

ASTM-D5229M. (1998). *Standard test method for moisture absorption properties and equilibrium conditioning of polymer matrix composite materials*. ASTM.

ASTM-D570. (2010). *Standard test method for water absorption of plastics*. ASTM.

Brunauer, S. E. (1940). *Journal of American Chemical Society, 62*, 1723.

BS-2782. (1983). *Part 4 Method 430A, determination of water absorption at 23 °C.*

CMH-17, R. G. (2011). Polymer matrix composites. guidelines, (Chapter 2). In *Composite materials handbook* (Vol. 1).

Crank, J. (1956). *Mathematic of diffusion*. Oxford University Press.

DVS. (2014). *Quantachrome instrument*. Dynamic Vapor Sorption Analyzers. Retrieved from http://www.quantachrome.com/vapor_sorption/aquadyne_dvs.html.

Galloway, J. M., & Miles, B. M. (1996). Moisture absorption and desorption predictions for plastic ball grid array packages. In *Proceedings of the intersociety conference on thermal phenomena* (pp. 180−186).

Henry, W. (1803). Experiments on the quantity of gases absorbed by water, at different temperatures, and under different pressures. *Philosophical Transactions of the Royal Society, 93*, 29–274.

IGASorp. (2014). *Hiden Isochema*. Dynamic Vapor Sorption Analyzer. Retrieved from http://www.hidenisochema.com/our_products/instruments/?id=12&title=igasorp.

ISO-62. (2008). *Plastic: Determination of water absorption*. International Standard Organisation.

IUPAC. (1972). Colloid and surface chemistry. *Pure and Applied Chemistry*. IUPAC Manual of Symbols and terminology, Appendix 2, Pt. 1.

JESD22-A120A. (2008). *Test method for the measurement of moisture diffusivity and water solubility in organic materials used in electronic devices*. JEDEC Standard.

JIS-K7209. (1984). *Methods for water and boiling water absorption of plastics*. JIS.

Langmuir, I. (1918). *The adsorption of gases on plane surface of glass, mica and platinum*. The Research Laboratory of the General Electric Company. pp. 1361–1402.

McBain, J. (1909). *Physical Chemistry, 38*, 471.

SEMI-G66. (1996). *Test method for the measurement of water absorption characteristics for semiconductor moulding compounds*.

Shi, Y. T. (2002). An effective method of characterising moisture desorption of polymer materials at high temperature. In *Electronic Packaging Technology Conference* (pp. 70–75). Singapore.

Stadie, N. P. (2013). *Synthesis and thermodynamic studies of physisorptive energy storage materials* (Ph.D. dissertation). California Institute of Technology. Retrieved from http://resolver.caltech.edu/CaltechTHESIS:09092012-010239493.

Steamtable. (2014). The Steamtable Online. Retrieved from http://www.steamtablesonline.com.

Teo, P. H. (2000). Investigation of under bump metallization systems for flip-chip assemblies. In *Electronic and Component Technology Conference* (pp. 33–39). Las Vegas, NV.

TLV. (n.d.). TLV steam calculator. Retrieved from http://www.tlv.com/global/TI/calculator/steam-table-temperature.html.

VSA. (2014). *TA instrument*. Vapour Sorption Analysis. Retrieved from http://www.tainstruments.com/product.aspx?id=239&n=1&siteid=11.

Wong, E. R. (2002). The mechanics and impact of hygroscopic swelling of polymeric materials in electronic packages. *Journal of Electronic Packaging, 124*, 122–126.

Wong, E. R. (2003). Moisture absorption & diffusion characterisation of packaging materials – advanced treatment. *Microelectronics Reliability, 43*, 2087–2096.

Advances in diffusion and vapour pressure modelling

5

5.1 The discontinuity of concentration

While the discontinuity of concentration across the interface of substances has been well accepted among the microelectronic assembly community, a theoretical argument has not been presented. Herein, we give a coherent argument of the discontinuity of concentration to serve as an introduction to later presentations on diffusion modelling, which has seen some advancement since the development of the 'wetness' method (Wong, 1998).

Based on experimental observations and Newton's law of cooling, Fourier (1822) proposed the well-known law:

$$J_{\mathrm{q}} = -k_{\mathrm{q}} \nabla T \tag{5.1}$$

where J_{q}(J/s m^2) is heat flux and k_{q}(J/s m K) is thermal conductivity. This suggests a linear relation between heat flux and the spatial gradient of temperature. Fourier's law as a fundamental law is affirmed by the zeroth law of thermodynamics, which states that the continuity of temperature at the interface of substances is independent of the chemical constituents, the molecular/atomic structure and the phases of the substances.

Under the condition of constant pressure, the first law of thermodynamics gives

$$\mathrm{d}H_{\mathrm{v}} = \mathrm{d}q = \rho c_p \mathrm{d}T \tag{5.2}$$

where H_{v}(J/m^3) is volumetric enthalpy, q(J/m^3) volumetric heat flow, ρ(kg/m^3) density and c_p(J/kg K) specific heat capacity at constant pressure. The multiplier ρc_p(J/m^3 K) is the volumetric heat capacity at constant pressure. Multiplying the RHS of Eqn (5.1) by $\rho c_p / \rho c_p$ and assuming ρc_p to be independent of spatial coordinate gives

$$J_{\mathrm{q}} = -\alpha \nabla H_{\mathrm{v}}, \quad \text{for} \quad p = \text{constant} \tag{5.3}$$

where $\alpha = k_{\mathrm{q}} / \rho c_p$(m^2/s) is thermal diffusivity. The assumption that ρc_p is independent of spatial coordinates implies ρc_p is also independent of temperature. Equation (5.3) suggests a linear relationship between heat flux and spatial gradient of volumetric enthalpy, which appears to be the exact equivalence of Fourier's law. In general, the volumetric heat capacity of substances differs. Figure 5.1 shows two substances, substance A and substance B, in physical contact, wherein $(\rho c_p)_B > (\rho c_p)_A$. At the interface where temperature is T', the volumetric enthalpies are $H_{\mathrm{v},A} = (\rho c_p)_A T'$ and $H_{\mathrm{v},B} = (\rho c_p)_B T'$, respectively; and $H_{\mathrm{v},B} > H_{\mathrm{v},A}$. That is, a discontinuity in the

Robust Design of Microelectronics Assemblies Against Mechanical Shock, Temperature and Moisture
http://dx.doi.org/10.1016/B978-1-84569-528-6.00005-8

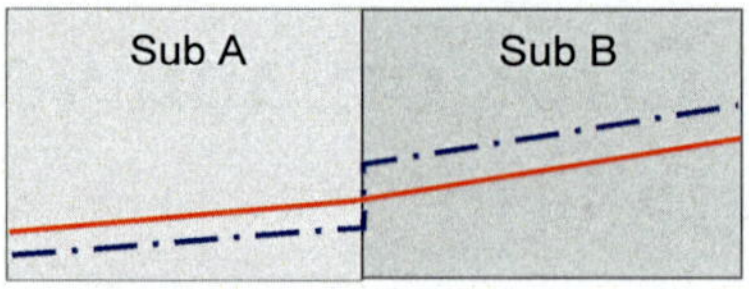

Figure 5.1 The continuity of temperature and discontinuity of volumetric enthalpy at the interface of two substances.

volumetric enthalpy develops at the interface of the two substances. Thus, unlike Eqn (5.1), Eqn (5.3) is not a fundamental equation but it is valid only for a body that has a homogeneous volumetric heat capacity (ρc_p).

Under the condition of constant pressure, the equation of conservation of heat energy is given by

$$\frac{\partial H_\mathrm{v}}{\partial t} + \nabla \cdot J_\mathrm{q} = \Theta_\mathrm{q} \tag{5.4}$$

where Θ_q(J/m^3 s) is internal heat source. Substituting with Eqns (5.2) and (5.3) gives

$$\frac{\partial(\rho c_p T)}{\partial t} + \nabla(-\alpha \nabla(\rho c_p T)) = \Theta_\mathrm{q} \tag{5.5}$$

assuming ρc_p to be time and spatially invariant and α spatially invariant leads to the well-known heat equation:

$$\frac{\partial T}{\partial t} = \alpha \nabla^2 T + \frac{\Theta_\mathrm{q}}{\rho c_p} \tag{5.6}$$

Conversely, substituting Eqn (5.3) in Eqn (5.4) and assuming α is spatially invariant gives the heat equation in enthalpy:

$$\frac{\partial H_\mathrm{v}}{\partial t} = \alpha \nabla^2 H_\mathrm{v} + \Theta_\mathrm{q} \tag{5.7}$$

While Eqn (5.6) is valid for a system with non-uniform volumetric heat capacity (ρc_p), Eqn (5.7) is only valid for a body that has a uniform volumetric heat capacity (ρc_p).

From experimental observations on diffusion of salt in a liquid, and drawing upon the inspiration of Fourier's law, Fick's law for diffusion (Fick, 1855) has been proposed:

$$J_\mathrm{m} = -D \nabla C \tag{5.8}$$

where J_m (kg/s m^2) is mass flux, D (m^2/s) mass diffusivity and C (kg/m^3) concentration of the solute in the substance. Similar to the volumetric heat capacity (ρc_p), different substances exhibit different volumetric solute capacity (C_{sat}). Therefore, just like volumetric enthalpy, discontinuity in concentration is developed at the interface of substances at equilibrium. Thus, unlike Fourier's law, Fick's law is not a fundamental law but is valid only for a system that has a uniform volumetric solute capacity (C_{sat}).

Substituting Eqn (5.8) into the equation of conservation of the mass of solute:

$$\frac{\partial C}{\partial t} + \nabla \cdot J_m = \Theta_m \tag{5.9}$$

where Θ_m (kg/m^3 s) is the internal source for the solute, and assuming spatial invariant of D gives the Fick's diffusion equation:

$$\frac{\partial C}{\partial t} = D\nabla^2 C + \Theta_m \tag{5.10}$$

The Fick's diffusion equation is analogous to the heat equation in enthalpy and is only valid for a body that has a uniform volumetric solute capacity (C_{sat}). This implies a body that is made of the same substance – same chemical constituents, molecular/atomic structures and phases. The fact that C_{sat} of substances is a function of temperature and pressure further constrains the body to be under the conditions of uniform temperature and pressure.

5.2 The fractional saturation

The above analysis has established the following correspondences between thermal and mass diffusion: (α, ρc_p, H_v) and (D, C_{sat}, C). The ratio of volumetric enthalpy (H_v) to volumetric heat capacity (ρc_p) returns the fundamental property in thermal diffusion – temperature:

$$T = \left(\frac{H_v}{\rho c_p}\right)_p \tag{5.11}$$

where the subscript p refers to the condition of constant pressure. By induction, the ratio of volumetric solute concentration (C) to volumetric solute capacity (C_{sat}) should return a corresponding fundamental property in mass diffusion:

$$w = \left(\frac{C}{C_{sat}}\right)_{T,\rho_s} \tag{5.12}$$

where the subscript T refers to the condition of constant temperature and ρ_s refers to the concentration/density of the solute in the source that has induced both C and

C_{sat} in the absorbant. The property, w, has been referred to as wetness for moisture as the solute; it is more appropriately referred to as fractional saturation since it is indeed a measurement of the degree of saturation. Physically, it is the quest for saturation that provides the driving force for mass diffusion, and the strength of the driving force is proportionate to the differential in the degree of saturation. This is in exact analogy with heat diffusion wherein temperature is a measurement of the degree of heat, and the quest for equality of the degree of heat provides the driving force for thermal diffusion. At equilibrium, physically connected substances should have equal degrees of heat and saturation. The continuities of temperature and fractional saturation are a natural conclusion of the equalities of heat and saturation, respectively. Nevertheless, a mathematical argument of the continuity of fractional saturation using the equalisation principle of chemical potential (μ) shall be presented in the following sections.

5.2.1 The chemical potential

Recognising the restriction of solute concentration, Gibbs (1948) proposed the equalisation principle of chemical potential (μ), which suggests that at equilibrium the chemical potential is continuous at the interface of substances, independent of the chemical constituents, the molecular/atomic structure and the phases of the substances.

The Gibbs fundamental equation of chemical thermodynamics for a system constituting of r species, with the ith species having n_i mole, is given by (Baierlein, 2003; Gibbs, 1948; Kittel, 1980)

$$\mathrm{d}G = -S\mathrm{d}T + V\mathrm{d}p + \sum_{i}^{r} \mu_i n_i \tag{5.13}$$

where S(J/K) and V are entropy and volume, respectively, of the system, and μ_i(J/mol) is the chemical potential of the ith species. At constant temperature and pressure, Eqn (5.13) is reduced to

$$\mathrm{d}G = \sum_{i}^{r} \mu_i n_i \tag{5.14}$$

which gives the definition of chemical potential as the Gibbs free energy per unit mole; that is,

$$\mu_i = \left(\frac{\partial G}{\partial n_i}\right)_{T,p,n_{j\neq i}} \tag{5.15}$$

where the subscripts T and p indicate the condition of constant temperature and pressure, and the subscript $n_{j\neq i}$ suggests the amount of all other constituents except the ith are kept constant. We shall focus on a system with a single diffusing solute, and dropping the subscript i in the following discussions. The chemical potential of the solute in the substance may be expressed as (Baierlein, 2003; Job, 2006)

$$\mu(C) = \mu^o(C^o) + RT \ln\left(\frac{\gamma_c C}{C^o}\right) \tag{5.16}$$

where $\mu^o(C^o)$ is the chemical potential at the reference solute concentration C^o, R(J/mol K) is the universal gas constant, $\gamma_c C/C^o$ is thermodynamic activity and γ_c is the activity coefficient, which is a function of C but approximates a constant for dilute solution. Assuming dilute solution, differentiating Eqn (5.16) gives

$$\nabla C = \frac{C}{RT}\nabla\mu \tag{5.17}$$

which upon substituting into Eqn (5.8) gives

$$J_m = -\frac{DC}{RT}\nabla\mu = -\frac{DC^o}{RT\gamma_c}\exp\left(\frac{\mu-\mu^o}{RT}\right)\nabla\mu \tag{5.18}$$

Equation (5.18) suggests a nonlinear flux–gradient relation for chemical potential. Substituting Eqns (5.17) and (5.18) into the equation of mass conservation, Eqn (5.9), yields a complex diffusion equation:

$$C\frac{\partial\mu}{\partial t} = D\nabla\cdot(C\nabla\mu) + RT\Theta_m \tag{5.19}$$

which is quite unlike the heat equation, Eqn (5.6).

5.2.2 The fractional saturation

Applying the equality of chemical potential at the interface of absorbants A and B, respectively, gives

$$\mu_A^o + RT \ln \frac{\gamma_{c,A} C_A}{C_A^o} = \mu_B^o + RT \ln \frac{\gamma_{c,B} C_B}{C_B^o} \tag{5.20}$$

Equation (5.20) suggests that under the condition of constant pressure, the ratio C_A/C_B is a constant (for any magnitudes of C_A and C_B). Substituting C with $w \cdot C_{sat}$ gives

$$\frac{C_A}{C_B} = \frac{w_A C_{sat,A}}{w_B C_{sat,B}} = \text{constant} \tag{5.21}$$

The equality must be satisfied for any magnitudes of w_A and w_B; this is possible only if $w_A = w_B$. Thus, the continuity of chemical potential (μ) implies the continuity of fractional saturation (w).

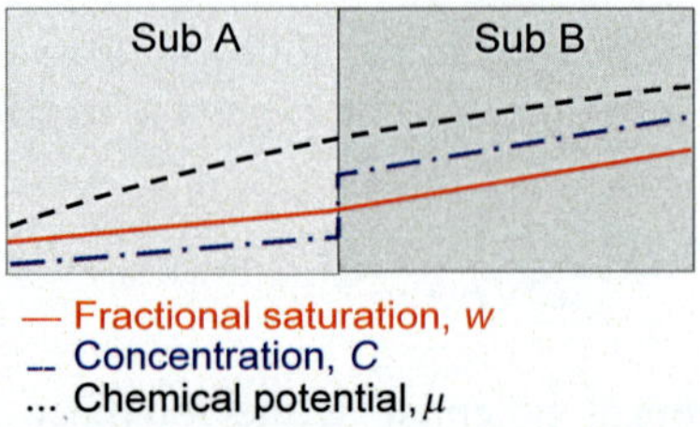

Figure 5.2 The profile of fractional saturation, concentration and chemical potential in a body of two contacting substances.

Stating again the fractional saturation, $w = C/C_{sat}$, it is clear that like concentration (C), there is also a linear flux−gradient relation for fractional saturation (w) if the saturated solute concentration (C_{sat}) in a substance of interest is independent of spatial coordinate. This will be true for a substance that has a uniform temperature and pressure. Figure 5.2 shows the continuous and linear characteristics of fractional saturation, the discontinuous but linear nature of concentration and the continuous but nonlinear characteristics of chemical potential.

Expressing the flux−fractional saturation gradient relation as

$$J_m = -k_m \nabla w \tag{5.22}$$

where k_m(kg/m s) is the conductivity of mass flux defined as the flux per unit gradient of fractional saturation, and substituting with $w = C/C_{sat}$, together with the condition that C_{sat} is independent of spatial coordinate, gives

$$J_m = -\frac{k_m}{C_{sat}} \nabla C \tag{5.23}$$

Comparing Eqn (5.23) with Eqn (5.8) suggests that $D = k_m/C_{sat}$, which is analogous to thermal diffusivity $\alpha = k_q/\rho c_p$.

The equation of conservation of mass, Eqn (5.9), expressed in terms of the fractional saturation, becomes

$$\frac{\partial(wC_{sat})}{\partial t} = \nabla(D\nabla(wC_{sat})) + \Theta_m \tag{5.24}$$

Under the condition of time-constant pressure and temperature, the saturated concentration of a solute in a substance (C_{sat}) is independent of time. Assuming the diffusivity D to be spatially invariant, then Eqn (5.24) gives

$$\frac{\partial w}{\partial t} = D\nabla^2 w + \frac{\Theta_m}{C_{sat}} \tag{5.25}$$

Equation (5.25) is applicable to the diffusion of any solute in any substance, be it a solid, a liquid, a gas or combination thereof. Equation (5.25) is a perfect analogue of the heat equation, Eqn (5.6). The perfect correspondence between temperature and

Table 5.1 The heat-diffusion correspondence

	Heat	Diffusion
Volumetric capacity	Heat capacity, ρc_p(J/m^3 K)	Solute capacity, C_{sat}(kg/m^3)
The discontinuous volumetric characteristic	Enthalpy, H(J/m^3)	Concentration, C(kg/m^3)
The continuous characteristic	Temperature, $T = H/\rho c_p$(K)	Fractional saturation, $C = C_{sat}$
Transient equation	$\frac{\partial T}{\partial t} = \alpha \nabla^2 T + \frac{\Theta_q}{\rho c_p}$	$\frac{\partial w}{\partial t} = D\nabla^2 w + \frac{\Theta_m}{C_{sat}}$
Conductivity	k_q(J/s m K)	$k_m = C_{sat}\, D$(kg/s m)
Diffusivity	$\alpha = k_q/\rho c_p$(m^2/s)	D(m^2/s)

fractional saturation facilitates the use of 'heat-diffusion' module in commercial software for modelling mass diffusion.

The heat-diffusion correspondence is summarised in Table 5.1. The two essential material properties required for solving the heat equation are the thermal diffusivity (α) and volumetric heat capacity (ρc_p); in the absence of heat generation, only thermal diffusivity (α) is needed. In practice, the three material properties (k_q, ρ, c_p) are more readily available and provided to the commercial heat-diffusion module. The thermal diffusivity ($\alpha = k_q/\rho c_p$) and volumetric heat capacity (ρc_p) are then evaluated from these properties. The two essential material properties required for solving the diffusion equation are D and C_{sat}, which are readily available. However, this information must be supplied in the form of (k_m, ρ, c_p) if the heat-diffusion analogy is applied to the commercial heat-diffusion module; that is, $k_m \to C_{sat}\, D$ and $\rho c_p \to C_{sat}$, which may be satisfied by unlimited combinations of ρ and c_p.

5.2.3 *The auxiliary C_{sat} and auxiliary fractional saturation*

It is worth noting that Eqn (5.12) is not restricted to source solute or absorbants of any particular phase. Dalton's law of partial pressure suggests that the partial pressure of a gaseous solute in a gaseous absorbant is independent of the elemental constitution of the absorbant; this is extended to the density of the solute assuming ideal gas behavior: $p_s = \rho_s R_s T$, where R_s is the constant of the gaseous solute. That is, all gaseous absorbants would have volumetric solute capacity equal to the density of the gaseous source solute ρ_s. Thus, C_{sat} in Eqn (5.12) may be substituted with ρ_s for gaseous solute in gaseous absorbant; the fractional saturation may be expressed as

$$\begin{aligned} w &= \frac{C}{C_{sat}} \\ &= \frac{C}{C_{sat}(\rho_s)} = \frac{C}{C_{sat}(p_s)} \quad \text{for gaseous source solute in solid absorbant,} \\ &= \frac{\rho}{\rho_s} = \frac{p}{p_s} \quad \text{for gaseous solute in gaseous absorbant} \end{aligned} \tag{5.26}$$

wherein it is understood that C, ρ and p in the above equation correspond to the source solute density of ρ_S.

It is suggested in Eqn (5.21) that the only requirement for the continuity of fractional saturation is that the ratio $C_{sat,A}(\rho_S)/C_{sat,B}(\rho_S)$ is a constant. For diffusion that occurs under a temporally constant source density ρ_S, Eqn (5.21) is automatically satisfied. In such a case, the continuity of the fractional saturation is independent of the characteristics of the sorption isotherm of the solute—absorbant system – the sorption isotherm may be linear or nonlinear and the two absorbants may even have different sorption isotherms.

However, this is not the case for diffusion under a time-varying source density $\rho_S(t)$. We shall examine the necessary characteristics of the sorption isotherms that will ensure the continuity of fractional saturation under such a condition. It is numerically cumbersome for a field variable to take the form $w = C/C_{sat}(\rho_S(t))$, in which its instantaneous magnitude is dependent on the instantaneous magnitude of its property. To overcome this difficulty, an auxiliary condition for the source density ρ_a may be defined. The auxiliary volumetric solute concentration is then $C_{sat}(\rho_a)$, and the auxiliary wetness, w_a, is defined as

$$w_a = \frac{C(\rho_s)}{C_{sat}(\rho_a)} \tag{5.27}$$

Substituting Eqn (5.27) into Eqn (5.21) gives

$$\frac{w_A C_{sat,A}(\rho_s)}{w_B C_{sat,B}(\rho_s)} = \frac{w_{a,A} C_{sat,A}(\rho_a)}{w_{a,B} C_{sat,B}(\rho_a)} \tag{5.28}$$

The condition for the continuity of w_a is therefore given by

$$\left(\frac{C_{sat,A}}{C_{sat,B}}\right)_{\rho_a} = \left(\frac{C_{sat,A}}{C_{sat,B}}\right)_{\rho_s} \tag{5.29}$$

that is, the sorption isotherms of solute in the two absorbants should be proportionate. If the sorption isotherms of the solute in the two absorbants are not only proportionate but linear, then the auxiliary fractional saturation w_a may be reduced to

$$\begin{aligned} w_a &= \frac{C(\rho_s)}{C_{sat}(\rho_a)} = w\frac{C_{sat}(\rho_s)}{C_{sat}(\rho_a)} = w\frac{\rho_s}{\rho_a} \\ &= w\frac{p_s}{p_a} \quad \text{for gaseous source solute} \\ &= \frac{p}{p_a} \quad \text{for gaseous solute in gaseous absorbant} \end{aligned} \tag{5.30}$$

For the case of water vapour as the source solute and the vapour saturation condition (100% relative humidity) selected as the auxiliary condition, the auxiliary fractional saturation is simply

$$w_a = w * \%RH \tag{5.31}$$

The application of the auxiliary fractional saturation will be demonstrated in the following examples.

5.2.4 The normalised concentration

The characteristic 'normalisation concentration':

$$p_\phi = \frac{C}{k_H} \tag{5.32}$$

where K_H(Kg/m^3 Pa) is Henry's coefficient, are used by some researchers (Jang, 2008; Xie, 2009). The first reported use of normalised concentration as a field variable for modelling moisture diffusion in microelectronic assembly may be traced to Galloway (1996), who used the inherent capability of the mass diffusion module in Abaqus theory manual (2014).

Henry's law states that (Henry, 1803): 'At a constant temperature, the amount of a given gas that dissolves in a given type and volume of liquid is directly proportional to the partial pressure of that gas in equilibrium with that liquid'. This also applies to gas–solid interfaces (Smith, 2007). Henry's law can be derived as follows. At equilibrium, the chemical potential of a solute in a substance (may be either solid, liquid or gas) is given by

$$\mu(C_{sat}) = \mu^o\left(C^o_{sat}\right) + RT\ \ln\left(\frac{\gamma_c C_{sat}}{C^o_{sat}}\right) \tag{5.33}$$

For solute in a gaseous substance, the ideal gas law of the solute gives $p_v \propto C$, and Eqn (5.33) may be reduced to

$$\mu(p_v) = \mu^o\left(p^o_v\right) + RT\ \ln\left(\frac{\gamma_p p_v}{p^o_v}\right) \tag{5.34}$$

The activity coefficients γ_p and γ_c are approximate constants for substances with dilute solute. For a solder or liquid substance in contact and in equilibrium with a gaseous substance, the equalisation principle of chemical potentials requires that $\mu(C_{sat}) = \mu(p_v)$. Thus, equating Eqns (5.33) and (5.34) gives

$$\frac{C_{sat}(p_v)}{p_v} = \frac{C^o_{sat}(p_v)}{p^o_v}\frac{\gamma_p}{\gamma_c} = k_H \tag{5.35}$$

It has been argued (in Section 4.3.3) that on the basis of the intermolecular forces decreasing with temperature, Henry's coefficient decreases with temperature (Smith, 2007) and may be expressed in the form of the Arrhenius equation:

$$k_{\mathrm{H}} = A_{\mathrm{H}}\mathrm{e}^{\Delta E/RT} \tag{5.36}$$

Substituting Eqn (5.35) into Eqn (5.32) gives

$$p_{\phi} = \frac{C}{C_{\mathrm{sat}}(p_{\mathrm{v}})}p_{\mathrm{v}} = w_{\mathrm{v}} \cdot p_{\mathrm{v}} \tag{5.37}$$

The continuity of $w_{\mathrm{v}} = C/C_{\mathrm{sat}}(p_{\mathrm{v}})$ has been established in Section 5.2.2 using the equalisation principle of chemical potentials; multiplying a continuity by a constant does not alter its continuity. Indeed, we may define as many forms of 'normalised concentration' as desired by multiplying the fundamental ratio, $C/C_{\mathrm{sat}}(p_{\mathrm{v}})$, with as many constants including the state property of water vapour, such as pressure, density, internal energy, enthalpy and entropy.

5.2.5 Applications of thermal–mass diffusion analogy

5.2.5.1 Moisture sorption in a dual-absorbant body with temporally constant and spatially uniform temperature T and source vapour pressure p_s

This basic condition is included for completeness and is illustrated in Figure 5.3. The inputs for the material properties, the initial condition and the boundary condition are:

- Diffusivity: $D(T)$;
- Volumetric moisture capacity: $C_{\mathrm{sat}}(T, p_{\mathrm{s}})$;
- Initial condition: $w = 0$, assuming the absorbants have been prebaked to remove residual moisture;
- Boundary condition: $w = 1$, $\nabla w = 0$.

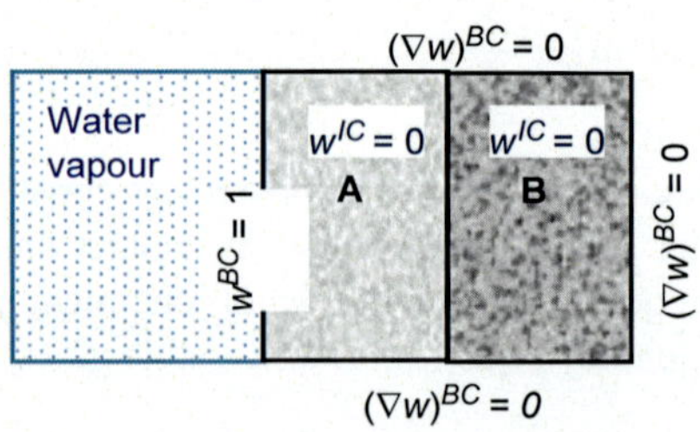

Figure 5.3 Moisture sorption under static temperature and vapour pressure.

5.2.5.2 *Moisture sorption in a single-absorbant or a multi-absorbant body with temporally constant and* spatially non-uniform temperature T*(x) and temporally constant source vapour pressure* p_s

Despite being a single-absorbant body, the spatially non-uniform temperature would lead to spatially nonuniform volumetric solute capacity in the absorbant, and rendering Fick's law inapplicable. The diffusion equation must be described using fractional saturation as the field variable. The inputs for the material properties, the initial condition and the boundary condition are:

- Diffusivity: $D(T(x))$;
- Volumetric moisture capacity: $C_{sat}(T(x), p_s)$;
- Initial condition: $w = 0$, assuming the absorbants have been prebaked to remove residual moisture;
- Boundary condition: $w = 1$, $\nabla w = 0$.

5.2.5.3 *Moisture sorption in a dual-absorbant body with temporally constant and spatially uniform temperature* T *and* temporary varying source vapour pressure $p_s(t)$

The condition is as illustrated in Figure 5.3 except that the humidity of the water vapour is now set to vary with time. The varying boundary condition does not allow the use of fractional saturation as the field variable, and the auxiliary fractional saturation w_a established in Section 5.2.3 should be used. Taking the vapour saturation condition (100% relative humidity) as the auxiliary condition and assuming linear sorption isotherm, the inputs for the material properties, the initial condition and the boundary condition are:

- Diffusivity: $D(T)$;
- Volumetric moisture capacity: $C_{sat}(T, p_a)$;
- Initial condition: $w_a = 0$, assuming the absorbants have been prebaked to remove residual moisture;
- Boundary condition: $w_a = \%RH(t)$, $\nabla w_a = 0$.

The use of auxiliary fractional saturation has enabled the time-varying source to be modelled simply as time-varying boundary condition.

5.2.5.4 *Moisture sorption in a multiphase body with temporally constant and spatially uniform temperature* T *and a* depleting source *that has an initial vapour pressure* p_s

In the previous three examples, the water vapour domain, and hence the vapour pressure, have been assumed to be spatially uniform such that the entire domain can be replaced by a boundary condition applied along the vapour–solid interface. If the vapour domain is not uniform, then it would have to be modelled as a domain of gas absorbant. This shall be illustrated using an example. Figure 5.4 shows a solid body of dual absorbants with an internal cavity that is placed in a small humidity

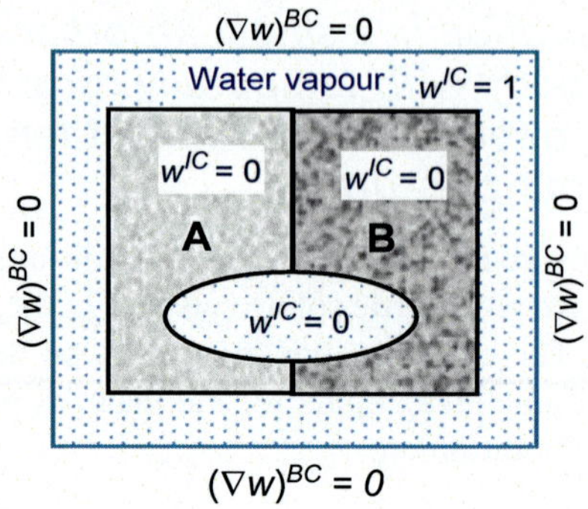

Figure 5.4 Moisture sorption in a multiphase body in an isolated chamber pre-filled with water vapour.

chamber. The entire system is maintained at a constant temperature T; the chamber is prefilled with vapour pressure p_s but is isolated from the external source. With ingression of water into the solid body and the internal cavity, the water vapour in the chamber will be depleted. The objective of this modelling is to find the decreasing vapour pressure in the chamber and the increasing vapour pressure in the internal cavity.

The model consists of the solid absorbants A and B, the gaseous absorbant in the chamber and the gaseous absorbant in the internal cavity. The inputs for the material properties, the initial conditions and the boundary condition are:

- Diffusivity: $D(T)$;
- Volumetric moisture capacity: $C_{sat}(T, p_s)$ for the solid absorbants and $\rho_s(T, p_s)$ for the gaseous absorbants, which may be looked up in a standard steam table;
- Initial condition: $w = 1$ for the gaseous absorbant in the chamber; and $w = 0$ for the solid absorbants and the internal gaseous absorbant assuming these absorbants have been pre-baked to remove residual moisture before being placed into the humidity chamber;
- Boundary condition: being isolated, there is no flux across the external boundary; that is, $\nabla w = 0$.

The magnitudes of the local water vapour pressure in the chamber and in the internal cavity may be evaluated using Eqn (5.30) as

$$\begin{aligned} p_{\text{chamber}}(x, y, t) &= p_s * w_{\text{chamber}}(x, y, t) \\ p_{\text{cavity}}(x, y, t) &= p_s * w_{\text{cavity}}(x, y, t) \end{aligned} \tag{5.38}$$

5.3 Diffusion under time-varying temperature and pressure

5.3.1 Dependence of C_{sat} on temperature and pressure

Fractional saturation is a function of temperature T and pressure (that is, total pressure P). This is because its denominator – the saturated solute concentration C_{sat} – is a function of temperature and pressure. The rates of change of C_{sat} with respect to temperature and pressure, respectively, are given by

$$\left(\frac{\partial C_{sat}}{\partial T}\right)_P = \left(\frac{\partial \mu}{\partial T}\right)_{P,n}\left(\frac{\partial C_{sat}}{\partial \mu}\right)_{P,T}$$
$$\left(\frac{\partial C_{sat}}{\partial p}\right)_T = \left(\frac{\partial \mu}{\partial p}\right)_{T,n}\left(\frac{\partial C_{sat}}{\partial \mu}\right)_{T,P} \quad (5.39)$$

The rate of change of chemical potential μ with respect to temperature and pressure, respectively, can be determined from Maxwell relations (Baierlein, 2003; Job, 2006; Kittel, 1980):

$$\left(\frac{\partial \mu}{\partial T}\right)_{P,n} = -\left(\frac{\partial S}{\partial n}\right)_{T,P} = -\overline{S}$$
$$\left(\frac{\partial \mu}{\partial p}\right)_{T,n} = \left(\frac{\partial V}{\partial n}\right)_{T,P} = \overline{V} \quad (5.40)$$

where $\overline{S}$ and $\overline{V}$ are molar entropy and molar volume, respectively. The function $(\partial C_{sat}/\partial \mu)_{T,P}$ can be evaluated from Eqn (5.17) as

$$\left(\frac{\partial C_{sat}}{\partial \mu}\right)_{T,P} = \frac{C_{sat}}{RT} \quad (5.41)$$

Substituting Eqns (5.40) and (5.41) into Eqn (5.39) gives

$$\left(\frac{\partial C_{sat}}{\partial T}\right)_P = -\overline{S}\,\frac{C_{sat}}{RT}$$
$$\left(\frac{\partial C_{sat}}{\partial p}\right)_T = \overline{V}\,\frac{C_{sat}}{RT} \quad (5.42)$$

The properties $\overline{S}, \overline{V} > 0$ for all substances; thus, C_{sat} of a substance increases with increasing pressure but decreases with increasing temperature. More specifically, the volume of a solid substance experiences little change while that of a gaseous substance experiences significant increase with increasing solute concentration. That is, we have $\overline{V}_{gas} > \overline{V}_{liquid} > \overline{V}_{solid} \approx 0$. The fact that a gaseous system has more microstates (Baierlein, 2003; Kittel, 1980) than a solid system suggests that $\overline{S}_{gas} > \overline{S}_{liquid} > \overline{S}_{solid}$. For the case of moisture diffusion in microelectronics assemblies, $C_{sat}(p_g)$ of solid microelectronic materials may be assumed to be independent of pressure and is less sensitive to temperature than $C_{sat}(p_g)$ of water vapour. In practice, $C_{sat}(T,p_g)$ of water vapour is readily available while $C_{sat}(T,p_g)$ of solid microelectronic materials can be established through sorption experiments as discussed in Section 4.3.

5.3.2 Diffusion modelling

There have been attempts to model moisture diffusion in a multi-material system experiencing time-varying temperature. The simplest approach assumes C_{sat} to be independent of temperature (Jang, 2008; Wong, 2002a), which is valid only if the range of temperature variation is relatively small. The more sophisticated method involves mapping the entire wetness field into $w(T) = C/C_{\text{sat}}(T)$ while enforcing wetness continuity along material interfaces using the conservation of concentration (Wong, 2002a,b):

$$w_{T+\Delta T} = w_T \frac{\sum_n C_{\text{sat}_T}}{\sum_n C_{\text{sat}_{T+\Delta T}}} \tag{5.43}$$

However, this method is numerically demanding and requires implementation of specially written macro-files onto the commercial FE software. Another reported method uses concentration as field variable while enforcing continuity of 'normalised concentration' along the material interfaces using the technique of Lagrange multiplier (Xie, 2009); however, this method requires a specially written numerical code. A simple method that is readily available on commercial FE software is presented below.

In the absence of diffusion, the concentration of solute in an isolated system must be conserved; this gives

$$\frac{\partial C_{\text{sat}}}{C_{\text{sat}}} = -\frac{\partial w}{w} \tag{5.44}$$

Substituting Eqn (5.44) into Eqn (5.42) gives

$$\left(\frac{\partial w}{\partial T}\right)_p = \overline{S}\frac{w}{RT}$$
$$\left(\frac{\partial w}{\partial p}\right)_T = -\overline{V}\frac{w}{RT} \tag{5.45}$$

That is, the fractional saturation in a substance increases with increasing temperature and decreases with increasing pressure, in exact opposite to the saturated concentration.

The dependence of C_{sat} on temperature and pressure renders the diffusion equation, Eqn (5.25), awkwardly difficult to solve when diffusion occurs under a varying temperature/pressure condition – the conservation of concentration requires that the field variable $w(T) = C/C_{\text{sat}}(T)$ be updated concurrent with the changing $C_{\text{sat}}(T)$. However, any active manipulation of field variable is computationally challenging (Wong, 2002a,b). This difficulty could be overcome through the use of an 'internal source' to update the fractional saturation surreptitiously.

An explanation of the scheme is presented below using a dual-absorbant model and for a single time increment. Figure 5.5 shows a dual absorbant in a temperature-humidity chamber whose condition changes linearly from condition 1: 60 °C/60% RH to condition 2: 85 °C/60% RH over a time increment Δt. The wetness field, w_1,

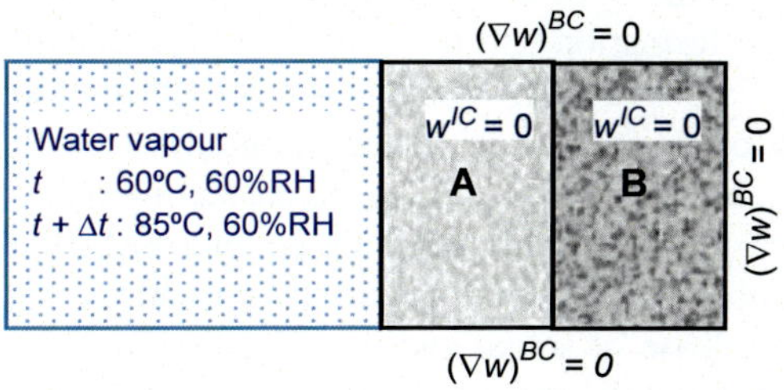

Figure 5.5 Moisture sorption in a dual-absorbant body exposed to time-varying temperature.

in the absorbants at time t, is known and the objective is to compute the wetness field, w_2, in the absorbants at time $t + \Delta t$.

The volumetric moisture capacity of the absorbants experiences a momentary change from $C_{\text{sat},1} = C_{\text{sat}}(60\ °\text{C},\ 60\%\ \text{RH})$ to $C_{\text{sat},2} = C_{\text{sat}}(85\ °\text{C},\ 60\%\ \text{RH})$ in the time increment Δt. In the absence of diffusion, the concentration of moisture in the absorbants must be conserved, which gives the corresponding fractional saturation in the absorbants at condition 2 as

$$w_2^* = \frac{C_{\text{sat},1} w_1}{C_{\text{sat},2}} \tag{5.46}$$

where the superscript * represents the condition without diffusion. The concentration of moisture to be 'added' to the absorbants in order to conserve the moisture concentration is given by

$$C_{1,2}^* = C_{\text{sat},2}\left(w_2^* - w_1\right) \tag{5.47}$$

which may be introduced to the individual constituting substance using the internal source function Θ_{m} in the diffusion equation Eqn (5.25). Over the time increment Δt, the concentration of moisture may be introduced at a constant rate given by

$$\Theta_{\text{m}} = \frac{C_{1,2}^*}{\Delta t} \tag{5.48}$$

The actual wetness in the absorbants at condition 2, w_2, is determined by solving the transient diffusion equation, Eqn (5.25), over the time increment Δt, using dynamic diffusivity that varies linearly from $D(60\ °\text{C})$ to $D(85\ °\text{C})$.

5.4 Advances in vapour pressure modelling

5.4.1 Introduction and review

The ultimate interest in modelling the diffusion of moisture into a microelectronic assembly is in modelling the risk of 'popcorn' cracking when the temperature in the microelectronic assembly is raised to the reflow temperature of the solder. A precondition for popcorn cracking is the formation and the expansion of a delamination during the solder reflow. The mismatched hygrothermal stresses during the solder reflow induce a delamination and cause it to open up. A vacuum is formed within the delamination; moisture that is absorbed in the polymeric packaging materials during moisture preconditioning moves toward the delamination and raises the pressure. When the microelectronic assembly is unable to contain the pressure, a violent rupture occurs accompanied by the release of sound energy – hence the name popcorn cracking. The instantaneous magnitude of vapour pressure within the delamination is a function of the instantaneous temperature and instantaneous wetness within the delamination given by Eqn (5.30). The wetness within the delamination is a function of the wetness of the surrounding polymeric material and the volume of the delamination; the latter is in turn dependent on the magnitude of pressure within the delamination and the structural stiffness of the microelectronic assembly. The evaluation of the instantaneous vapour pressure within the delamination therefore requires the coupled field analysis of transient mass diffusion and structural deformation.

The prevalence of popcorn cracking and the challenge of evaluating the magnitude of the instantaneous vapour pressure within the delamination of a microelectronic assembly during solder reflow have attracted a great deal of interest. The first reported attempt on modelling vapour pressure in an integrated circuit (IC) package during solder reflow was made by Kitano (1988), who has rigorously captured the full physics of the event. A schematic of his methodology is shown in Figure 5.6. The dynamic development of vapour pressure within the delamination is modelled through time integration. The vapour pressure at each time step is evaluated using

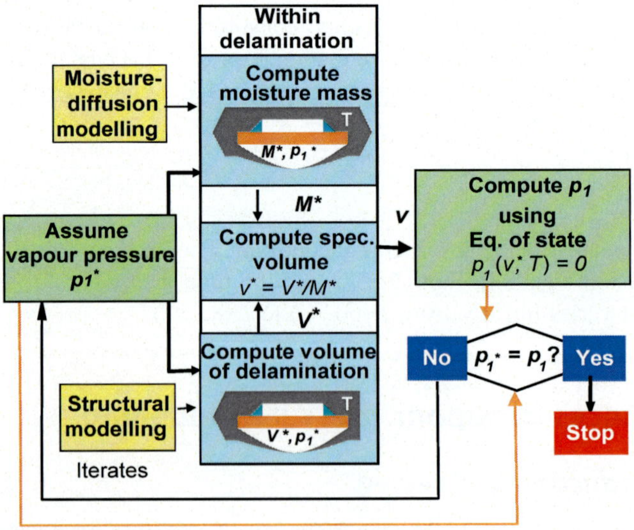

Figure 5.6 Kinato's methodology.

fixed-point iteration technique. This means that at each time (or temperature) increment, a value of vapour pressure (p_1^*) is assumed and used to evaluate the mass of transported moisture into the delamination (M^*) and the deformed volume of the delamination (V^*); the specific volume of water vapour is then evaluated as $v^* = V^*/M^*$; and the vapour pressure, p_I, is obtained by the equation of the state of water vapour, $p_I(v^*,T) = 0$. The two values of vapour pressure are compared, and if $p_1 \neq p_1^*$, then p_1^* becomes the assumed vapour pressure and the entire analysis is repeated. Kitano implemented the methodology using a finite difference scheme assuming a one-dimensional (1-D) system. Tay (1996) subsequently implemented the methodology into a 2-D finite element analysis procedure. Despite its rigour, the methodology suffers from computational inefficiency due to the expensive iteration process – each iteration involves a time-consuming moisture-diffusion analysis and a structural analysis.

Bhattacharyya (1988) assumes an ideal gas for water vapour and a microelectronic assembly that is fully saturated with moisture after moisture conditioning and retains its saturation during solder reflow:

$$p_r = \frac{k_b N_a T_m \rho_{v,m} \exp\left[-2740\left(\frac{1}{T_r} - \frac{1}{T_m}\right)\right]}{M_o} \tag{5.49}$$

where p_r is vapour pressure in the delamination at reflow; k_b is the Boltzmann constant = 1.38,065 × 10^{-23} J/K; $N_a = 6.022 \times 10^{23}$ is Avogadro's number; T_m and T_r are temperatures (K) at moisture conditioning and solder reflow, respectively; $\rho_{v,m}$ is density of water vapour at moisture conditioning; and $M_o = 18.02 \times 10^{-3}$ kg/mol is molecular weight of water. Equation (5.49) may be written more concisely as

$$p_r = p_{v,m} \exp\left[-2740\left(\frac{1}{T_r} - \frac{1}{T_m}\right)\right] \tag{5.50}$$

where $p_{v,m}$ is the pressure of water vapour in the moisture-conditioning chamber.

Sawada (1993) gave two equations:

$$\begin{aligned} p_r &= G_1 C_{sat}\left(p_{v,m}\right) T_r \\ p_r &= p_g(T_r)\%\text{RH} \end{aligned} \tag{5.51}$$

and suggested the correct vapour pressure is the lower of the two equations, where G_1 is an empirical fitting constant; $C_{sat}(p_{v,m})$ is volumetric solute capacity of moulding compound at moisture conditioning; $p_g(T_r)$ is the saturation pressure of water vapour at T_r; %RH is the relative humidity at moisture conditioning. The first Eqn (5.51) is purely empirical and the assumed linearity between p_r and T_r does not agree with the Arrhenius relation of water vapour $p_g = p_o e^{-E_v/RT}$ as reported in Eqn (5.1) of Chapter 4. The second Eqn (5.51) is based on the thermodynamic property of water vapour and assumes a microelectronic assembly that is fully saturated with moisture after moisture conditioning and retains its saturation during solder reflow.

Accounting for the non-saturation of moisture in the IC package after moisture conditioning and assuming non-diffusion of moisture during solder reflow, Wong (1998) proposed:

$$p_{\mathrm{r}} = p_{\mathrm{v,m}} \exp\left(\frac{-E_{\mathrm{v}}}{R}\left(\frac{1}{T_{\mathrm{r}}} - \frac{1}{T_{\mathrm{m}}}\right)\right) w_{\mathrm{v,m}} \tag{5.52}$$

where $w_{\mathrm{v,m}}$ is wetness along the delamination interface after moisture conditioning and $E_{\mathrm{v}} = 38.83$ kJ/mol is activation energy of $p_{\mathrm{g}} = p_{\mathrm{o}} \mathrm{e}^{-E_{\mathrm{v}}/RT}$ for the temperature range between 100 °C and 250 °C, and $E_{\mathrm{v}}/R = 4670$ K. Comparing Eqns (5.52) and (5.50), it seems that Bhattacharyya has assumed an unusually low activation energy given by $E_{\mathrm{v,bhatt}} = 2740R = 22.78$ kJ/mol.

Accounting for the diffusion of moisture during solder reflow, Wong (2005) modified the equation to

$$p_{\mathrm{r}} = p_{\mathrm{g}}(T_{\mathrm{r}})\ \%\mathrm{RH}\ w_{\mathrm{v,r}} \tag{5.53}$$

where $w_{\mathrm{v,r}}$ is the wetness along the delamination interface during solder reflow. Note that $p_{\mathrm{g}}(T_{\mathrm{r}})\%\mathrm{RH} = p_{\mathrm{v,m}} \exp\left(\frac{E_{\mathrm{v}}}{R}\left(\frac{1}{T_{\mathrm{m}}} - \frac{1}{T_{\mathrm{r}}}\right)\right)$. Equations (5.52) and (5.53) are effectively identical except for the difference in the definition of wetness. Equation (5.53) is similar to the second of Eqn (5.51), but it accounts for the non-saturation of the water around the delamination and the diffusion of moisture during reflow.

Assuming an ideal gas and a rectangular delamination area, Lim (1998) derived a rather simple equation:

$$p_{\mathrm{r}} = \sqrt{\frac{\pi^2 R_{\mathrm{w}} h^3 m E T_{\mathrm{r}}}{4 f_1 a^4 A}} \tag{5.54}$$

where $R_{\mathrm{w}} = 461.5$ J/kg K is the gas constant of water vapour; a and A are the length and the area of the delamination; h and E are the thickness and the elastic modulus of the delaminated packaging material; f_1 is a function of the aspect ratio of the delamination and also the Poisson ratio of the packaging material. Unfortunately, the most challenging task of evaluating the mass of moisture within the delamination m was not included.

Tee (1999) proposed to first determine a characteristic saturation temperature T_1 assuming the absorbed moisture resides in the micro-voids as saturated vapour:

$$\rho_{\mathrm{g}}(T_1) = \frac{C}{f_{\mathrm{v}}} \tag{5.55}$$

where f_v is the volume fraction of void in a solid substance; and hence the vapour pressure is computed using one of the following three equations:

$$p_r = \begin{cases} \dfrac{Cp_g(T_m)T_r}{\rho_g(T_m)T_m f_v} & \text{if } T_1 \le T_m \\[2ex] \dfrac{p_g(T_1)T_r}{T_1} & \text{if } T_m \le T_1 \le T_r \\[2ex] p_g(T_r) & \text{if } T_1 \ge T \end{cases}$$

Xie (2009) proposed a simpler set of equations:

$$\rho = \frac{C}{f_v} \tag{5.56}$$

$$p_r = \begin{cases} \dfrac{CRT_r}{M_\alpha f_v} & \text{if } \quad \rho \le \rho_g(T_r) \\[2ex] p_g(T_r) & \text{if } \quad \rho \ge \rho_g(T_r) \end{cases}$$

The computation of vapour pressure using Eqns (5.55) and (5.56) requires knowledge of the volume fraction of the pores, which is not readily available. However, the main criticism of Eqns (5.55) and (5.56) is that capillary condensation occurs in a pore at a partial vapour pressure $p_{g,pore}$ and density $\rho_{g,pore}$ that are much smaller than the corresponding values (p_g and ρ_g) in the open environment. The partial pressure (and partial density) at which condensation occurs in a pore has been given in Section 4.3.2, which is

$$\frac{p_{g,pore}}{p_g} = \exp\left(-\frac{2\gamma v_m}{rRT}\right) \quad \text{and} \quad \frac{\rho_{g,pore}}{\rho_g} = \exp\left(-\frac{2\gamma v_m}{rRT}\right) \tag{5.57}$$

where r is pore radius; $v_m = 18.02 \times 10^{-6}$ m^3/mol M is volume of water; γ (N/m) is surface tension of water vapour, which is a function of temperature. The vapour saturation density ρ_g and saturation vapour pressure p_g in Eqns (5.55) and (5.56) must correspond to the saturation condition in the pores, which requires knowledge of the exact radius of the pores.

Assuming moisture diffusion into the delamination can be modelled as that through a semi-infinite medium with moisture concentration C_d in the bulk and nil moisture concentration in the delamination, and using ideal gas for equation of state and plate theory for vapour pressure and volume of delamination, Alpern (2000) obtained the following three equations with closed-form solutions:

$$
\begin{aligned}
m &= C_d A\sqrt{4/\pi D_r t_r} \\
p_r V &= nRT_r, \\
V &= k_w(q)\frac{p_r A^3}{D_e}
\end{aligned}
\tag{5.58}
$$

where D_r and t_r are diffusivity and duration, respectively, at reflow temperature; D_e is the flexural rigidity of a plate; $k_w(q)$ is a function of the aspect ratio of the delaminated area. Equation (5.58) may be reduced to (Wong, 2005):

$$
p_r = \begin{cases}
\sqrt{384\dfrac{D_e C_d}{a_r^4}\dfrac{R}{M_{H_2O}}T_r\left(\dfrac{D_r t_r}{\pi}\right)^{1/4}} & \text{for clamped edge} \\[2ex]
\sqrt{384\dfrac{D_e C_d}{a_r^4}\dfrac{R}{M_{H_2O}}T_r\dfrac{1+\nu}{7+\nu}\left(\dfrac{D_r t_r}{\pi}\right)^{1/4}} & \text{for hinged edge}
\end{cases}
\tag{5.59}
$$

for axisymmetric delamination area. However, the assumption of water vapour as ideal gas without due consideration for saturation would lead to overestimation of the magnitude of vapour pressure.

The new capability of modelling moisture diffusion in multiphase substances under time-varying temperature presented in this section has facilitated the modelling of vapour pressure in a delamination during solder reflow. Three techniques are presented below.

5.4.2 The upper-bound method

Delamination within a microelectronic assembly during solder reflow occurs along a weak interface, and this is typically very well defined; for example, along the die-attach and die-pad for lead-frame IC packaging and along the die-attach and die-flag for PBGA packaging. The formation of delamination during solder reflow must lead to a depletion of moisture along the solid substances enclosing (or interfacing with) the delamination. The conservation of concentration gives

$$w_{a,2}(T_2, \Delta p) = C_2(\Delta p)/C_{sat,aux}(T_2, p_g) \tag{5.60}$$

The equality of wetness gives

$$w_{\mathrm{a,delam}} = w_{\mathrm{a,int}} = \frac{w_{\mathrm{a,o}}}{1 + \frac{\rho_{\mathrm{v}} V_{\mathrm{delam}}}{\Delta V_{\mathrm{sub}} C_{\mathrm{sat,sub}}}} \tag{5.61}$$

where $C_{\mathrm{sat,sub}}$ is volumetric solute capacity of surrounding substances, ΔV_{sat} is affected volume of the substances; $w_{\mathrm{a,o}}$ and $w_{\mathrm{a,int}}$ are the auxiliary wetness along the substances interfacing with the delamination before and after the formation of delamination; V_{delam} and $w_{\mathrm{a,delam}}$ are the volume and the auxiliary wetness of the delamination. The maximum magnitude of $w_{\mathrm{a,delam}}$ happens when $V_{\mathrm{delam}} \rightarrow 0$, which is the case of zero delamination volume. That is,

$$w_{\mathrm{a,delam}}^{\mathrm{max}} = \lim_{V_{\mathrm{delam}} \rightarrow 0} w_{\mathrm{a,int}} = w_{\mathrm{a,o}} \tag{5.62}$$

This is shown in Figure 5.7. The assumption of nil V_{delam} suggests a microelectronic assembly of high flexural rigidity. In case the wetness along the interface is not uniform, then an average value of $w_{\mathrm{a,int}}$ should be used; that is,

$$w_{\mathrm{a,delam}}^{\mathrm{max}} = \lim_{V_{\mathrm{delam}} \rightarrow 0} \overline{w}_{\mathrm{a,int}} \tag{5.63}$$

Taking the auxiliary condition as the saturated vapour condition, the upper-bound magnitude of vapour pressure in a delamination is given by

$$p_{\mathrm{delam}}^{\mathrm{upper}}(T_{\mathrm{r}}) = p_{\mathrm{g}}(T_{\mathrm{r}}) * \operatorname*{limit}_{V_{\mathrm{delam}} \rightarrow 0} \overline{w}_{\mathrm{a,int}} \tag{5.64}$$

Therefore, the evaluation of the upper-bound magnitude of vapour pressure within a delamination as a function of the solder-reflow temperature requires only moisture-diffusion analysis and prior knowledge of where a delamination will most likely occur within a microelectronic assembly.

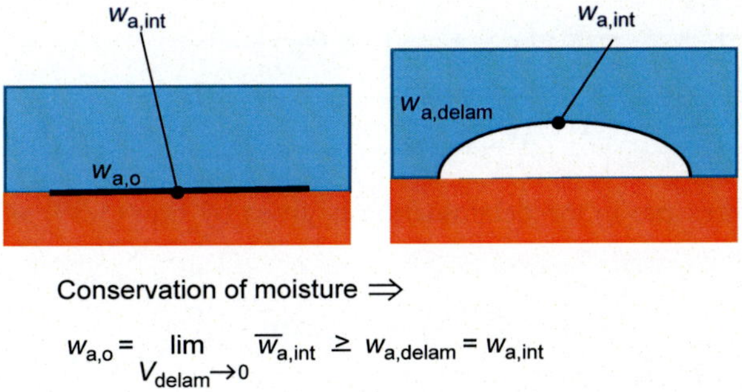

Figure 5.7 Maximum wetness when there is nil volume of delamination.

5.4.3 The multi-physics methods (direct-pressure and delta-pressure)

Assuming 1-D diffusion, the wetness w_m at the substance–delamination interface at the end of moisture conditioning, is given by (Crank, 1956):

$$w_m = 1 - \frac{4}{\pi}\sum_{n=0}^{\infty}\frac{(-1)^n}{2n+1}\exp\left(\frac{-\pi^2(2n+1)^2 D t_m}{4h^2}\right) \tag{5.65}$$

where t_m is duration of moisture conditioning and h is the length of the diffusion path. Consider only $n = 0$,

$$w_m \approx 1 - \frac{4}{\pi}\exp\left(\frac{-\pi^2 D t_m}{4h^2}\right) \tag{5.66}$$

During reflow, the moisture diffuses out of the microelectronic assembly and this is given (consider only $n = 0$) approximately by

$$w_r \approx \frac{4}{\pi}\exp\left(\frac{-\pi^2 \overline{D} t_r}{4h^2}\right) w_m \tag{5.67}$$

where $\overline{D}$ is the average diffusivity for the reflow and is given by

$$\overline{D} = \frac{\int_0^{t_r} D(T(t))\,\mathrm{d}t}{t_r} \tag{5.68}$$

Equation (5.31) gives $w_{a,o} = w_r{*}\%\mathrm{RH}$; substituting this and the structural deformation relation, $V_{delam} = kp_{delam}/h^3$, where k is a constant of proportionality, into Eqn (5.67) gives

$$w_{a,delam} = \frac{\frac{4}{\pi}\exp\left(\frac{-\pi^2\overline{D}t_r}{4h^2}\right)\left[1 - \frac{4}{\pi}\exp\left(\frac{-\pi^2 D t_m}{4h^2}\right)\right]}{1 + \frac{k\rho_v p_{delam}}{\Delta V_{sub} C_{sat,sub} h^3}}\%\mathrm{RH} \tag{5.69}$$

The vapour pressure in the delamination is given by Eqn (5.30) as

$$p_{delam}(T_r) = p_g(T_r)\frac{\frac{4}{\pi}\exp\left(\frac{-\pi^2\overline{D}t_r}{4h^2}\right)\left[1 - \frac{4}{\pi}\exp\left(\frac{-\pi^2 D t_m}{4h^2}\right)\right]}{1 + \frac{k\rho_v(T_r) p_{delam}(T_r)}{\Delta V_{sub} C_{sat,sub}(T_r) h^3}}\%\mathrm{RH} \tag{5.70}$$

Equation (5.70) captures the coupled physics of moisture diffusion and structural deformation. However, the determination of ΔV_{sub} has many uncertainties. In practice, it is more reliable to solve the couple-field problem by numerical analysis. Two procedures – the direct-pressure method and the delta-pressure method – are presented below.

5.4.3.1 The direct-pressure method

The direct-pressure method is shown in Figure 5.8. It involves first evaluating $w_{a,2}^*$ by moisture-diffusion analysis assuming $V = V_1$, hence evaluating p_2^* using Eqn (5.33); this is followed by evaluating the volume $V_2^*(p_2^*, T)$ using structural analysis. The increased volume (from V_1 to V_2^*) reduces $w_{a,2}^*$ to $w_{a,2} = w_{a,2}^* V_1/V_2^*$ and hence, p_2^* to p_2. The volume V_2 is then updated using structural analysis; and the procedures repeat for the next time increment. No iteration is needed in each time increment; however, there is small risk of error propagation due to the use of V_1 in evaluating $w_{a,2}$.

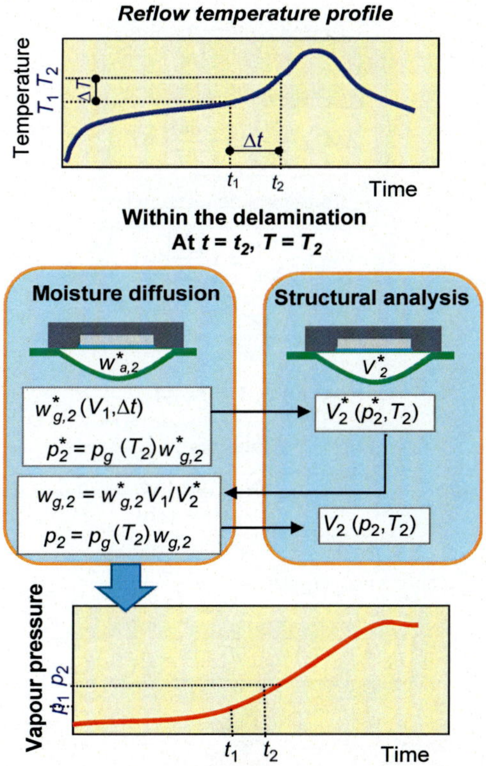

Figure 5.8 The direct-pressure method for evaluating the dynamic vapour pressure in a delamination during solder reflow (Note: w_a and w_g are used interchangeably).

5.4.3.2 The delta-pressure method

The delta-pressure method is illustrated in Figure 5.9. It involves expressing the incremental mass of moisture in the delamination ΔM_1 and the incremental volume of the delamination ΔV_1 in terms of Δp and ΔT through perturbation analysis in moisture diffusion and structural deformation, respectively. The temperature increment ΔT is defined by the reflow profile $T(t)$; thus, the output from the perturbation analysis gives $\Delta M_1(\Delta p)$ and $\Delta V_1(\Delta p)$, respectively. The mass of moisture in the delamination and the volume of delamination at $t = t_2$ are given by $M_2 = M_1 + \Delta M_1(\Delta p)$ and $V_2 = V_1 + \Delta V_1(\Delta p)$, respectively. The concentration and wetness of moisture in the delamination are $C_2(\Delta p) = M_2/V_2$ and $w_{g,2}(T_2,\Delta p) = C_2(\Delta p)/C_{sat}(T_2,p_g)$, respectively. The vapour pressure in the delamination is given by Eqn (5.30) as

$$p_1 + \Delta p = p_g(T_r)w_{a,2}(T_2, \Delta p) \tag{5.71}$$

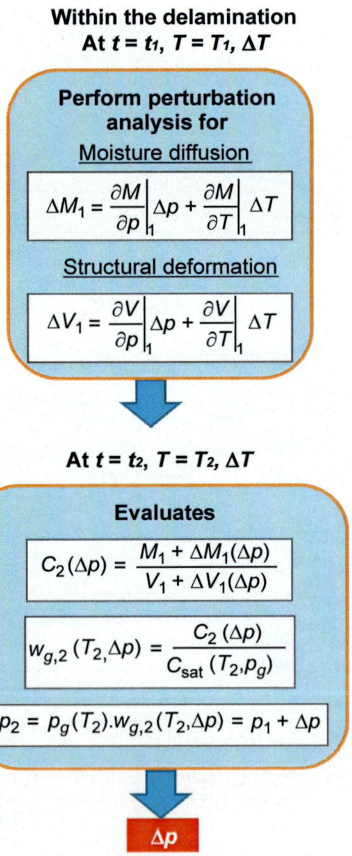

Figure 5.9 The delta-pressure method for evaluating the dynamic vapour pressure in a delamination during solder reflow.

Solving Eqn (5.71) gives Δp, and the procedures are repeated for the next time increment. The only iteration in each time increment is solving the nonlinear equation, Eqn (5.71), which is computationally much simpler than the 'moisture diffusion + structural' analysis of the Kitano's method (see Figure 5.6).

5.4.4 Benchmarking

5.4.4.1 Kitano's small outline J-lead vehicle, static reflow temperature

In view of the diversity of the various techniques proposed, a common vehicle was selected as a benchmark to evaluate the validity and accuracy of these techniques. The small outline J-lead (SOJ) vehicle used by Kitano (1988) was selected for the following reasons: (1) the package dimensions, material properties, moisture preconditioning and reflow temperature are readily available (Kitano, 1988; Tay, 1996); and (2) the magnitude of the vapour pressure has been evaluated rigorously by Kitano (1988) and validated separately by Tay (1996).

The SOJ vehicle was subjected to 85 °C/85% RH moisture conditioning for 34 h prior to constant temperature vapour phase reflow at 215 °C for 40 s. The package dimensions are given in Figure 5.10; the material properties relevant to moisture conditioning information are shown in Table 5.2. Delamination is assumed to occur along the entire die pad to the moulding-compound interface. Using Eqn (5.68), the wetness along the interface has been evaluated as $w_{v,m} = 0.63$. The results of benchmarking are summarised in Table 5.3.

Compared to the benchmark vapour pressure of 0.98 MPa obtained by Kitano (1988) and Tay (1996), the equations of Bhattacharyya (1988) underestimated the magnitude of vapour pressure by 62% while Sawada (1993) overestimated the magnitude of vapour pressure by 78%. The underestimation by Bhattacharyya is attributed to the unusually low value of activation energy assumed: $E_{v,bhatt} = 2740\,R = 22.78$ kJ/mol. The overestimation by Sawada is attributed to the assumption that the IC package is fully saturated with moisture prior to solder reflow. The

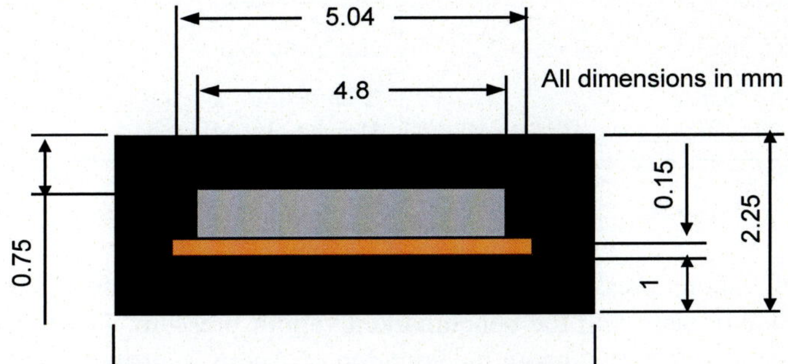

Figure 5.10 Dimensions of SOJ vehicle used in the benchmarking exercise.

Table 5.2 Properties of materials used in the benchmarking exercise

Properties	Data
Henry's coefficient $k_H = k_o \exp\left(\frac{\Delta E}{RT}\right)$	$k_o = 4.96 \times 10^{-4}$ kg/m^3 $\Delta E = 3.87 \times 10^4$ J/mol $E_v = \Delta E$ C_{sat}(85 °C/85% RH) = 10.8 kg/m^3
Diffusivity $D = D_o \exp\left(\frac{-E_d}{RT}\right)$	$D_o = 0.472 \times 10^{-4}$ m^2/s $E_d = 4.84 \times 10^4$ J/mol
Storage modulus	$T_g = 140$ °C E below $T_g = 10$ GPa E above $T_g = 1$ GPa
Moisture conditioning; wetness, concentration along delamination interface	85 °C/85% RH for 34 h; $w_{v,m} = 0.63$, $C_d = 6.8$ kg/m^3
Reflow condition	$T_r = 215$ °C, $t_r = 40$ s $p_g(T_r) = 2.1$ MPa
Delamination	Full-pad delamination

Table 5.3 Results of benchmarking (computation was performed using a single core desktop computer in year 2003)

Method	Vapour pressure
Kitano (1988), Tay (1996)	0.98 MPa (max)
Bhattacharyya Eqn (5.49)	0.37 MPa
Sawada Eqn (5.51)	1.79 MPa
Wong Eqn (5.52)	0.98 MPa
Lim Eqn (5.54)	Not evaluable
Tee Eqn (5.55)	Not evaluable
Xie Eqn (5.56)	Not evaluable
Alpern Eqn (5.59)	Clamped, 9.2 MPa Hinged, 3.9 MPa

magnitude of vapour pressure computed using the wetness method (Wong, 1998) agreed remarkably well with the benchmarked vapour pressure. The high magnitude of vapour pressure evaluated using the equation of Apern is attributed to the assumption of ideal gas behaviour for vapour pressure and hence it has no saturation pressure.

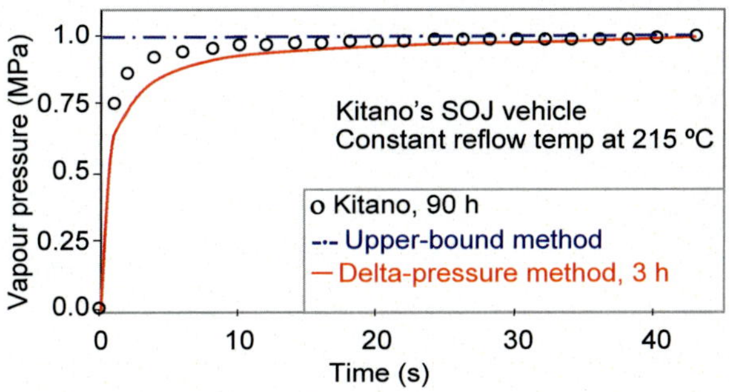

Figure 5.11 Comparisons of the upper-bound, delta-pressure and Kitano's methods using SOJ package, static reflow temperature.

The equation of Lim (1998) could not be evaluated due to lack of clarity in the computation of mass m. The equations of Tee (1999) and Xie (2009) could not be evaluated because of lack of clarity in the volume fraction and the radius of the pores.

The dynamic development of vapour pressure in the delamination for the entire reflow duration computed using the upper-bound method, the delta-pressure method and the Kitano method are displayed in Figure 5.11. In view of the relatively short reflow duration, the wetness along the delamination interface experiences no changes. The upper-bound method therefore yields $p_r = 0.98$ MPa for the entire duration of the reflow. Both Kitano's method and the delta-pressure method have been implemented onto ANSYS, and the analysis is fully automated to provide the same basis for assessing the computation efficiency. A full-pad delamination is assumed for both methods. Because of the formation of the delamination and the depletion of moisture along the edges of the interface, both Kitano and delta-pressure methods show a gradual increase in vapour pressure in the delamination for the first 5 s and approach the upper-bound solution with increasing time. There is only a marginal difference between the three methods after 10 s, and the difference between the magnitudes of vapour pressure obtained is merely 2% after 40 s. The computational times for the Kitano and delta-pressure methods are 90 and 3 h, respectively. The exceptional efficiency of the delta-pressure method derives from the elimination of the iterative analysis of moisture diffusion and structural deformation.

5.4.4.2 *Kitano's SOJ vehicle, dynamic reflow temperature*

In the second analysis, the same SOJ vehicle is subjected to a vapour-phase reflow profile described by (Kitano, 1988):

$$T_r = 22\,°\mathrm{C} + 0.916 t_r + 23.63\sqrt{t_r}, \quad \text{for} \quad 0 \le t_r \le 40\ \mathrm{s} \tag{5.72}$$

Kitano's method is found to be numerically unstable unless extremely fine time-integration steps are used and the analysis is abandoned. By contrast, the delta-pressure method is able to complete the analysis in 3 h. The dynamic development

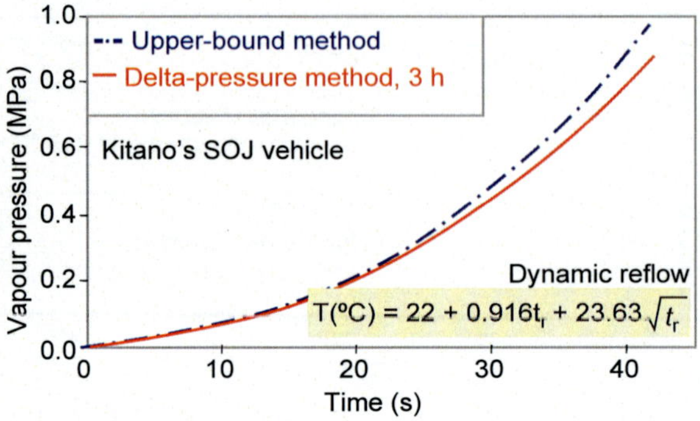

Figure 5.12 Comparisons of the upper-bound and the delta-pressure methods using SOJ package, dynamic reflow temperature.

of vapour pressure in the delamination for the entire reflow duration computed using the upper-bound method and the delta-pressure method are depicted in Figure 5.12. Good agreement is obtained between the upper-bound and the delta-pressure methods. The deviations at the higher temperature are attributed to the increasing deflection of moulding compound at higher temperature due to the reduction in its elastic modulus and the higher thermal expansion mismatch.

5.4.4.3 Low-profile wire bond BGA

In this analysis, a low-profile wire bond BGA (LPBGA) is used as a vehicle to accentuate the effect of the deflection of the moulding compound on the computed vapour pressure. A cross-section of the package is shown in Figure 5.13. The moulding compound is assumed to have the same properties as those of Kitano's SOJ vehicle. The package has been moisture preconditioned at 85°C/85% RH for 168 h prior to linear heating to 210 °C in a thermal mechanical analyser (TGA). The temperature at popcorn is found to be near 210 °C. Delamination occurs between the moulding compound and die passivation interface. The development of vapour pressure computed using the upper-bound and the delta-pressure methods is shown in Figure 5.14. The delta-pressure method successfully completes the analysis in 5 h.

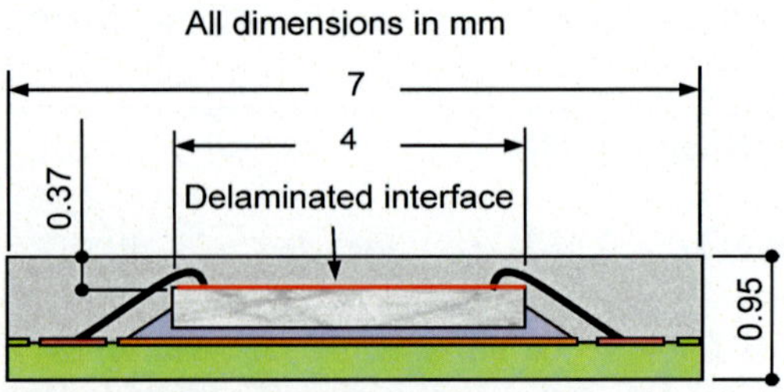

Figure 5.13 The low-profile BGA vehicle.

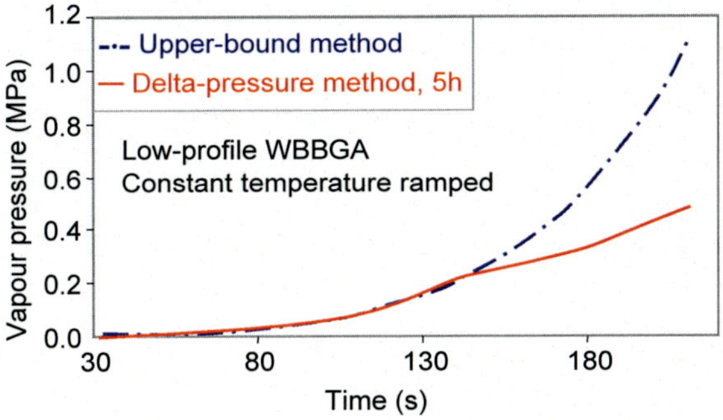

Figure 5.14 Comparisons of the upper-bound and the delta-pressure methods using LPBGA, linear ramping of temperature.

The solution from the upper-bound method agrees well with the delta-pressure method until $T = 140$ °C when the solution from the delta-pressure method deviates abruptly from the upper-bound solution. This is attributed to the fact that a step reduction of the elastic modulus of the moulding compound is defined for the moulding compound at its glass transient temperature ($T_g = 140$ °C), giving rise to a step increase in the delamination volume and hence an abrupt reduction in the computed vapour pressure. The abruptness of the deviation has been accentuated by the thin nature of the moulding compound that encourages large deflections. This example highlights the importance of multi-physics modeling for microelectronic assemblies that are pronged to large delamination volumes.

5.4.5 Robust design analysis

The vapour pressure–modelling techniques offer a power tool for designing and optimising the package construction, packaging materials and reflow profile against 'popcorn' cracking. These are demonstrated through the following parametric analysis of Kitano's SOJ vehicle (i.e. moisture conditioning at 85 °C/85% RH followed by static reflow at 215 °C for 40 s).

5.4.5.1 Elastic modulus of moulding compound

The effect of the elastic modulus of moulding compound on the magnitude of vapour pressure computed using the upper-bound and the delta-pressure methods are shown in Figure 5.15. The upper-bound method predicts a constant vapour pressure of 1 MPa independent of the elastic modulus of the moulding compound. This is because the technique inherently assumes an infinite modulus for the moulding compound. The delta-pressure method suggests a monotonically decreasing vapour pressure with decreasing modulus of the moulding compound due to its increasing deflection. There is a threshold value of elastic modulus ($\sim$0.65 GPa in this case) above which the package becomes so stiff that there is negligible delamination volume. Hence, increasing the modulus leads to little increase in vapour pressure. This parametric analysis

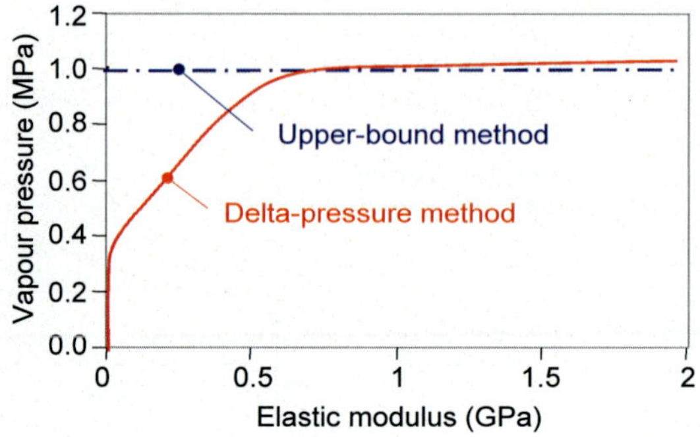

Figure 5.15 Parametric analysis for the modulus of moulding compound.

suggests that moulding compound of low modulus that promotes large deflection volume is beneficial to popcorn resistance; however, to be effective, the selected modulus must be below a threshold value, which is a function of the die pad and the thickness of the molding underneath.

5.4.5.2 *Thickness of moulding compound*

The effect of the thickness of moulding compound underneath the die pad on the magnitude of vapour pressure computed using the upper-bound and the delta-pressure methods is shown in Figure 5.16. The upper-bound method predicts an increasing vapour pressure with decreasing thickness of the moulding compound. This is attributed to the increasing wetness along the delamination during moisture conditioning (see Eqn (5.66)). However, the delta-pressure method yields a decreasing vapour pressure with decreasing moulding compound thickness. This suggests that the effect of increasing delamination volume during solder reflow outweighs the effect of increased wetness along the interface during moisture conditioning. While it appears that a thinner package is desirable, a more complete analysis should include the

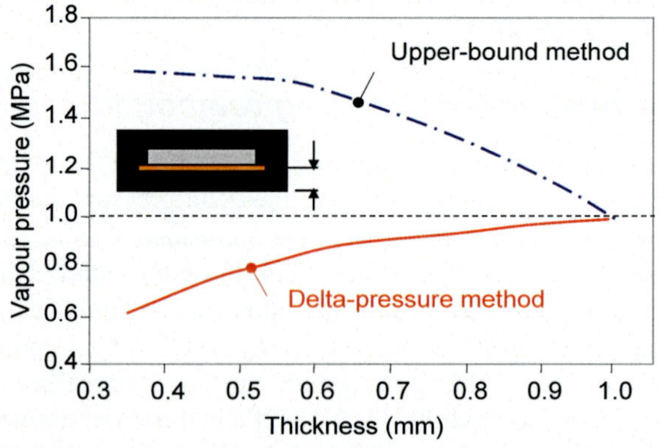

Figure 5.16 Parametric analysis for the thickness of moulding compound below die pad.

reduction in interfacial strength due to the higher wetness and the reduced fracture resistance associated with the thinner package.

5.4.5.3 Increased reflow temperature

The effect of increasing the peak reflow temperature on the package with high flexural stiffness is captured in the Arrhenius equation of saturation pressure for water vapour:

$$p_g = p_o \exp\left(\frac{-E_v}{RT}\right) \tag{4.1bis}$$

where $R = 8.314$ J/K mol and ($p_o = 30.05$ GPa, $E_v = 38.83$ kJ/mol) for temperature between 100 and 250 °C. The nonlinear increase in vapour pressure with temperature leads to rapid increase in vapour pressure at higher temperature. For example, the saturation vapour pressure doubles as the temperature is raised from 220 °C to 260 °C. The upper-bound vapour pressure in a delamination increases proportionately with increasing saturation pressure since:

$$p(T_r) = p_g(T_r) \cdot w_g \tag{5.26}$$

For thin assemblies, the increase in vapour pressure with increasing reflow temperature will be more moderate because of the increasing delamination volume due to the reducing modulus of the moulding compound and increasing mismatched thermal expansion.

References

Alpern, P. (2000). A simple model for the mode I popcorn effect for IC packages. *Microelectronics Reliability, 40*, 1503–1508.

Baierlein, R. (2003). *Thermal physics*. Cambridge University Press.

Bhattacharyya, B. (1988). Moisture absorption and mechanical performance of surface mountable plastic packages. In *38th electronic component and technology conference* (pp. 49–58).

Crank, J. (1956). *Mathematic of diffusion*. Oxford University Press.

Fick, A. (1855). *Edinburg and Dublin Philosophical Magazine and Journal of Science*, 30–39.

Fourier, J. (1822). *Theorie analytique de la chaleur*. Firmin-Didot Pere et Fils.

Galloway, J. M. (1996). Moisture absorption and desorption predictions for plastic ball grid array packages. In *Intersociety conference on thermal phenomena* (pp. 180–186).

Gibbs, J. (1948). *The collected works of J. Willard Gibbs*. New Haven: Yale University Press.

Henry, W. (1803). Experiments on the quantity of gases absorbed by water, at different temperatures, and under different pressures. *Philosophical Transactions on the Royal Society, 93*, 29–274.

Jang, C. P. (2008). Advanced thermal-moisture analogy scheme for anisothermal moisture diffusion problem. *ASME Transactions on Journal of Electronic Packaging, 130*, 011004.

Job, G. H. (2006). Chemical potential – a quantity in search of recognition. *European Journal of Physics, 27*, 353–371.

Kitano, M. N. (1988). Analysis of package cracking during reflow soldering process. In *26th international reliability physics symposium* (pp. 90–95).

Kittel, C. K. (1980). *Thermal physics* (2nd ed.). W. H. Freeman.

Lim, J. (1998). Vapour pressure analysis of popcorn cracking in plastic IC packages by fracture mechanics. In *Electronics packaging technology conference* (pp. 36–42).

Sawada, K. (1993). Package deformation and cracking mechanism due to reflow soldering. In *Japan international electronics manufacturing technology symposium* (pp. 295–298).

Smith, F. (2007). Avoid common pitfalls when using Henry's law. *Chemical Engineering Progress, 103*, 33–39.

Tay, A. L. (1996). Moisture diffusion and heat transfer in plastic IC packages. *IEEE Transactions on CPMT-Part A, 19*, 186–193.

Tee, T. (1999). Modeling of whole field vapour pressure during reflow for flip chip BGA and wire bond PBGA packages. In *1st electronic materials and packaging workshop* (pp. 38–45).

TLV. (n.d.). TLV: A steam table specialist. Retrieved from Calculator: Saturated Steam Table: http://www.tlv.com/global/TI/calculator/steam-table-temperature.html

Wong, E. (2002). Comprehensive treatment of moisture induced failure – recent advances. *IEEE Transactions on Electronic Packaging Manufacturing, 25*, 223–230.

Wong, E. K. (2002). Advanced moisture diffusion modeling & characterisation for electronic packaging. In *52nd electronic components & technology conference* (pp. 1297–1303).

Wong, E. K.-M.-W. (2005). Advances in vapour pressure modeling for electronic packaging. *IEEE Transactions on Advanced Packaging, 29*, 751–759.

Wong, E. T. (1998). Moisture diffusion and vapour pressure modeling for electronic packaging. In *48th electronic components & technology conference, Seattle* (pp. 1372–1378).

Xie, B. F. (2009). Direct concentration approach of moisture diffusion and whole-field vapour pressure modeling for reflow process – part I: theory and numerical implementation. *ASME Transactions on Journal of Electronic Packaging, 131*, 031010.

Abaqus theory manual.(2014). Retrieved from http://abaqus.ethz.ch:2080/v6.11/pdf_books/THEORY.pdf.

Part Three

Robust design against drop impact

The physics of failure of portable electronic devices in drop impact

6

6.1 Product drop testing

The drop testing of portable electronic devices (PEDs) is routinely performed by original equipment manufacturers; however, these are rarely reported. Extensive drop testing of commercial PEDs has been carried out (Lim, 2003, 2004; Seah, 2002, 2004) under a collaborative project between the Institute of Microelectronics, National University of Singapore and the University of Cambridge using the facilities at the impact laboratory of National University of Singapore. The objective of the testing is to gather information on the types and range of loading that can be experienced by the printed circuit board (PCB) assemblies of commercial products. The information provides a basis for designing a subsystem test that can reproduce as closely as possible the conditions that components and interconnects are subjected to in an actual mobile device.

6.1.1 The mechanics of drop impact

6.1.1.1 Drop impact of a simple subject

The severity of impact of a simple subject against a stationery target may be measured using a number of parameters: impulse, impact force, acceleration and strain. Impulse measures the change in momentum of the object upon impact: $I = m\Delta v$. This change in momentum is a function of the mass of the subject, the velocity of the subject prior to impact and the rebound velocity of the subject after impact. The situation of maximum impulse occurs when the rebound velocity equals the impact velocity – the situation of perfect transformation of kinetic energy to strain energy and back to kinetic energy when no energy is channelled to fracturing or permanently deforming the subject. The magnitude of impulse is therefore not a good indication of the risk of damage to the subject.

Experimentally, the impulse of an impact can be conveniently measured using a force transducer (load cell) giving $I = \int_0^t F\mathrm{d}t$. The impact force, F, measures the intensity of mechanical interaction between the subject and the target. A large impact force could lead to excessive stress in the region of the subject that is in contact with the ground leading to local failure. For the same magnitude of impulse, the magnitude of impact force could be lowered by increasing the impact duration, which could be achieved by reducing the rigidity of the subject either globally or, more effectively, at the region of contact (e.g. by adding rubber padding to the corners of the product).

The impact force, F, induces contact stresses in the subject local to the contact domain. Local failure could occur if the magnitude of contact stresses exceeds the

Robust Design of Microelectronics Assemblies Against Mechanical Shock, Temperature and Moisture
http://dx.doi.org/10.1016/B978-1-84569-528-6.00006-X

strength of the material that the subject is made of. The impact force also acts on the subject to decelerate it. The magnitude of deceleration (or acceleration) can be measured using an accelerometer. Acceleration leads to inertial force. If the subject is perfectly rigid (and assuming the subject experiences no rotation) such that the entire body of the subject is assumed to experience uniform acceleration, a, then the magnitude of the inertial force at distance h from the free end of the subject is given by

$$f_h = \int_0^h \rho_x A_x a \, dx \tag{6.1}$$

where ρ_x and A_x are the density and cross-sectional area, respectively, of the segment at distance x from the free end. The magnitude of the inertial force increases from nil at the free end of the subject to F at the impacting end of the subject. The inertial force induces direct compressive stress in the cross-section of the subject given by $\sigma_h = f_h / A_h$, assuming the compress stress is uniformly distributed throughout the cross-section. Failure could occur if the direct compressive stress exceeds the material strength of the subject. It is clear that the compressive stress on the subject depends as much on the construction of the object as the magnitude of acceleration. Hence, while acceleration is most frequently cited as an indication of the severity of impact, it is not necessarily an absolute indication of the risk of damage to the subject concerned.

If the subject is elastic, then finite time is needed for the stress wave (and the accompanying acceleration) to propagate from the point of contact to the free end of the device. The resultant inertial force could also lead to deformation of the subject in the form of bending, torsion and even buckling. Strain measurement provides a reliable indication of the magnitude of these deformations and the risk of damage.

6.1.1.2 *Drop impact of portable electronic devices*

Consider a PED that is impacted against a rigid target at a horizontal orientation as shown in Figure 6.1. The impact force results in rapid deceleration of the device housing, including those points and edges of the PCB assembly that are attached to the device housing. Being relatively flexible, the propagation speed of deceleration wave in the PCB assembly is substantially slower than that in the device housing. As a result, the PCB assembly continues to move downward when its edges have

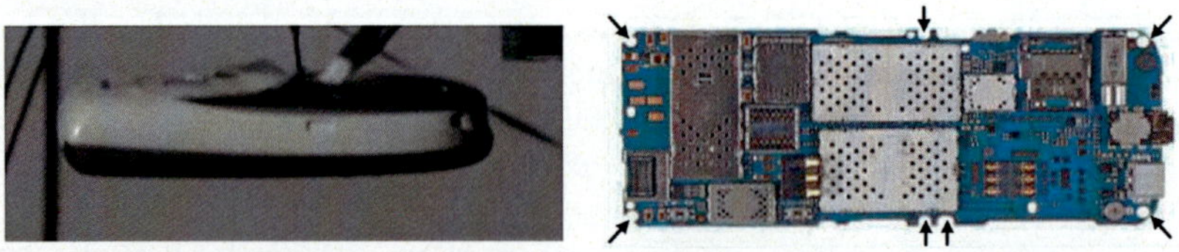

Figure 6.1 Horizontal impact of a PED and the attached sites of PCB assembly to the device housing.

come to a stop. This results in bending of the PCB assembly. Eventually, the entire PCB assembly, including the attached electronic components, would come to a stop at the maximum bending of the PCB assembly. The solder joints that interconnect the IC components to the PCB could experience three sources of loading: (1) the inertial load of the IC component that acts to pull the IC component away from the PCB; (2) the differential bending of the IC component and the PCB that has to be accommodated by the solder joints; and (3) the restrictive spacing between the PCB assembly and the device housing that promotes inter-knocking, which could induce secondary bending of the PCB.

6.1.2 Selected candidates and experimental setup

Table 6.1 tabulates the specifications for the PEDs characterised in the survey drop test. The test vehicles cover mobile phones and personal digital assistants (PDAs).

The three modes of loading on the solder joints mentioned in the last section are best measured using miniature accelerometers and strain gauges mounted on the PCB assembly. In-plane and out-of-plane (with respect to the PCB) accelerometers were mounted on selected IC packages to measure the inertial loading on solder joints. Strain gauges were mounted at the edges of ball grid array (BGA) processor packages to measure the magnitude of flexural deformation of the PCB around the IC packages. Two to four strain gauges were mounted on the board assembly of each mobile product. The directions of measurement are shown in Figure 6.2. Figure 6.3(a) shows a strain gauge mounted next to a BGA component and a miniature accelerometer mounted on the package, and Figure 6.3(b) shows sensors with thin measurement leads that need to be carefully routed out of the product housing during the test set-up. Results from the survey may be taken as estimates of surface strains in the central region (where most BGA components are located) of a 1-mm-thick mobile device board assembly. It is important to note that the measured strain is a summation of two components of strain: the membrane strain component, which is uniform over the thickness of the PCB; and the bending strain component, which varies linearly with thickness with nil magnitude at its neutral plane and maximum magnitudes on its outer surfaces.

Each product was subjected to drops from a height of 1 m onto a load cell that measures the impact force and triggers the data capture. The product is impacted on its faces and edges at orientations of 0° and 45° from the impact surface. Figure 6.4(a) illustrates these orientations. The orientation control is achieved using a specially designed gripper, shown in Figure 6.4(b), that maintains the product's orientation through much of the free fall and releases the product just before impact, thus ensuring that impact occurs in the desired orientation.

6.1.3 Analysis of drop test data

Figures 6.5 and 6.6 show examples of peak magnitudes of strain and acceleration, respectively, taken on the PCB assembly of four PEDs in eight impact orientations.

Table 6.1 List of portable electronic devices that have been drop tested

Model		Dimensions L × W × T (mm)/vol (cm^3)	Mass (g)	Announced date
Nokia 8250	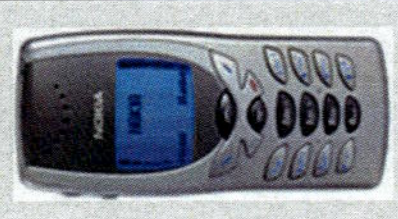	102.5 × 45 × 19/73	69	January 2000
Nokia 8310		97 × 43 × 19/66	84	January 2001
Nokia 3650	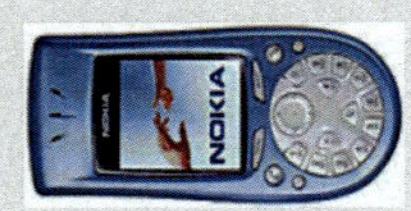	130 × 57 × 26/139	130	1st Q 2003
Sony Ericsson T68i		101 × 48 × 19.5	84	April 2002
Panasonic GD88		97 × 49 × 23	103	November 2002
Siemens S57		101 × 46 × 30	85	December 2002

HP Compaq iPaq H3850 PDA		133 × 84 × 16	190	November 2001
Palm M105	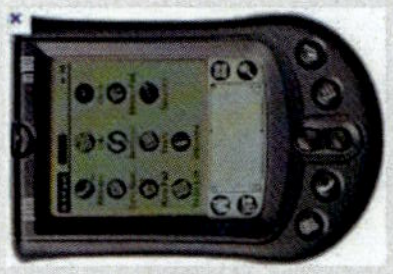	118 × 79 × 18	125	March 2001
Palm M500	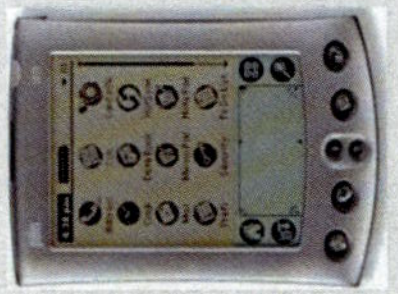	114 × 79 × 12	118	March 2001
Nokia 3110 classic		108.5 × 45.7 × 15.6/72	87	February 2007
Nokia 2630		105 × 45 × 9.9/45	66	May 2007
Nokia 5610		98.5 × 48.5 × 17/75	111	Augest 2007

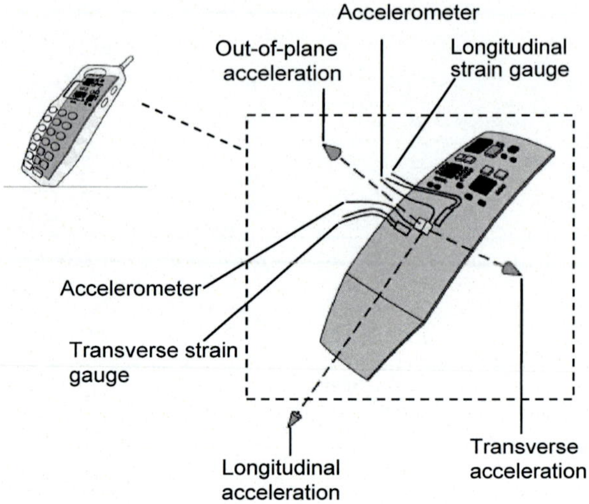

Figure 6.2 Sketch of board assembly showing measurement directions of sensors.

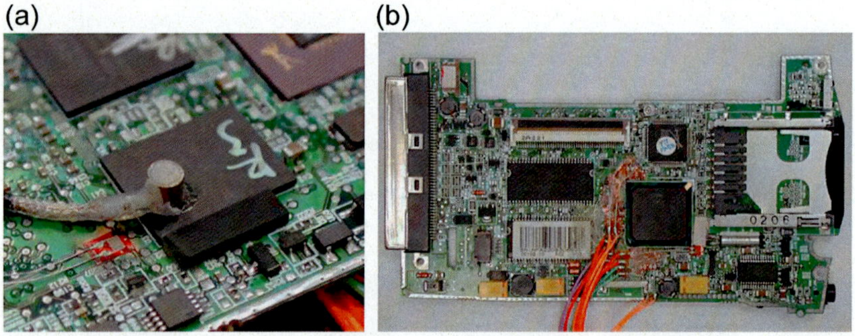

Figure 6.3 Sensor mounting for product test. (a) Strain gauge and accelerometer mounted on board assembly. (b) Strain gauges around package, with lead wires connected to gauges.

The peak impact force and duration and the peak in-plane acceleration, peak out-of-plane acceleration, peak longitudinal strain and peak transverse strain taken on the PCB assembly of a selection of PEDs are analysed and the following observations were made.

6.1.3.1 *Impact force and impact duration*

For all devices, the highest magnitude of impact force is observed for the vertical impact orientation; this is followed by the horizontal impact orientation. The result is not surprising as impacting in these orientations tends to be more intense as reflected in the shorter impact duration. By contrast, impacting at angled orientation tends to induce angular rotation of the device with residual momentum leading to secondary impact. The magnitude of momentum (and hence, impact force) in individual impact is therefore lowered.

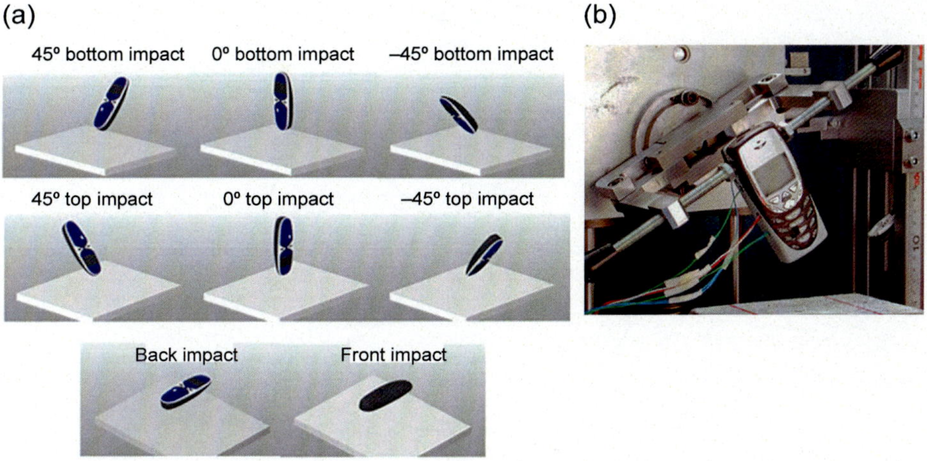

Figure 6.4 Product drop test (a) orientations, and (b) gripper that allows orientation control.

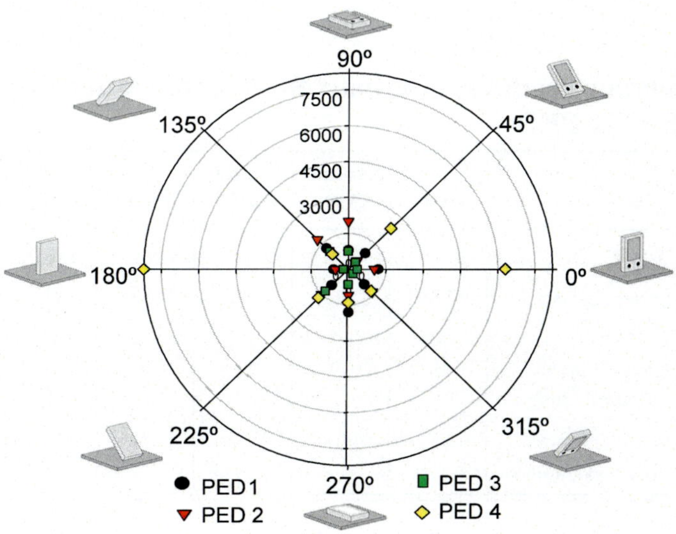

Figure 6.5 Peak magnitudes of PCB strain (in microstrains) in four PEDs for various impact orientations.

The average readings of impact force, impact duration and their products over eight orientations for the six devices are tabulated in Table 6.2. Notice that m105 has the highest magnitude when it has only the second highest weight after H350. The reason for this becomes clear when it is noted that the impact duration of H350 is almost twice that of m105. Figure 6.7 shows good correlation between the average nominal impulse and the mass of the devices.

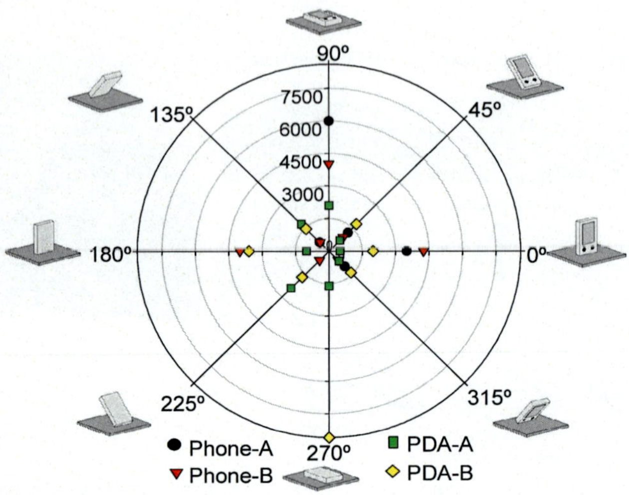

Figure 6.6 Peak magnitudes of acceleration (in g's) in four PEDs for various impact orientations.

Table 6.2 Average readings of impact force, impact duration and impulses for six PEDs

		T68i	Nokia 8310	Nokia 8250	m500	m105	H3850
	Mass (g)	85	90	90	106	130	190
(a)	Average impact force (N)	627	879	1171	1350	1653	1173
(b)	Average impact duration (ms)	1.49	1.09	1.11	1.08	1.11	2.04
(a)×(b)	Average nominal impulse (N.ms)	936	956	1305	1458	1838	2389

6.1.3.2 Acceleration

The peak out-of-plane acceleration of the PCB ranges from 580 to 8600 g for five of the six devices. The average readings of peak in-plane accelerations for the vertical, horizontal and angled orientations are 2200, 3100 and 1200 g, respectively. The high reading of out-of-plane acceleration of the PCB at 2200 g in the vertical impact orientation is perplexing. There is no evidence of lateral rigid-body acceleration of the devices – the devices have virtually nil lateral velocity before impact and negligible lateral velocity after impact. This leaves only one possible scenario: that the PCB

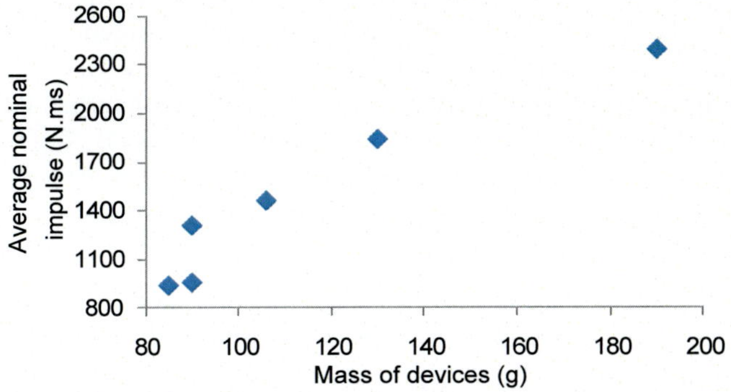

Figure 6.7 Correlation between nominal impulse and mass of devices.

assembly has undergone an out-of-plane bending deformation, which could be together with the entire device or independent of the housing of the device. Assuming the bending deformation can be given by $w = A \sin \omega t$, the magnitude of acceleration is then $\omega^2 A$. For a bending frequency of 600 Hz, it takes only 1.5 mm of bending of PCB to arrive at an acceleration of 2200 g.

The peak longitudinal acceleration of the PCB ranges from 510 to 5700 g for the six devices. The average readings of peak in-plane accelerations for the vertical, horizontal and angled orientations are 2800, 2900 and 1200 g, respectively. The high reading of longitudinal acceleration of the PCB at 2900 g in the horizontal impact orientation is even more perplexing, and there is no good explanation for this.

The average out-of-plane and longitudinal accelerations over eight orientations for the devices are tabulated in Table 6.3. These are compared with the nominal

Table 6.3 Average readings of nominal and measured accelerations for six PEDs

		T68i	Nokia 8310	Nokia 8250	m500	m105	H3850
(a)	Mass (g)	85	90	90	106	130	190
(b)	Nominal acceleration (g)	753	996	1328	1129	1297	630
(c)	Average out-of-plane acceleration (g)	1754	1396	1700	2929	1338	
(d)	Average longitudinal acceleration (g)	1408	2470	1397	2123	1713	2908
(c)/(b)		2.3	1.4	1.3	2.6	1.0	
(d)/(b)		1.9	2.5	1.1	1.9	1.3	4.6

accelerations of the devices, which are obtained by dividing the average measured impact force by the respective mass of the devices. The measured acceleration ranges between 1 and 4.3 times that of the nominal acceleration for the six devices. The larger magnitude of the measured accelerations is a result of the resonance of the PCB assembly and possibly also the accelerometer. It highlights the frequently futile experience of trying to interpret the state of the PCB assembly from the measured acceleration.

6.1.3.3 Strains

The peak longitudinal strain of the PCB among the six devices ranges from 150 to 8200 microstrain, while the peak transverse strain of the PCB ranges from 150 microstrain to 3150 microstrain. The average readings of peak longitudinal strain for five devices (excluding m500) and peak transverse strain for six devices for the vertical, horizontal and angled impact orientations are (810×10^{-6}, 690×10^{-6}), (1140×10^{-6}, 1910×10^{-6}) and (940×10^{-6}, 590×10^{-6}), respectively, where the first and second readings in the brackets show longitudinal and transverse strains.

The dominance of bending strain over membrane strain is obvious when noting that the magnitude of PCB strain registered in horizontal impact is two to three times that registered in vertical impact. Indeed, it is likely that bending strain dominates even in the vertical impact orientation as devices are frequently observed to undergo bending even in vertical impact. The transverse strain registered during vertical bending is a result of anticlastic curvature due to the Poisson's ratio effect.

The exact magnitudes of membrane and bending strains can be extracted by mounting two strain gauges on the surfaces of PCB directly opposite each other. If the strains on the two surfaces are given by p and q, respectively, then the membrane strain is given by $(p + q)/2$ and the bending strain is given by $|p - q|/2$.

The average longitudinal and transverse strains over eight orientations for the devices (except m500) are shown in Table 6.4; the two strains are plotted against each other in Figure 6.8. A very good linear correlation between the two strains is observed. The fitting gradient of near one is interesting considering that all the impact orientations (less the horizontal impact) landed on the longitudinal edges of the devices and none landed on the transverse of the devices. The only logical explanation is that the measured PCB strain consists of predominantly bending strain, whose orientation (be it longitudinal or transverse) is less sensitive to the specific edges of the device in impact.

Table 6.4 **Average readings of peak PCB strains for six PEDs**

	T68i	Nokia 8310	Nokia 8250	m500	m105	H3850
Average longitudinal strain ($\times 10^{-6}$)	1220	1056	433	3171	620	1375
Average transverse strain ($\times 10^{-6}$)	1118	860	373	1164	531	1345

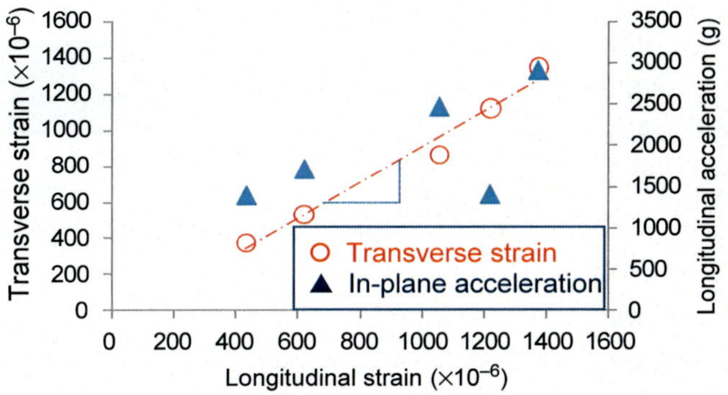

Figure 6.8 Correlation between longitudinal strain, transverse strain and longitudinal acceleration.

There is also good correlation between the longitudinal strain and the longitudinal acceleration for four of the devices, which is expected in view of the fact that the bending of the PCB is inertially induced.

6.1.4 Detailed analysis of the dynamic responses of PCB assemblies

The dynamic behaviour of the board assembly within a mobile device subjected to drop impact is very complex due to the effects of the product housing, PCB mountings and other interior components. Figure 6.9 shows sensor data (impact force, longitudinal strain and out-of-plane accelerations on PCB assembly) captured for a mobile phone dropped in a 'back-impact' orientation. Perfect horizontal impact is near impossible as indicated by the two impulses of impact ① & ⑤. Given the lower magnitude and the longer duration of its initial impact pulse, the phone has most likely experienced an initial impact on one of its edges followed by a second impact on the opposite edge/face. Upon the first impact, marked as ① in Figure 6.9(b), the housing decelerates rapidly at the point of contact, but the signal only travels to the points of support of the PCB at the housing after a fraction of a millisecond ②. The centre region of the PCB begins to decelerate ③ while it continues to move downward, resulting in a positive fibre strain on the bottom surface of the PCB ④. The first impact on the edge results in rotation of the phone – part of which is in mid air – that has led to impact by the opposite edge, which can also be a surface, of the phone with a larger impulse ⑤. The rotation and the resulting second impact are transparent to the PCB, which completes its downward elastic deflection and begins on its upward elastic deflection – negative strain on the bottom surface of the PCB ⑥. However, it appears that second impact could have happened in the region of the phone that is more closely packed, and as a result, the impact force seems to have been transferred directly to the centre of the PCB, giving rise to the high magnitude of upward acceleration of the PCB ⑦and a corresponding 'boost' in its upward elastic deflection ⑧. The sharper peak of the second impact corroborates the theory of a harder impact – a result of more rigid contact

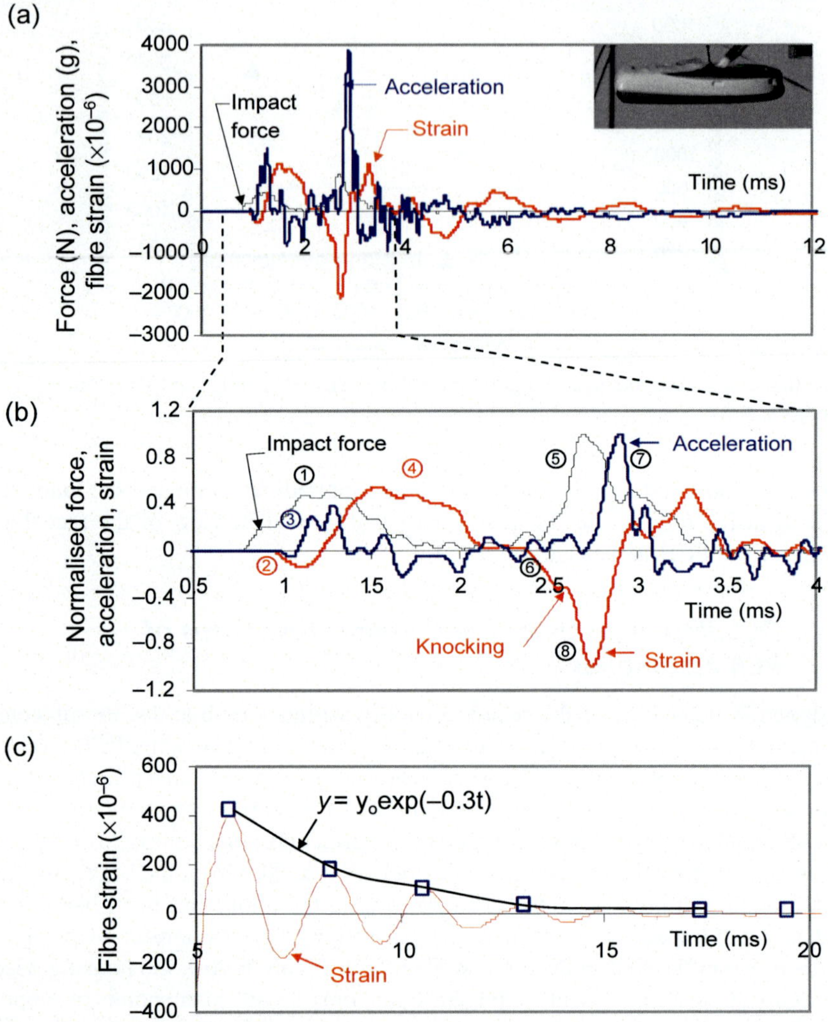

Figure 6.9 The impact response of a typical mobile phone experiencing horizontal drop impact between the duration 0–12 ms; (b) the normalised response data between the duration 0.5–4 ms; and (c) the damping characteristic of the fundamental frequency of the PCB assembly.

due to the more closely packed components within the phone at the second impact surface.

The response of the PCB assembly consists of the fundamental frequency and other higher frequencies. The higher frequency components damp off rather quickly as evident in Figure 6.9(a). The frequency and the damping characteristic of the PCB at its fundamental mode may be estimated from the free vibration of the PCB after filtering off the higher frequency components as shown in Figure 6.9(c). The fundamental frequency of the PCB has been evaluated as 440 Hz. Its damping ratio is estimated from the decay constant, which is $\lambda = 0.3\ \text{ms}^{-1} = 300\ \text{s}^{-1}$. The damping ratio is given by

$$\zeta = \frac{\lambda}{\omega_n} \approx 0.11 \tag{6.2}$$

The use of fibre strain instead of deflection of the PCB assumes linearity between these two properties. Assuming simple beam theory, the fibre strain on the PCB is given by

$$\varepsilon = -\frac{h}{2}\frac{\mathrm{d}^2 z}{\mathrm{d}x^2} \tag{6.3}$$

where h is thickness of PCB. The deflection may be expressed by $z = z_{\text{o}} f(\text{x})$; hence, $z'' = z_{\text{o}} f''(\text{x})$, which implies $\varepsilon \propto z_{\text{o}}$.

The clattering impact is evident in a phone impacting in a 45° orientation as seen in the high-speed video sequence of Figure 6.10(a). The fibre strains on the PCB are measured at four locations shown in Figure 6.10(b). Two patterns ① & ② of transient strain oscillations can be observed interrupted by a short pause. The first is generated by the initial impact at one end of the phone that causes it to rebound, while the second arises from a *clattering impact* at the other end after a brief rotation of the phone through the air. The four strains measured at widely spaced locations are in phase, making it likely that the flexural vibration occurs in a half-sine fundamental mode shape, having a frequency of 330 Hz. The PCB strains damp off rapidly, and each impact may be approximated to produce a single loading cycle.

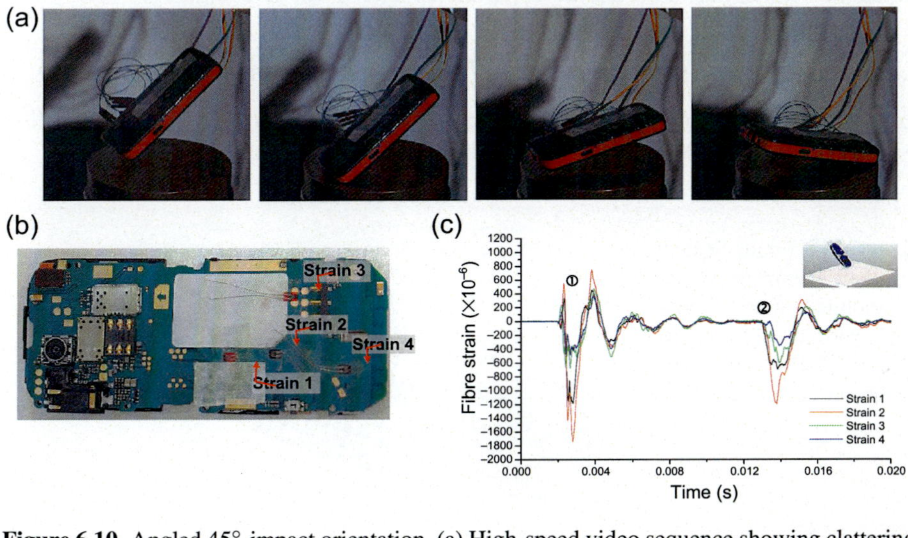

Figure 6.10 Angled 45°-impact orientation. (a) High-speed video sequence showing clattering-impact frequency of the PCB assembly; (b) strain gauge locations; and (c) strain measurements.

For angled orientations, the fundamental vibration mode generally dominates the strain response. For horizontal orientations, sharp-bending spikes of high amplitude and frequency occasionally occur, such as seen in the impact response of phone in Figure 6.9. These spikes are attributed to knocking of the PCB by the housing (directly or through intermediate devices) in the region other than those designed to be supported by the housing. Such spikes are deemed highly damaging not only because of the resulting high magnitudes of acceleration and bending strain but also because the high frequency (high strain rate) event tends to promote brittle fracture in the intermetallic compound. It is advisable for engineers of mobile devices to design with the aim of preventing these spikes.

It should be noted that the results discussed in this section are findings from a rather limited product drop-impact survey – compared to the wide diversity of mobile electronics available today. Selecting different products or using a larger number of sensors placed at more varied locations may well result in different survey results. However, the study is useful in providing a basic understanding of what happens under field conditions.

6.2 The physics of failure

Section 6.1 has identified two physics principles that could be responsible for the failure of the solder joints that interconnect the IC components to the PCB: (1) the inertial load of the IC component, $F = ma$, that acts to tear it from the PCB; and (2) the bending of the PCB assembly due to inertia and the resultant differential bending of the IC component and the PCB lead to severe deformation of the solder joints at the peripheral of the IC component as shown in Figure 6.11; secondary bending may be induced by knocking of electronic components against the housing.

6.2.1 The dominant physics of failure – experimental evidence

Experiments have been performed to ascertain the dominant physics that is responsible for the failure of the solder joints in the drop impact of PEDs.

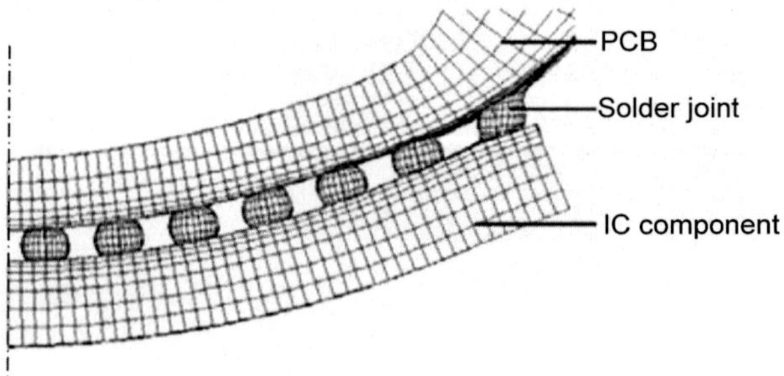

Figure 6.11 Finite element plot (half-symmetry, symmetry plane on left) with magnified displacements showing the effect of PCB bending on solder joint deformation.

6.2.1.1 Inertial load of IC component

Two experiments are performed to investigate the effects of inertial loading of IC component alone by removing as much as possible the effects of PCB bending. Figure 6.12 shows the set-up of the first experiment. A PCB assembly is sandwiched between two thick aluminum plates that prevent its bending. An opening in the center of the bottom aluminum plate exposes the IC component to allow its inertial loading acting on the solder joints attaching the IC component to the PCB. A peak acceleration of 4000 g is introduced by a shock table and the same magnitude of acceleration is measured on the IC component, which is twice what it would be if the PCB were allowed to flex freely.

Figure 6.13 shows the set-up of the second experiment. The board assembly is cut to a size slightly larger than the IC component and is attached to the underside of

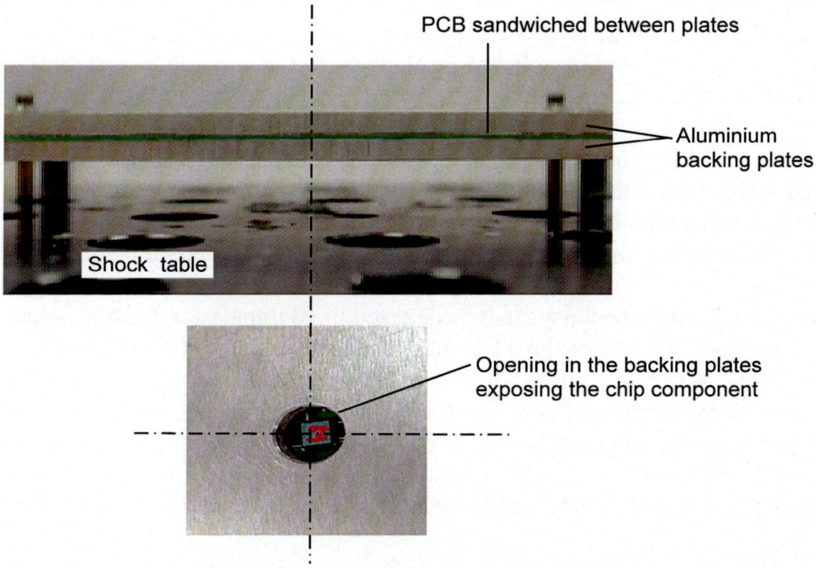

Figure 6.12 First experiment to investigate the inertia of IC component.

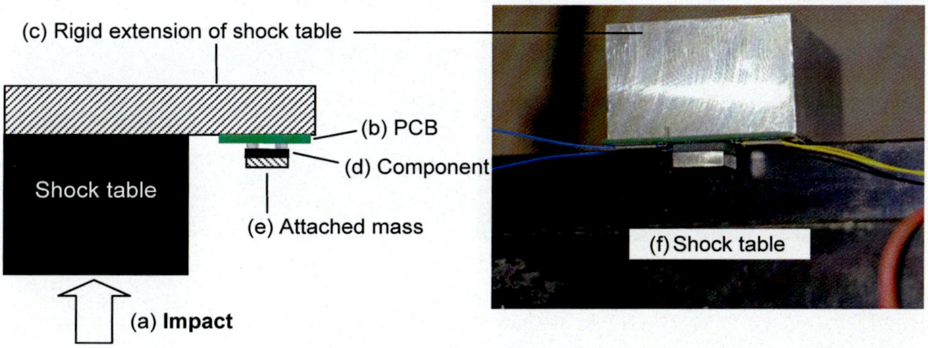

Figure 6.13 Second experiment to investigate the inertia of IC component.

a rigid extension that overhangs the shock table. An inertial mass made of steel of thickness 3 mm – equivalent to six times the inertia of the IC component – is bonded to the IC component. The overhanging extension is rigid enough to prevent flexing of itself or the PCB to ensure that the loading on the solder joints is purely inertial. An acceleration shock pulse of amplitude 3000 g and duration 0.5 ms is then applied to the shock table, and the same acceleration is measured on the PCB.

The electrical connectivity of the solder joints is monitored *in situ* through a daisy chain connection. In both experiments, the daisy chain remains intact after more than 30 drops, which is much longer than a freely flexing board would have survived. These two experiments have practically ruled out the inertia of IC component as the physics of failure.

6.2.1.2 Bending of PCB

The effect of PCB bending on joint failure is most easily observed by monitoring concurrently the bending strains in the board and the electrical connectivity of the joints, in a standard laboratory board-level shock test. Here, a PCB assembly with a single central BGA component is supported at four corners (Figure 6.14) on a board-level shock test fixture and subjected to half-sine mechanical shocks of amplitude 1500 g and duration 0.5 ms. The electrical connectivity (resistance) of the daisy-chained solder ball joints is monitored *in situ* during the test using a high-speed data logger. A strain gauge mounted adjacent to the component measures the local strains near the component. An accelerometer measures the local acceleration near the board centre.

An open circuit in the daisy chain is detected in 10 impacts, indicating solder joint failure. Figure 6.15 shows the PCB strain and electrical resistance graphs for an impact shock performed immediately after this first detection of failure. The PCB strain waveform consists of two superposed vibration mode components, namely a fundamental mode component with a half-period of approximately 2.5 ms, and a higher bending mode component with a period of 0.8 ms. At time = 2 ms, before the impact, the already-failed daisy chain was electrically closed (indicated by low

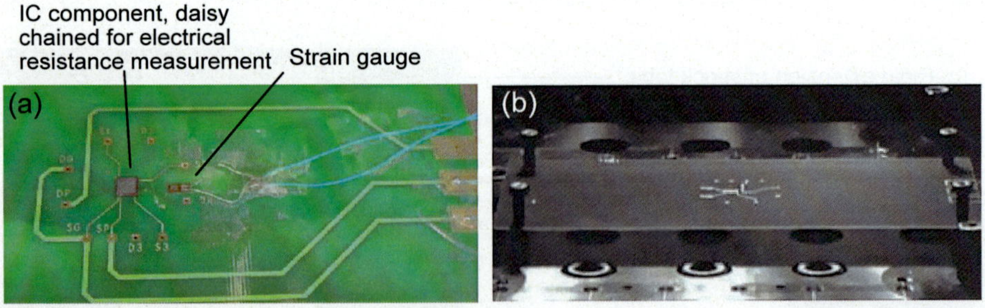

Figure 6.14 Board-level drop test set-up. (a) Strain gauge mounted near component. (b) Board mounted on shock test fixture, with component facing downward.

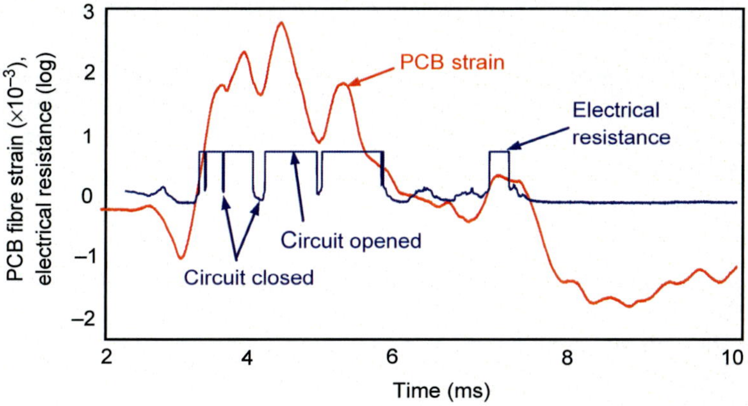

Figure 6.15 PCB strain and electrical resistance waveforms from a board-level shock test.

resistance) due to contact between the cracked surfaces maintaining the electrical connectivity. The impact shock sets the PCB into dynamic oscillations, and the daisy chain reopens when the PCB develops a bending strain of roughly 1000 microstrains at an approximate time of 3.2 ms. Subsequently, there is intermittent opening and closing of the daisy chain roughly corresponding to the peaks and troughs of the higher bending component. The one-to-one correspondence of the two waveform patterns suggests PCB bending as a key physics of failure in drop impact of PEDs. It should be noted that the electrical connectivity does not appear to close/open at a specific level of measured strain. Several reasons may explain this phenomena, namely (1) wear/deformation of the fracture surfaces when they contact each other; (2) the effect of bending components in the transverse axis, which may be slightly out of phase from the ones measured; or (3) the fact that the strain gauge is located a short distance away from the joint and may not perfectly reflect the events occurring at the joint.

6.2.2 The dominant physics of failure – analytical evidence

A relative dominance of the two postulated physics of failure principles may be evaluated quantitatively using simple analytical equations.

6.2.2.1 Inertia load of IC component

Assuming (1) the inertia load due to the IC component acts evenly over all joints and (2) each solder ball joint is modelled as a cylinder with stress evenly distributed over its cross-section, the stress due to inertia of the IC component is given by

$$\sigma_g \approx \frac{ma_g}{nA_3} \tag{6.4}$$

where m is the combined mass of the IC component and its solder joints, a_g the acceleration experienced by the mass during the drop shock, n the number of joints on the IC component and A_3 the cross-sectional area of a single solder joint. Using a typical $10 \times 10 \times 1\ \text{mm}^3$ IC component with mass $m = 0.3$ g, measured peak acceleration magnitude $a_g = 4000$ g, and $nA_3 = 25\ \text{mm}^2$, the average stress in the joints due to inertial loading is estimated to be $\sigma_g = 0.44$ MPa, which is way below the strength of a typical solder joint.

6.2.2.2 *Bending of PCB*

The analytical solution for the stresses in the solder joint due to bending of PCB is discussed in detail in Chapter 11, but we shall take the analytical equation here without proof for the current discussion. The maximum peeling stress in the solder joint, $\sigma_{p,\max}$, is given by

$$\sigma_{p,\max} = \sigma_{a,\max} + \left(m_{\Phi,\max} + F_{s,\max}\frac{h_3}{2}\right)\frac{d_3}{2I_3} \tag{6.5}$$

where $\sigma_{a,\max}$, $F_{s,\max}$ and $m_{\Phi,\max}$ are axial stress, shear stress and bending moment, respectively, on the critical solder joint and are given by

$$\sigma_{a,\max} \approx \frac{4\alpha_z\varepsilon_2(L)}{\overline{D}_e h_2}\left[\sin\alpha_z p\cosh\alpha_z p - \sin\alpha_z(2L-p)\sinh\alpha_z p\right] \tag{6.6}$$

$$F_{s,\max} \approx \frac{q\varepsilon_2(L)}{\lambda_x}\left(1 - e^{-\beta p}\right) \tag{6.7}$$

$$m_{\Phi,\max} \approx \frac{2E_3 I_3 \varepsilon_2(L)}{\alpha_z h_2} \tag{6.8}$$

where $\varepsilon_2(L)$ is the magnitude of fibre strain at the edge of the IC component; the characteristic constants are given by $\alpha_z = \sqrt[4]{\overline{D}_e/4\lambda_z}$ and $\beta = \sqrt{\lambda_x/\kappa_s}$; and the compliances are given by

$$\overline{D}_e = \frac{12}{E_1 h_1^3} + \frac{12}{E_2 h_2^3}, \quad \lambda_x \approx \sum_{i=1}^{2}\frac{1}{E_i h_i}\left(4 + \frac{3h_3}{h_i}\right)$$

$$\lambda_z = \frac{13h_1}{32E_1} + \frac{13h_2}{32E_2} + \frac{pqh_3}{E_3 A_3}$$

$$\kappa_s \approx \frac{1}{8G_1}(h_1 - 2h_3) + \frac{1}{8G_2}(h_2 + 2h_3) + \frac{pqh_3}{G_3 A_3} + \frac{h_3^3 pq}{12E_3 I_3}$$

Table 6.5 Assumed dimensions and mechanical properties of PCB assembly

	Dimensions				Mechanical properties	
Component	**Planer**	h	p, q	d	E	ν
#1, IC component	10 × 10	1			25	0.3
#2, PCB		1.25			24	
#3, Solder joints		0.15	0.3, 0.3	0.168	25	

Notes: Dimensions in mm; E (GPa).

Table 6.6 Evaluated compliances, characteristic constants and forces/moment

Compliances				Characteristic constants		Forces/moment		
D_e	λ_x	λ_z	κ_s	α_z	β	$\sigma_{a,max}$	$F_{s,max}$	$m_{\Phi,max}$
7.36×10^{-4}	3.23×10^{-4}	6.18×10^{-5}	1.19×10^{-4}	1.31	1.65	$1960\ \varepsilon_2$	$362\ \varepsilon_2$	$1.2\ \varepsilon_2$

Note: D_e ($N^{-1}.mm^{-1}$), λ_x ($N^{-1}.mm$), λ_z, κ_s ($N^{-1}.mm^3$), α_z, β (mm^{-1}), $\sigma_{a,max}$ (N mm^{-2}), $F_{s,max}$ (N), $m_{\Phi,max}$ (N mm).

where p, q are the pitches in the x and y directions, respectively; h, d, I, E, G, ν are height, diameter, second moment of area, stretch modulus, shear modulus and Poisson's ratio, respectively; L is the half-length of the IC component; the subscripts 1,2,3, represent the IC component, PCB, and solder joints, respectively; λ_x, λ_z and κ_s are compliances in the x, z and x–z directions, respectively. The assumed dimensions and mechanical properties of the PCB assembly are given in Table 6.5; note that the total area of solder joints is identical to that used in the above section; that is, $nA_3 = 25\ mm^2$. The evaluated compliances, characteristic constants and forces/moment are shown in Table 6.6.

The maximum peeling stress in the solder joint has been evaluated as $\sigma_{p,max} = 6.3 \times 10^4\ \varepsilon_2(L)\ N\ mm^{-2}$. Table 6.4 shows it is possible for the PCB to acquire a fibre strain magnitude of 0.001 in the drop impact of a PED. Substituting $\varepsilon_2(L) = 0.001$ into the above equation gives $\sigma_{p,max} = 63$ MPa.

Comparing $\sigma_{p,max} = 63$ MPa with $\sigma_g = 0.44$ MPa, it is clear that PCB bending is the dominant physics of failure for board-level solder joints in the drop impact of a PED.

References

Lim, C. A. (2003). Drop impact survey of portable electronic products. In *53rd Electron. compon. technol. conf.* (pp. 113–120).

Lim, C. M. (2004). Drop impact study of handheld electronic products. In *Fifth international symposium on impact engineering. United Kingdom.*

Seah, S. L. (2002). Mechanical response of PCBs in portable electronic products during drop impact. In *4th electron. packag. technol. conf.* (pp. 120–125) (Singapore).

Seah, S. W. (2004). *Experiments and failure drivers in drop impact of portables*. San Jose: JEDEX.

Subsystem testing of solder joints against drop impact

7

7.1 Board-level testing

The failure of solder joints interconnecting the integrated circuit (IC) component to the printed circuit board (PCB) in a PED experiencing drop impact has raised a number of difficult questions for the microelectronic assembly industries: What design features and materials of the IC component and PCB are responsible for the robustness of the PCB assembly in product drop impact? How can the designs of IC components and the PCB assembly be assessed for product drop-impact failure? The first question will be addressed in later sections. The second question was more urgent for practical reasons, triggering an industry-wide effort to establish a test standard.

The most obvious method for assessing the design of the IC component is to drop test it in the particular PED in which the IC component will be assembled. However, this is not possible because it would be too late in the development cycle of the PED. Also, IC components are designed with universal applications in mind, independent of the design of the portable device. As a starting point, a board-level test method that is independent of the portable device and is able to assess qualitatively (or comparatively) the reliability of IC components against drop impact is essential.

7.1.1 Development of test methods: a historical perspective

Some of the earliest work on drop-impact testing was performed by Juso, Yamaji, Kimura, Fujita, and Kada (1998) and Yamaji et al. (2000) of Sharp, who mounted PCB assemblies onto a test phone, which was then dropped repetitively in the longitudinal and transverse orientations from a height of 1 m. Such tests are also routinely performed in-house by manufacturers of mobile devices as part of product qualification. However, microelectronic and PCB assembly manufacturers desire a board-level test method that is independent of the portable product. Peng et al. (2001) of Auburn University and Mishiro et al. (2002) of Fujitsu performed drop tests of bare PCB assemblies. Peng et al. conducted experiments where the bare PCB was guided in the inner bore of a tube, whose diameter was marginally larger than the width of the PCB, to enable vertical impact of the PCB assembly. The main drawback for this test is the inability to control the impact velocity of the PCB assembly with the target due to friction between the PCB and the bore of the tube. Mishiro et al. simply released the PCB assembly from a horizontal orientation to fall unguided. The main drawback of this test is the inability to control the impact angle.

A common drawback of drop testing bare boards without fixtures is that the bending modes of the PCB are not easily controlled. These drawbacks led to the fixed and

Robust Design of Microelectronics Assemblies Against Mechanical Shock, Temperature and Moisture
http://dx.doi.org/10.1016/B978-1-84569-528-6.00007-1

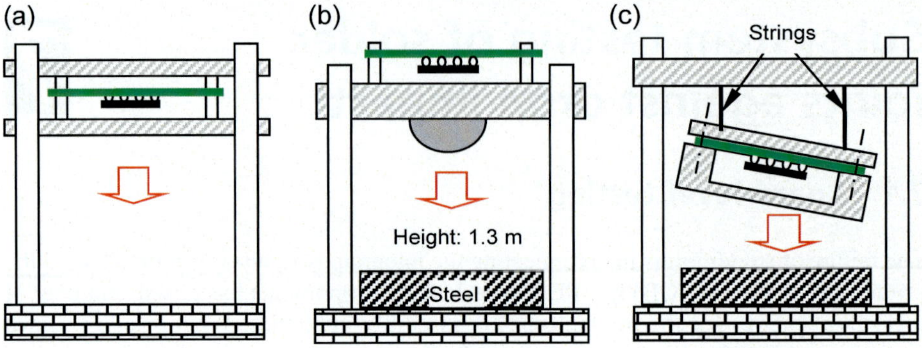

Figure 7.1 Board-level drop test setups of (a) Ericsson, (b) Toshiba and (c) NEC.

orientation-controlled drop test methods. Zhu (2001) of Ericsson used a setup where the PCB assembly was suspended along its edges in a guided fixture during drop so as to impact horizontally onto the floor (Figure 7.1(a)). Sogo and Hara (2001) of Toshiba reported an improved fixture with a hemisphere attached to its bottom (Figure 7.1(b)) to ensure consistent impacts. Hirata, Yoshida, and Morizaki (2001) of NEC suspended the fixture on strings such that the orientation of impact could be adjusted (Figure 7.1(c)). Qiang et al. (2002) of Yokohama University studied the positions of the IC components on the PCB using the setup of Figure 7.1(a); they reported a more severe loading for IC packages that experienced symmetric bending.

At this juncture, it is important to highlight the technical difference between a drop-impact test (ASTM-D5276, 2009) and a mechanical shock test (ASTM-D6537, 2014; IPC/JEDEC-9703, 2009). In a drop-impact test, the test specimen is released from a defined height to impact upon a defined target; the key test parameters are impact velocity and the contact stiffness of the target. The drop testing of PEDs described in Chapter 6 and the three test methods shown in Figure 7.1 are examples of drop-impact tests; the latter may be referred to as board-level drop-impact test. The reproducibility of the board-level drop-impact test depends on the consistency of the drop-height, the local stiffness of the target and the angle of impact. It is assumed that by controlling these parameters, the test specimen will experience a reproducible dynamic experience, which is best reflected in the form of accelerated-time history. This being the case, it makes more sense to control directly the acceleration−time characteristic of the impact. This leads to the board-level mechanical shock test, which prescribes an acceleration−time characteristic for the test fixture.

Goyal, Buratynski, and Elko (2000) of Lucent used mechanical shock tests to investigate the effectiveness of support damping, as shown in Figure 7.2(a) They characterised the response of the PCB mounted on damped supports at six points along its edges to half-sine shock pulses of durations ranging from 1.5 to 16 ms. The use of shock tests to assess the robustness of board-level interconnects was reported by Kim et al. (2001) of Hyundai. However, the test by Kim et al. is believed to be ineffective because the PCB assembly was attached rigidly to the fixture, as shown in Figure 7.2(b), and was thus prevented from bending.

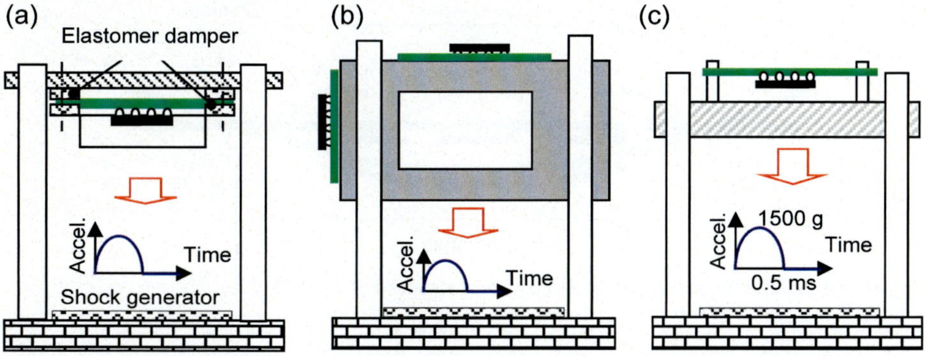

Figure 7.2 Board-level shock test setups of (a) Lucent, (b) Hyundai and (c) JESD22B111.

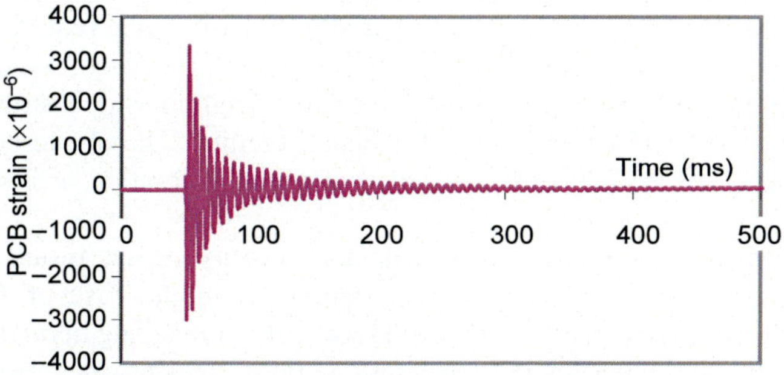

Figure 7.3 PCB strain response of JEDEC test standard JESD22B113.

Motivated by the need for an industry standard, Syed et al. of Amkor led a working group for a new JEDEC test standard JESD22B111 (JESD22B111, 2003), which prescribes a half-sine acceleration–time of amplitude 1500 gravitational acceleration over a duration of 0.5 ms (equivalent to 1000 Hz) for the test fixture, as shown in Figure 7.2(c). The desired acceleration–time shock characteristic is achieved through manipulation of the drop-height and the contact stiffness of the shock generator. The PCB assembly is allowed to flex under its own inertia. Figure 7.3 shows a typical strain response of the PCB assembly in the JESD22B111 test. The mechanical shock has excited a transient bending deflection in the PCB assembly, which is followed by natural vibration of the PCB assembly with decaying amplitude.

Because the velocity of impact in the JESD22B111 test is derived purely from gravitational acceleration, it may be described more precisely as a board-level drop-shock test (BLDST). Specialised mechanical shock testers are available (Langmont, n.d.), which introduce additional acceleration (on top of gravitational acceleration) to the test fixture through either compressed air or compressed springs.

By prescribing a half-sine acceleration–time characteristic of the impact, JESD22B111 BLDST should offer better reproducibility for the board-level drop-impact tests shown in Figure 7.1. But, in practice, it is nearly impossible to achieve

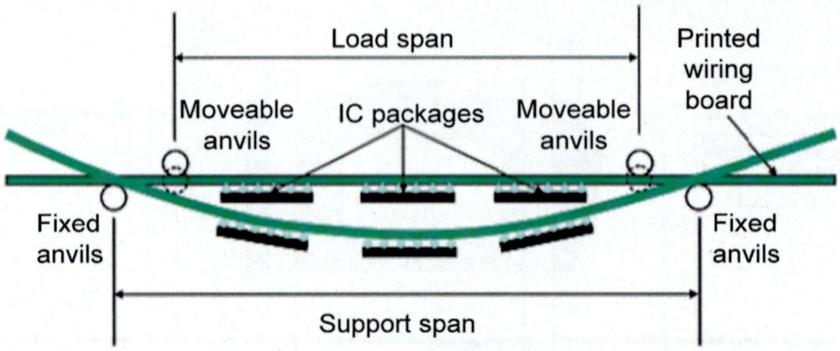

Figure 7.4 Schematic test setup of JEDEC test standard JESD22B113.

a perfect half-sine acceleration–time characteristic however the drop-height and the contact stiffness are manipulated, which raises questions about the reproducibility of the test method. Another concern with the JESD22B111 BLDST is the bulkiness of the test fixture. A mechanised hoist is required to hoist it to the required test height, which slows down the throughput rate of testing. The impact is acoustically deafening and generates ground-shock that could affect the neighbouring instrument so that testing needs to be performed in a well-isolated facility. The above concerns with the JESD22B111 BLDST have motivated continued exploration of board-level test methods.

The principal element of a mechanical shock is velocity impact. Instead of a moving test fixture impacting against a stationary target, as in the case of JESD22B111 BLDST, a velocity impact may be achieved by a striker impacting upon the test fixture at rest, imparting it with the desired acceleration–time characteristics. However, given the weight of the test fixture, a striker of significant mass and velocity is required; this test method will suffer from similar setbacks as the JESD22B111 BLDST.

It was highlighted in Chapter 6 that cyclic bending of PCB assemblies is the dominant physics of failure for solder joints interconnecting the IC component to the PCB in the drop impact of PEDs. The JESD22B111 BLDST simply attempts to duplicate this physics by setting up cyclic bending in the PCB assembly by imparting a mechanical shock to the test fixture. There are many simpler methods to induce cyclic bending in PCB assemblies; the simplest approach is to mechanically deflect the PCB assembly with a mechanical actuator, as described in the JEDEC test standard JESD22B113, a schematic of which is shown in Figure 7.4. The acceleration of a mechanical actuator is limited by its inertia; for a sinusoidal displacement, this is given by $\ddot{z}_o = z_o\omega^2$. The versatile Instron model 5848 MicroTester is used as an illustration; the design envelop is shown in Figure 7.5, which suggests a limiting acceleration of 0.01 gravitational acceleration. In view of the high sensitivity of the fatigue resistance of solder joints with strain rate (evidence given in Chapter 8), it is essential that the solder joints in a board-level test are fatigued at a similar strain rate as that in the PCB assembly of a PED experiencing a drop impact. Chapter 6 suggests that the PCB assembly in a PED experiences a nominal acceleration of a few hundred gravitational acceleration and a fibre strain of a few thousand microstrains, when drop impact from a height of 1 m. This is way beyond the capability of a standard mechanical tester.

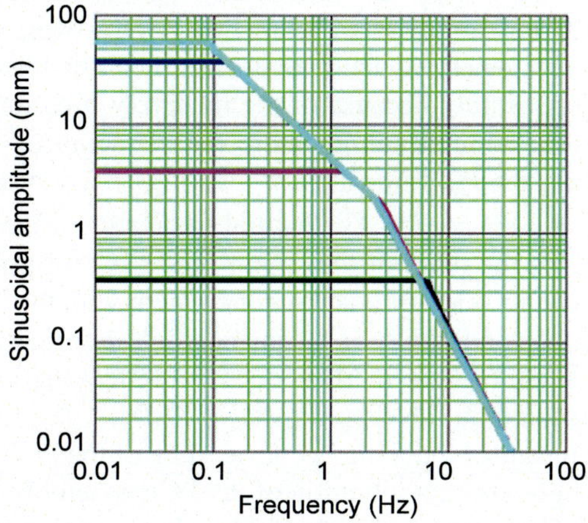

Figure 7.5 Amplitude–frequency envelope of Instron model 5848 MicroTester.

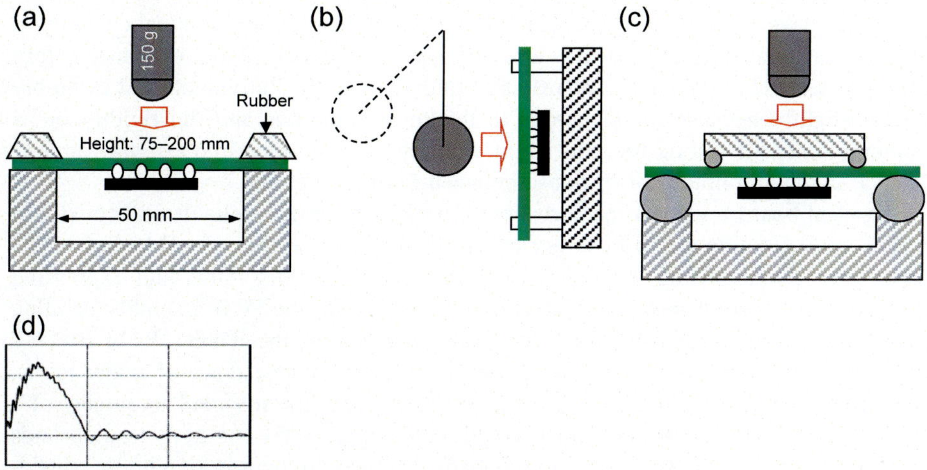

Figure 7.6 Impact bending test setups of (a) Hitachi, (b) CALCE and (c) Motorola. (d) The strain response of the PCB assembly of Hitachi.

The simplest means of imparting high acceleration to a PCB assembly is by striking the PCB assembly directly at high speed. This requires much less energy than introducing mechanical shock to the bulky test fixture. Yaguchi, Yamada, and Yamamoto (2003) and Nagano, Yaguchi, Terasaki, and Yamamoto (2006) of Hitachi reported using an impact bending test method in which the PCB assembly, clamped at two ends, was impacted at its centre by a falling plunger (see Figure 7.6(a)). Varghese and Dasgupta (2003), of the Center for Advanced Life Cycle Engineering (CALCE) at University of Maryland, delivered the impact using a swinging steel ball (see Figure 7.6(b)). Reiff and Bradley (2005) of Motorola used a method similar to that of Hitachi but introduced a fixture between the impactor and the PCB to enforce four-point bending of the PCB assembly (see Figure 7.6(c)). However, one drawback is that the bending frequencies generated tend to be significantly lower than the natural frequency of the PCB assembly because of the relatively high mass of the impactor, and also that of the four-point bending fixture for the setup of Motorola. This can be visualised by treating the PCB impactor as a spring–mass system whose resonant frequency is given by $\omega_n = \sqrt{k/m}$. The natural frequency of these test methods can be increased by reducing the mass of the striker (and the four-point bending fixture) while increasing its striking velocity. The bigger concern with the 'strike' test methods is that the PCB assembly undergoes only half-bending deformation, as displayed in Figure 7.6(d) for the setup of Hitachi. This does not reflect the bi-directional fatigue experienced by solder joints in the drop impact of PEDs.

The most efficient means of inducing cyclic bending of the PCB assembly at its natural frequency is through resonant vibration of the PCB assembly using a mechanised shaker (ASTM-D999, 2008; ISO-5347-22, 1997), when a large amplitude of response can be induced from a small amplitude of input. A resonant vibration test method was proposed by Marjamäki, Mattila, and Kivilahti (2006) of Helsinki University of Technology, in which a PCB assembly held in a test fixture is resonated using an electromagnetic shaker. However, this apparently ideal test method has practical challenges. Large resonant amplification occurs in a very narrow frequency band, with a steep fall-off on either side of the band. The resonant frequency of the individual PCB assembly differs slightly because of deviations in thickness, modulus and the rigidity of its mounting onto the test fixture; this slight difference is translated into a significant difference in the amplitudes of response between PCB assemblies.

An ideal board-level test method should, besides offering a reliable and consistent means of evaluating the robustness of the design of board-level solder joints, be capable of characterising the fatigue resistance of the solder joints and expressing the results in a useful form, such as $\Delta\sigma = C_o N^{\beta_o}$. With the PCB assembly bending at a decaying amplitude following each mechanical shock, the JESD22B111 BLDST is incapable of offering a viable expression for the fatigue resistance of solder joints. Similarly incapable is the resonant vibration method; at the resonant frequency of a few hundred hertz, the solder joints would have experienced hundreds to thousands of fatigue cycles as the PCB assembly is being ramped up the test amplitude; considering the fact that the fatigue life of the solder joints is only in the hundreds to low thousands, much of the fatigue life of the solder joints would have been consumed during the ramp-up. The high-speed cyclic bending method proposed by Seah, Wong, Mai, Rajoo, and Lim (2006) appears to provide the ultimate solution.

7.1.2 The high-speed cyclic bending test

Figure 7.7 shows the high-speed cyclic bend tester, which is a result of collaboration with Instron. Four-point bending of a PCB assembly is achieved through two sets of anvils: a set of fixed anvils and a set of displacing anvils. The displacing anvils are given cyclic displacement by a cam through a lever, whose level arm could be adjusted to vary the amplitude of displacement. The cam is driven by a high-power servo motor,

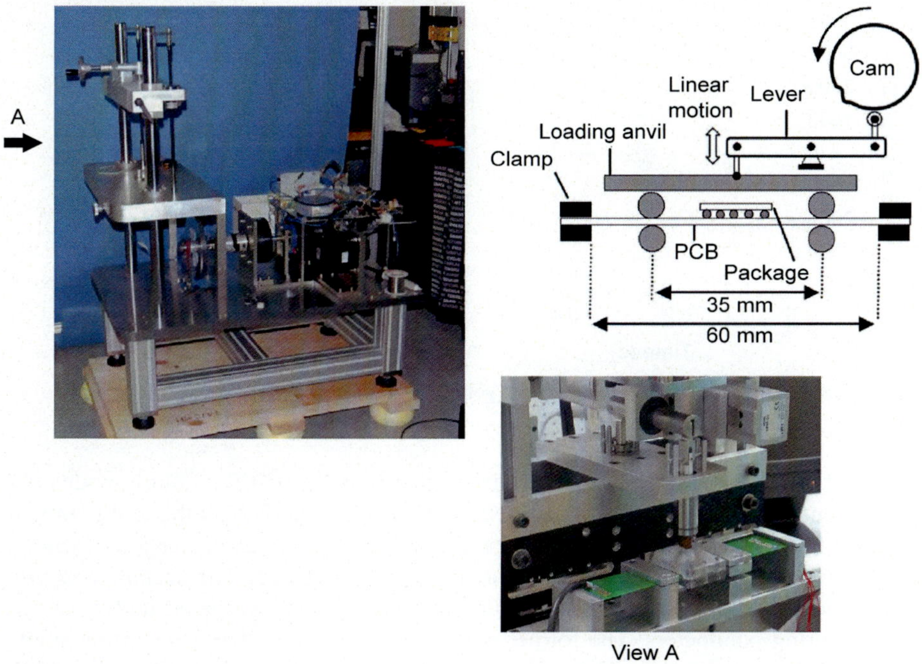

Figure 7.7 High-speed bend tester.

which has a top speed of 1500 rpm, or 25 revolutions per second (rps). A sinusoidal profile is built into the cam over a segment of 20°. This gives a maximum linear cyclic frequency of 450 Hz. It takes finite time (and hundreds of revolutions) for the high-power motor to be accelerated to its operating speed. During this period, the motor is disconnected with the cam. Upon reaching the operating speed, the motor is then engaged with the cam through a frictional clutch. The inertia of the cam, the level and the moving anvils are minimised so that the anvil is brought to the design frequency in a single cycle. This is illustrated in Figure 7.8, which shows the fibre strain of a PCB assembly for three cycles; each cycle is a near-perfect sinusoidal waveform of 100 Hz. The interval between each cycle is the result of the level moving over the circular contour of the cam. Besides offering consistent bending amplitude, frequency and waveform, the high-speed cyclic bending test (HSCBT) also offers a very high rate of test throughput; for example, at 20 rps, it takes less than a minute to accomplish 1000 cycles of bending. An equivalent test using the JESDB111 BLDST would require hours.

7.1.3 Experimental studies on correlations between HSCBT and BLDST

Two studies were performed to evaluate the ability of high-speed bend testing to reproduce the solder interconnect failures encountered in a board-level shock test

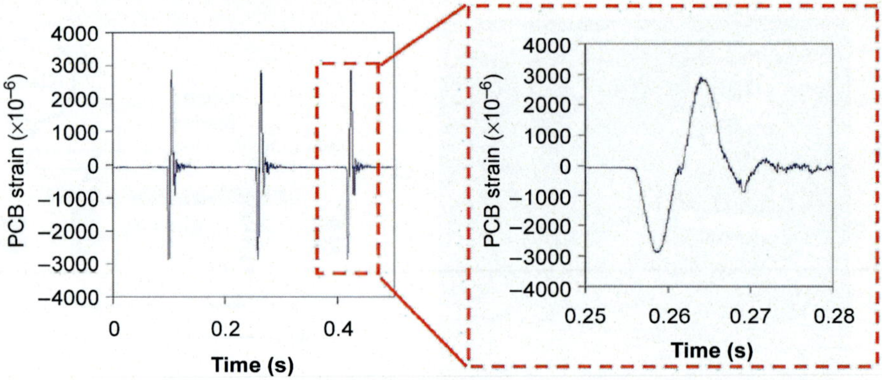

Figure 7.8 PCB bending strain waveforms generated by the tester.

(Wong et al., 2009). In these studies, the cycle-to-failure and the failure modes obtained using the HSCBT are compared with the drop-to-failure and the failure modes obtained from BLDST. The first study involves testing various solder joint material systems, using various solder material and pad combinations. The second study involves the testing of a limited number of material systems prepared under various manufacturing parameters. The former is referred to as the materials variation study, whereas the latter is the manufacturing variation study. The test specimen, test conditions and test matrices are described in the following sections.

7.1.3.1 Test specimen

The test specimen is shown in Figure 7.9. The IC component is simulated by a piece of laminate material. The PCB and the IC component are fabricated from the same laminate panel; hence, both have identical pad finishes. This minimises the inconsistency derived from variations in the fabrication processes. The board dimensions, solder joint layout and solder joint dimensions are shown in Figure 7.9(a)–(c), respectively. The four corner joints are designed to experience a significantly higher stress magnitude than the inner joints, ensuring early but controlled failure there. Symmetry ensures identical loading on these joints. The corner joints are separated from each other by several rows and columns of inner joints to ensure independent failure of each corner joint; that is, the failure of one corner joint does not significantly alter the loads experienced by the remaining corner joints. Thus, the design allows for four independent data values to be obtained from a single test specimen. The inner pads are connected into a daisy-chain for electrical monitoring to verify that the inner joints do not fail before the corner joints.

7.1.3.2 Test conditions

The BLDST was performed on an AVEX SM 110 mechanical shock tester. The test specimen is bolted at four points and held along its entire width with a span of

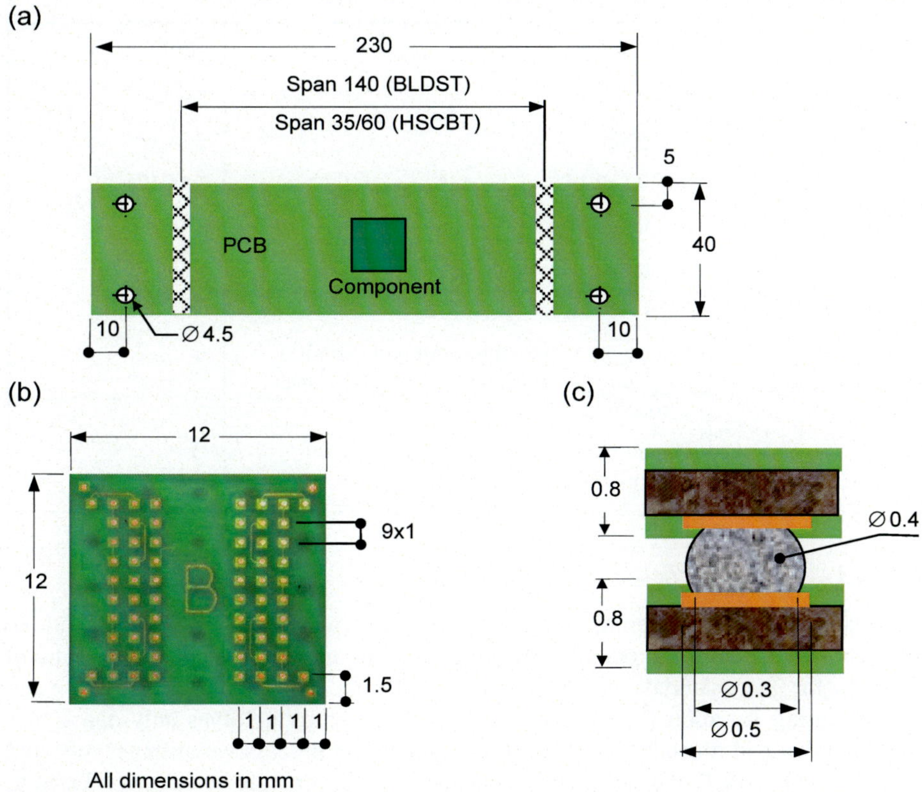

Figure 7.9 Test specimen: (a) PCB; (b) IC component; and (c) solder joint.

140 mm, as shown in Figure 7.9(a). A half-sine shock pulse of amplitude 800 g and duration 0.5 ms was applied to the supports. This shock pulse is selected for generating a peak longitudinal PCB strain of 2000 μstrain, measured adjacent to the IC component, which results in failures within 10–100 shocks, depending on the solder joint material system or manufacturing variation. The fundamental mode frequency is approximately 200 Hz for the specimen mounted in this configuration. The number of drops to complete electrical failure/open in each individual corner joint is recorded. Failure is defined as the point when the electrical resistance of the corner joint exceeds 100 Ω.

The HSCBT was performed on the high-speed cyclic bend tester shown in Figure 7.7. The specimen is clamped by a set of fixed anvils along its entire width at a span of 60 mm and is cycled by a set of moving anvils that has a span of 35 mm, as shown in Figure 7.9(a), such that the specimen experiences almost pure bending within the inner span. The test specimens are flexed using a sinusoidal bending waveform at a frequency of 100 Hz and a peak fibre strain of 2000 μstrain, similar to the peak response in the BLDST. The number of flexing cycles to failure is recorded.

Table 7.1 **Test matrix for the materials variation study**

Parameter	Description	Symbol
Solder alloy	Sn37Pb	SnPb
	Sn3.5Ag	SnAg
	Sn1.0Ag0.1Cu	SAC101
	Sn3.0Ag0.5Cu	SAC305
	Sn1.0Ag0.1Cu with 0.02% Ni & 0.05% In	SAC101(d)
	Sn3.0Ag0.5Cu with 0.05% Ni	SAC305(d)
Solder pad finish (from vendor Z)	Electroless nickel immersion gold (ENIG); Ni (5 μm max thickness), Ag (0.1 μm max thickness), P (6–8%)	E
	Organic solderability preservative over copper (OSP)	O

7.1.3.3 Test matrix

The test matrix for the materials variation study is shown in Table 7.1. Six solder alloys and two pad finishes for a total of 12 material systems are investigated. The simplified symbols for the materials are also given in Table 7.1. The test matrix for the manufacturing variation is shown in Table 7.2. The study involves only one solder alloy and two pad finishes; the effect of manufacturing process variations is reflected in (1) laminates (PCB and component) that are sourced from three vendors (referred to as X, Y and Z); (2) three thicknesses of immersion gold for electroless nickel plating; and (3) three thermal ageing conditions, for a total of 11 manufacturing systems. Twenty data points were collected for each system tested under the BLDST and HSCBT.

7.1.4 The consolidated experimental data

The cumulative distribution of drops-to-failure (for BLDST) and cycles-to-failure (for HSCBT) for the materials variation and manufacturing variation studies is shown in Figure 7.10. The symbols for the materials variation graphs are to be read as [Solder alloy]_[Pad finish], in which pad finish symbols *E* and *O* represent electroless nickel immersion gold (ENIG) and organic solderability preservative over copper (OSP), respectively. The symbol for the manufacturing variation graphs is to be read as [Vendor]_[Pad finish][#]. The # numeric represents the manufacturing variation. With reference to Table 7.2, #1 and #2 represent thermal ageing at 150 °C for 100 and 200 h, respectively, and #3 and #4 represent thick and thin immersion gold, respectively.

A cursory examination of the graphs in Figure 7.10 shows good agreement between BLDST and HSCBT in the performance ranking of the various materials systems. The

Table 7.2 Test matrix for the manufacturing variation study

Parameter	Description/variation		
Solder alloy	SAC101(d)		
	Pad finish	**Vendor**	**Variation**
Manufacturing variation of pad	ENIG	X	No conditioning
		Z	Thermal ageing at 150 °C for: Baseline – 0 h #1, 100 h #2, 200 h Gold thickness: Baseline – 0.1 μm max #3, thick at 0.15 μm #4, thin at 0.03 μm max
	OSP	X	No conditioning
		Y	No conditioning
		Z	Thermal ageing at 150 °C for: Baseline – 0 h #1, 100 h #2, 200 h

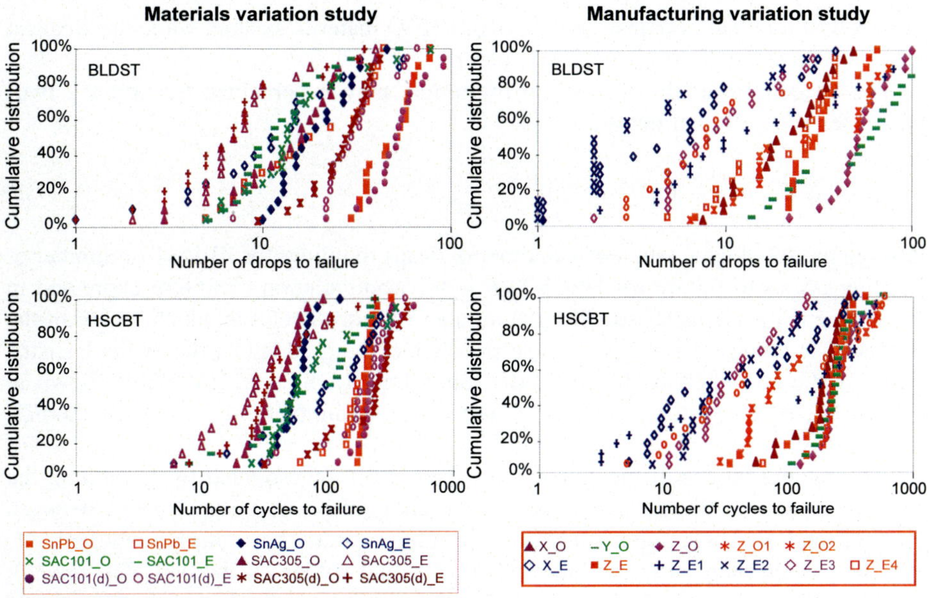

Figure 7.10 Cumulative distribution of the characteristic life of BLDST and HSCBT for materials variation and manufacturing variation studies.

Table 7.3 Summary of characteristic life and brittleness index of BLDST and HSCBT for the materials variation study

	BLDST		HSCBT	
Material system	**Characteristic life**	**Brittleness index**	**Characteristic life**	**Brittleness index**
SnPb_O	57	0.0	226	0.0
SnPb_E	27	0.0	212	0.1
SnAg_O	24	0.6	66	0.8
SnAg_E	19	0.5	165	0.3
SAC101_O	20	0.8	121	0.6
SAC101_E	16	0.4	118	0.3
SAC305_O	20	0.9	50	0.9
SAC305_E	11	0.8	54	0.5
SAC101(d)_O	64	0.0	297	0.0
SAC101(d)_E	39	0.4	250	0.2
SAC305(d)_O	31	0.3	297	0.3
SAC305(d)_E	11	0.8	62	0.7

most robust joints in both tests are the SAC101_O material system, while the weakest joints are those made using SAC305 solder.

The characteristic life of a group of sample may be obtained by fitting a two-parameter Weibull distribution:

$$F(x, \alpha, \beta) = 1 - \exp[-(x/\alpha)]^{\beta} \tag{7.1}$$

through the set of data, where x is number of drops to failure for BLDST or number of bending cycles to failure for HSCBT, F is the accumulation of failure expressed in fraction, α is the characteristic life (the number of drops/cycles at which 63.2% of the product has failed) and β is the shape parameter (the gradient of the Weibull distribution). The characteristic life of solder joints for BLDST and HSCBT is shown in Tables 7.3 and 7.4 for materials variation and manufacturing variation studies, respectively.

The fracture modes of the solder joints have been investigated by examining the residual solder on the metal pad. For easy comparison of failure modes between HSCBT and BLDST, four categories of failure modes have been identified: (1) 'ductile', when there is more than 50% of residual solder on the pad; (2) 'mixed', when there is solder residue on 10–50% of the pad surface; (3) 'brittle', when less

Table 7.4 Summary of characteristic life and brittleness index of BLDST and HSCBT for the manufacturing variation study

	BLDST		HSCBT	
Material system	**Characteristic life**	**Brittleness index**	**Characteristic life**	**Brittleness index**
X_O	24	0.5	207	0.1
Y_O	70	0.3	282	0.2
Z_O	64	0.0	297	0.0
Z_O1	45	0.4	199	0.3
Z_O2	13	0.8	157	0.4
X_E	9	0.9	84	0.6
Z_E	39	0.4	250	0.2
Z_E1	24	0.3	214	0.5
Z_E2	7	0.7	60	0.7
Z_E3	13	0.6	56	0.7
Z_E4	29	0.3	292	0.2

than 10% of the pad surface is covered with solder; and (4) 'pad peel', if the copper pad lifts off from the PCB. Examples of these failure mode classifications are shown in Figure 7.11. Note that the mixed failure mode category includes specimens that exhibit clusters of bulk solder on a predominantly brittle fracture surface, as shown in Figure 7.11(b).

The failure modes observed in the HSCBT and BLDST tests for both the materials variation and manufacturing variation studies are summarised in Figure 7.12. There is good agreement in the failure mode distribution patterns between the two test methods for both variation studies. For a more quantitative comparison, a brittleness index is introduced to describe the degree of brittleness of a particular solder joint material system. Ignoring the pad peel failure mode, which rarely occurs, and assuming that 50% of mixed failure modes are assignable to brittle failure, the brittleness index is defined as

$$\text{Brittleness index} = \text{Brittle occurrence} + 1/2 \text{ Mixed failure occurrence} \quad (7.2)$$

A brittleness index of 1 and 0 corresponds to 100% and 0% brittle failure, respectively, of samples in a material system. The brittleness index of solder joints for BLDST and HSCBT is listed in Tables 7.3 and 7.4 for materials variation and manufacturing variation, respectively.

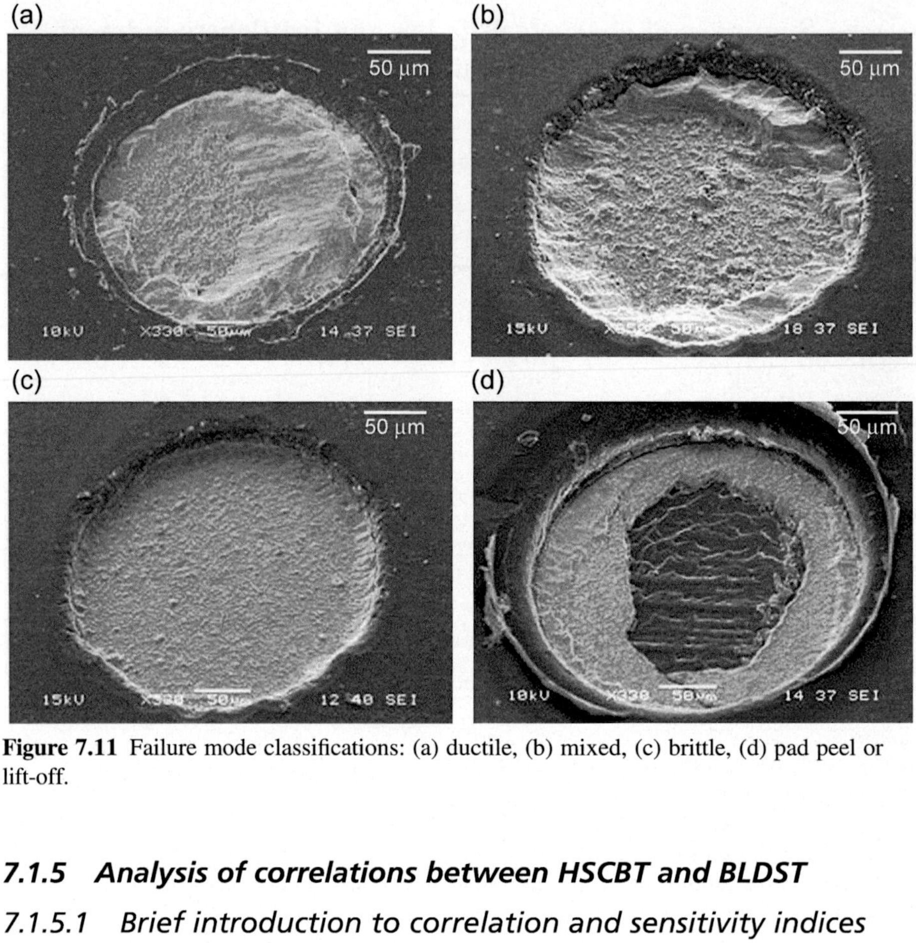

Figure 7.11 Failure mode classifications: (a) ductile, (b) mixed, (c) brittle, (d) pad peel or lift-off.

7.1.5 Analysis of correlations between HSCBT and BLDST

7.1.5.1 Brief introduction to correlation and sensitivity indices

The correlation coefficient between two sets of data, x_i and y_i, is defined as

$$\rho_{x,y} = \frac{\frac{1}{n}\sum_{i=1}^{n}(x_i - \overline{x})(y_i - \overline{y})}{\sigma_x \sigma_y} \tag{7.3}$$

The values of correlation coefficient $\rho_{x,y}$ range from -1 to 1. A higher absolute value indicates a stronger correlation; a positive value indicates a similar trend between the two sets of data. The operation can be performed using the standard Microsoft Excel command 'Correl(x, y)'.

In the presence of a strong correlation between two sets of data, the sensitivity of one set of data with respect to the other may be analysed by assuming a linear relationship between the two sets of data and computing the gradient. The Microsoft Excel function 'Slope(y, x)' computes the gradient of data set y with respect to data set x using the equation

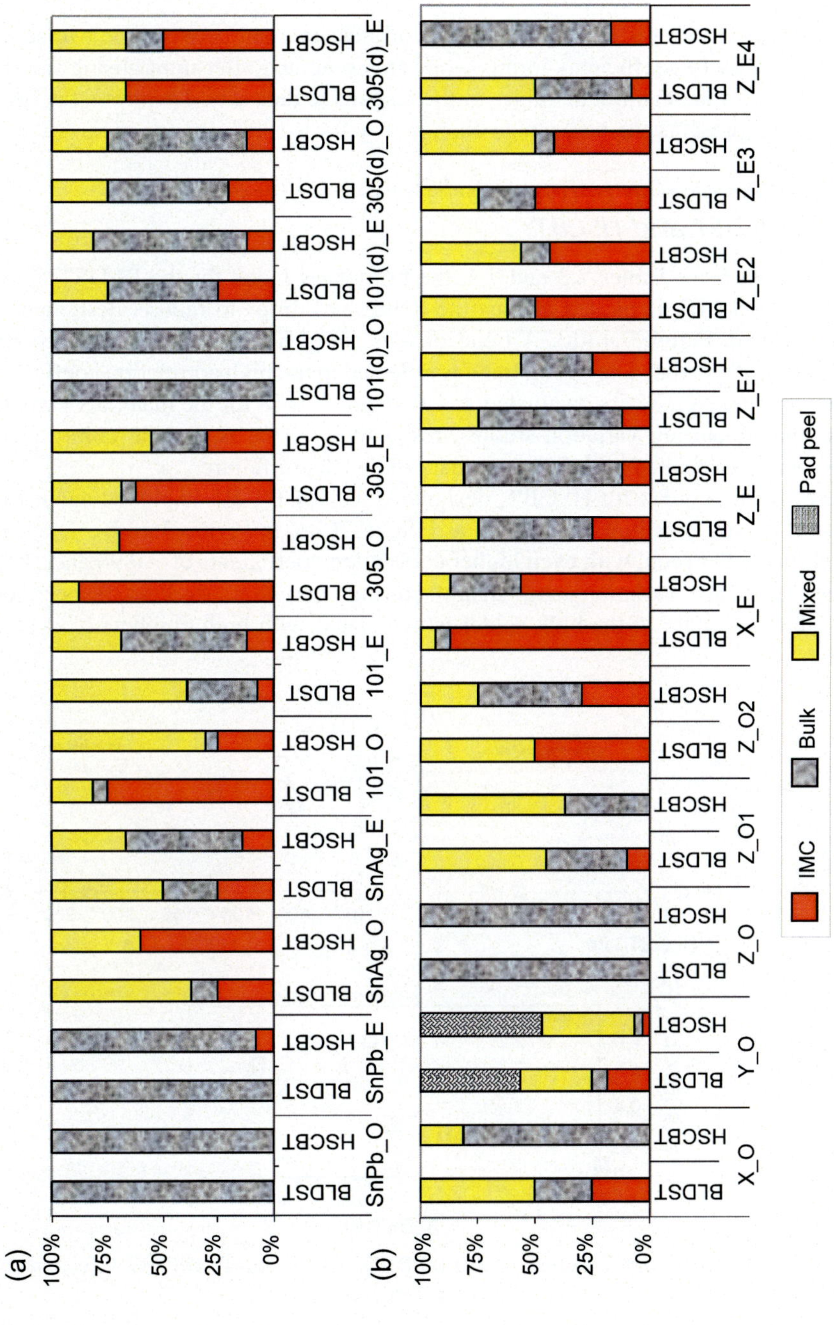

Figure 7.12 Distribution of failure modes for the (a) materials variation study and (b) manufacturing variation study.

$$G_{x,y} = \frac{n\sum xy - (\sum x)(\sum y)}{n\sum x^2 - (\sum x)^2} \tag{7.4}$$

To account for the difference in the units of measurement between the two sets of data, the sensitivity coefficients in this work are computed after normalising the data sets x and y with the maximum values of the respective data sets. Therefore, the sensitivity index represents the percentage change of y for a 100% change in x.

7.1.5.2 Correlation between brittleness indices of BLDST and HSCBT

Using the data from Tables 7.3 and 7.4, the brittleness index for the BLDST, designated as BLDST-BI, is plotted against the number of drops to failure, designated as BLDST-Life, in Figure 7.13(a). A trend of increasing BLDST-Life with decreasing BLDST-BI is observed. The strong trend is reflected in the high correlation coefficients between the two parameters, evaluated as −0.76 and −0.78 for the materials variation and the manufacturing variation studies, respectively. That is, ductile failure in the bulk solder is associated with high numbers of shocks to failure.

The brittleness index for HSCBT, designated as HSCBT-BI, is plotted against the number of cycles to failure, designated as HSCBT-Life, in Figure 7.13(b). The same trend as BLDST is seen, with even higher correlation coefficients of −0.90 and −0.86 for the materials variation and manufacturing variation studies, respectively. Hence, ductile failure in the bulk solder is associated with high numbers of cycle to failure.

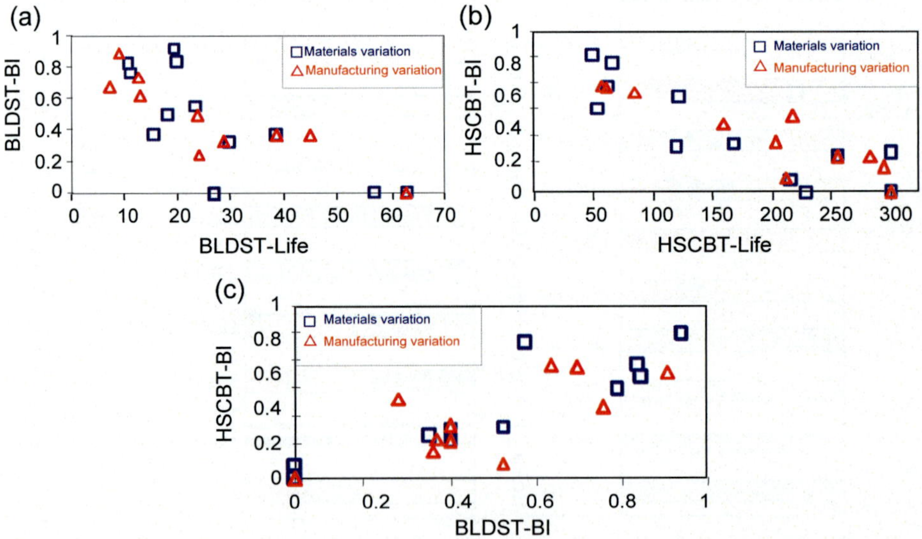

Figure 7.13 Brittleness index plotted against (a) BLDST-Life and (b) HSCBT-Life; and (c) brittleness index for the two tests plotted against each other.

Plotting BLDST-BI versus HSCBT-BI, as shown in Figure 7.13(c), it is no surprise that the two sets of data correlate well. The correlation coefficients for the materials variation and the manufacturing variation are 0.94 and 0.74, respectively. Thus, the HSCBT is capable of duplicating the failure mode of the BLDST.

The good agreement between the brittleness of the two test methods suggests they share the same failure driving force – namely, the high-speed flexing of PCB that induces bending stresses in the solder joints at a high strain rate.

7.1.5.3 Correlation between fatigue life of BLDST and HSCBT

HSCBT-Life is now plotted against BLDST-Life in Figure 7.14(a) for the materials variation and Figure 7.14(b) for the manufacturing variation. For the purpose of clarity, the OSP finish is bracketed. The correlation and sensitivity coefficients for the data are

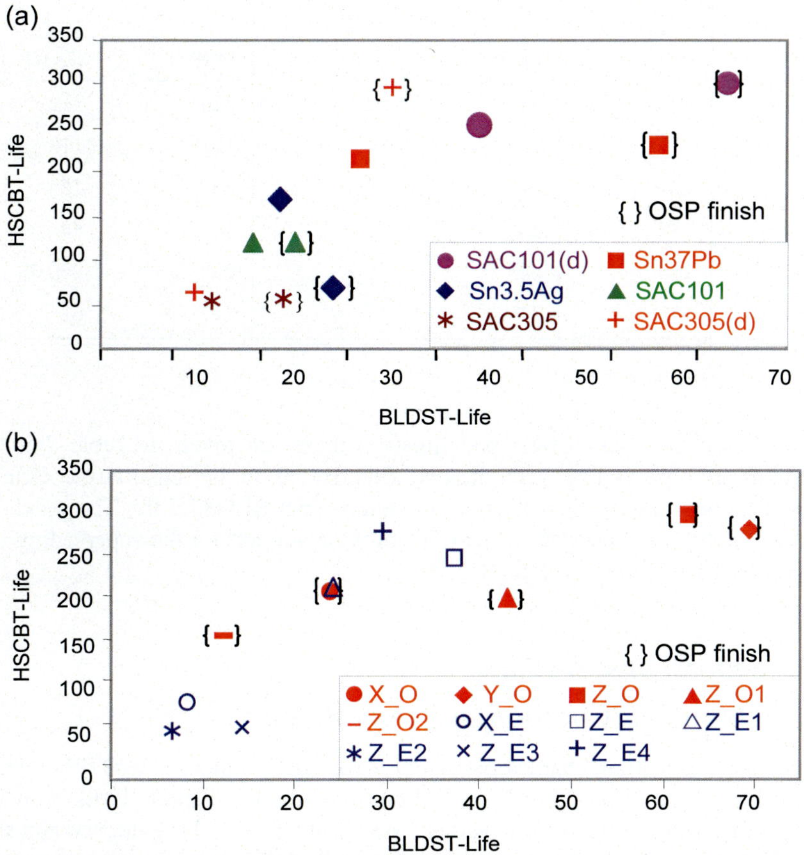

Figure 7.14 HSCBT-Life plotted against BLDST-Life for the (a) materials variation study and (b) manufacturing variation study.

Table 7.5 $\rho_{x,y}$ correlation coefficients and sensitivity coefficients for HSCBT-Life versus BLDST-Life

Materials variation			Manufacturing variation		
	$\rho_{x,y}$	$G_{x,y}$		$\rho_{x,y}$	$G_{x,y}$
OSP[a]	0.95	120%	OSP	0.95	160%
ENIG	0.95	180%	ENIG	0.91	50%

[a]The data for SAC305(d)_O was an outlier and therefore was excluded.

Table 7.6 Summary of the flow stress of solder alloy, BLDST-Life and HSCBT-Life

Solder alloys	Flow stress (MPa)	BLDST-Life		HSCBT-Life	
		OSP	ENIG	OSP	ENIG
PbSn	80	57	27	226	212
SAC101	110	20	16	121	118
SnAg	125	24	19	66	165
SAC305	135	20	11	50	54
SAC101(d)	110	64	39	297	250
SAC305(d)	135	31	11	297	62

segregated into OSP and ENIG pad finishes; these are given in Table 7.5. Good correlations are observed between the two tests. Based on the sensitivity coefficients, HSCBT-Life has a higher sensitivity to variations than BLDST-Life. The good correlation and the higher sensitivity of HSCBT provide support for the superseding of the JESDB111 BLDST with the HSCBT.

7.1.6 General observations

7.1.6.1 Materials variation and flow stress

The dynamic stress-strain characteristics of the four solder alloys – Sn37Pb, Sn3.5Ag, Sn1.0Ag0.1Cu and Sn3.0Ag0.5Cu – will be presented in Chapter 12; the flow stress of these solder alloys at 10% strain and at a strain rate of 50 s^{-1} is extracted and shown in Table 7.6 alongside the BLDST-Life and the HSCBT-Life of the solder joints. The BLDST-Life and HSCBT-Life of the solder joints are plotted against their flow

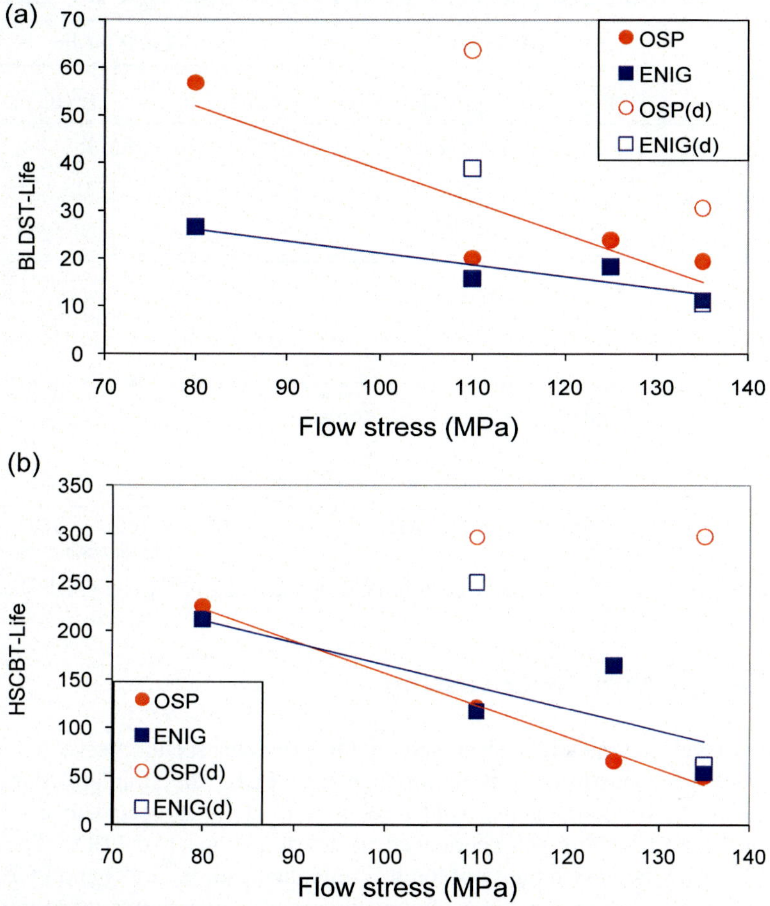

Figure 7.15 (a) BLDST-Life and (b) HSCBT-Life plotted against flow stresses of solder alloys.

stresses in Figure 7.15. It is observed that the BLDST-Life and HSCBT-Life of the solder joints decrease with increasing flow stress of the solder alloys. Because increasing the flow stress of the solder joint increases the peeling stress and reduces the effective plastic strain in the joint, this indicates that the fatigue driving force is the peeling stress and not the plastic strain.

7.1.6.2 Manufacturing variation: thermal ageing

Table 7.7 lists the BLDST-Lift and HSCBT-Life after different durations of ageing at 150 °C and Figure 7.16 plots the normalised BLDST-Life and HSCBT-Life against these thermal ageing durations. The results suggest that thermal ageing could degrade the fatigue life of solder joints very significantly.

Table 7.7 **Thermal ageing versus BLDST-Life and HSCBT-Life**

Thermal ageing time at 150 °C (h)	BLDST-Life		HSCBT-Life	
	OSP	ENIG	OSP	ENIG
0	64	39	297	250
100	45	24	199	214
200	13	7	157	60

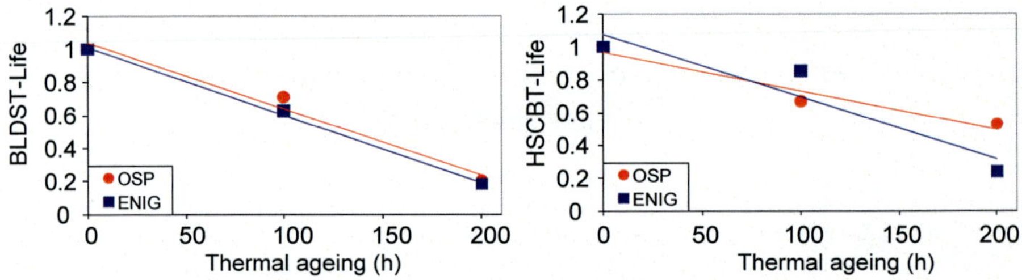

Figure 7.16 Detaching of IC component from PCB by knocking followed by ball shear.

7.2 Component-level testing

The board-level test methods discussed in the previous section serve to evaluate qualitatively the robustness of IC components on PCB (the structure, dimensions and materials of IC components, solder joints and PCB) subjected to a relatively high strain rate of loading as experienced in the drop impact of PEDs. These tests are useful during the design phase of the IC component, solder joints and PCB. However, with designs frozen, it is not economically sensible to continue using the board-level test as a quality assurance test in manufacturing.

The area array interconnection of ball grid array (BGA) packages to PCB is facilitated by the arrays of solder balls on the metal (typically Cu) pads of BGA packages. The attachment of solder balls to the metal pads is accomplished through melting and reflowing the solder balls over the metal pads, forming a metallurgical bond in the form of an intermetallic compound (IMC) between the solder ball and the metal pad. The manufacturing quality of the metallurgical bond is typically assessed using JEDEC standard JESD22B117A, a ball shear test (JESD22-B117A, 2006) in which a chosen solder ball is sheared with a tool of a width marginally larger than the solder ball and at a speed less than 1 mm/s. Because the quasi-static shear strength of the IMC is typically much higher than that of the solder alloy that makes up the solder ball, the tool is expected to smear through the solder ball without fracturing the IMC. Fracture of IMC is a reflection of an extremely poor solder-reflow process. This test method works well for solder balls that are destined for applications in which the solder alloy is the weakest element and the strength of IMC is not critical, as in the creep fatigue of solder joints. At the strain rate experienced by the solder joints of a PCB assembly in a PED experiencing drop impact, the fracture strength of IMC becomes critical and the JESD22B117A ball shear test becomes unviable. A new test method for assessing the quality of the ball-attachment process is needed.

While the JESD22B111 BLDST or the other simpler board-level test methods may be used for this purpose, the necessity to attach the IC component to a test board renders the these board-level test methods economically unjustifiable. An ideal quality assurance test method for the solder ball on the IC component is one that retains the simplicity of the JESD22B117A ball shear test and is able to emulate the strain rate experienced by solder joints in product drop impact. Such an ideal test method has wider applications. The quality of the IMC between the solder ball and the metal pad is dependent not only on the solder-reflow process but also on the quality of the surface finishing of the metal pads (e.g., ENIG, organic surface preservative) in the laminate substrate. The ideal test method can be used by IC component manufacturers to control the incoming quality of laminate substrates and by the substrate manufacturers to control its manufacturing quality. This also applies to PCB manufacturers and PCB assembly houses.

7.2.1 Development of test methods

A solder ball on an IC component may be deformed through three principal actions: shearing, stretching or twisting. The last two require holding the minute ball using a mechanical vice; this demands good alignment and the application of optimum holding force so as not to induce damage to the IMC structure of the solder ball. Shearing is the simplest test method. It requires only a shear tool moving over a solder ball at a speed that will induce brittle fracture in the solder ball. This typically needs a shearing speed of a few hundred millimetres per second, which is way beyond the capability of standard mechanical testers that are screw or hydraulic driven. By contrast, an impactor can acquire such speed with ease.

The earliest report of ball impact shearing test (BIST) is probably by Shoji, Yamarmoto, Kajiwara, Morita, Sato, and Date (2002) of Hitachi. The shearing tool is built into the end of a miniature pendulum, as shown in Figure 7.17. The linear velocity of the shearing tool just before and just after the impact are measured optically; the difference in the kinetic energy is taken to be consumed in plastically deforming and fracturing the solder ball. Date, Shoji, Fujiyoshi, Sato, Tu (2004a,b), also of Hitachi, used the same method to study the fracture energy and the fracture modes of solder ball-on-pad for solder balls of SnPb, SnAg, SnAgCu, SnZn and SnZnBi alloys. Similar tests were also reported by Ou, Xu, Tu, Alam, Chan (2005) of the University of California, Los Angeles, for solder balls of compositions Sn1Ag, SAC105 and Sn1Ag0.5Cu1In. The angular positions of the pendulum are recorded and the difference in the potential energy of the pendulum before and after the impact is assumed to have consumed in plastically deforming and fracturing the solder ball.

In view of the large deflection of the pendulum, it is highly unreliable to evaluate the potential energy of the pendulum through the angular position of the pendulum. The direct measurement of the velocity of the shearing tool as reported by researchers

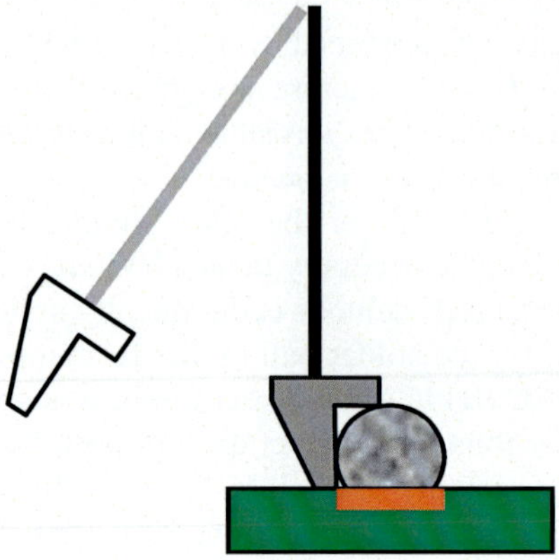

Figure 7.17 Pendulum ball impact shearing test.

of Hitachi offers better accuracy. The main source of error for the pendulum method is that not all the loss of the potential energy of the pendulum goes into plastically deforming and fracturing the solder ball; a fraction of the energy goes into deforming the pendulum elastically, which if unaccounted for can result in significant measurement errors.

The following analysis serves to provide the guidelines for designing a pendulum impactor. The design is guided by two considerations. The first consideration is the need to limit the magnitude of the elastic energy in the pendulum impactor after impact so as to limit measurement error. The fracture energy of the solder ball for a half-sine impact pulse of shear amplitude $F_{\rm i}$ and a shear displacement d is given by

$$\mathrm{FE} = \frac{2F_{\rm i}d}{\pi} \tag{7.5}$$

Assuming a pendulum of uniform cross-section, the strain energy of the pendulum of length L is given by

$$\mathrm{SE} = \frac{F_{\rm i}^2 L^3}{6EI} \tag{7.6}$$

Constraining the strain energy of the pendulum to be at 10% of the fracture energy of the solder ball gives

$$\frac{L^3}{EI} = \frac{6d}{5\pi F_{\rm i}} \tag{7.7}$$

The second design consideration is the need to limit the input energy of the pendulum impactor so as to have a high measurement resolution. The kinetic energy of the pendulum with a mass, m, and a tip speed, v, is given by

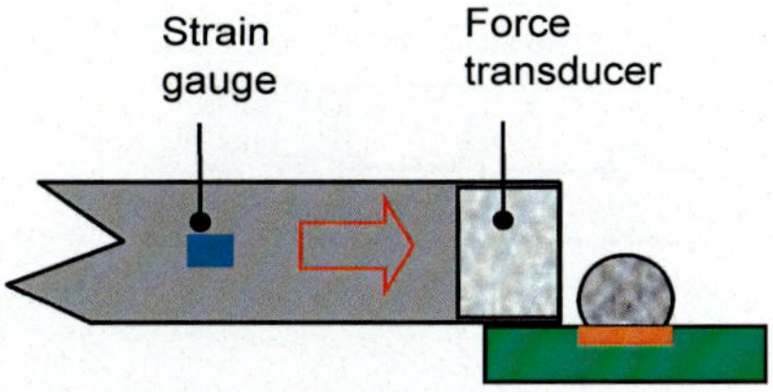

Figure 7.18 Instrumented ball impact shearing test.

$$KE = \frac{mv^2}{6} \tag{7.8}$$

The kinetic energy of the pendulum must be larger than the fracture energy of a solder ball, but it must not be too much larger as to lose sensitivity; an ideal kinetic energy could be between three and six times the fracture energy of a solder ball. Equating the kinetic energy with three times the loss in the potential energy (ignoring friction),

$$PE = mgh \tag{7.9}$$

where h is the drop-height at the centre of mass of the pendulum, gives

$$h = \frac{v^2}{18g} \tag{7.10}$$

Setting $v = 1\ \mathrm{m\,s^{-1}}$, the rest of the parameters can be evaluated iteratively.

Williamson et al. (2002) reported the use of a sophisticated high-speed water jet tester that sends a narrow stream of high-speed water jet (at 317 m/s) to impinge upon the back of a BGA package; they reported the number of dislodged solder balls per shot as 1, 0.5, 3.3 and 3.3, for SnPb, SnAg, SnAgCu and SnAgCuSb, respectively. While the simplicity of this test method is appealing, the high-speed water jet tester is not commercially available. Williamson et al. also reported using the modified split-Hopkinson pressure bar technique to measure the fracture strength of the above solder balls through strain gauging the pressure bar. However, the complexity of the Hopkinson bar setup limits its application to a laboratory rather than a manufacturing environment. It was left to the commercial companies, Dage and Instron, to offer a commercial ball impact shear tester.

BIST using a tester instrumented with either a strain gauge or a force transducer, as shown in Figure 7.18, was reported by Newman (2005) of Sun Microsystems, Wong et al. (2005, 2006) of the Institute of Microelectronics, Yeh, Lai, Chang, and Chen (2005, 2007) and Lai, Chang, and Yeh (2007) of ASE, Valota, Losavioa, Renarc, and Vicenzo (2006) of ST Microelectronics, Song, Lee, Newman, Clark, and Sykes (2007) and Song, Lee, Newman, Sykes, and Clark (2007) of HKUST, and Zhao, Caers, De Vries, Wong, and Rajoo (2007) of Philips.

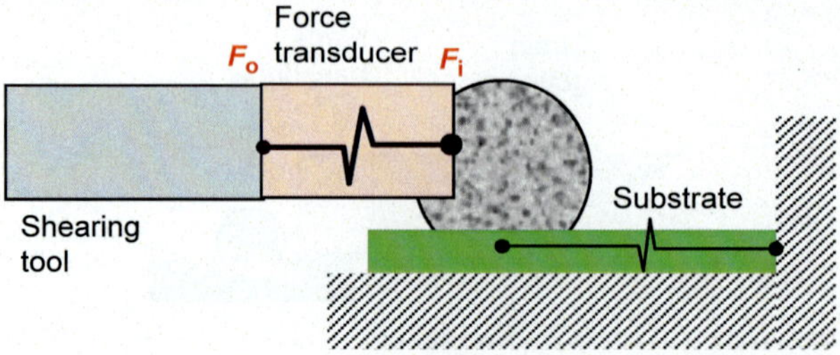

Figure 7.19 Schematic of ball impact shear test (BIST).

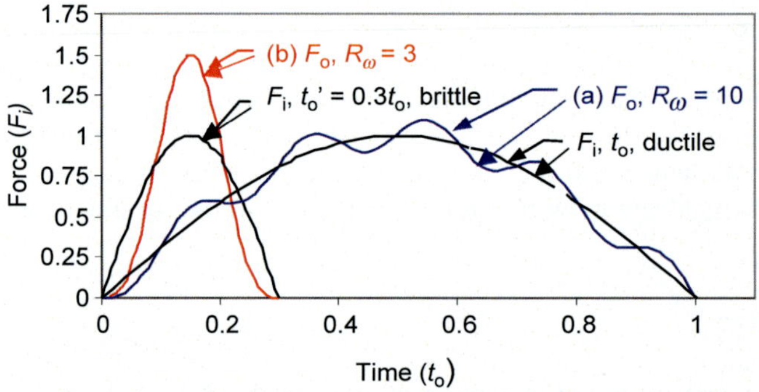

Figure 7.20 Dynamic response of strain gauge to impact pulses.

The shearing tool is an elastic body, whose resonance leads to complication in the measurement of impact force. The placement of a force transducer at the tip eliminates the resonance of the shearing tool. This feature was first adopted by the Micro Impactor of Instron. However, even a force transducer has resonant frequency and a piezoelectric transducer offers the highest natural frequency. A piezoelectric transducer was used in the first generation of Micro Impactor. Figure 7.19 shows a schematic of the BIST, where only the deformation of the elements in the horizontal direction is considered. The impact gives rise to a force F_i between the shearing tool and the solder ball. However, the compliance of the force transducer results in the transducer registering a false magnitude of F_o. Assuming the impulse from the impact to be half-sine with frequency Ω, and approximating the duration of impact as

$$t_o = \frac{1}{2\Omega} \approx \frac{\text{pad diameter}}{\text{shear speed}} \tag{7.11}$$

a shear speed of 1 m/s and a pad diameter of 0.3 mm would give $t_o \approx 0.3$ ms and $\Omega \approx 1.7$ kHz. Commercial piezoelectric transducers have resonant frequencies of approximately $\omega \approx 30$ kHz. This gives the frequency ratio, $R_\omega = \omega/\Omega \approx 18$. In the case of brittle fracture, fracture would occur before the shearing tool smashes through the entire solder ball. Figure 7.20 illustrates two responses of $F_o - t$ for ductile ball

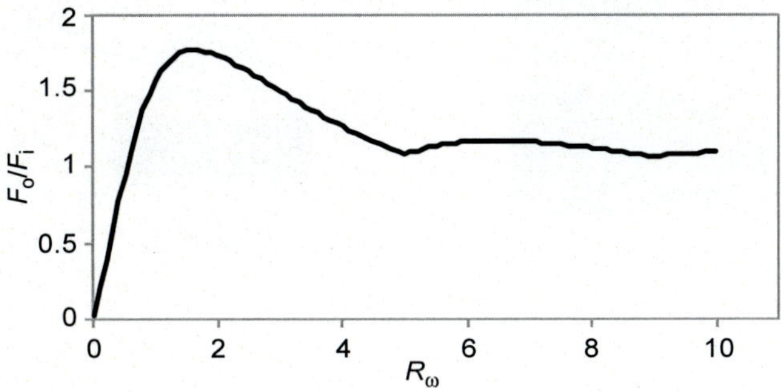

Figure 7.21 Amplification of impact force.

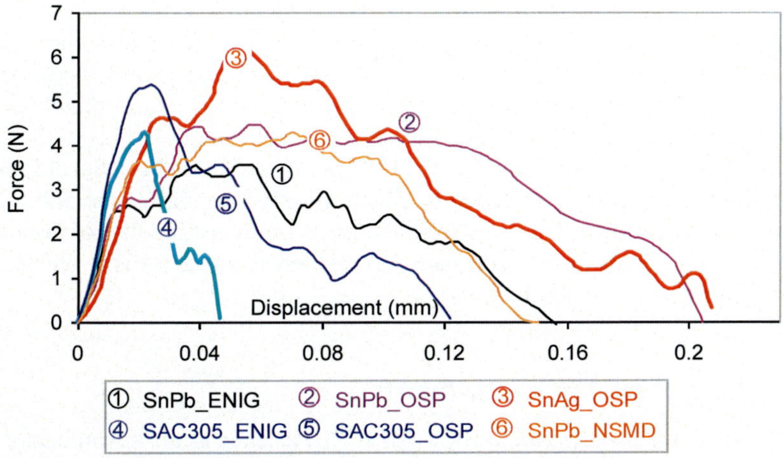

Figure 7.22 Typical force—displacement characteristics of solder joint material systems.

shearing ($R_\omega = 10$, impact duration $= t_o$) and brittle ball shearing ($R_\omega = 3$, impact duration $= 0.3t_o$), respectively. The case of ductile shearing returns a measured force F_o that is a near duplication of the impact force, F_i, but with rippling, which gives rise to minor overestimation in the magnitude of the impact force magnitude. The case of brittle shearing returns a measured force F_o, whose magnitude is 50% higher than the impact force F_i. Figure 7.21 shows the amplification of impact force, F_o/F_i, as a function of frequency ratio, R_ω. Thus, errors due to amplification can be reduced by increasing R_ω, which may be achieved through lowering the speed of impact and by using a transducer with higher natural frequency.

Rippling is an inherent feature of the instrumented BIST, and high-frequency filtering is used routinely to suppress excessive rippling. Figure 7.22 (Wong et al., 2005) shows typical force—displacement characteristics of various solder joint material systems after filtering. The displacement of the shearing tool is measured using a linear variable differential transducer. The test results have highlighted the fragility of the SAC305 solder joint for impact application.

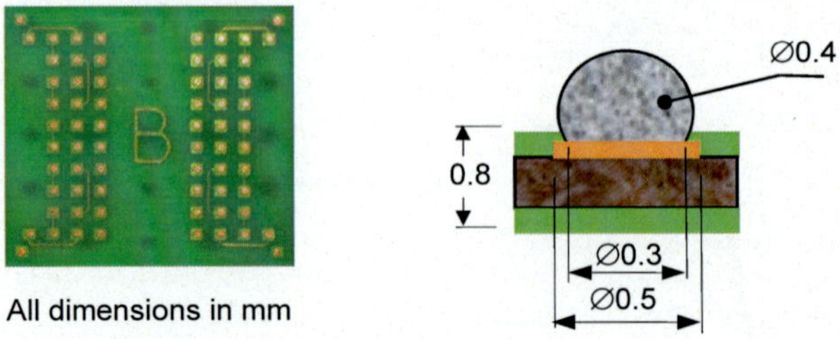

Figure 7.23 Schematic of test specimen for BIST.

Newman (2005), Song, Lee, Newman, Clark, et al.(2007) and Song, Lee, Newman, Sykes, et al.(2007) have evaluated the fracture of solder balls using high-speed pulling. Compared to ball impact-shear tests, ball impact-pull tests exhibit a lower ductile-to-brittle transition test speed. This is not at all surprising in view of the fact that pulling introduces predominantly mode I (opening mode) loading and that mode I fracture toughness of most materials is lower than its mode II (in-plane shear mode) fracture toughness. However, the ball impact-pull test suffers from some drawbacks that have limited its adoption: (1) gripping of the solder ball prior to impact-pulling could potentially compromise the integrity of the IMC because of the substantial deformation of the ball by the grip that is needed to enable positive griping of the ball; (2) it is impossible to extract fracture energy from the impact-pull test because of the substantial slippage between the grip and the ball; and (3) there is a need for very precise alignment of the grip to the solder ball.

7.2.2 The correlation experiments

The brittleness of the solder ball in BIST (BIST-BI) is taken as the key indicator of the quality of the IMC structure in a solder ball (on pad). However, the assessment of brittleness tends to be subjective and may differ between quality inspectors. The instrumental BIST could generate a number of characteristic parameters: the peak load, total fracture energy and energy to peak load. The aim of this correlation experiment is to establish the characteristic parameters that would give the most reliable indication of the brittleness of the solder ball (on pad).

Another aim of this correlation exercise is to assess the ability of BIST to duplicate BLDST. The experimental data for BLDST reported in Section 7.1.3 will be used in this correlation study; hence, the description of experiments will only be given for the BIST (Wong et al., 2008).

7.2.2.1 Test specimen

Figure 7.23 shows a schematic of the ball-on-substrate specimen used for the ball impact shear tests. The solder balls are 0.4 mm in diameter on a solder-mask-defined pad with 0.3 mm opening diameter.

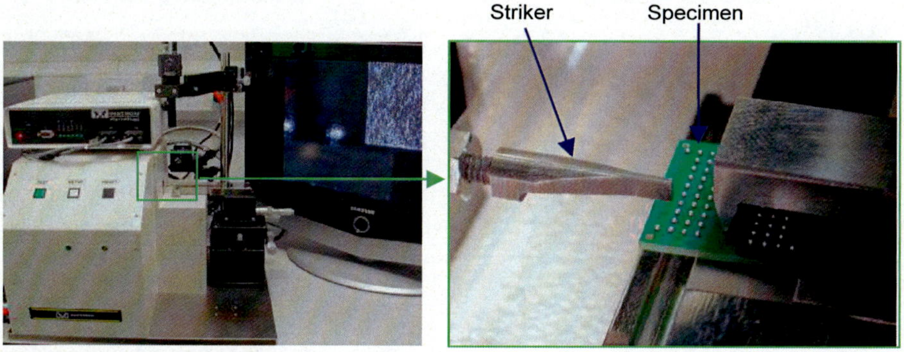

Figure 7.24 Instron microimpactor.

7.2.2.2 Test conditions

The BIST is performed on the Instron MicroImpact tester shown in Figure 7.24. A piezoelectric transducer with a resonant frequency of 35 kHz is attached to the tip of a shearing tool, which is driven by a set of flexural springs whose strain energy can be adjusted to vary the impact speed. The displacement of the shearing tool is measured with a linear velocity displacement transducer. BIST is performed at two shearing speeds: 0.5 and 1 m/s. The peak force, energy-to-peak force (area under the force–displacement curve up to the point of peak force) and fracture energy (area under the force–displacement curve) are recorded.

7.2.2.3 Test matrix

There are two test matrices: materials variations and manufacturing variations; these have been detailed in Tables 7.1 and 7.2, respectively. Between 16 and 20 data points are obtained for each test combination. A small matrix made of four SAC solders with increasing silver contents – SAC105, SAC205, SAC305 and SAC405 – on an OSP pad finishing is prepared for BIST at 1 m/s.

7.2.3 The consolidated experimental data

The cumulative distributions of measured characteristics for the material and manufacturing variations for the shear speed of 0.5 m/s are shown in Figure 7.25. The failure mode distributions are shown in Figure 7.26, in which the designation BIST-0.5 and BIST-1.0 represents shearing speeds of 0.5 and 1.0 m/s, respectively. The characteristic parameters and the brittleness indices for the material and manufacturing variations studies are listed in Tables 7.8 and 7.9, respectively. The cumulative distributions and failure modes for the four SAC solders with different silver content are summarised in Figure 7.27.

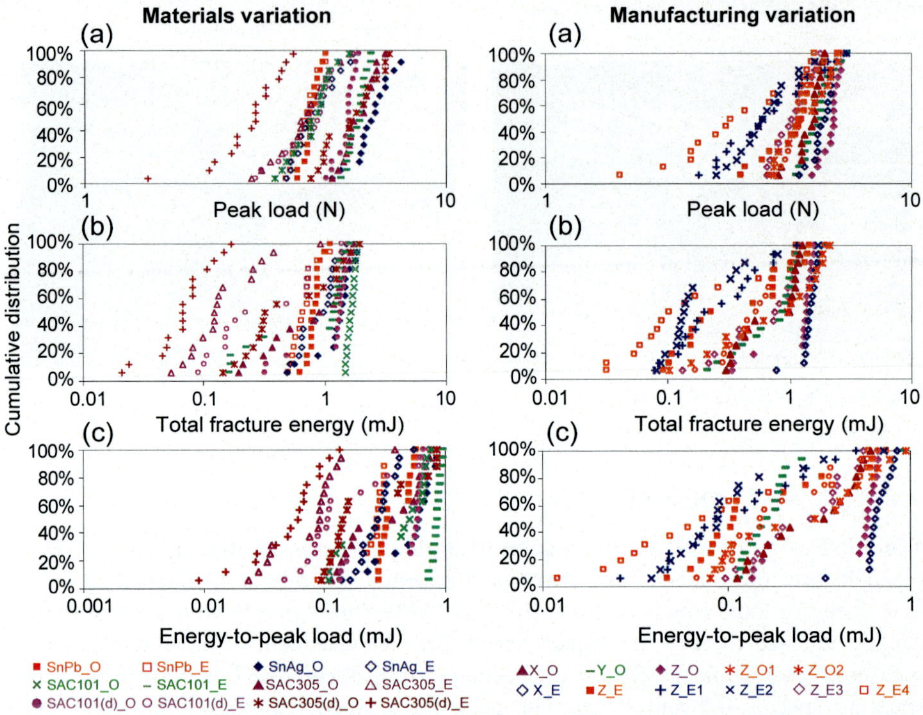

Figure 7.25 Cumulative distribution of the characteristic parameters of BIST: (a) peak fracture load, (b) total fracture energy and (c) energy-to-peak load for materials variation and manufacturing variation.

7.2.4 The analysis of correlations among the characteristic parameters of BIST

The brittleness index, peak load, total fracture energy and energy-to-peak load are designated as BIST-BI, BIST-Load, BIST-Energy and BIST-EPL, respectively. BIST-BI plotted against BIST-Load, BIST-Energy and BIST-EPL for shear speeds of 0.5 and 1.0 m/s are shown in Figure 7.28. Correlation coefficients for the materials variations, manufacturing variations and combinations thereof are listed in Table 7.10.

In general, the brittleness index decreases with increased BIST-Load, BIST-Energy and BIST-EPL. Figure 7.28 suggests that the brittle index exhibits a very distinct trend with BIST-Energy, a less distinct trend with BIST-EPL and a weak trend with BIST-Load. The correlation data in Table 7.10 suggest that the weak correlation between BIST-BI and BIST-Load is restricted to the materials variation study; it is especially apparent at the lower shearing speed. The remarkably good correlation between BIST-BI and BIST-Energy is only disturbed by the outlier data points of Sn37Pb solder in the materials variation study, which is attributed to the much lower flow stress of Sn37Pb solder compared to the other solders.

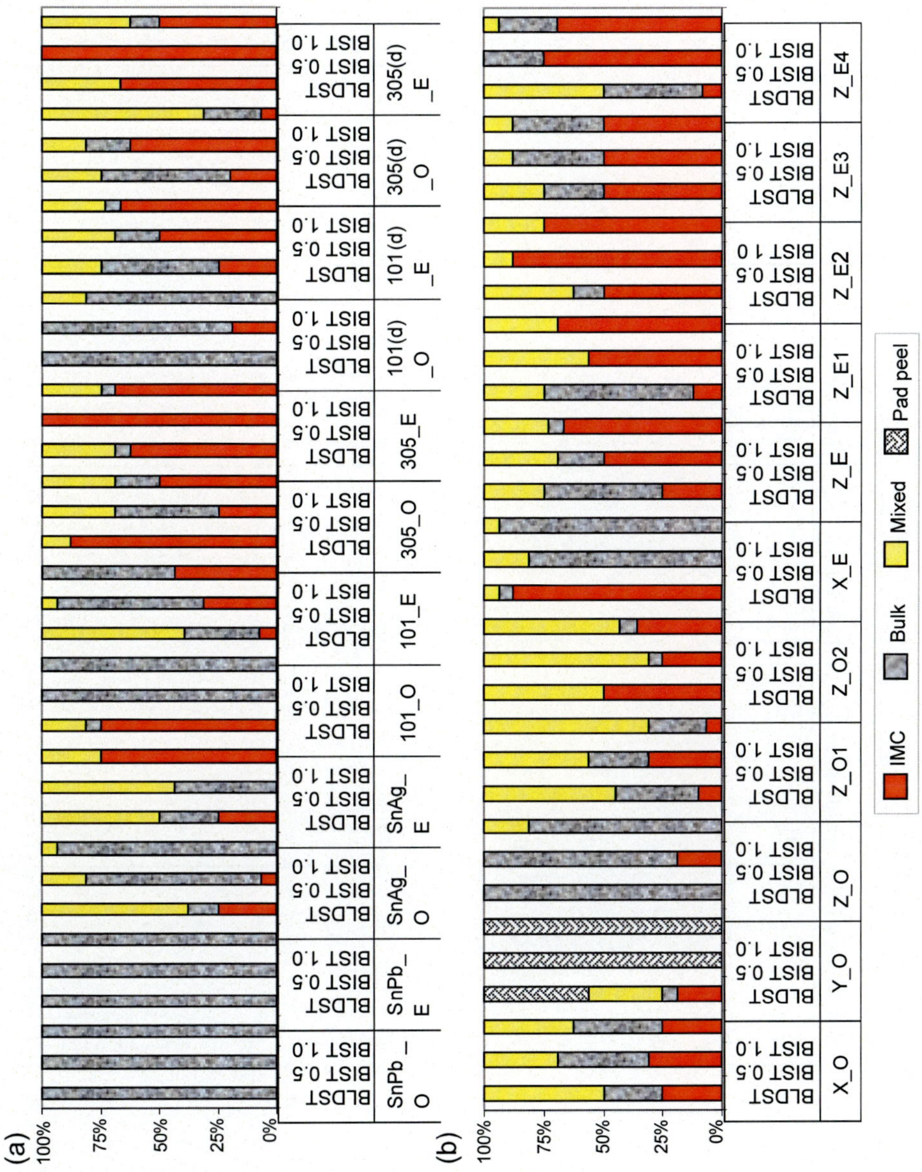

Figure 7.26 Distribution of failure modes for (a) materials variation and (b) manufacturing variation.

Table 7.8 **Summary of characteristic parameters and brittleness indices for the materials variation study**

	BLDT		BIST Shear speed: 0.5 m/s Shear speed: 1.0 m/s			
Solder joint	**Characteristic life**	**Brittle index**	**Peak load**	**Fracture energy**	**Energy-to-peak load**	**Brittle index**
SnPb_O	57	0.0	4.9 5.5	0.9 0.8	0.5 0.2	0.0 0.0
SnPb_E	28	0.0	4.8 5.6	0.7 0.9	0.3 0.3	0.0 0.0
SnAg_O	24	0.6	7.5 9.4	1.5 1.9	0.7 0.6	0.2 0.0
SnAg_E	19	0.5	5.3 5.4	1.1 0.3	0.3 0.1	0.3 0.9
SAC101_O	20	0.8	5.1 8.3	1.7 1.9	0.5 0.6	0.0 0.0
SAC101_E	16	0.4	6.6 5.6	1.2 0.9	0.9 0.3	0.3 0.4
SAC305_O	20	0.9	6.8 7.5	1.1 0.8	0.5 0.3	0.4 0.7
SAC305_E	11	0.8	5.0 6.6	0.2 0.5	0.1 0.2	1.0 0.8
SAC101(d)_O	64	0.0	6.2 7.9	1.4 1.7	0.6 0.5	0.2 0.1
SAC101(d)_E	39	0.4	5.2 5.9	0.6 0.4	0.2 0.2	0.7 0.8
SAC305(d)_O	31	0.3	6.7 7.2	0.8 1.1	0.4 0.4	0.7 0.4
SAC305(d)_E	11	0.8	3.6 5.3	0.1 0.4	0.1 0.2	1.0 0.7

Table 7.9 Summary of characteristic parameters and brittleness indices for the manufacturing variation study

Solder joint	BLDT		BIST Shear speed: 0.5 m/s Shear speed: 1.0 m/s			
Solder alloy: SAC101(d)	Characteristic life	Brittle index	Peak load	Fracture energy	Energy-to-peak load	Brittle index
X_O	24	0.5	5.4	0.9	0.4	0.5
			6.3	0.8	0.3	0.4
Y_O	70	0.61[a]	5.8	0.9	0.2	[a]
			6.5	0.8	0.2	[a]
Z_O	64	0.0	6.2	1.5	0.6	0.2
			7.9	1.8	0.5	0.1
Z_O1	45	0.4	5.5	1.1	0.4	0.5
			6.9	1.1	0.3	0.4
Z_O2	13	0.8	5.2	0.7	0.3	0.6
			6.3	0.6	0.2	0.6
X_E	9	0.9	5.9	1.6	0.7	0.1
			6.5	1.5	0.3	0.0
Z_E	39	0.4	5.2	0.6	0.2	0.7
			5.9	0.4	0.2	0.8
Z_E1	24	0.3	4.7	0.4	0.2	0.8
			4.0	0.3	0.1	0.8
Z_E2	7	0.7	4.5	0.3	0.2	0.9
			4.7	0.3	0.1	0.9
Z_E3	13	0.6	5.0	0.8	0.3	0.6
			5.9	0.8	0.2	0.6
Z_E4	29	0.3	4.1	0.3	0.2	0.7
			5.3	0.5	0.2	0.7

[a]Predominantly pad peeling.

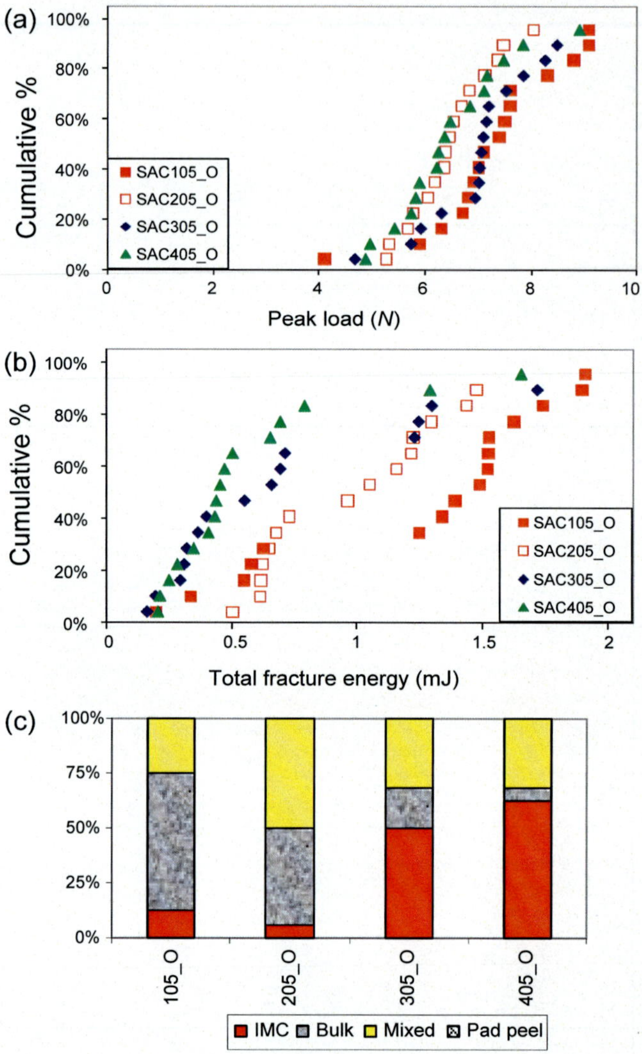

Figure 7.27 Silver content. (a) Cumulative distribution for peak load. (b) Cumulative distribution for total fracture energy. (c) Distribution of failure modes.

Increasing the shearing speed from 0.5 to 1 m/s has resulted in shifting of the data for BIST-BI versus BIST-Load to the right (Figure 7.28(a)) and shifting of the data for BIST-BI versus BIST-EPL to the left (Figure 7.28(c)). In comparison, the data of BIST-BI versus BIST-Energy appear to be insensitive to shearing speed (Figure 7.28(b)).

Referring to Figure 7.27, the data for the four solders bunched up in the case of BIST-Load, suggesting its indifference to the content of silver. The data for

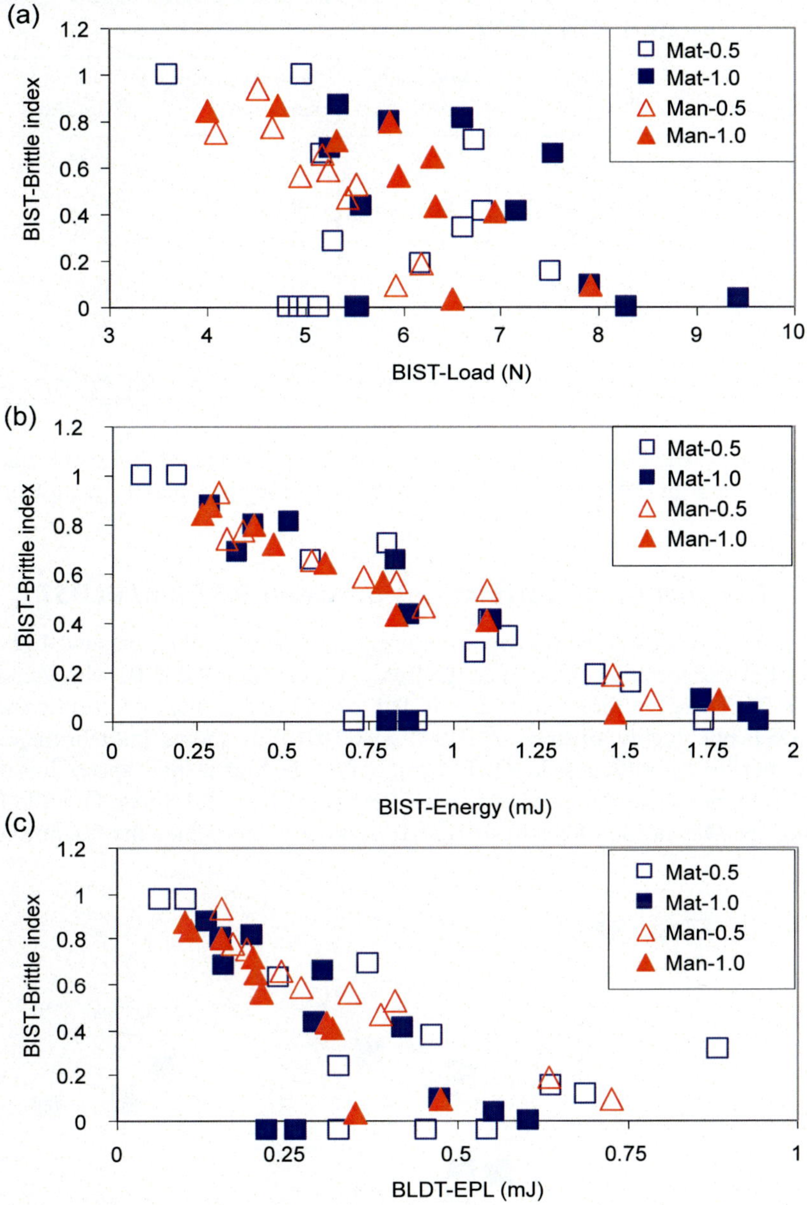

Figure 7.28 BIST-BI plotted against (a) BIST-Load, (b) BIST-Energy and (c) BIST-EPL.

BIST-Energy present a clear trend of decreasing fracture energy with increasing silver content. The trend of reducing fracture energy is supported by the observed increased brittleness of the fracture joints.

Table 7.10 The correlations of BIST-Brittle index with BIST-Load, BIST-Energy and BIST-EPL

Characteristics	Shear speed	Materials variation	Manufacturing variation	Combined
BIST-Brittle index vs BIST-Load	0.5 m/s	−0.26	−0.89	−0.45
	1.0 m/s	−0.43	−0.81	−0.59
BIST-Brittle index vs BIST-Energy	0.5 m/s	−0.77	−0.97	−0.83
	1.0 m/s	−0.78	−0.97	−0.85
BIST-Brittle index vs BIST-EPL	0.5 m/s	−0.64	−0.98	−0.75
	1.0 m/s	−0.70	−0.94	−0.79

In summary, BIST-Energy is the most reliable indicator and BIST-Load is the least reliable indicator of failure mode in BIST for both the materials and the manufacturing variations.

7.2.5 The analysis of correlations between BIST and BLDST

Figure 7.29 shows BIST-BI (at two shear speeds: 0.5 and 1.0 m/s) versus BLDST-BI for material variance that shows a vague trend of increasing BIST-BI with increasing BLDST-BI. There are three outliers: SAC101_O and SnAg_O exhibit brittle fracture in BLDST but ductile fracture in BIST; SAC101(d)_E shows mild brittleness in BLDST but strong brittleness in BIST, albeit only at the higher shear speed. It is postulated that higher shearing speed is needed for SAC101_O and SnAg_O while lower shearing speed is needed for SAC101(d)_E in order to reproduce the fracture model

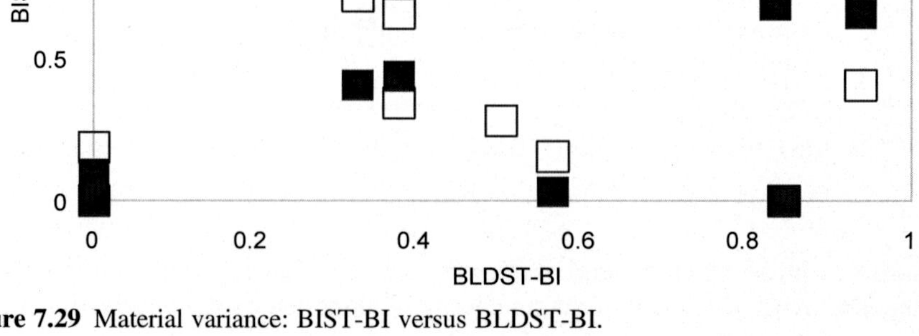

Figure 7.29 Material variance: BIST-BI versus BLDST-BI.

Table 7.11 **Material variance: correlations between BIST-BI and BLDST-BI**

Shear speed	All data	Excluding SAC101_O and SnAg_O
0.5 m/s	0.46	0.72
1.0 m/s	0.49	0.82

of BLDST. Table 7.11 shows moderate correlations for the two test methods, which are improved dramatically when the two data points of SAC101_O and SnAg_O are excluded.

The BIST-Load and BIST-Energy for the materials variations at the shearing speed of 0.5 m/s are plotted against BLDST-Life in Figure 7.30, where the data symbols for OSP finish are given in brackets. The correlation and sensitivity coefficients are shown

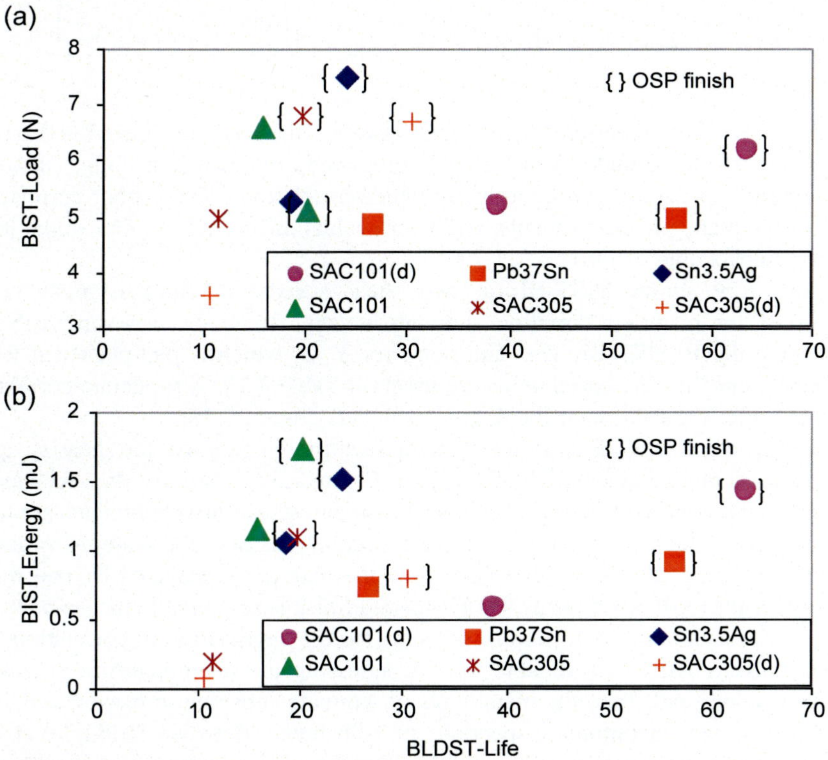

Figure 7.30 BIST (0.5 m/s) versus BLSDT for materials variation: (a) fracture load and (b) fracture energy.

Table 7.12 Material variance: correlations of BIST (0.5 m/s)-Load and BIST-Energy with BLDST-Life

BIST-Load			BIST-Energy		
Range	$\rho_{x,y}$	$G_{x,y}$	Range	$\rho_{x,y}$	$G_{x,y}$
Full	0.11	–	Full	0.27	–
Segregation by finishes					
OSP	−0.39	–	OSP	−0.25	–
ENIG	0.17	–	ENIG	0.24	–

Table 7.13 Manufacturing variance: correlations between BIST-BI and BLDST-BI

Shear speed	All data	Excluding X_E
0.5 m/s	−0.03	0.49
1.0 m/s	−0.02	0.48

in Table 7.12. No universal correlations between BLDST-Life and BIST-Load or BIST-Energy are found even after segregation by pad finishing, although BIST-Energy appears to give marginally better correlations. These poor correlations have ruled out any prospect of using BIST for evaluation and selection of solder joints for drop-impact applications.

Figure 7.29 shows BIST-BI (at two shear speeds: 0.5 and 1.0 m/s) versus BLDST-BI for materials variation showing a vague trend of increasing BIST-BI with increasing BLDST-BI. The data point for Y_O, which is pad-peeled in BIST, has been excluded. The correlations are given in Table 7.13. A moderate correlation appearing after the data point for X_E is excluded (Figure 7.31).

BIST-Load and BIST-Energy for the manufacturing variation at the shearing speed of 0.5 m/s are plotted against BLDST-Life in Figure 7.32, where the data for thermal ageing are given within brackets. The correlation and sensitivity coefficients are listed in Table 7.14, which show somewhat better correlations than the material variation. A much better correlation is observed when the data are segregated by pad finish, but only for the OSP finish. Thus, the ENIG pad finish is responsible for the poor correlation. The data point of X_E is responsible for the degradation in correlation.

In summary, BIST is a reliable test for assessing the manufacturing process of reflowing solder balls onto the metal pads of laminate substrate; however, it is not a reliable test for incoming inspection of substrate. Between BIST-Load and BIST-Energy, the latter is a more reliable characteristic parameter, and it also has a higher sensitivity.

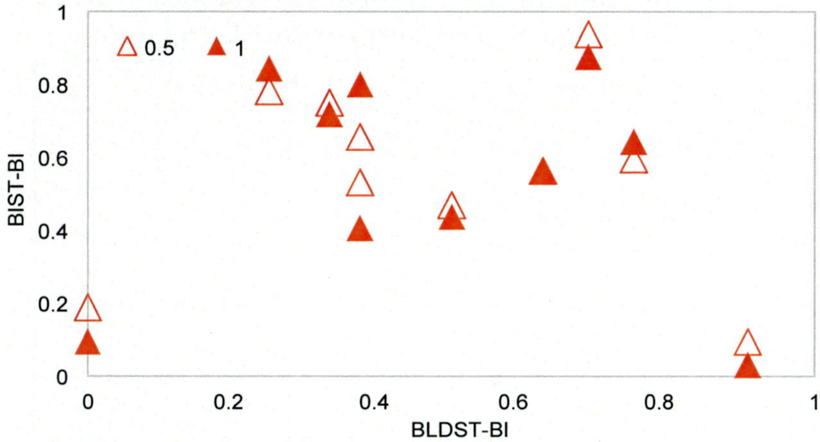

Figure 7.31 Manufacturing variation: BIST-BI versus BLDST-BI.

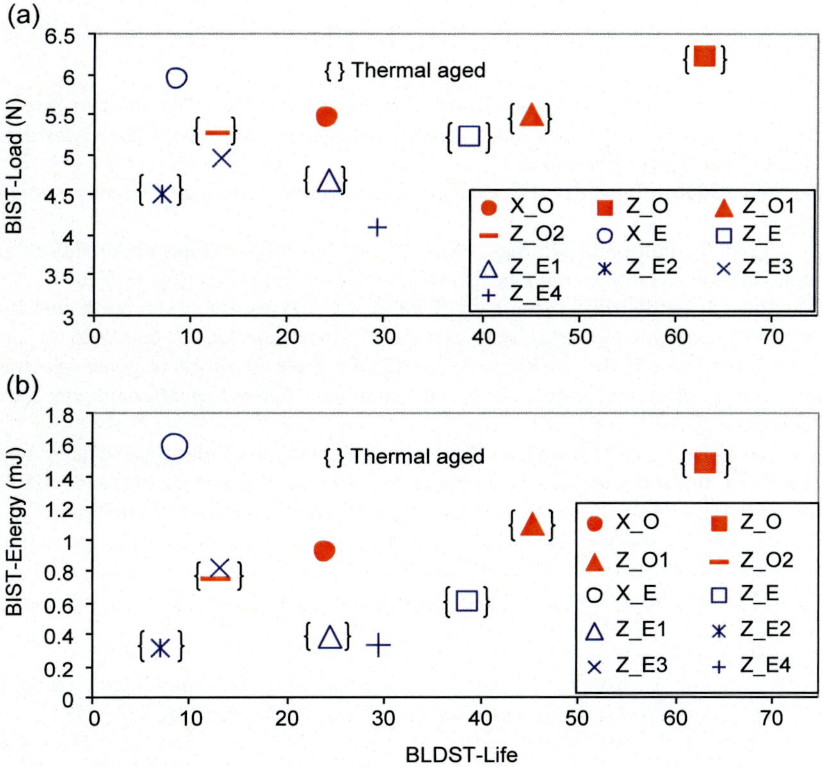

Figure 7.32 BIST (0.5 m/s) versus BLDST for manufacturing variation: (a) fracture load and (b) fracture energy.

Table 7.14 Manufacturing variation: correlations of BIST (0.5 m/s)-Load and BIST-Energy with BLDST-Life

BIST-Load			BIST-Energy		
Range	$\rho_{x,y}$	$G_{x,y}$	Range	$\rho_{x,y}$	$G_{x,y}$
Full	0.41	–	Full	0.31	–
Segregation by finishes					
OSP	0.93	18%	OSP	0.99	59%
ENIG	−0.25	–	ENIG	−0.41	–
For thermal-aged specimens only					
OSP	0.93	18%	OSP	0.99	61%
ENIG	0.94	16%	ENIG	0.95	56%

NB: Sensitivity coefficients were computed only for those cases whose correlation coefficients are greater than 0.6.

References

ASTM-D5276. (2009). *Standard test method for drop test of loaded containers by free fall.*

ASTM-D6537. (2014). *Standard practice for instrumented package shock testing for determination of package performance.*

ASTM-D999. (2008). *Standard test methods for vibration testing of shipping containers.* ASTM.

Date, M., Shoji, T., Fujiyoshi, M., Sato, K., & Tu, K. N. (2004a). Impact reliability of solder joints. In *54th electronic components and technology conference* (pp. 668–674).

Date, M., Shoji, T., Fujiyoshi, M., Sato, K., & Tu, K. N. (2004b). Ductile-to-brittle transition in Sn–Zn solder joints measured by impact test. *Scripta Material, 51*, 641–645.

Goyal, S., Buratynski, E. K., & Elko, G. W. (2000). Role of shock response spectrum in electronic product suspension design. *International Journal of Microcircuits Electron Packaging, 23*, 182–189.

Hirata, I., Yoshida, Y., & Morizaki, I. (2001). Drop-simulation of electronic boards mounted with CSPs. In *1st international conference for electronic packaging* (pp. 422–426).

IPC/JEDEC-9703. (2009). *Mechanical shock test guideline for solder joint reliability.*

ISO-5347-22. (1997). *Methods for the calibration of vibration and shock pick-ups – Part 22: Accelerometer resonance testing – General methods.*

JESD22-B117A. (2006). *Solder ball shear.* JEDEC Standard.

JESD22B111. (2003). *Board level drop test method of components for handheld electronic products.* JEDEC.

Juso, H. A., Yamaji, Y., Kimura, T., Fujita, K., & Kada, M. (1998). Board level reliability of CSP. In *48th electronic components and technology conference* (pp. 525–531).

Kim, J.-Y., Kang, I. S., Park, M. G., Kim, J. H., Cho, S. J., Park, L. S., et al. (2001). Characterization of wafer level package for mobile phone application. In *51st electronic components and technology conference* (pp. 1398–1401).

Lai, Y.-S., Chang, H.-C., & Yeh, C.-L. (2007). Evaluation of solder joint strengths under ball impact test. *Microelectronics Reliability, 47*, 2179–2187.

Langmont. (n.d.). Retrieved from http://www.lansmont.com/shocktest/default.htm.

Marjamäki, P. M., Mattila, T. T., & Kivilahti, J. K. (2006). A comparative study of the failure mechanisms encountered in drop and large amplitude vibration tests. In *56th electronic components and technology conference* (pp. 95–101).

Mishiro, K., Ishikawa, S., Abe, M., Kumai, T., Higashiguchi, Y., & Tsubone, K.-i. (2002). Effect of the drop impact on BGA/CSP package reliability. *Microelectronics Reliability, 42*, 77–82.

Nagano, K., Yaguchi, A., Terasaki, T., & Yamamoto, K. (2006). Fatigue strength of BGA type solder joints between package and printed wiring board of portable device. *JSME Series A, 49*, 220–228.

Newman, K. (2005). BGA brittle fracture – alternative solder joint integrity test methods. In *55th electronic components and technology conference* (pp. 1194–1201).

Ou, S., Xu, Y., Tu, K. N., Alam, M. O., & Chan, Y. C. (2005). Micro-impact test on lead-free BGA balls on Au/electrolytic Ni/Cu bond pad. In *55th electronic components and technology conference* (pp. 467–471).

Peng, H., Johnson, R. W., Flowers, G. T., Ricketts, A.-G., Yeager, E. K., Konarski, M. M., et al. (2001). Underfilling fine pitch BGAs. *IEEE Transactions on Electronics Packaging Manufacturing, 24*, 293–299.

Qiang, Y., Kikuchi, H., Ikeda, S., Shiratori, M., Kakino, M., & Fujiwara, N. (2002). Dynamic behaviour of electronics package and impact reliability of BGA solder joints. In *Inter society conference on thermal phenomena* (pp. 953–960).

Reiff, D., & Bradley, E. (2005). A novel mechanical shock test method to evaluate lead-free BGA solder joint reliability. In *55th electronic components and technology conference* (pp. 1519–1525).

Seah, S., Wong, E. H., Mai, Y. W., Rajoo, R., & Lim, C. T. (2006). High-speed bend test method and failure prediction for drop impact reliability. In *56th electronic components and technology conference* (pp. 1003–1008).

Shoji, T., Yamarmoto, K., Kajiwara, R., Morita, T., Sato, K., & Date, M. (2002). Study of interface reaction phenomena between Pb-free solder and Cu. In *16th JIEP annual meeting* (pp. 97–98).

Sogo, T. H., & Hara, S. (2001). Estimation of fall impact strength for BGA solder joint. In *1st international conference for electronic packaging* (pp. 369–373).

Song, F., Lee, S. W. R., Newman, K., Clark, S., & Sykes, B. (2007). Comparison of joint strength and fracture energy of lead-free solder balls in high speed ball shear/pull tests and their correlation with board level drop test. In *Electronic packaging technology conference* (pp. 450–458).

Song, F., Lee, S. W. R., Newman, K., Sykes, B., & Clark, S. (2007). High-speed solder ball shear and pull tests vs board level mechanical drop tests: correlation of failure mode and loading speed. In *57th electronic component technology conference* (pp. 1504–1513).

Valotaet, A., Losavioa, A., Renarc, L., & Vicenzo, A. (2006). High speed pull test characterization of BGA solder joints. In *7th EuroSime.*

Varghese, J., & Dasgupta, A. (2003). Test methodology for impact testing of portable electronic products. In *ASME international mechanical engineering congress & exposition.* pp. IMECE2003-41844.

Williamson, D. M., Palmer, S. J. P., Kennedy, C. F., Siviour, C. R., Siviour, G. B., Walley, S. M., et al. (2002). *Spall, quasi-static and high strain rate shear strength data for electronic solder materials.* Internal report Cavendish Laboratory No. SP 1113.

Wong, E., Mai, Y.-W., Rajoo, R., Tsai, K. T., Liu, F., Seah, S. K. W., et al. (2006). Micro impact characterisation of solder joint for drop impact application. In *56th electronic component technology conference* (pp. 64–71).

Wong, E., Rajoo, R., Mai, Y.-W., Seah, S. K. W., Tsai, K. T., & Yap, L. M. (2005). Drop impact: fundamentals and impact characterisation of solder joints. In *55th electronic component technology conference* (pp. 1202–1209).

Wong, E., Rajoo, R., Seah, S., Selvanayagam, C. S., Driel, V. W., Caers, J., et al. (2008). Correlation studies for component level ball impact shear test and board level drop test. *Microelectronic Reliability, 48*, 1069–1078.

Wong, E., Seah, S., van Driel, W., Caers, J., Zhao, X., Owens, N., et al. (2009). High-speed cyclic bend tests and board-level drop tests for evaluating the robustness of solder joints in printed circuit board assemblies. *Journal of Electronic Materials, 38*(6), 884–895.

Yaguchi, A., Yamada, M., & Yamamoto, K. (2003). Reliability evaluation of solder joints in ball-grid-array-type packages by impact bending test. *Journal of Japan Institute of Electronics Packaging, 6*, 314–321.

Yamaji, Y., Yamasaki, H., Juso, H., Ohsono, M., Kimura, T., Fujita, K., et al. (2000). A proposal: the assessing method of the CSPs mechanical reliability on board. *International Journal of Microcircuits Electron Packaging, 23*, 138–144.

Yeh, C.-L., Lai, Y.-S., Chang, H.-C., & Chen, T.-H. (2005). Correlation between package-level ball impact test and board-level drop test. In *7th electronic packaging technology conference* (pp. 270–275).

Yeh, C.-L., Lai, Y.-S., Chang, C.-H., & Chen, T.-H. (2007). Empirical correlation between package-level ball impact test and board-level drop reliability. *Microelectronics Reliability, 47*, 1127–1134.

Zhao, X. J., Caers, J. F. J. M., De Vries, J. W. C., Wong, E. H., & Rajoo, R. (2007). A component level test method for evaluating the resistance of Pb-free BGA solder joints to brittle fracture under shock impact. In *57th electronic component technology conference* (pp. 1522–1529).

Zhu, L. (2001). Submodeling technique for BGA reliability analysis of CSP packaging subjected to an impact loading. In *International electronic packaging technology conference* (pp. IPACK2001-15873). Kauai, HI.

Fatigue resistance of solder joints: strain-life representation

8

8.1 Introduction

8.1.1 Fatigue as the failure driving force

Portable electronic devices (PEDs) are expected to survive multiple drops, with most original equipment suppliers specifying a number between 30 and 50 drops. The robustness requirement of a product has to be supported by the components that make up the product. The solder joints in printed circuit board (PCB) assemblies used for portable products are therefore required to survive multiple shocks; the JEDEC test standard JESD22-B111 suggests 30 drop-shocks. At each drop-shock, the PCB assembly undergoes cyclic flexural deformation at a relatively high amplitude of fibre strain registered on the PCB, but the amplitude decays rapidly, as shown in Figure 8.1.

The board-level solder joints, which experience the same nature of fatigue history, may fail by fatigue; however, the solder joints may also fail by overstress due to the relatively high amplitude of the initial response. Evidence of the fatigue failure of solder joints is provided in Figure 8.2, which was obtained from a PCB assembly subjected to multiple drop-shocks. Marks of fatigue striations, which are the result of the grow-and-stop of fatigue crack in each load reversal, can be observed in the fractured specimen under a scanning electron microscope. Fatigue striations in the fractograph indicate fatigue damage initiated from the left edge of the solder joint and propagated towards the right for approximately two-thirds of the diameter, where an overstress fracture occurred, as indicated by the dimples that characterise ductile fracture. Note that fatigue striations on solder joints are

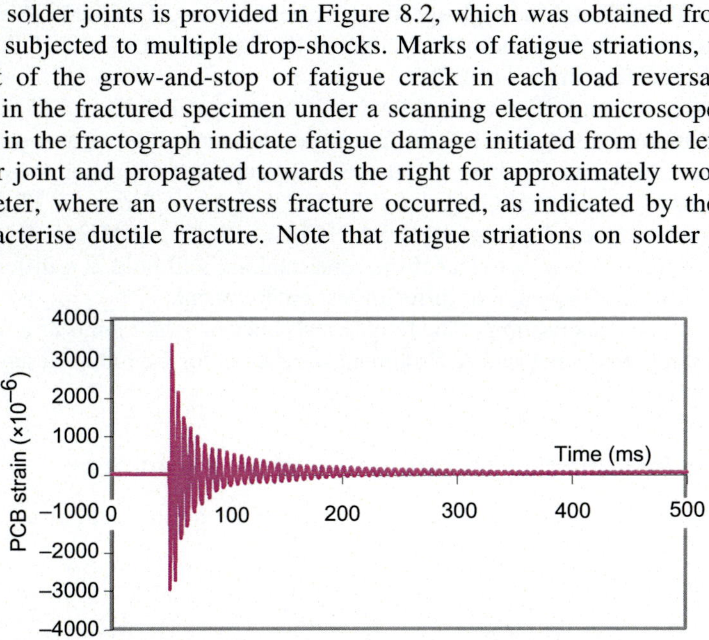

Figure 8.1 PCB strain response for JEDEC test standard JESD22B113.

Robust Design of Microelectronics Assemblies Against Mechanical Shock, Temperature and Moisture
http://dx.doi.org/10.1016/B978-1-84569-528-6.00008-3

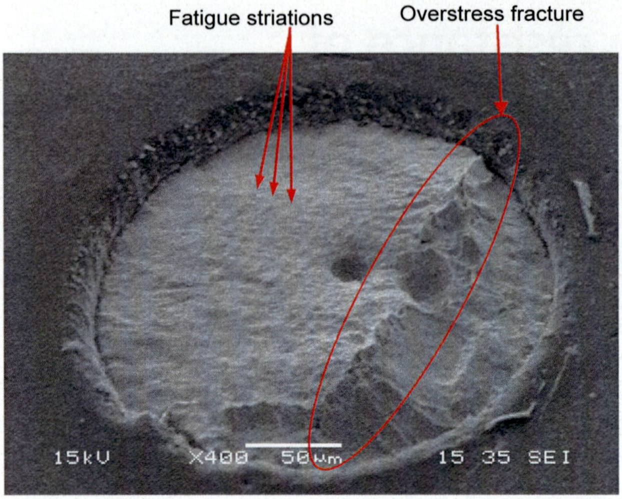

Figure 8.2 Evidence of fatigue failure in solder joint.

rarely observed; the softness of the solder material makes it susceptible to deformation in the subsequent load reversals.

As reported in Chapter 7, there is good correlation between the board-level drop-shock test (BLDST) and the high-speed cyclic bending test (HSCBT). This provides further support for the hypothesis that fatigue is the driving force for the failure of solder joints in the drop impact of PEDs.

When assessed at a PCB strain amplitude of 2×10^{-3} at 100 Hz and 25 °C in a laboratory environment using HSCBT, the fatigue performance of the organic solderability preservative (OSP) pad finish is found to be far superior to that of the electroless nickel immersion gold (ENIG) pad finish. The fatigue performance of the solder alloys is ranked in the following order: SAC101(d) > SnPb > SAC305(d) > SnAg > SAC101 > SAC305. It is not clear if the same ranking will hold at a different strain amplitude, different frequency or different test environment.

A thorough characterisation of the fatigue resistance of solder joints as a function of strain amplitude, frequency and environment is essential for the robust design of PEDs against drop impact.

8.1.2 An introduction to fatigue resistance equations

The fatigue resistance of materials can be expressed in the form of power-law:

$$F = aN^b \tag{8.1}$$

where F is the amplitude of the driving force; N is the fatigue life, expressed as the number of cycles to failure; and a and b are material constants. Cyclic stress amplitude has been found to be the driving force for fatigue failure requiring a large number of cycles (Wöhler, 1855)—that is, high cycle fatigue (Basquin, 1910):

$$\sigma = a_\sigma N^{b_\sigma} \tag{8.2}$$

Cyclic plastic strain amplitude has been found to be driving force for fatigue failure requiring a small number of cycles (Coffin, 1954)—that is, low cycle fatigue:

$$\varepsilon_\mathrm{p} = a_\mathrm{p} N^{b_\mathrm{p}}. \tag{8.3}$$

Expressing σ as $E\varepsilon_\mathrm{e}$, Eqns (8.2) and (8.3) may then be plotted on the same graph. This is shown in Figure 8.3 (Dowling, 2012), in which the solid lines represent the respective domains of Eqns (8.2) and (8.3). The intersection of the two solid lines gives the transitional (i.e., between the low and high cycles) fatigue life N_t:

$$N_\mathrm{t} = \left(\frac{Ea_\mathrm{p}}{a_\sigma}\right)^{\frac{1}{b_\sigma - b_\mathrm{p}}}; \tag{8.4}$$

The corresponding strain amplitudes are:

$$\varepsilon_\mathrm{e} = \varepsilon_\mathrm{p} = a_\mathrm{p}\left(\frac{Ea_\mathrm{p}}{a_\sigma}\right)^{\frac{b_\mathrm{p}}{b_\sigma - b_\mathrm{p}}} \tag{8.5}$$

Assuming that Eqn (8.2) is extendable to low cycle fatigue and Eqn (8.3) is extendable to low cycle fatigue (the dotted lines in Figure 8.3), then a fatigue life equation based on the total strain amplitude as the driving force may be defined as follows:

$$\varepsilon = \varepsilon_\mathrm{e} + \varepsilon_\mathrm{p} = \frac{a_\sigma}{E} N^{b_\sigma} + a_\mathrm{p} N^{b_\mathrm{p}} \tag{8.6}$$

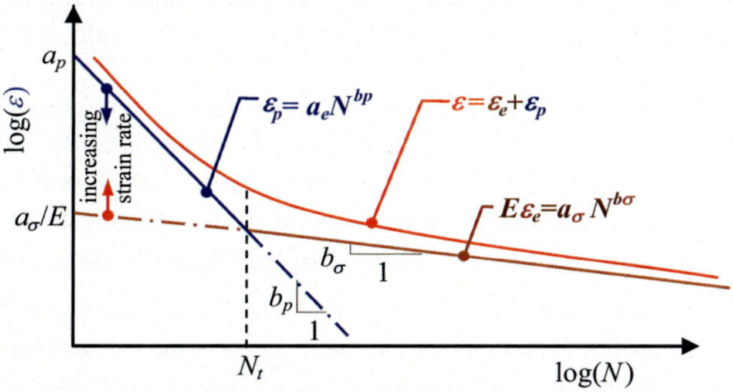

Figure 8.3 Strain range versus cyclic life.

The assumed extendibility of Eqns (8.2) and (8.3) implies that fatigue failure is concurrently dictated by two driving forces: the cyclic stress and the cyclic plastic strain. Substituting Eqn (8.2) into Eqn (8.3) gives

$$\varepsilon_{\mathrm{p}} = a_{\mathrm{p}}\left(\frac{\sigma}{a_{\sigma}}\right)^{b_{\mathrm{p}}/b_{\sigma}}. \tag{8.7}$$

The cyclic stress and cyclic plastic strain are not independent but are governed by the cyclic stress-strain relationship. Expressed in the form of a Ramberg-Osgood equation,

$$\varepsilon_{\mathrm{p}} = \left(\frac{\sigma}{K}\right)^{1/n}. \tag{8.8}$$

Substituting into Eqn (8.7) gives

$$\left(\frac{\sigma}{K}\right)^{1/n} = a_{\mathrm{p}}\left(\frac{\sigma}{a_{\sigma}}\right)^{b_{\mathrm{p}}/b_{\sigma}}, \tag{8.9}$$

which suggests the following compatibility relationships:

$$b_{\sigma} = nb_{\mathrm{p}} \quad \text{and} \quad a_{\sigma} = Ka_{\mathrm{p}}{}^{n}. \tag{8.10}$$

By assuming the extendibility of Eqns (8.2) and (8.3) at high strain rate fatigue and expressing the material constant K in the Ramberg-Osgood equation to be a function of strain rate, for example,

$$K = K_{\mathrm{o}}\dot{\varepsilon}^{\alpha}, \tag{8.11}$$

where α is a positive real number, Eqn (8.10) is readily extended to high strain rate fatigue. Hence, for the same magnitude of allowable plastic strain amplitude, the allowable amplitude of stress (or elastic strain) increases with increasing strain rate. However, for the same magnitude of allowable stress (or elastic strain), the allowable amplitude of plastic strain decreases with increasing strain rate. This is shown in Figure 8.4.

Therefore, the assumed compatibility of the failure drivers suggests that the fatigue life of a material will not be affected by an increase in strain rate if (a) the plastic strain amplitude remains unchanged despite an increase in the stress amplitude or (b) the stress amplitude remains unchanged despite a reduction in the plastic strain amplitude. By contrast, in the absence of the compatibility of the failure drivers, we shall expect to see the following:

- A reduction in fatigue life with increasing strain rate for fatigue under constant plastic strain amplitude if stress is the failure driver; and
- An increase in fatigue life with increasing strain rate for fatigue under constant stress amplitude if plastic strain is the failure driver.

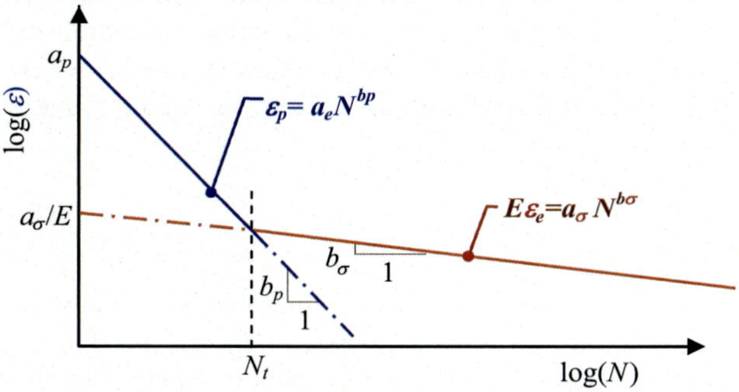

Figure 8.4 Cyclic stress–strain.

Equations (8.2) and (8.3) are strictly valid for the fatigue cycle about nil mean stress or strain. However, these may be extended to cases of mean stress and strain using simple assumptions. At $N = 1$, $\sigma \rightarrow a_\sigma$ and $\varepsilon_p \rightarrow a_p$; that is, the coefficient a_σ and a_p correspond to the amplitudes of stress and strain, which will lead to failure in a single cycle. Assuming the presence of mean stress σ_m and mean strain ε_m leads to a reduction of σ and ε_p at $N = 1$ to $\sigma|_{N=1} = a_\sigma - \sigma_m$ and $\varepsilon_p|_{N=1} = a_p - \varepsilon_m$. Assuming this reduction applies to other values of N, then Eqns (8.2) and (8.3) may be modified for the presence of mean stress and mean strain as follows:

$$\sigma = a_\sigma \left(1 - \frac{\sigma_m}{a_\sigma}\right) N^{b_\sigma} \quad (8.12)$$

$$\varepsilon_p = a_p \left(1 - \frac{\varepsilon_m}{a_p}\right) N^{b_p} \quad (8.13)$$

8.2 Design of test specimens

The inability of the JEDEC JESD22-B111 BLDST method to introduce cyclic bending of constant amplitude to the test specimen has excluded it from being a viable test method for establishing the fatigue resistance equation of solder joints. The HSCBT, which is capable of imposing constant amplitude cyclic bending to the PCB assembly over a wide spectrum of frequencies, is ideal for characterising and establishing the fatigue resistance of solder joints.

The standard fatigue characterisation experiments (ASTM-D7774, 2012; ASTM-D7791, 2012) use test specimens of simple geometry, which are selected so that there is a simple distribution of stress/strain in the cross-section within the ‘gauge domain’ of the test specimen when subjected to an applied load (uniaxial, bending, torsion) of cyclic nature. Engineering components in practical applications are frequently made of complex geometry; many result in singularity of stress/strain, making the exact evaluation of their magnitude impossible. In such a case, the only viable approach is to perform a fatigue test using an engineering test specimen that is similar in geometrical construction to that of the engineering component of interest. However, singularity at the point of fatigue failure renders strain measurement impossible; instead, a representative strain measurement is taken at a point where there is a relatively small strain gradient. The fatigue resistance equation takes the form

$$\Delta\varepsilon_{\text{rep}} = aN^{b} \tag{8.14}$$

Considering the criticality of the fatigue resistance of intermetallic compound (IMC) structure in solder joints destined for drop-impact applications, the use of engineering test specimens that have a similar IMC structure is beyond debate. The test specimen for characterising the fatigue resistance of solder joints is shown in Figure 8.5, which is identical to that used in HSCBT described in Chapter 7. The test specimen is supported at two spans, such that a uniform bending moment is generated within the inner span and the fibre strain on the PCB provides a consistent and representative strain measurement for the failure driving force. The fatigue resistance equation of solder joints in a PCB assembly may be expressed as

$$\Delta\varepsilon_{\text{pcb}} = CN^{-\beta} \tag{8.15}$$

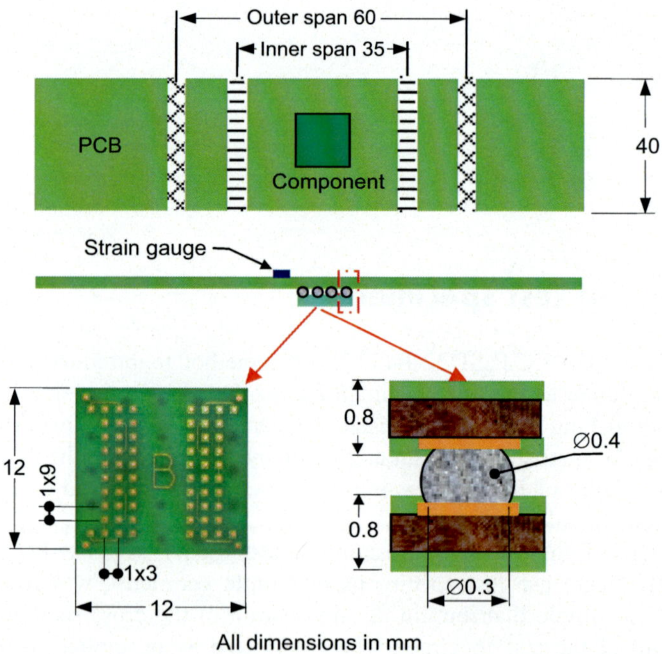

Figure 8.5 Test specimen for characterising the fatigue resistance of solder joints at high cyclic frequency.

where ε_{pcb} and β are the fatigue resistance coefficient and exponent, respectively. The four corner joints of the test specimen are designed to experience a higher magnitude of stress/strain compared to the inner solder joints and will fail independently. The electrical resistance at the four corner joints is monitored individually; failure is defined as the point when the electrical resistance of the corner joint exceeds 10 Ω for the duration of 1 ms.

It must be noted that the fatigue resistance equation established using an engineering test specimen is a function of its geometrical construction, dimensions and elastic stiffness. The relations for transferring the coefficients of the fatigue resistance equation to engineering specimens of other geometrical constructions, dimensions and elastic stiffness can be established both analytically and numerically.

Four aspects of the fatigue resistance of solder joints are investigated: solder joint materials, cyclic frequency, test environmental conditions and room-temperature aging.

8.3 Fatigue resistance equations: materials

8.3.1 *Test matrix*

Six out of the 12 solder joint material systems investigated in the board-level subsystem test in Chapter 7 are given further investigations (Wong et al., 2009). The material systems are shown in Table 8.1. A standard HSCBT is used to cycle test specimens at 100 Hz at five constant PCB strain amplitudes: 1200, 1500, 1800, 2100 and 2400 microstrains. The magnitude of strain amplitude is selected based on the range of measured strains of PCB assembly in the drop impact of PEDs, as reported in Chapter 2. Each test leg consists of between 12 and 20 data points.

8.3.2 *Consolidated results and analysis*

The cumulative failure distributions for the 33 test legs were evaluated; those of SnPb_OSP and SnPb_ENIG solder joints are shown in Figure 8.6. Assuming Weibull distribution, the characteristic life, N, and shape parameters, γ, of each test leg

Table 8.1 Solder material system

Parameter	Description	Symbol
Solder alloy	Sn37Pb Sn1.0Ag0.1Cu Sn1.0Ag0.1Cu with 0.02% Ni and 0.05% In	SnPb SAC101 SAC101(d)
Pad finish	Organic solder preservative Electroless nickel immersion gold	OSP ENIG

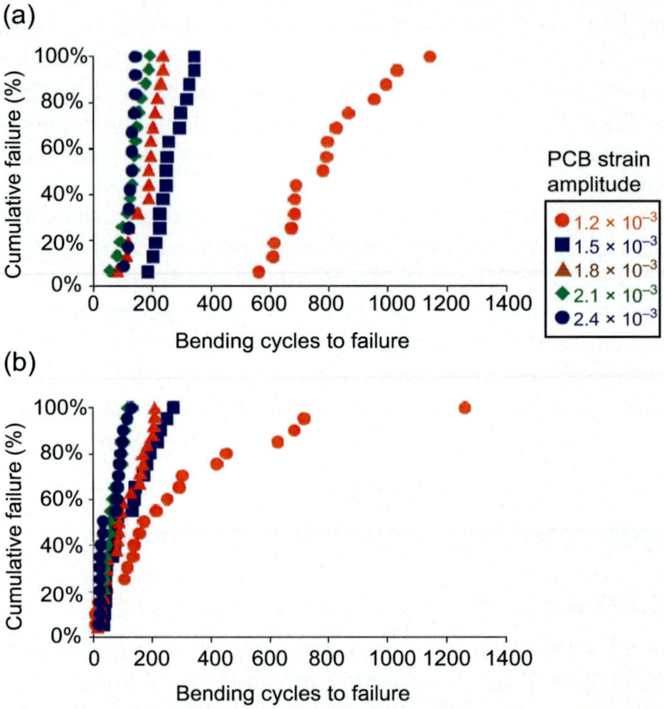

Figure 8.6 Cumulative failure of (a) SnPb_OSP and (b) SnPb_ENIG solder joint material systems when subjected to PCB bending at 100 Hz and 5 PCB strain amplitudes.

were evaluated. The quality of the data was assessed using two criteria: Weibull shape parameter and R^2 of the linearised Weibull distribution, $Y = \gamma X + c$. To aid the assessment, the z-value (the magnitude of deviation from the mean expressed in unit of standard deviation) of these quality indicators was evaluated. These values are given in Table 8.2. The values of γ and R^2 that are of suspicious quality (z-value > 1) and the corresponding fatigue life were underlined; data of suspicious quality were examined in detail and the outliers discarded in subsequent analysis.

The mean characteristic life of three solder alloys and those of the two pad finishes are listed in Table 8.3. Among the two pad finishes, the OSP pad finish outperformed the ENIG pad finish by almost 2:1. Among the three solder alloys, SAC101(d) distinctively outperformed the other two solder alloys, with SAC101 marginally outperforming SnPb solder alloy. When the fatigue life of solder joints was evaluated using BLDST and HSCBT at a single PCB strain amplitude of 2000 microstrains in Chapter 7, the fatigue life of the OSP pad finish was found to be far superior to that of the ENIG pad finish; and the fatigue life of solder alloys was ranked in the

Table 8.2 The fatigue life of solder joint material systems at various PCB strain amplitudes

Solder joint material system	PCB strain amplitude, ε_{pcb} ($\times 10^{-3}$)	Characteristic life, N (cycles)	Quality of data			
			γ, Weibull		R^2	
			Abs	z	Abs	z
SnPb_OSP	1.2	830	5.6	−0.08	0.93	−0.03
	1.5	273	6.0	0.04	0.94	0.08
	1.8	190	3.5	−0.78	0.88	−1.64
	2.1	141	3.3	−0.82	0.96	0.67
	2.4	132	10.9	1.65	0.96	0.92
SnPb_ENIG	1.2	258	0.7	−1.54	0.92	0.33
	1.5	127	1.6	0.28	0.89	−0.30
	1.8	88	1.5	0.07	0.94	0.71
	2.1	69	2.0	1.24	0.95	0.86
	2.4	61	1.4	−0.05	0.83	−1.60
SAC101_OSP	1.0	1013	2.3	1.28	0.97	0.82
	1.2	855	2.3	1.30	0.98	0.95
	1.5	292	1.3	−0.50	0.97	0.94
	1.8	210	1.2	−0.76	0.95	0.44
	2.1	288	1.6	0.06	0.86	−1.03
	2.4	312	1.6	−0.03	0.86	−1.20
	2.7	55	0.9	−1.35	0.87	−0.91
SAC101_ENIG	1.2	278	2.4	1.15	0.92	0.65
	1.8	144	1.9	0.50	0.87	−0.86
	2.1	48	0.8	−1.02	0.93	1.06
	2.4	52	1.0	−0.63	0.87	−0.85
SAC101(d)_OSP	1.2	1348	3.0	−0.75	0.96	0.46
	1.5	411	4.3	0.38	0.97	0.54
	1.8	317	2.7	−1.07	0.81	−1.79
	2.1	250	3.8	−0.01	0.95	0.36
	2.4	215	5.5	1.46	0.96	0.42

Continued

Table 8.2 **Continued**

Solder joint material system	PCB strain amplitude, ε_{pcb} ($\times 10^{-3}$)	Characteristic life, *N* (cycles)	Quality of data			
			γ, Weibull		R^2	
			Abs	*z*	Abs	*z*
SAC101(d)_ENIG	1.0	867	3.3	1.93	0.90	−0.53
	1.2	774	1.9	0.23	0.98	1.54
	1.5	280	1.4	−0.36	0.95	0.86
	1.8	231	1.2	−0.57	0.90	−0.50
	2.1	114	1.1	−0.69	0.91	−0.22
	2.4	122	1.2	−0.54	0.88	−1.15

Note: Underlined data were omitted in the subsequent evaluations.

Table 8.3 **The mean fatigue life of solder joint material systems and pad finishes**

Solder joint material system	Mean characteristic life, *N* (cycles)	Pad finish	Mean characteristic life, *N* (cycles)
SnPb	217	OSP	436
SAC101	328	ENIG	234
SAC101(d)	448		

order: SAC101(d) > SnPb > SAC305(d) > SnAg > SAC101 > SAC305. These two test results agree on the following:

- Superiority of OSP pad finish over the ENIG pad finish
- Superiority of SAC101(d) over SAC101 and SnPb solder alloys

However, the two test results disagree on the relative superiority of SAC101 and the SnPb solder alloys. The different conclusion is attributable to the significant difference in the magnitude of the fatigue exponents of these two solder alloys, which will be discussed in the next section. SnPb solders have relatively superior fatigue performance at high PCB strain, while SAC101 solder has relatively superior fatigue performance at low PCB strain. In view of the higher damage associated with higher PCB strain, it is suggested that a high magnitude of PCB strain should be used for evaluating the fatigue resistance of solder joints, if the fatigue test could afford to be performed at only one PCB strain amplitude.

8.3.3 Fatigue resistance equations

The ε_{pcb}–N data of the six solder joint material systems, plotted in normal scale and in logarithmic scale, respectively, are shown in Figure 8.7. The solder joints with OSP pad finish are shown with 'open' symbols, whereas those with ENIG finish are indicated by solid symbols. The superiority of the OSP pad finishing over the ENIG pad finishing is apparent.

The near-linear nature of the data expressed in the logarithmic scale suggests a power-law relation that has endorsed the validity of Eqn (8.15). The optimum values of C and β that best fit the experimental data of the individual solder joint material system have been evaluated using the method of regression; these are tabulated in Table 8.4. Because of the use of engineering test specimens, the fatigue resistance coefficient C depends not only on the solder joint material systems but also on the

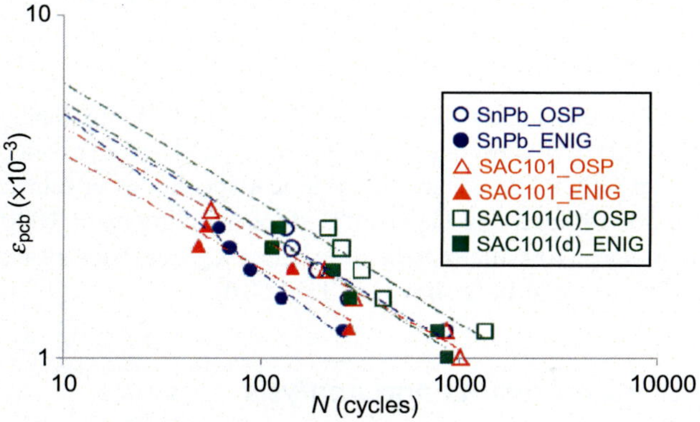

Figure 8.7 ε_{pcb}–N characteristics of solder joint material systems.

Table 8.4 The constants of $\varepsilon_{pcb} = CN^{-\beta}$ as extracted from experimental data

		$\varepsilon_{pcb} = CN^{-\beta}$						
		All data			Exclude data at $\varepsilon_{pcb} = 1.2 \times 10^{-3}$			Only data at $\varepsilon_{pcb} = 1.2 \times 10^{-3}$ & $\varepsilon_{pcb} = 2.4 \times 10^{-3}$
Solder alloy	**Pad finish**	C ($\times 10^{-3}$)	β	R^2	C ($\times 10^{-3}$)	β	R^2	β
SnPb	OSP	11.7	0.35	0.90	41.9	0.60	0.96	0.20
	ENIG	15.0	0.46	0.96	29.7	0.62	0.98	0.32
SAC101	OSP	9.8	0.32	0.98				
	ENIG	8.5	0.34	0.88				
SAC101(d)	OSP	14.3	0.35	0.88	109	0.71	0.99	0.19
	ENIG	13.1	0.37	0.95				

geometry and dimensions of the solder joints, the layout of the solder joints on the component/PCB, and the dimensions and elastic constants of the PCB and the component. Conversely, the fatigue resistance exponent β is a characteristic of the solder joint material system alone. It is interesting to note that the fatigue resistance exponent β for the six solder joint material systems falls within the range of 0.32–0.46. If excluding SnPb_ENIG, the fatigue resistance exponent β for the other five solder joint material systems falls within the range of 0.32–0.37. This suggests that, except for the SnPb_ENIG solder joint, the relative fatigue resistance of solder joint material systems is insensitive to the amplitude of the PCB strain. Hence, the relative performance of the solder joints may be assessed with reasonable confidence using just one single PCB strain amplitude.

8.4 Fatigue resistance equations: frequency

8.4.1 Test matrix

Two out of the above six material systems, SAC101_OSP (Wong, Seah, Caers, & Lai, 2014) and SAC101(d)_ENIG (Wong, Seah, & Shim, 2014), were given further consideration. A standard HSCBT was used to cycle test specimens at four cyclic bending frequencies: 30, 50, 100 and 150 Hz at PCB strain amplitudes of 1000, 1250, 1500, 1800, 2100, 2400 and 2700 microstrains. Each test leg consisted of between 12 and 32 data points. The test matrix is shown in Table 8.5.

8.4.2 Consolidated results and analysis

The cumulative failure distributions for these test legs were evaluated. The SAC101_OSP and SAC101(d)_ENIG solder joints at bending frequencies of 30 and 100 Hz are shown in Figure 8.8 and Figure 8.9 as illustrations.

Following the procedures used in Section 8.3, Weibull distribution was assumed and the characteristic life, N, and shape parameters, γ, were evaluated for individual test leg; the quality of the data was assessed using the same criteria: Weibull shape parameter and the R^2 of the linearised Weibull distribution. These values are given in Tables 8.6 and 8.7 for SAC101_OSP and SAC101(d)_ENIG, respectively. The values of γ and R^2 that were of suspicious quality (z-value > 1) and the corresponding fatigue life were underlined. Data of suspicious quality were examined in detail; the outliers were discarded in the construction of the PCB ε_{pcb}–N curves.

Table 8.5 Cyclic frequency

Solder joints	Cyclic frequency
SAC101_O SAC101(d)_ENIG	30 Hz, 50 Hz, 100 Hz, 150 Hz

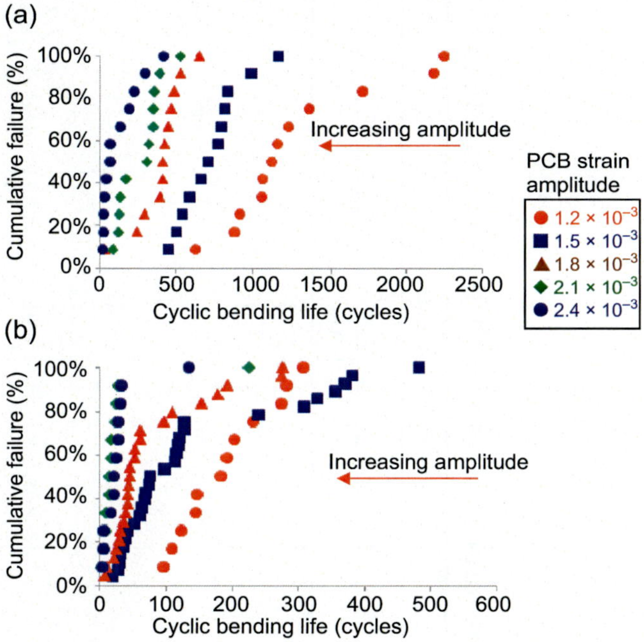

Figure 8.8 Cumulative failure of SAC101_OSP solder joints when subjected to cyclic bending (a) at 30 Hz and (b) at 150 Hz.

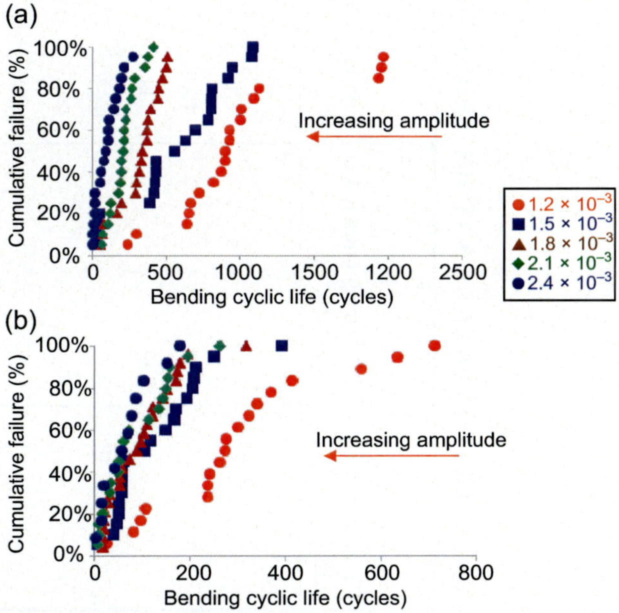

Figure 8.9 Cumulative failure of SAC101(d)_ENIG solder joints when subjected to bending (a) at 30 Hz and (b) at 150 Hz.

Table 8.6 The fatigue life of SAC101_OSP at four cyclic frequencies at various PCB strain amplitudes

Cyclic frequency, f (Hz)	PCB strain amplitude, ε_{pcb} ($\times 10^{-3}$)	Characteristic life, N (cycles)	Quality of data			
			γ		R^2	
			Abs	z	Abs	z
30	1.2	1375	2.9	0.22	0.93	0.22
	1.5	776	4.1	1.20	0.98	1.62
	1.8	450	3.4	0.58	0.90	−0.67
	2.1	300	1.8	−0.71	0.90	−0.89
	2.4	<u>121</u>	<u>1.1</u>	−<u>1.29</u>	0.92	−0.28
50	1.2	998	2.2	1.31	0.98	1.57
	1.5	559	1.5	0.07	0.91	−0.86
	1.8	361	1.7	0.46	0.91	−0.78
	2.1	303	1.1	−0.49	0.94	0.35
	2.4	<u>195</u>	<u>0.6</u>	−<u>1.35</u>	0.93	−0.28

100	1.0	1013	2.3	1.28	0.97	0.82
	1.2	855	2.3	1.30	0.98	0.95
	1.5	292	1.3	–0.50	0.97	0.94
	1.8	210	1.2	–0.76	0.95	0.44
	2.1	288	1.6	0.06	0.86	–1.03
	2.4	312	1.6	–0.03	0.86	–1.20
	2.7	55	0.9	–1.35	0.87	–0.91
150	1.2	204	2.8	1.79	0.97	1.02
	1.5	139	1.2	–0.44	0.92	–0.20
	1.8	77	1.2	–0.46	0.93	0.10
	2.1	18	1.3	–0.37	0.95	0.66
	2.4	24	1.2	–0.52	0.86	–1.57

Note: Underlined data were omitted in the subsequent evaluations.

Table 8.7 **The fatigue life of SAC101(d)_ENIG at four bending frequencies at various PCB strain amplitudes**

Cyclic frequency, f (Hz)	PCB strain amplitude, ε_{pcb} ($\times 10^{-3}$)	Characteristic life, N (cycles)	Quality of data			
			γ		R^2	
			Abs	z	Abs	z
30	1.2	1139	1.9	0.74	0.91	−0.08
	1.5	673	0.8	−0.99	0.85	−1.37
	1.8	383	1.5	0.12	0.89	−0.48
	2.1	237	2.2	1.17	0.96	1.05
	2.4	107	0.7	−1.04	0.95	0.89
50	1.2	768	1.3	0.03	0.92	0.08
	1.5	441	1.8	1.35	0.92	−0.26
	1.8	169	0.7	−1.41	0.90	−1.55
	2.1	220	1.1	−0.29	0.94	0.96
	2.4	158	1.4	0.32	0.93	0.77

100	1.0	867	3.3	1.91	0.90	−1.58
	1.2	774	1.9	0.28	0.98	1.14
	1.5	292	1.3	−0.36	0.96	0.42
	1.8	252	1.1	−0.58	0.95	0.13
	2.1	126	1.0	−0.70	0.96	0.65
	2.4	132	1.2	−0.55	0.92	−0.77
150	1.2	337	1.3	0.99	0.92	−0.54
	1.5	138	1.2	0.23	0.90	−1.30
	1.8	102	1.3	0.84	0.93	−0.15
	2.1	85	1.0	−0.76	0.96	0.90
	2.4	68	0.9	−1.30	0.96	1.09

Note: Underlined data were omitted in the subsequent evaluations.

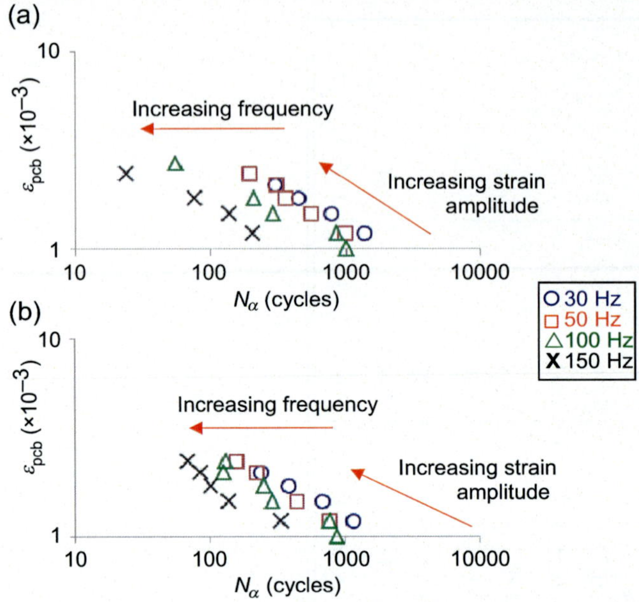

Figure 8.10 ε_{pcb}–N characteristics versus bending frequency for (a) SAC101_OSP and (b) SAC101(d)_ENIG solder joint systems.

8.4.3 Fatigue resistance equations

The ε_{pcb}–N characteristics of SAC101_OSP and SAC101(d)_ENIG solder joints, expressed in terms of logarithmic scale, are shown in Figure 8.10. The fatigue lives of both solder joint materials decrease with increasing bending frequency of PCB. For example, increasing the bending frequency from 30 to 150 Hz reduces the fatigue life of the SAC101_OSP and the SAC101(d)_ENIG solder joint systems by 80% and 70%, respectively. Increasing frequency suppresses plastic strain while raises the level of stress in the solder joints; the hypothesis of strain as a failure driver would predict increased fatigue life, while the hypothesis of stress as a failure driver would predict reduced fatigue life. The experimental evidence supports the hypothesis of stress as the failure driver.

Expressing the PCB strain-solder life relation in the form of $\varepsilon_{pcb} = CN^{-\beta}$, the constants C and β for the two material systems at four cyclic frequencies were extracted using the method of linear regression; these are given in Table 8.8. It was noted that the coefficient C increases as the frequency is increased from 30 to 50 Hz but decreases as the frequency is increased further to 150 Hz. By contrast, the exponent β fluctuates within a much smaller range with no clear trend. It was postulated that the fatigue resistance exponent β is independent of the cyclic frequency; taking the average of β over the four bending frequencies, the mean values of fatigue exponent $\overline{\beta}$ for SAC101_OSP and SAC101(d)_ENIG solder joints were evaluated as 0.36 and 0.40, respectively.

Table 8.8 **The constants of $\varepsilon_{pcb} = CN^{-\beta}$ as extracted from experimental data**

Material systems	Cyclic frequency, f (Hz)	$\varepsilon_{pcb} = CN^{-\beta}$	
		C ($\times 10^{-3}$)	β
SAC101_OSP	30	16.7	0.36
	50	24.5	0.44
	100	9.8	0.32
	150	6.6	0.31
SAC101(d)_ENIG	30	14.6	0.35
	50	22.9	0.44
	100	14.9	0.39
	150	13.5	0.42

Using these mean values, the fatigue resistance coefficient C was revaluated using the method of regression; the new constants C and and $\overline{\beta}$ for the two material systems are listed in Table 8.9. Figure 8.11 shows that the fatigue resistance coefficient C decreases linearly, at an almost identical rate, with increasing cyclic frequency of PCB for the two solder joint material systems.

The following unified equation for creep-fatigue was introduced in Chapter 2:

$$\varepsilon_{pcb} = C_o c(T,f) N^{-\beta_o s(T,f)} \tag{8.16}$$

Table 8.9 **The constants of $\varepsilon_{pcb} = CN^{-\beta}$ assuming a mean value of constant β**

Material systems	Bending frequency, f (Hz)	$\varepsilon_{pcb} = CN^{-\overline{\beta}}$	
		$\overline{\beta}$	C ($\times 10^{-3}$)
SAC101_OSP	30	0.36	16.3
	50		15.3
	100		12.2
	150		8.3
SAC101(d)_ENIG	30	0.40	19.6
	50		17.6
	100		15.7
	150		12.0

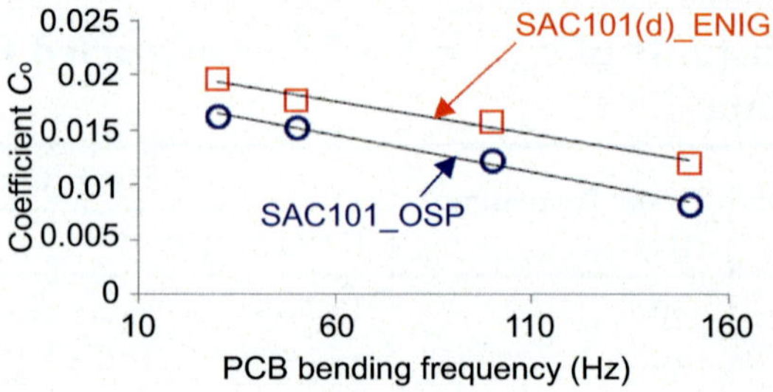

Figure 8.11 Coefficient C_o versus PCB bending frequency.

where the function $c(T,f)$ accounts for the reduction of allowable plastic strain due to creep, while the function $s(T,f)$ accounts for increased creep with increasing stress. The unified equation for rate-induced reduction in fatigue life takes the form

$$\varepsilon_{pcb} = C_o r(f) N^{-\beta_o} \tag{8.17}$$

The linearity of C_o versus frequency, shown in Figure 8.11, suggests the rate function has the form

$$r_{linear}(f) = 1 - r_1(f - f_{ref}) \tag{8.18}$$

where r_1 is a positive constant and f_{ref} is the frequency at and below which the rate effect is not active; the fatigue exponent is assumed to be independent of strain rate. Taking $f_{ref} = 1$ Hz, the unified equation for SAC101_OSP and SAC101(d) ENIG solder joints using regression analysis gives $\{C_o = 0.0176,\ r_1 = 0.00344,\ \beta_o = 0.353\}$ and $\{C_o = 0.0177,\ r_1 = 0.00241,\ \beta_o = 0.383\}$, respectively. These results suggest that SAC101_OSP solder has a stronger dependence on strain rate.

By defining the transformation relation as $N_{ref} = r(f)^{-1/\beta_o} N$ or $\varepsilon_{pcb,ref} = r(f)^{-1}\varepsilon_{pcb}$, Eqn (8.17) is transformed to the reference conditions:

$$\varepsilon_{pcb} = C_o N_{ref}^{-\beta_o} \tag{8.19}$$

and

$$\varepsilon_{pcb,ref} = C_o N^{-\beta_o} \tag{8.20}$$

respectively. Plots of ε_{pcb} versus N at the reference condition of $f_{ref} = 1$ Hz for SAC101_OSP solder joints are shown in Figure 8.12, which shows the collapse of all data into one curve at $f_{ref} = 1$ Hz.

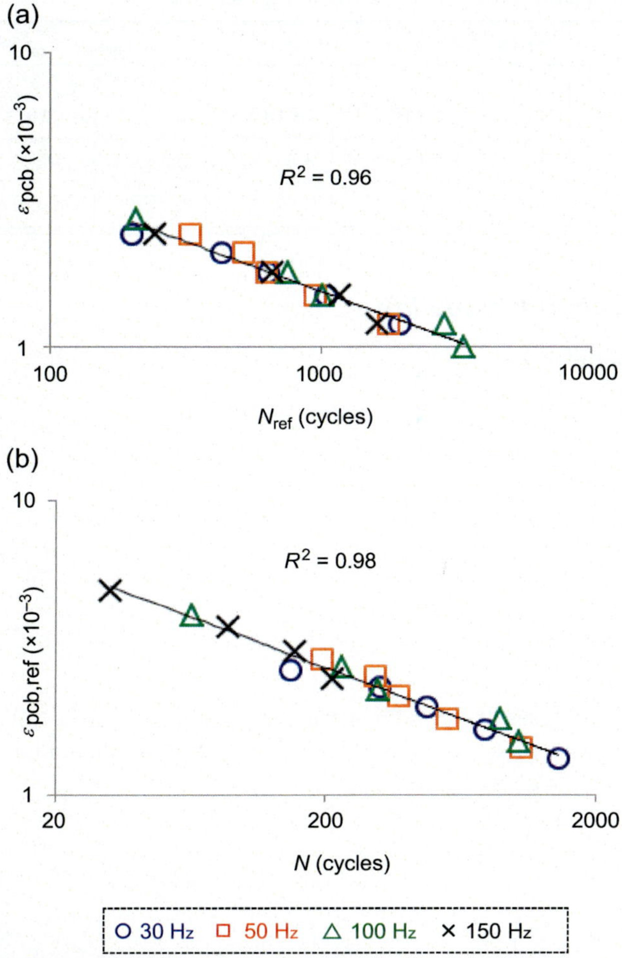

Figure 8.12 ε_{pcb} versus N at $f_{ref} = 1$ Hz: (a) ε_{pcb} versus N_{ref} and (b) $\varepsilon_{pcb,ref}$ versus N for SAC101_OSP solder joint.

For ease of subsequent manipulation, the rate function is assumed to take the power-law form

$$r_{power}(f) = \left(\frac{f}{f_{ref}}\right)^{-q} \tag{8.21}$$

Taking $f_{ref} = 1$ Hz, the coefficients have been evaluated by regression as $\{C_o = 0.050,\ q = 0.341,\ \beta_o = 0.345\}$ and $\{C_o = 0.0332,\ q = 0.211,\ \beta_o = 0.377\}$ for SAC101_OSP and SAC101(d)_ENIG solder joints, respectively. These are shown in Table 8.10.

Table 8.10 The coefficients of unified equations

	$\varepsilon_{pcb} = C_o[1-r_1(f-1)]N^{-\beta_o}$			$\varepsilon_{pcb} = C_o f^{-q} N^{-\beta_o}$		
Material systems	β_o	C_o ($\times 10^{-3}$)	r_1 ($\times 10^{-3}$)	β_o	C_o ($\times 10^{-3}$)	q
SAC101_OSP	0.353	17.6	3.41	0.345	0.050	0.341
SAC101(d)_ENIG	0.383	17.7	2.44	0.377	0.0332	0.211

8.4.4 *Equivalent fatigue life*

Rearranging Eqn (8.17) to

$$C_o N^{-\beta_o} = \frac{\varepsilon_{pcb}}{r(f)} \tag{8.22}$$

and substituting ($\varepsilon_{pcb,1}, f_1$) and ($\varepsilon_{pcb,2}, f_2$), respectively, into Eqn (8.22) and equating them, we obtain

$$\frac{r(f_1)}{r(f_2)} \frac{\varepsilon_{pcb,2}}{\varepsilon_{pcb,1}} = 1 \tag{8.23}$$

Equation (8.23) gives the condition for the equivalent fatigue life. For reasons of simplicity, let $r(f) = r_{power}(f) = f^{-q}$. Substituting into Eqn (8.23) gives $\left(\frac{f_2}{f_1}\right)^q \frac{\varepsilon_{pcb,2}}{\varepsilon_{pcb,1}}\bigg|_{f_1=f_2} = \left(\frac{f_2}{f_1}\right)^q \frac{\varepsilon_{pcb,2}}{\varepsilon_{pcb,1}}\bigg|_{\varepsilon_{pcb,1}=\varepsilon_{pcb,2}}$

$$\Rightarrow \frac{\varepsilon_{pcb,2}}{\varepsilon_{pcb,1}} = \left(\frac{f_2}{f_1}\right)^q \tag{8.24}$$

Substituting $q = 0.341$ for the SAC101_OSP solder joint and increasing the cyclic frequency by 10 times is equivalent to increasing the PCB strain amplitude by 2.2 times; the latter is much easier to achieve experimentally.

8.5 Fatigue resistance equations: environment

8.5.1 *Tester for environmental testing*

Mechanical test machine manufacturers, such as Instron and MTS, provide a temperature chamber as an accessory to their range of mechanical testers. An example of a general-purpose temperature chamber is shown in Figure 8.13. It also shows a schematic of a bending test setup in the temperature chamber.

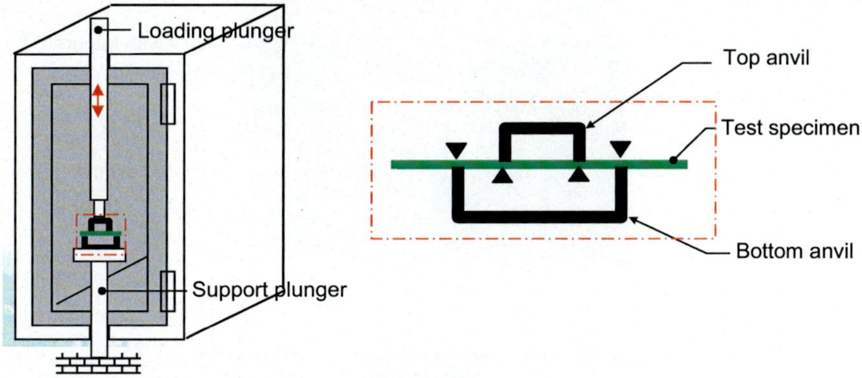

Figure 8.13 A general-purpose environmental chamber.

However, a general-purpose environmental chamber is too large to be accommodated in a high-speed cyclic bend tester. The long plungers introduce false displacements due to thermal expansion; the larger chamber presents a larger thermal inertia. An ideal environmental chamber is one that is integrated with the loading anvils, as illustrated in Figure 8.14. The environmental chamber encloses only a controlled volume over the domain of interest—in this case, the section of PCB that is assembled with the integrated circuit component. The environmental chamber doubles up as a loading anvil applying vertical displacements to the test specimen, while the two ends of the test specimen are prevented from vertical displacement by the fixed pivots. The integrated chamber is made in two pieces, each with a cavity, and the two pieces are mechanically fastened together. The desired test environment in the chamber is established by piping in the appropriate environment from external source. The cavities are sealed along their faces to maintain the desired environment within the controlled volume; the cavities in Figure 8.14 are open in the front view in order to show the domain of interest for illustrative purposes. Figure 8.15 shows an integrated environmental chamber in operation with a PCB assembly, complete with instrumentation for temperature monitoring of the chamber, electrical resistance across solder joints, and strain gauging of PCB. This setup is used to study the environmental effects of solder joints when fatigued at high cyclic frequencies.

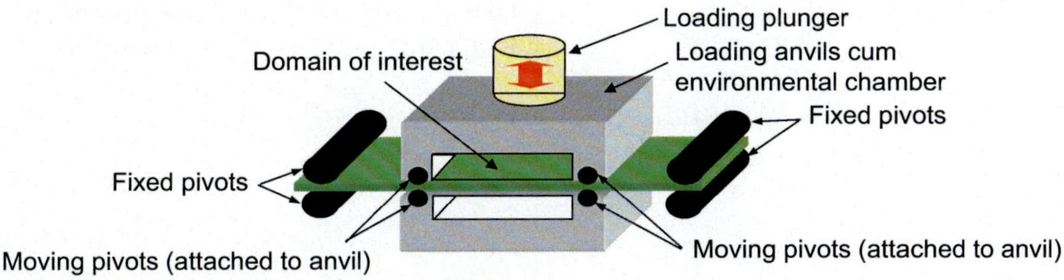

Figure 8.14 A schematic representation of an integrated environmental chamber: symmetric bending of a PCB assembly.

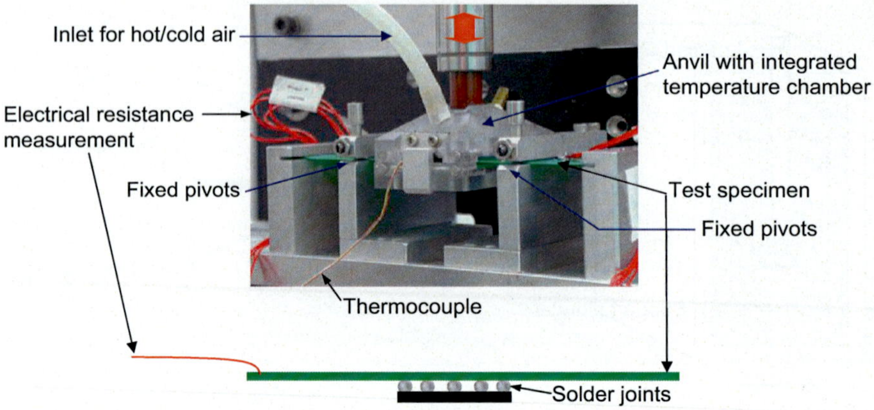

Figure 8.15 An integrated environmental chamber in operation: symmetric bending of a PCB assembly.

8.5.2 The effects of temperature on the fatigue performance of solder joint material systems

PEDs are susceptible to drop impact at a wide range of temperatures. So far, the fatigue performance of solder joint material systems has been evaluated in a laboratory's ambient environment. The ductility of solder joints decreases with decreasing temperature; the hypothesis of stress as the fatigue failure driver would predict decreasing fatigue performance with decreasing temperature. This hypothesis is investigated in this section.

8.5.2.1 Test matrix

Six solder alloys and two pad finishes were evaluated at two test temperatures. These are tabulated in Table 8.11. The test temperature of 22 °C is the ambient temperature of the test laboratory, while dry nitrogen was used for the lower test temperature.

Table 8.11 Test matrix for test temperatures

Solder alloys	Designation	Pad finish (PCB strain)	Test temperature
Sn–37Pb	SnPb	OSP (2×10^{-3})	22 °C (ambient)
Sn–1.0Ag–0.1Cu	SAC101	ENIG (1.2×10^{-3})	0 °C (dry nitrogen)
Sn–3.0Ag–0.5Cu	SAC305		
Sn–1.0Ag–0.1Cu–0.02Ni–0.05In	SAC101(d)		
Sn–0.6Cu–0.05Ni + Ge	SN100-0.6		
Sn–0.8Cu–0.05Ni + Ge	SN100-0.8		

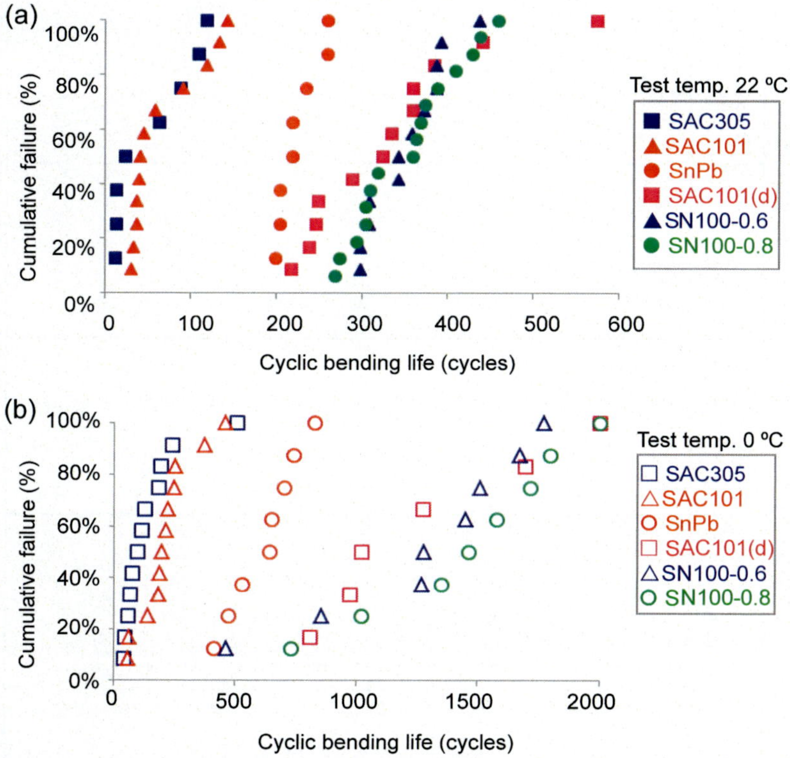

Figure 8.16 Cumulative percent-failures of solider joints with OSP pad finish subjected to PCB bending strain of 2×10^{-3} at 100 Hz and at temperatures of (a) 22 °C, ambient environment and (b) 0 °C, dry nitrogen.

The fatigue life of the solder joints was evaluated only at one PCB strain: 2×10^{-3} for OPS pad finish and 1.2×10^{-3} for ENIG pad finish. The magnitudes of PCB strain were selected so that most of the solder joints will have experienced failure within 2000 bending cycles. A cyclic bending frequency of 100 Hz is used for all test specimens. Between 8 and 16 data points have been collected for each solder joint.

8.5.2.2 Consolidation of data

The cumulative percent-failure plots for the six solders with OSP pad finish at 22 °C and at 0 °C are shown in Figure 8.16. The characteristic life, first-fail life and shape factors for the solder joints, assuming Weibull distribution, are given in Table 8.12. The same results for the ENIG pad finish are plotted in Figure 8.17 and in Table 8.13. The solder joints have not been monitored for failure beyond 2000 bending cycles. Because no failure of Sn100-0.8 was observed up to 2000 bending cycles at a test temperature of 0 °C, no data were reported for this alloy at 0 °C. For the same reason, only three data for Sn100-0.6 were reported at 0 °C.

Table 8.12 Fatigue characteristics of solder joints on OSP pad finish at two test temperatures[a]

		Characteristic life, N		First-fail life, $N_{1\%}$		Shape factor, γ	
Solder alloys	Temperature	Absolute	Ratio (0 °C/ 22 °C)	Absolute	Ratio (0 °C/ 22 °C)	Absolute	Ratio (0 °C/ 22 °C)
SAC305	22 °C	46 ⑥	3.0	13 ⑥	3.5	1.1	1.5
	0 °C	136 ⑥		45 ⑥		1.6	
SAC101	22 °C	70 ⑤	3.2	33 ⑤	1.8	2.0	1.0
	0 °C	221 ⑤		61 ⑤		1.9	
SnPb	22 °C	228 ④	2.8	200 ④	2.1	10.9	0.5
	0 °C	630 ④		415 ④		5.0	
SAC101(d)	22 °C	347 ③	3.6	219 ③	3.7	4.2	0.8
	0 °C	1257 ③		810 ①		3.4	
SN100-0.6	22 °C	364 ②	3.5	300 ①	1.6	9.2	0.3
	0 °C	1289 ②		465 ③		2.4	
SN100-0.8	22 °C	369 ①	4.0	270 ②	2.7	6.8	0.5
	0 °C	1465 ①		732 ②		3.4	
Ave	22 °C	237	3.5	173	2.4	5.8	0.53
	0 °C	830		422		3.0	
Coefficient of variation			0.13		0.35		0.56

Note: Circled superscript numbers show the ranking of the solder joints.
[a]22 °C at laboratory ambient, 0 °C using dry nitrogen.

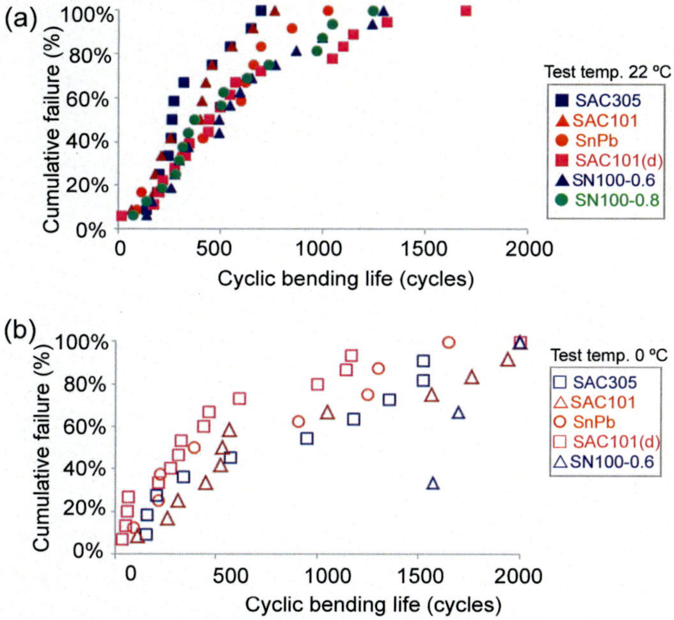

Figure 8.17 Cumulative percent-failures of solider joints with ENIG pad finish subjected to PCB bending strain of 1.2×10^{-3} at 100 Hz and at temperatures of (a) 22 °C, ambient environment and (b) 0 °C, dry nitrogen.

8.5.2.3 Analysis: OSP pad finish

In Tables 8.12 and 8.13, the relative ranking of the solder joints based on characteristic life and first-fail life, respectively, at the two test temperatures is indicated (enclosed in circles) besides their respective fatigue life. The fatigue life of the individual solder joints at the test temperature of 0 °C was normalised against the same at 22 °C; this is reflected as the 'Ratio (0 °C/22 °C)'. The average characteristics and the coefficient of variation (COV) taken over the six solder joints are tabulated in the last rows of the tables.

The following observations and analyses were made:

- The data are generally of good quality, as reflected in the relatively high magnitudes of shape factors for most of the test legs.
- Judging from either the characteristic life or the first-fail life, the solder joints exhibited near-identical rankings in fatigue performance at both test temperatures. The results suggest that evaluation of the fatigue performance of solder joint material systems can be performed at just one temperature—typically, the room temperature of the laboratory. The unwelcoming prospect of having to perform evaluation at multiple test temperatures is thus averted.
- The characteristic life of the six solder joints improved when tested at the lower temperature. The ratio of improvement ranges from 2.8 times for SnPb to 4 times for Sn100-0.8, with an average improvement of 3.5 times for the six solder joints. This ratio of improvement has a COV of 0.13, suggesting good uniformity in improvement among the six solder joints.

Table 8.13 Fatigue characteristics of solder joints on ENIG pad finish at two test temperatures[a]

		Characteristic life, N		First-fail life, $N_{1\%}$		Shape factor, γ	
Solder alloys	Temperature	Absolute	Ratio (0 °C/ 22 °C)	Absolute	Ratio (0 °C/ 22 °C)	Absolute	Ratio (0 °C/ 22 °C)
SAC305	22 °C	359 ⑥	2.2	140 ②	1.1	2.3	0.5
	0 °C	804 ④		156 ③		1.2	
SAC101	22 °C	384 ⑤	2.2	71 ⑤	1.6	1.7	0.7
	0 °C	837 ③		112④		1.3	
SnPb	22 °C	504 ④	1.2	95 ③	0.9	1.5	0.7
	0 °C	587 ⑤		90⑤		1.1	
SAC101(d)	22 °C	616 ①	0.7	18 ⑥	2.0	1.1	0.9
	0 °C	435 ⑥		36 ⑥		0.9	
SN100-0.6	22 °C	603 ②	>2.8	145 ①	10.9	1.8	–
	0 °C	>1693 ②		1575 ②		–	
SN100-0.8	22 °C	542 ③	>3.7	75 ④	>26.7	1.5	–
	0 °C	>2000 ①		>2000 ①		–	
Ave	22 °C	501	>2.1	91	>7.2	1.6	0.67
	0 °C	>1059		>394		1.1	
Coefficient of variation			>0.51		>1.43		0.29

Circled superscript numbers show the ranking of the solder joints.
[a]22 °C at laboratory ambient, 0 °C using dry nitrogen.

- A similar trend was observed for the first-fail life of the solder joints, even though the ratio of improvement is lower at 2.4 times. The lower ratio of improvement points to an increased spread of data caused by early failure when tested at the lower temperature. The higher value of COV at 0.35 suggests relatively larger variation (compared to that of characteristic life) in the ratios of improvement among the six solder joints. The observed early failure when tested at the lower temperature is reflected in the deterioration in the Weibull shape factors of the solder joints. The average ratio of degradation is 0.53.

8.5.2.4 Analysis: ENIG pad finish

The following observations and analyses were made:

- Compared to the data for the solder joints with OSP pad finishes, the data of solder joints with ENIG pad finishes were generally of poorer quality, as reflected in the relatively low values of shape factors and the lower first-fail life.
- Much less consistency in the ranking pattern was observed between the characteristic life and the first-fail life of the solder joints and between the characteristic life of the solder joints at the two test temperatures. Most alarming is the dramatic change of order of the SAC101(d)_ENIG solder joints at the two test temperatures—from first rank at 25 °C to sixth rank at 0 °C. The results highlighted the risk of evaluating and selecting solder joints with ENIG pad finishes at a single test temperature.
- Unlike the solder joints with OSP pad finishes, there was no across-the-board improvement in the characteristic life among the six solders when tested at the lower temperature. Two distinctly different responses were observed from the six solder joints at the two test temperatures. Although there were significant improvements (more than 2.8 times) for the two Sn100 solder joints when tested at the lower temperature, degradation in fatigue life was observed for the SAC101(d) solder joint; marginal improvements were found for other three solder joint material systems.
- The fatigue performance of SAC101(d) solder joints decreased at −10 °C, dry nitrogen, in an opposite trend from the other five solder joints. This is attributed to its inclination to brittle fracture at the lower test temperature, as is evident in Figure 8.18.

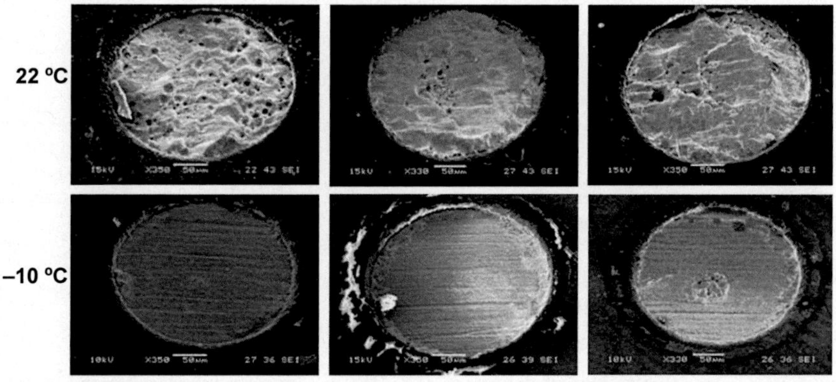

Figure 8.18 Prevailing fracture modes of SAC101(d)_ENIG solder joint cyclic bend at 100 Hz at 22 °C and at −10 °C.

Table 8.14 **Test matrix for PCB strain-life versus test environment**

Solder joint	Designation	PCB strain	Test environment
Sn–0.8Cu–0.05Ni + Ge on OSP pad finish	SN100-0.8_OSP	3×10^{-3}	22 °C (laboratory ambient) 22 °C (dry nitrogen) 50 °C (dry nitrogen)

The superior fatigue performance at the lower temperature contradicts the hypothesis of stress as the fatigue failure driver. However, besides temperature, the compositions of the environmental gases are also different between the two test conditions. It is unclear if the dry nitrogen used at 0 °C was responsible for the unexpected enhancement of fatigue life.

8.5.3 The effects of test environment on fatigue performance of solder joints

8.5.3.1 Test matrix

The fatigue life of the Sn100-0.8_OSP solder joint was evaluated at a PCB strain amplitude of 3×10^{-3} at 100 Hz in three test environments: 22 °C, laboratory ambient; 22 °C, dry nitrogen; and 50 °C, dry nitrogen, as given in Table 8.14. Eight data points were collected for each test leg.

8.5.3.2 Consolidation of test results

The cumulative percent-failure plots for the three test environments are shown in Figure 8.19. The Weibull characteristic life and shape parameters of the solder joints are listed in Table 8.15. Also included in the table are the Weibull characteristics of the

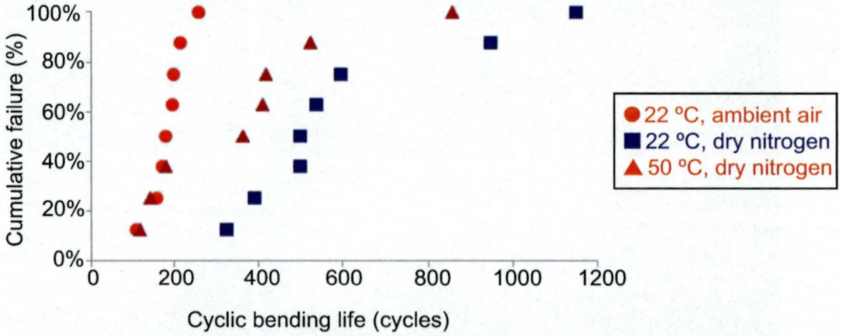

Figure 8.19 Cumulative failure of the Sn100-0.8_OSP solder joint when subjected to cyclic bending at 3×10^{-3} PCB strain amplitude at 100 Hz in three test environments.

Table 8.15 Fatigue characteristics of Sn100-0.8_OSP in three test environments

S/No	Test environment	PCB strain amplitude ($\times 10^{-3}$)	Characteristic life, N (cycles)		Shape parameter, γ
			Absolute	Ratio	
A	22 °C, ambient air	3	189	–	4.6
B	22 °C, dry nitrogen	3	607	B/A = 3.2	2.8
C	50 °C, dry nitrogen	3	344	C/B = 0.6	1.6
[a]D	22 °C, ambient air	2	369	–	6.8
[a]E	0 °C, dry nitrogen	2	1465	E/D = 4.0	3.5
[b]F	0 °C, dry nitrogen	3	460	F/B = 0.76	

[a]Extracted from Table 8.12.
[b]Extrapolated from E assuming a fatigue exponent of 0.35.

same solder joint evaluated at a PCB strain amplitude of 2×10^{-3} at 100 Hz at the two test conditions: 22 °C (ambient) and 0 °C (dry nitrogen); these are extracted from Table 8.12 and listed under 'D' and 'E', respectively, in Table 8.15. The fatigue life for 0 °C (dry nitrogen) is extrapolated to PCB strain amplitude of 2×10^{-3} by assuming a fatigue exponent of 0.35. This is listed under 'F' in Table 8.15. The computed fatigue life varies between 380 and 532 for the exponent range between 0.3 and 0.4.

8.5.3.3 Analysis

It is noted that the fatigue life of the solder joint is enhanced by 3.2 times by simply changing the test environment from ambient air to nitrogen while keeping the test temperature at 22 °C, highlighting the role of dry nitrogen in enhancing the fatigue life of the solder joints. The physics of the dry nitrogen test environment is believed to be similar to that of the vacuum test environment: the elimination of oxidation along the faces of the slip bands facilitates their intrusion and extrusion, thus eliminating local stress/strain lock-up that leads to crack initiation.

Raising the test temperature from 0 °C to 22 °C in a dry nitrogen environment increases fatigue life; this is in agreement with the hypothesis of stress being the failure driver. However, raising the test temperature from 22 °C to 50 °C under dry nitrogen almost halves the fatigue life of the solder joint, which contradicts the hypothesis. Unlike the SAC101(d)_ENIG solder joint, which shows brittle fracture at 0 °C (dry nitrogen; see Figure 8.18), Figure 8.20 shows that fatigue crack propagation is contained entirely within the ductile bulk solder, even for the solder joint that failed at 0 °C. Although the crack path has migrated towards the solder–pad interface, it remains predominantly away from the layers of brittle IMCs. The only logical explanation is that the fatigue resistance of the solder joints has been degraded at 50 °C, and the magnitude of degradation is more than the reduction in the fatigue failure driving force.

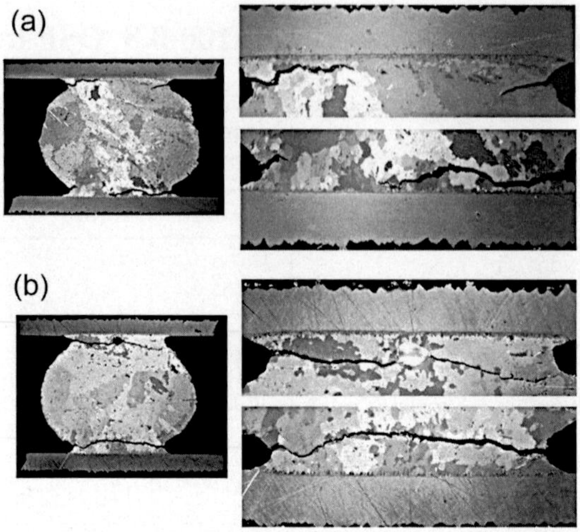

Figure 8.20 Cross-section of the Sn100-0.8_OSP solder joint at test conditions: (a) 2×10^{-3} PCB strain amplitude, 0 °C, dry nitrogen and (b) 3×10^{-3} PCB strain amplitude, 22 °C, dry nitrogen.

8.5.4 The effects of the test environment on fatigue resistance equations

The fatigue equation of solder joints has been evaluated at laboratory ambient condition. This section investigates the fatigue resistance equation at 0 °C under dry nitrogen.

8.5.4.1 Test matrix

Three solder joints from the previous experiment, including the SAC101(d)_ENIG solder joint that shows 'normal' characteristics in the previous test, are selected for these experiments. The solder joints are subjected to cyclic bending at 100 Hz at two test temperatures: 22 °C and −10 °C; the former is the equilibrium ambient temperature of the test laboratory and the latter is achieved using dry nitrogen. The fatigue life of the solder joints was evaluated at four PCB strain amplitudes: 1.2×10^{-3}, 1.5×10^{-3}, 1.8×10^{-3} and 2.1×10^{-3}. The test matrix is summarised in Table 8.16. Between 12 and 20 data points have been collected for each solder joint.

8.5.4.2 Consolidation of test results

The cumulative percent-failure plots for the SnPb_OSP solder joint, the SAC101_ENIG solder joint and the SAC101(d)_ENIG solder joint at the four PCB strain amplitudes and two test temperatures are shown in Figures 8.21−8.23, respectively. The corresponding characteristic life and shape parameters of all solder joints are tabulated in Table 8.17. The coefficients and exponents of the fatigue

Table 8.16 Test matrix for PCB strain-life versus test temperatures

Solder joints	Designation	PCB strain	Test temperature
Sn–37 Pb with OSP pad finish	SnPb_OSP	1.2×10^{-3} 1.5×10^{-3} 1.8×10^{-3} and 2.1×10^{-3}	22 °C (ambient) 0 °C (dry nitrogen)
Sn–1.0Ag–0.1Cu with ENIG pad finish	SAC101_ENIG		
Sn–1.0Ag–0.1Cu–0.02Ni–0.05In with OSP pad finish	SAC101(d)_ENIG		

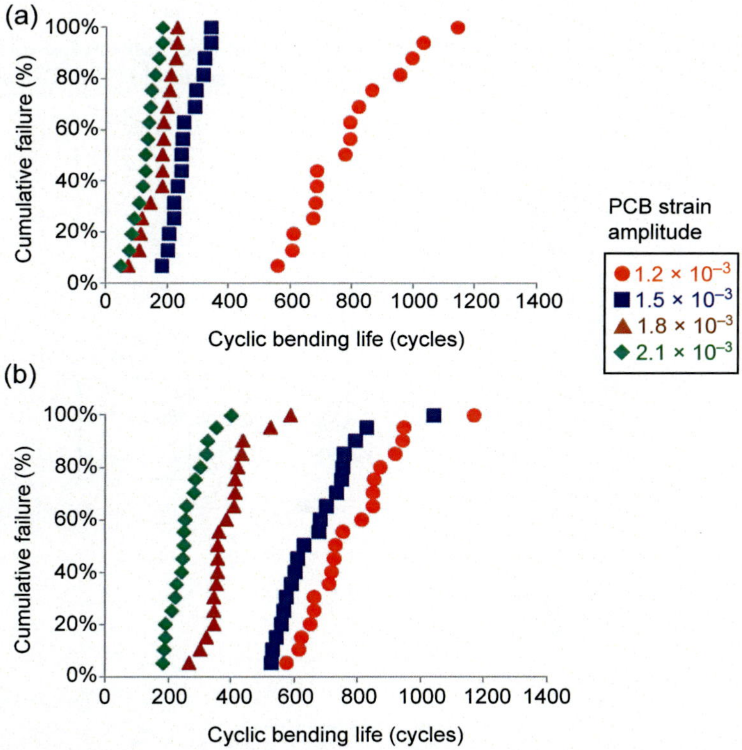

Figure 8.21 Cumulative failure of SnPb_OSP solder joint when subjected to cyclic bending at 100 Hz at four PCB strain amplitudes and at temperatures of (a) 22 °C, ambient environment and (b) −10 °C, dry nitrogen.

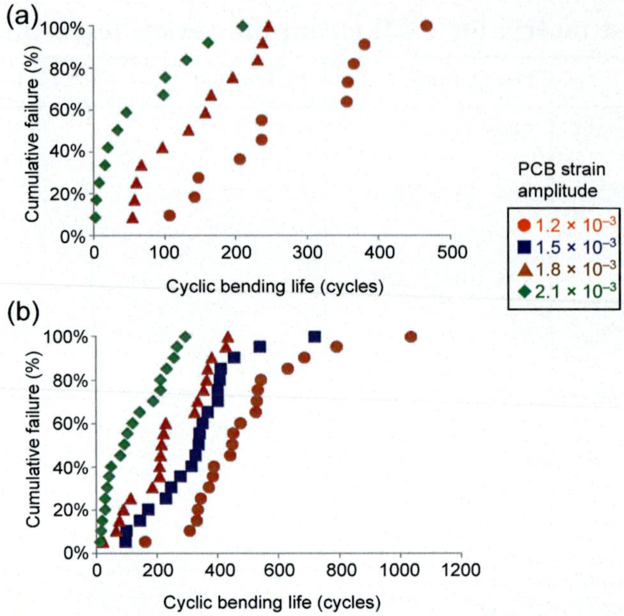

Figure 8.22 Cumulative failure of SAC101_ENIG solder joint when subjected to cyclic bending at 100 Hz at four PCB strain amplitudes and at temperatures of (a) 22 °C, ambient environment and (b) −10 °C, dry nitrogen.

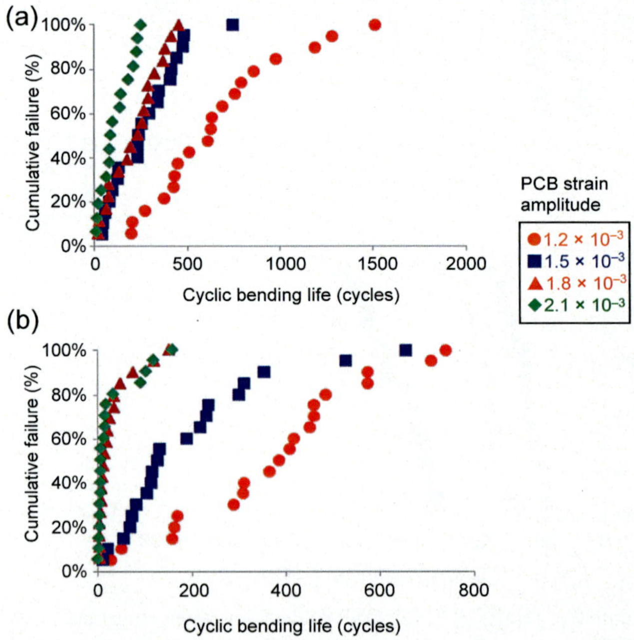

Figure 8.23 Cumulative failure of SAC101(d)_ENIG solder joint when subjected to cyclic bending at 100 Hz at four PCB strain amplitudes and at temperatures of (a) 22 °C, ambient environment and (b) −10 °C, dry nitrogen.

Table 8.17 **PCB strain-life of SnPb_OSP, SAC101_ENIG and SAC101(d)_ENIG at two test temperatures**[a]

Test environment	PCB strain amplitude, ε_{pcb} ($\times 10^{-3}$)	Characteristic life, N (cycles)			Shape parameter, γ		
		SnPb_OSP	SAC101_ ENIG	SAC101(d)_ ENIG	SnPb_ OSP	SAC101_ ENIG	SAC101(d) _ENIG
22 °C, ambient	1.2	830	278	708	5.6	2.4	2.1
	1.5	273	–	280	6.0	–	1.4
	1.8	190	144	231	3.5	1.9	1.2
	2.1	141	48	114	3.3	0.8	1.1
−10 °C, dry nitrogen	1.2	821	519	405	6.5	2.9	1.3
	1.5	707	356	188	6.4	2.2	1.1
	1.8	407	259	27	6.4	1.5	1.0
	2.1	275	115	21	5.3	1.1	0.8

[a]The data for 22 °C, ambient, were extracted from Table 8.2.

Table 8.18 The constants of $\varepsilon_{pcb} = CN^{-\beta}$ as extracted from experimental data[a]

Solder joints	Test environment	$\varepsilon_{pcb} = CN^{-\beta}$		
		$C(\times 10^{-3})$	β	R^2
SnPb_OSP	22 °C, ambient	11.7	0.35	0.90
	−10 °C, nitrogen	63	0.58	0.90
SAC101_ENIG	22 °C, ambient	8.5	0.34	0.88
	−10 °C, nitrogen	24	0.48	0.95
SAC101(d)_ENIG	22 °C, ambient	13.1	0.37	0.95
	−10 °C, nitrogen	4.6	0.22	0.97

[a]The data for 22 °C, ambient, were extracted from Table 8.4.

equations are given in Table 8.18. Note that the characteristic life and the coefficients for the test environment of 22 °C (ambient) were extracted from Tables 8.2 and 8.4, respectively.

8.5.4.3 Analysis

The values of the shape factor in Table 8.17 suggest good data quality for the SnPb_OSP solder joint but poor data quality for the other two solder joints. The characteristic lifetimes of the three solder joints are given in the ε_{pcb}–N_α plots of Figure 8.24. Changing the test environment from 22 °C (ambient) to −10 °C (dry nitrogen) resulted in translation of the ε_{pcb}–N characteristics of the SnPb_OSP and the SAC101_ENIG solder joints to the right along the N axis; thus, the positive effect of dry nitrogen applies across the entire range of PCB strains.

By contrast, changing the test environment from 22 °C (ambient) to −10 °C (dry nitrogen) has resulted in translation of the ε_{pcb}–N characteristics of the SAC101(d)_ENIG solder joint to the left along the N axis – more so for the larger amplitude of PCB strain. This is reflected in the reduced magnitude of fatigue coefficient and fatigue exponent as shown in Table 8.18. The reduced fatigue life is attributed to the transition to brittle failure of the SAC101(d)_ENIG solder joints at the lower temperature, despite the presence of dry nitrogen as shown in Figure 8.18; the tendency for brittle failure increases with increasing PCB strain amplitude.

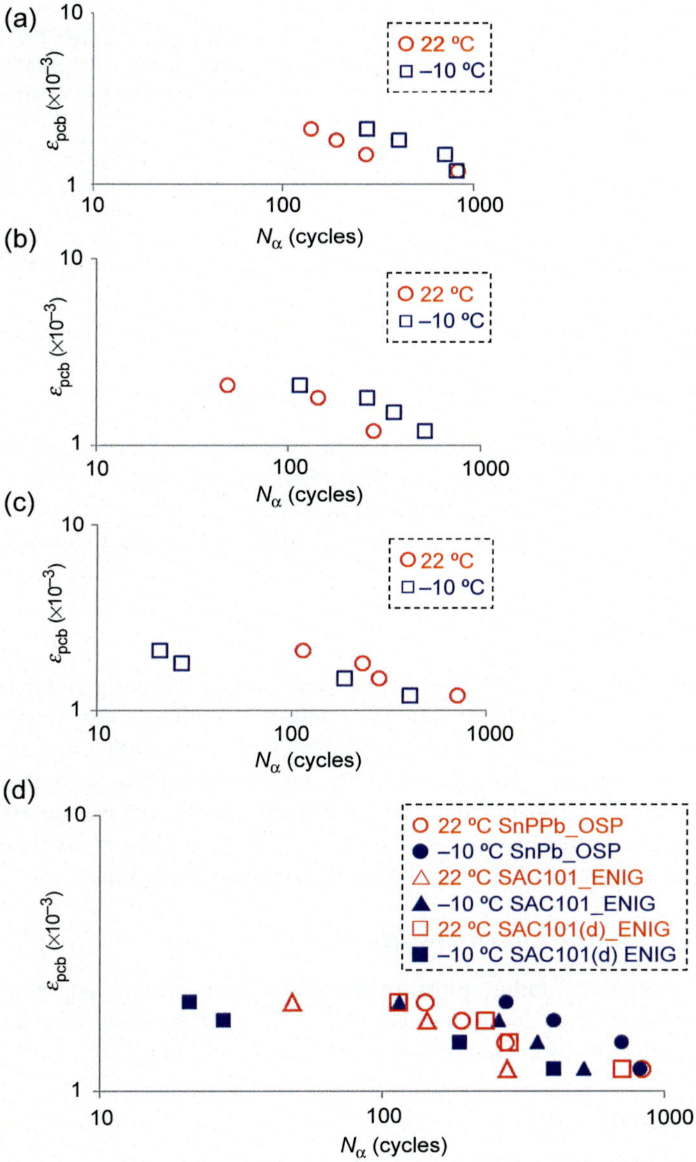

Figure 8.24 $\varepsilon_{pcb}-N_\alpha$ characteristics versus test temperature for (a) SnPb_OSP, (b) SAC101_ENIG, (c) SAC101(d)_ENIG and (d) all three solder joints.

Table 8.19 Test matrix for room-temperature aging

Solder alloys	Designation	Pad finish	Aging condition	Test condition
Sn–37Pb Sn–3.0Ag–0.5Cu Sn–1.0Ag–0.1Cu –0.02Ni–0.05In Sn–0.6Cu –0.05Ni + Ge Sn–0.8Cu –0.05Ni + Ge	SnPb SAC305 SAC101(d) SN100-0.6 SN100-0.8	OSP	Laboratory ambient for • less than 8 h • 10 days	BLDST (JEDEC condition H equivalent)

8.5.5 The effects of room-temperature aging

Newman (2005) reported a transition of the failure mode of solder joints towards ductile fracture with increased duration of aging at room temperature when tested under the conditions of ball impact shearing and ball impact pulling. The effects of aging on the board-level drop-shock resistance of solder joints are investigated in this section.

8.5.5.1 Test matrix

Five solder joint alloys on an OSP pad finish were studied. The fatigue resistance of the solder joints was evaluated using a BLDST with test conditions equivalent to Condition H of the JEDEC standard: a prescribed shock pulse of 2900 g and a duration of 0.3 ms. The test samples were manufactured and tested on two separate occasions: one on the same day of manufacture as the control vehicle and another 10 days after manufacture while stored in ambient laboratory conditions. The test matrix is tabulated in Table 8.19. The number of drop-shocks to failures was recorded.

8.5.5.2 Consolidation of results

The cumulative percent-failure plots for the solder joints without and with aging are shown in Figure 8.25(a) and (b), respectively. The corresponding characteristic life and shape parameters of the solder joints are shown in Table 8.20.

8.5.5.3 Analysis

The effects of room-temperature aging vary among the five solder joint material systems. Aging up to 10 days has virtually no effect on the BLDST resistance of three solder joint material systems: SnPb_OSP, Sn100-0.6_OSP and Sn100-0.8_OSP. However, the BLDST resistance of the SAC101(d)_OSP solder joint appears to be susceptible to degradation by room-temperature aging—by up to 40% after 10 days.

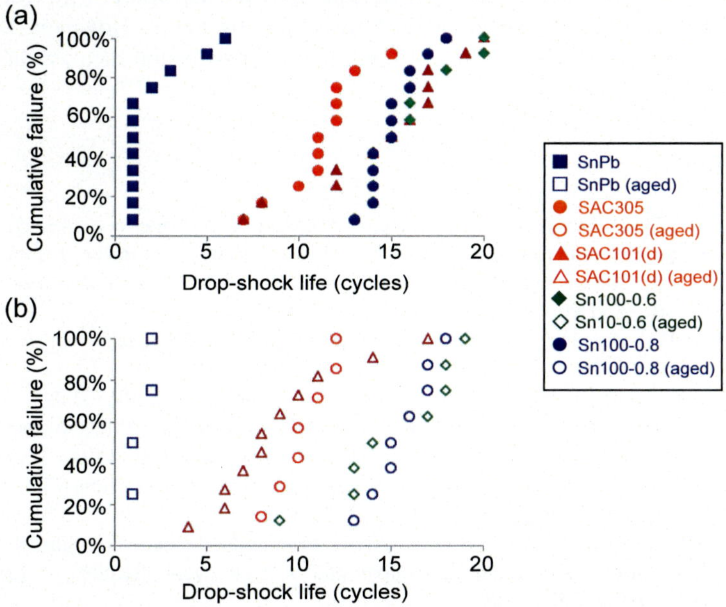

Figure 8.25 Cumulative failure of solder joint material systems in BLDST (a) without aging and (b) with 10 days of aging at room temperature before test.

Table 8.20 BLDST resistance of solder joints with and without room-temperature aging

Serial number	Solder joint	Aging	Characteristic life, *N* (cycles)		Shape parameter, γ
			Absolute	Ratio	
A	SnPb	✗	12.1	–	4.6
B	SnPb	✓	12.1	B/A = 1.0	5.4
C	SAC305	✗	1.9	–	1.4
D	SAC305	✓	3.9	C/D = 2.0	1.5
E	SAC101(d)	✗	15.1	–	3.4
F	SAC101(d)	✓	9.3	E/F = 0.6	2.9
G	Sn100-0.6	✗	16.3	–	7.7
H	Sn100-0.6	✓	18.6	G/H = 1.1	3.4
I	Sn100-0.8	✗	15.4	–	12.3
J	Sn100-0.8	✓	17.2	I/J = 1.1	7.9

The results also point to a higher risk of data spread after aging. The test data for SAC305_OSP suggest improved BLDST resistance by aging. However, the above observations are not conclusive due to the limited experimental data available.

References

ASTM-D7774. (2012). *Standard test method for flexural fatigue properties of plastics*. ASTM.

ASTM-D7791. (2012). *Standard test method for uniaxial fatigue properties of plastics*. ASTM.

Basquin, O. (1910). The exponential law of endurance tests. In *ASTM annual meeting* (Vol. 10, pp. 625–630). ASTM.

Coffin, L. J. (1954). A study of cyclic thermal stresses in a ductile metal. *Transactions of the ASME, 76*, 931–950.

Dowling, N. (2012). *Mechanical behaviour of materials* (4th ed.). Prentice Hall.

Newman, K. (2005). BGA brittle fracture – alternative solder joint integrity test methods. In *55th electronic components and technology conference* (pp. 1194–1201).

Wöhler, A. (1855). Theorie rechteckiger eiserner Brückenbalken mit Gitterwänden und mit Blechwänden. *Zeitschrift für Bauwesen, 5*, 121–166.

Wong, E., Seah, S., Caers, J., & Lai, Y.-S. (2014). Frequency-dependent strain-life characteristics of Sn1.0Ag0.1Cu solder on cupper pad at high cyclic frequency. *International Journal of Fatigue, 59*, 43–49.

Wong, E., Seah, S., Driel, W., Caers, J., Owens, N., & Lai, Y.-S. (2009). Advances in the drop-impact reliability of solder joints for mobile applications. *Microelectronics Reliability, 49*(2), 139–149.

Wong, E., Seah, S., & Shim, V. (2014). Frequency-dependent low cycle fatigue of Sn1Ag0.1Cu(In/Ni) solder joints subjected to high frequency loading. *Journal of Electronic Materials, 43*(2), 586–593.

Fatigue crack growth in solder joints at high strain rate

9

9.1 Introduction

Attempts to quantify the creep-fatigue crack growth in solder joints due to temperature cycling have been reported using an interrupted fatigue test (Darveaux, 2002). The crack growth in the solder joints is monitored by cross-sectioning the joints at intervals and averaging the crack growth rate over multiple samples of solder joints. This destructive test is expensive to perform in view of the large sample size required; each sample requires cross-sectioning, polishing and quantification of crack length. The interrupted test also introduces inaccuracies due to the discontinuous monitoring of data. Moreover, because of the loss of history in individual solder joints, no useful insights into the nature of crack propagation are obtained.

Although low-sensitivity resistance measurements are commonly used in the industry for the detection of complete failures of solder joints, high-sensitivity measurements for characterising the fatigue leading up to final failure are rarely performed. Zheng and Constable (1996) performed high-sensitivity resistance monitoring on a circuit board subjected to low-amplitude vibration. Because the circuits being monitored consisted of multiple joints, they reported that cracking occurred catastrophically at the end of the fatigue life; the observed gradual resistance changes were attributed to an aggregate of metallurgical changes caused by cyclic loading. Using high-sensitivity resistance tracking over a single solder joint, Seah et al. (2008) and Caers et al. (2010) reported progressive crack growth in solder joints subjected to high strain rate loading. Their studies have highlighted the feasibility of continuous tracking of crack growth in a solder joint through in situ monitoring of the change of electrical resistance across the solder joint.

9.2 Establishment of continuous crack growth tracking capability

The methodology for establishing continuous crack growth tracking in a solder joint is shown in Figure 9.1.

9.2.1 High-sensitivity in situ electrical resistance measurement

The test specimen is identical to that used for characterising the fatigue resistance equation, as shown in Figure 8.6. However, dedicated test structures, which allow very high measurement accuracy of the electrical resistance across the individual corner solder joints, have been added. The test structure is shown in Figure 9.2.

Robust Design of Microelectronics Assemblies Against Mechanical Shock, Temperature and Moisture
http://dx.doi.org/10.1016/B978-1-84569-528-6.00009-5

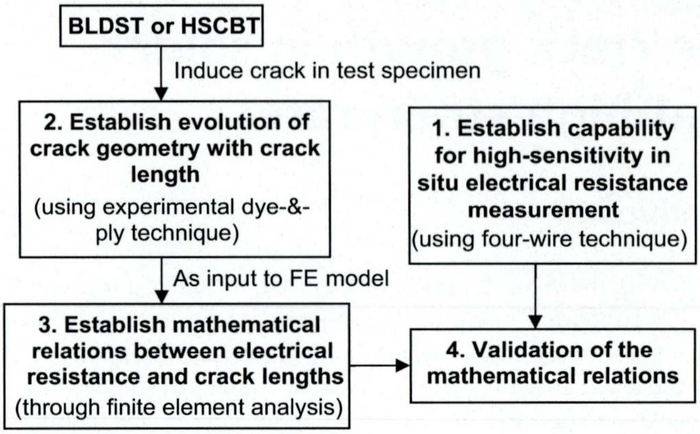

Figure 9.1 Methodology for establishing continuous crack growth tracking in a solder joint.

The presence of a crack restricts the flow of current, which is reflected in the increase of electrical resistance. Electrical resistance was measured using a four-wire (Kelvin connection) technique, where a stable source current of 200 mA was passed through each corner circuit, and the voltage drop across the corner joint was measured across two points close to the joint. To realise the required high measuring resolution, the current source was provided by a Keithley 2410 source meter that has an accuracy of 0.07% for the current reading; voltage measurements were made with an HP 8.5- digit reference multimeter that has a resolution of 0.05 μV. The voltage measurement has a sampling rate of 1000 data points per second, which is adequate for a medium transient dynamic event, such as the resonant vibration of a printed

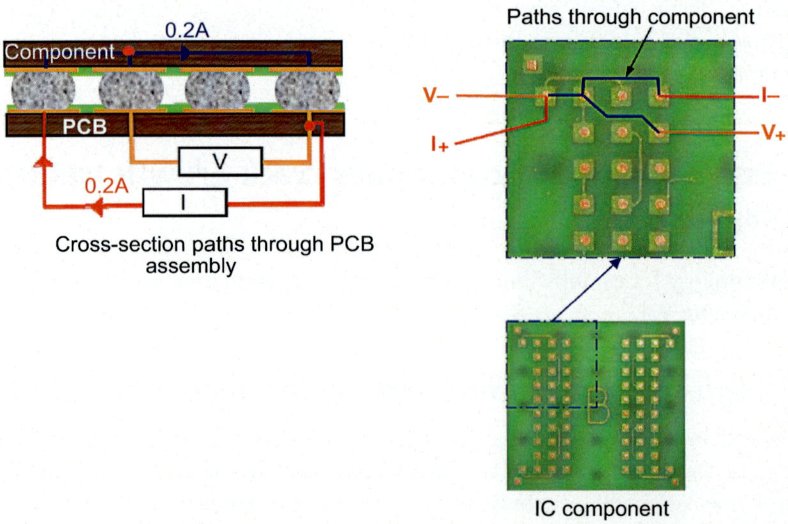

Figure 9.2 Test structure in test specimen for four-wire electrical resistance measurement.

circuit board (PCB) assembly. However, if a higher sampling rate is desired for highly transient event, such as impact, then a 16-bit oscilloscope with higher sampling rate could be used.

The design measures the voltage drop (and hence, the electrical resistance) across a solder joint, including a short length of Cu trace leading to the joint pads. It should be noted that even a short length of Cu trace accounts for more than 95% of the total electrical resistance due to the small cross-section and the significant length of the trace relative to the joint. However, this does not matter because it is the incremental change in electrical resistance that is of interest. The minimum detectable resistance change is approximately 20 μΩ, which is the noise floor for the measurement bandwidth of 5 kHz used for the dynamic bending tests.

In a typical experiment, the current is applied for several minutes prior to the start of cyclic bending to stabilise temperature drifts and resistance fluctuations caused by Joule heating of the Cu traces. Cyclic bending is then applied, during which an oscilloscope tracks changes in potential drop across the joint caused by the fatigue process. The test is continued until a completely open electrical circuit occurs, indicating a 100% cracking.

Undesirable fluctuations in measured voltage (on the order of microvolts) have been observed, which can be attributed to joule heating (electrical heating in the form of I^2R), the body heat of the individual when handling the test specimen, and environmental temperature fluctuations. To minimise joule heating, the source current was limited to 200 mA, without significantly compromising the resolution of the electrical voltage/resistance measurement. Background measurement noise of up to 20 μV in amplitude is prevalent, which can be easily handled with appropriate filtering. Electrical noise, caused possibility by electromagnetic induction, also may be noticed from the unrestrained flapping of the cables, which should be restrained when performing the test.

9.2.2 Evolution of crack geometry with crack length

Cracks of progressively larger size in the solder joints can be obtained through an interrupted test generated by either board-level drop-shock test (BLDST) or board-level high-speed cyclic bend test (HSCBT). At different intervals, samples were taken out and the crack surfaces were marked by dye. The cracked sample was immersed in Dykem Steel Red/Steel Blue dye for 5 min, followed by oven baking at 150 °C for 15 min to allow the dye penetrate the cracked faces and to evaporate the residual dye. The use of a hot-air blower is not advised because this will blow off the dye before it has fixed onto the crack faces. To enhance the dye-staining of the cracked surfaces, the cracked samples were pre-bent with the aid of an auxiliary board and a splitter before immersing into the dye, as shown in Figure 9.3. The outer solder joint will generally experience more than one crack; a large crack (referred to as the main crack) may originate from the outer edge of the solder joint near the solder–PCB interface, whereas a smaller crack (referred to as a secondary crack) may originate from the inner edge of the solder joint near the solder–component interface. For this reason, each sample was subjected to two pre-bends: one opened up

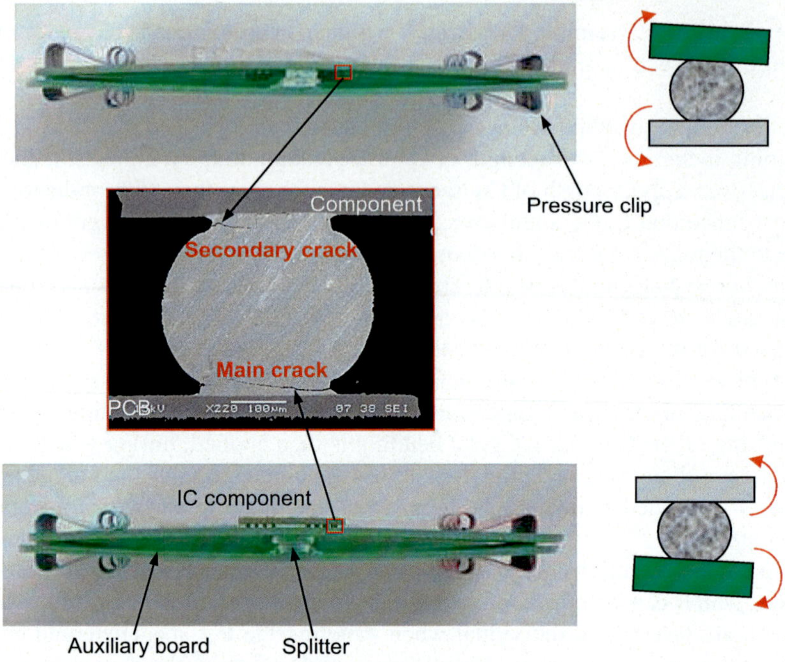

Figure 9.3 Pre-bent samples before immersing into dye.

the main crack and the other opened up the secondary crack. The magnitude of pre-bending was controlled so as not to induce artificial extension of existing cracks in the solder joints.

To characterise the planar geometry of the main crack, the component together with the solder joints have to be removed from the PCB to reveal the residual stained surface of the main crack on the PCB. Prying the component away from the PCB will cause the solder joint to undergo excessive plastic deformation, hence distorting the shape of the crack front. A technique was introduced here to take advantage of the fact that the fracture mode of the solder joint is highly sensitivity to strain rate. The component was dislodged from the PCB by light-impact hammering, as illustrated in Figure 9.4; this revealed the crack front of the main crack with virtually no artificial deformation. The relevant solder joint was then removed from the component to reveal the stained surface of the secondary crack while minimizing by impact shearing plastic deformation of the crack front.

The cracks run in the direction approximately parallel to the metal pad. Figure 9.5 shows the typical evolution of a main crack. The crack initiates along the edge, over approximately half the perimeter of the solder joint; its front assumes a curvature similar to that of the circular metal pad (a). The crack advances at different rate along its crack front, with a higher rate closer to the middle of the crack front and a lower rate nearer the edges (b), resulting in the gradual straightening of the crack front (reduced curvature) as the crack advances (c). The crack front becomes almost fully straightened

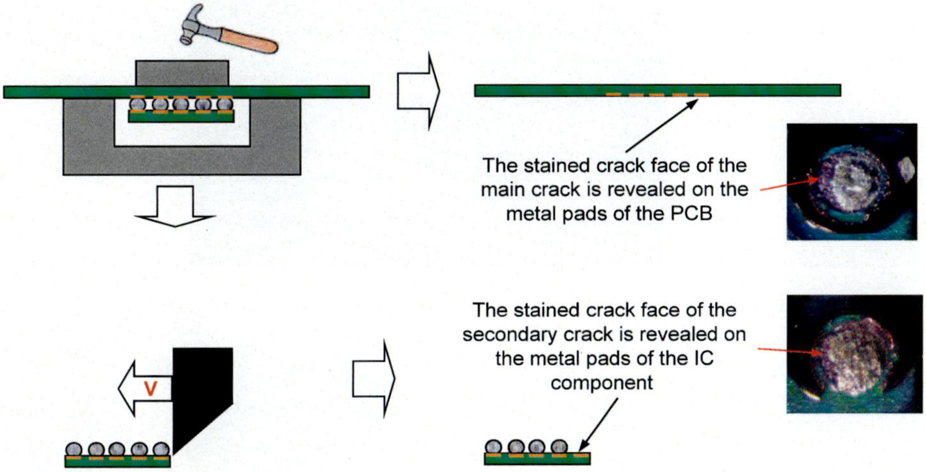

Figure 9.4 Dislodging of the IC component from the PCB by impact hammering and the solder joint from ICC component by impact shearing.

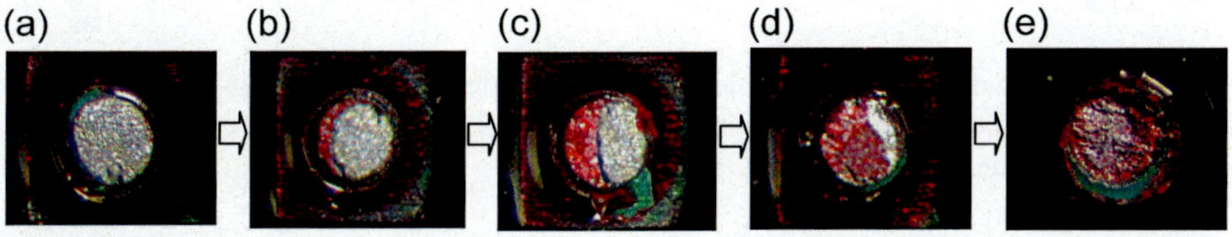

Figure 9.5 Evolution of the crack profile in the solder joint as observed from the dye-and-break experiments.

when the area of the crack grows to approximately three quarters of the cross-sectional area (d). This pattern of crack growth continues until the solder joint is fully broken (e).

9.2.3 Relating electrical resistance to crack length

Given the geometries of a solder joint, including the geometry of the crack, and the electrical conductivity of the materials involved, the electrical resistance across the solder joint can be evaluated through numerical modelling. The aim of this activity is to establish the theoretical relationship between the incremental electrical resistance across the solder joint and the size of cracks in the solder joint. The electrical model for a four-point resistance measurement circuit is shown in Figure 9.6(a). Because only the incremental electrical resistance is needed, and it is assumed that the electrical resistance of the copper traces and the bulk solder joints are independent of the current density, it is sufficient just to model a particular solder joint. This is shown in Figure 9.6(b), in which the boundary conditions for the model are also shown. The model includes the solder joint, the metal pads from the integrated circuit (IC) component (top) and the PCB (bottom) and short segments of copper traces. Again, because

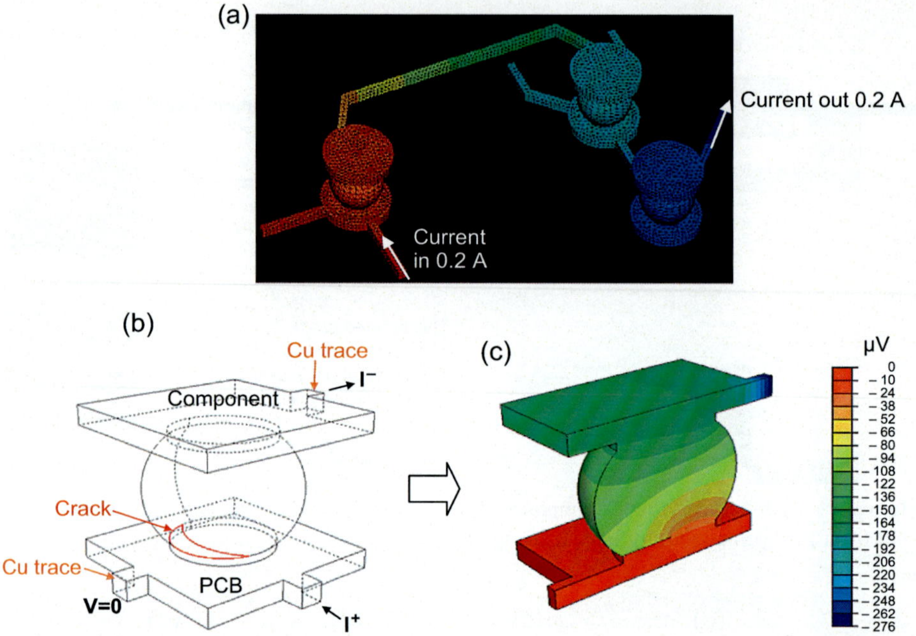

Figure 9.6 Electrical model of solder joint with pre-existent crack. (a) Four-point resistance circuit; (b) solder joint of interest showing the boundary condition; and (c) contour of electrical potential in the solder joint of interest.

only the incremental change in the electrical resistance is required, the intermetallic compounds (IMCs) are not included in the model. A pre-existing planar crack oriented parallel to the metal pad was introduced into the solder joint adjacent the PCB pad. The crack faces were modelled as two surfaces that have no mutual electrical connectivity. Figure 9.6(c) shows the electrical potential contour in the cross-section of the solder joint along a plane symmetric to the crack front. Much of the potential drop between the top and bottom pads occurred in the vicinity of the crack, where potential discontinuity was observed over the crack surfaces; high potential gradient developed around the remaining area of the solder joint due to the constricted current flow.

The crack fronts were modelled to have similar shapes as those observed in the experimental dye-and-break tests. The mathematical description of the crack front is given in Figure 9.7. The metal pad is represented by a circle of radius r_{pad}. The crack front is defined by the arc of a larger circle that intercepts with the circumference of the pad. The area enclosed by the two arcs gives the crack area, A_{crack}, and the offset between the two arcs gives the crack length, l_{crack}. A_{crack} is a function of l_{crack} and vice versa. Advancement of the crack front is modelled by increasing the offset while increasing the diameter of the intercepting circle. The relationship between the offset and the diameter of the intercepting circle is defined by experimental observations of Figure 9.5. Two intersecting circles, Circle 1 and Circle 2, are shown in Figure 9.7 to illustrate the progression of the crack at two crack fronts.

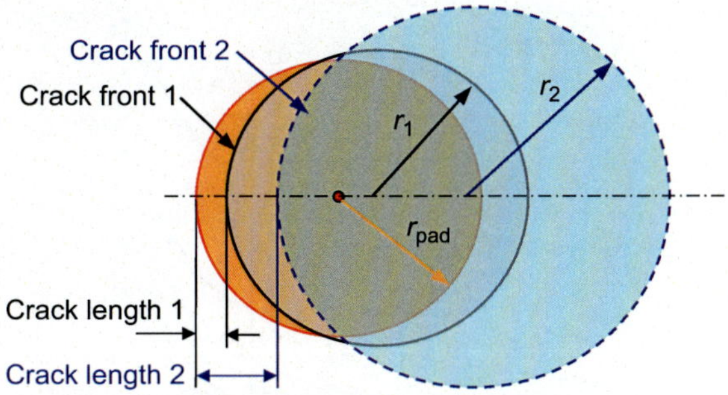

Figure 9.7 Description of the planar crack front in the solder joint.

Designating the potential drop across the solder joint as ΔV, which is a function of the crack size, the electrical resistance across the solder joint is simply $R = \Delta V/I$. For the same current input (0.2 A), the potential difference and hence the electrical resistance increases with increasing crack size. Setting the electrical resistance across a pristine solder joint as nil, the incremental electrical resistance, ΔR, across the solder joint as a function of the fractional fractured area of the crack (the area of crack as a fraction of the cross-sectional area of the solder–pad interface) is shown in Figure 9.8. The exponential nature of the relationship suggests increasing sensitivity of measurement with increasing crack size. As an illustration, a 100-μΩ incremental resistance may imply a growth of fractional fracture area from 30% to 45% or from 90% to 92%. Given that the crack length, l_{crack}, is a function of the crack area, A_{crack}, the horizontal axis may be expressed in terms of crack length.

A secondary crack that occurs diagonal to the main crack is frequently observed. The effects of the secondary crack on the electrical resistance across the solder joint have been modelled and are depicted in Figure 9.9. The horizontal axis is again the

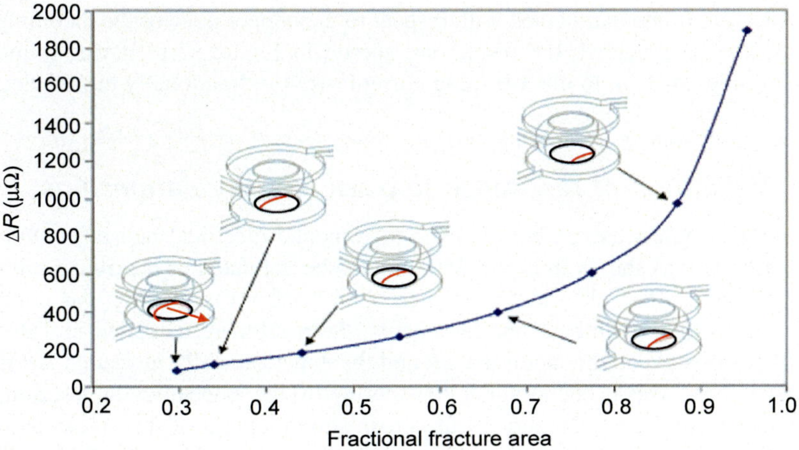

Figure 9.8 Electrical resistance versus crack size (main crack).

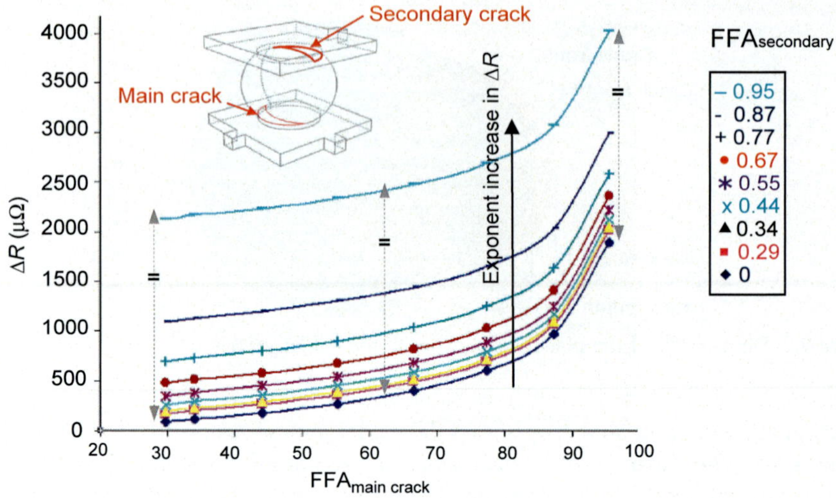

Figure 9.9 Electrical resistance versus crack size (main and secondary cracks).

fractional fracture area of the main crack, designated as FFA_{main}; the secondary crack is also modelled as a fraction of the solder–pad area, designated as $FFA_{secondary}$ and ranging from 0 to 0.95. For any particular size of the main crack, the electrical resistance increases exponentially with increasing size of the secondary crack, as expected. However, it is interesting to note that the absolute electrical resistance due to the secondary crack is almost independent of the size of the main crack, implying a simple additive relation for the electrical resistance attributed to the two cracks. As an illustration, Figure 9.8 gives ΔR equal to 180 and 400 μΩ for $FFA_{main} = 0.44$ and 0.67, respectively; the sum of resistance equals to 580 μΩ. Referring to Figure 9.9, taking $FFA_{main} = 0.44$ and $FFA_{secondary} = 0.67$ gives approximately 580 μΩ. That is, the electrical resistances from the two cracks behave as if they are in series connection in the electrical circuit.

The location of the main crack with respect to the copper trace of the incoming current was also investigated; the results are shown in Figure 9.10. It was found that cracks forming adjacent to the incoming current result in fractionally higher electrical resistance.

9.2.4 Validation of resistance to crack length relationships

The theoretical relationships between the incremental electrical resistance and fractional crack area, as shown in Figure 9.9, have to be validated or otherwise calibrated with experiments. The electrical resistance obtained from modelling was compared against that from experimental measurements; the results are tabulated in Table 9.1. The difference between the modelled ΔR and the experimentally measured ΔR ranges from 2% to 30%. This is encouraging considering (a) the inaccuracy in describing the shape and size of the cracks and (b) the gross simplification of the cracks as planar cracks, when in reality the crack could be deviated from the horizontal plane.

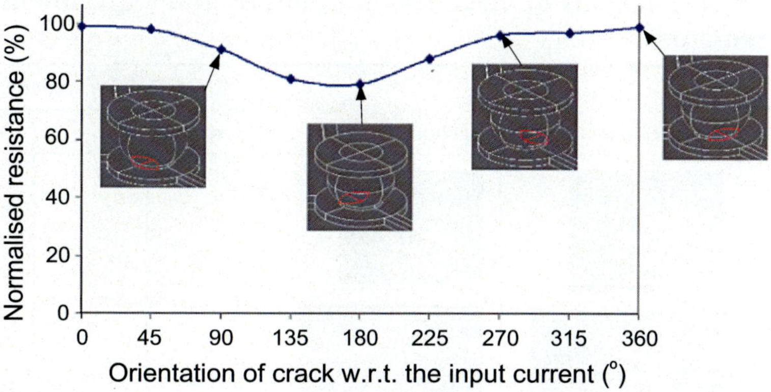

Figure 9.10 Electrical resistance versus crack orientation with respect to the Cu trace.

9.3 Crack propagation characteristics: board-level drop shock test

With the relationship between the crack size and the electrical resistance established, it is now possible to investigate in situ the characteristics of crack growth in the solder joint of interest. The electrical resistance in the corner joint of a test specimen with eutectic SnPb_OSP solder joints when subjected to consecutive BLDST of amplitude 800 g and duration = 0.5 ms was studied. The characteristics of electrical resistance for consecutive drop numbers 1, 3, 5 and 7 of the test specimen are shown in Figure 9.11. The resistance in the circuit is approximately 15 mΩ. A momentary increase in electrical resistances of magnitude between 30 and 36 μΩ was observed at 200 ms, which is in synch with the maximum bending of the test specimen. Much of this increased resistance was recovered with the unbending of the test specimen. However, residual resistance ranging from 12 to 25 μΩ was observed. The recoverable resistance, which has an average magnitude of 16 μΩ, can be attributed to the elastic deformation of the local traces in the resistance measurement circuit. The relationship is given by:

$$R + \Delta R = \rho \frac{L(1+\varepsilon)}{A(1-2\nu\varepsilon)} = \frac{1+\varepsilon}{1-2\nu\varepsilon} \Rightarrow \frac{\Delta R}{R} = \frac{\varepsilon(1+2\nu)}{1-2\nu\varepsilon} \tag{9.1}$$

where ρ is electrical resistivity and $R = \rho L/A$. Substituting $R = 15\ \text{m}\Omega$, $\Delta R = 16\ \mu\Omega$ and $\nu = 0.3$ gives the average stretching strain in the traces as 700 microstrains. The lower magnitude of strain compared to the amplitude of the PCB strain is expected in view of the fact that not all the local traces in the resistance circuit experience the maximum magnitude of the PCB strain.

While it is tempting to attribute the residual resistance to the plastic deformation of the copper traces in the resistance measurement circuit, this cannot explain the

Table 9.1 Comparisons of electrical simulation and experimental measurements

			Electrical resistance (μΩ)		
	Main crack	Secondary crack	Exp	FE	% Diff
1	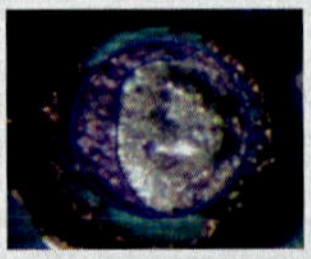		220	142	−35
2	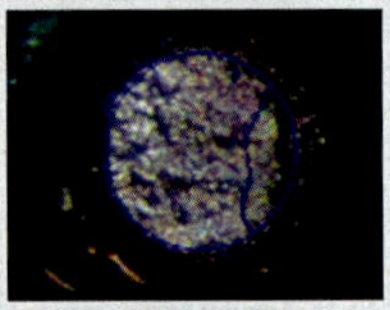	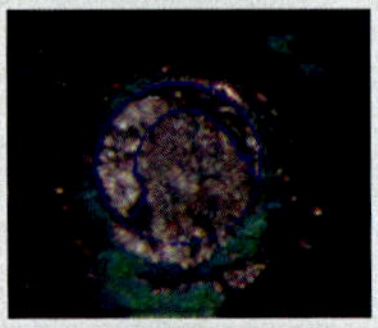	925	910	−2
3	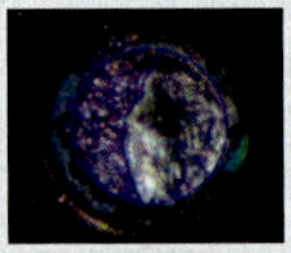		250	325	30
4	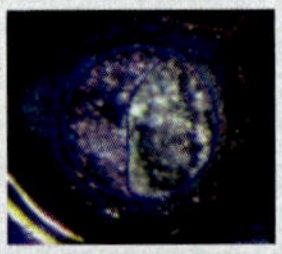	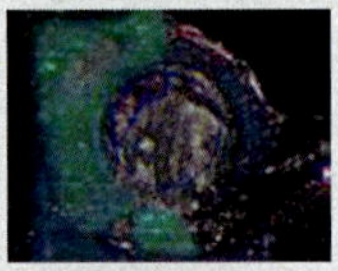	220	235	7
5	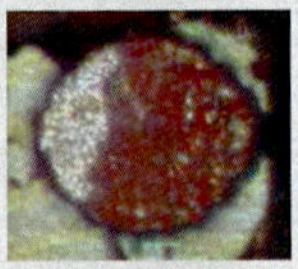	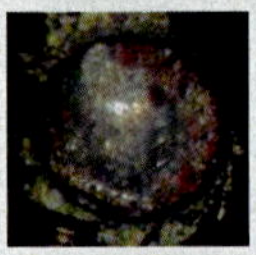	500	530	6
6	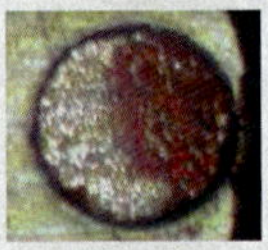	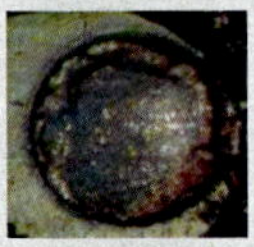	170	210	24
7			200	160	−20

Exp, experiment; FE, finite element; % Diff, percentage difference.

observed increase in the residual resistance with increasing number of drops at the same severity-increased plastic deformation of the copper traces is only possible if the test specimen is subjected to subsequent drop-shocks at increased severity. However, the presence of cracks in the solder joint will provide a satisfactory explanation for the observed residual resistance. It now appears that a single drop of the specimen

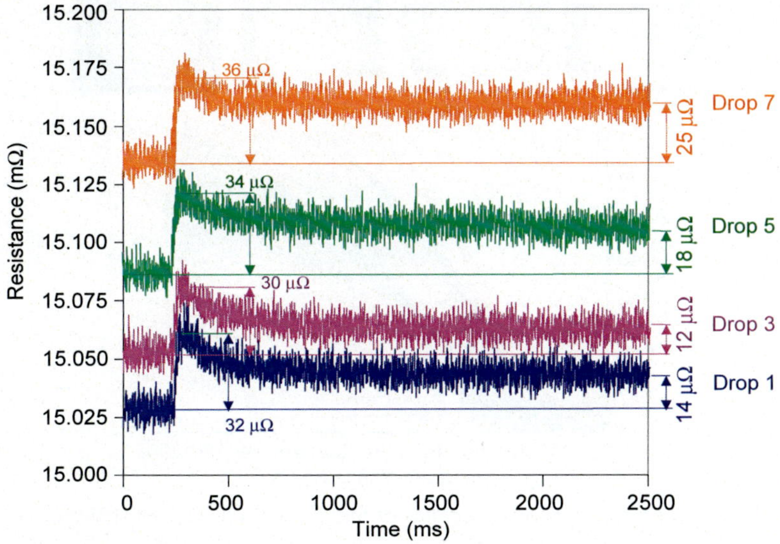

Figure 9.11 Electrical resistance characteristics of SnPb_OSP subjected to BLDST.

is enough to induce a detectable crack at approximately 5% FFA_{main}, which progressed to approximately 20% FFA_{main} at the seventh drop.

9.4 Crack propagation characteristics: high-speed cyclic bending test

Chapter 7 reported that solder joints exhibiting different failure mode have very distinct crack propagation path: predominantly within the bulk solder for ductile solder joints, such as SnPb_OSP and SAC101(d)_OSP, and predominantly within the IMC structures for brittle solder joints such as SAC305_ENIG and a combination of bulk solder and IMC in SAC101_OSP. It is expected that solder joints which have distinctively different failure modes would exhibit different crack propagation characteristics.

9.4.1 Crack propagation in ductile solder joints

Figure 9.12 shows four typical characteristics of crack propagation in SnPb_OSP solder joints subjected to the high-speed cyclic bending test at PCB strain amplitude of 1800 microstrains and frequency of 100 Hz. The fatigue crack is expressed in terms of crack length (%), which is the length of the crack observed in a sectional view though the centre of the joint expressed as a percentage of the pad diameter. The scanning electron micrograph of the sectional view of the solder joints, obtained from cross-sectioning the joints after testing, shows that the cracks occur entirely within the bulk solder material.

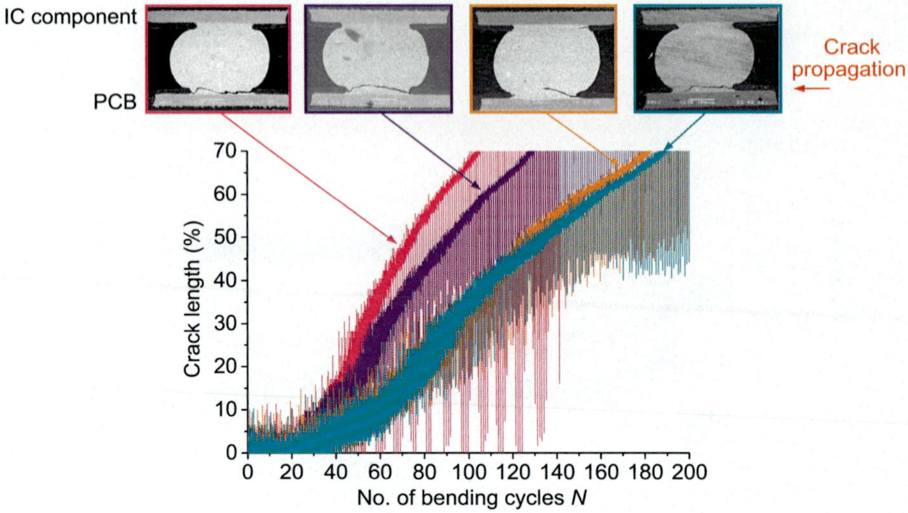

Figure 9.12 Crack propagation characteristics of SnPb_OSP subjected to HBSCT.

Detailed analysis of a particular SnPb solder is presented in Figure 9.13. A magnified view of the crack growth at the circuit board interface is shown in Figure 9.13(b). The sharp peaks and troughs correspond to the sinusoidal bending pulses of HSCBT. For each pulse, bending in the direction of an upward arch causes the crack to advance, as indicated by the small peaks in the waveform. Bending in the reverse (downward arch) direction causes the fracture surfaces to come back into contact, resulting in a momentary drop in electrical resistance, indicated by the deep and sharp troughs in the waveform.

It should be noted that troughs due to crack closure have much higher magnitudes and frequencies than peaks due to crack opening, although both result from the same 100 Hz sinusoidal bending pulse. This difference points to a significant limitation in frequency-domain measurements, which extract narrow bandwidths around certain frequencies of interest. In Figure 9.13(c), the high-frequency peaks and troughs have been filtered out for a clearer view of the crack growth trend from the start of cycling until failure at cycles-to-failure $N_f = 248$. The curve presents an S-shaped pattern that is typical of cracks occurring in bulk solder. The rate of crack growth, d*a*/d*N*, is shown in Figure 9.13(d). The rate of crack growth increases steadily from 0.3 μm/cycle at the initiation of crack to the peak rate of 1.7 μm/cycle, which corresponds to a crack length of 38%; from here, the rate of crack growth decreases steadily to 0.6 μm/cycle at 90% crack length before fracture.

This pattern of crack growth is due to two opposing effects on the crack driving force. The rate of crack propagation is dependent on crack driving force, which may be expressed simplistically as $K = Y\sqrt{\sigma a}$, where K is the stress intensity factor, Y is the geometrical parameter, σ is the remote stress and a is the crack length. The crack driving force K is effectively dependent on the product of σ and a, which are inversely

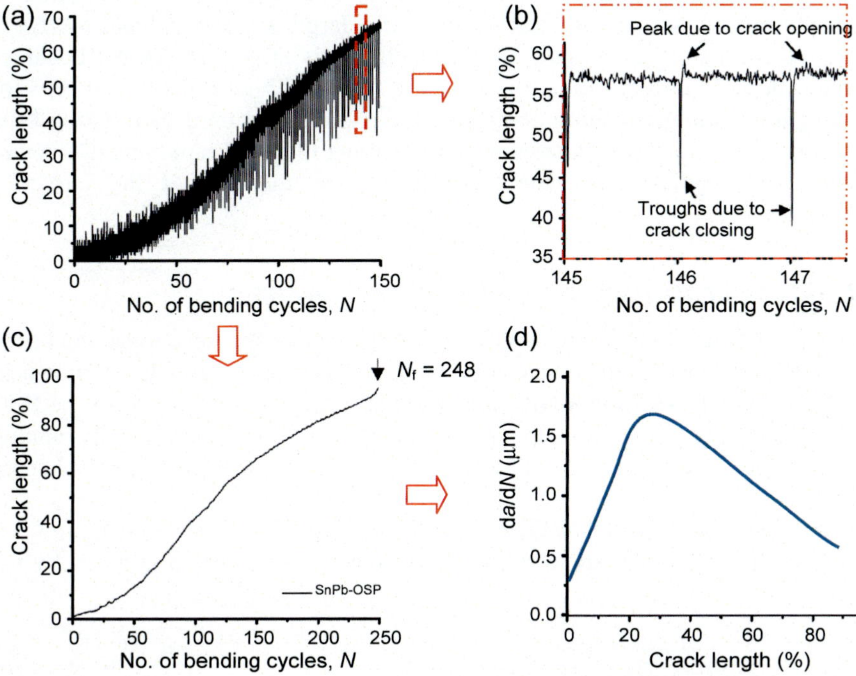

Figure 9.13 Detailed analysis of the crack propagation characteristics of SnPb-OSP subjected to HBSCT.

related: increasing a increases the compliance of the system and hence decreases σ. The crack driving force is effectively given by:

$$K + \Delta K = Y\sqrt{(\sigma + \Delta\sigma)(a + \Delta a)} \tag{9.2}$$

which after some manipulation becomes:

$$K + \Delta K = K\sqrt{1 + \frac{\Delta a}{a} + \frac{\Delta\sigma}{\sigma}} \tag{9.2a}$$

Expanding $\sqrt{1 + \frac{\Delta a}{a} + \frac{\Delta\sigma}{\sigma}}$ using the binomial series and ignoring the higher order term gives

$$\Delta K = \frac{K}{2}\left(\frac{\Delta a}{a} + \frac{\Delta\sigma}{\sigma}\right) \tag{9.3}$$

The rate of crack driving force is given by

$$\frac{\Delta K}{\Delta a} = \frac{K}{2}\left(\frac{1}{a} + \frac{\Delta\sigma}{\sigma\Delta a}\right) \tag{9.4}$$

wherein $1/a$ is always positive and $\Delta\sigma/\Delta a$ is always negative for the condition of fixed displacement. At a small crack length, $1/a$ is large, $1/\sigma$ is small and $\Delta K/\Delta a$ is positive. At a large crack length, $1/a$ is small, $1/\sigma$ is large and $\Delta K/\Delta a$ becomes negative. The equilibrium point for the solder joint of Figure 9.13 is at 38% of crack length. As fatigue loading continues, the crack grows until there is too little solder to maintain a mechanical connection, resulting in a final rapid rate of fracture propagation, as indicated by the sharp upturn at the end of the curve of Figure 9.13(c).

9.4.2 Crack propagation in brittle solder joints

SAC305 and SAC101 solders on Cu-OSP and ENIG pad finishes tend to exhibit brittle to mixed failure modes with much more unpredictable crack growth, as shown in Figure 9.14 for four SAC305_ENIG solder joints. Cracks running through the IMC (samples a and b) gave rise to steeper crack propagation. Conversely, cracks running through the ductile solder (sample d and part of sample c) yielded more gradual crack propagation.

The rate of crack growth $\Delta K/\Delta a$ is given by Eqn (9.4). Referring to Figure 9.15, the moment m is synonymous to σ and the stiffness of the system is given by the

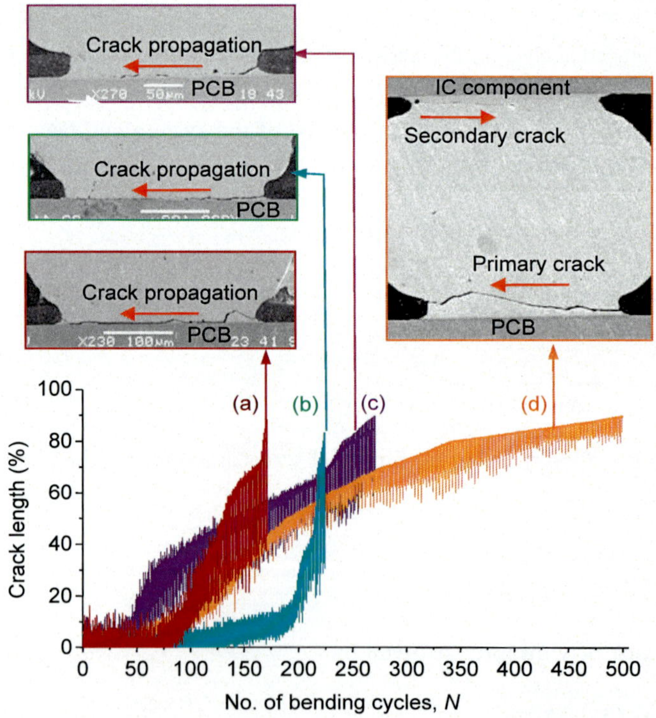

Figure 9.14 Crack propagation characteristics of the SAC305_ENIG solder joint.

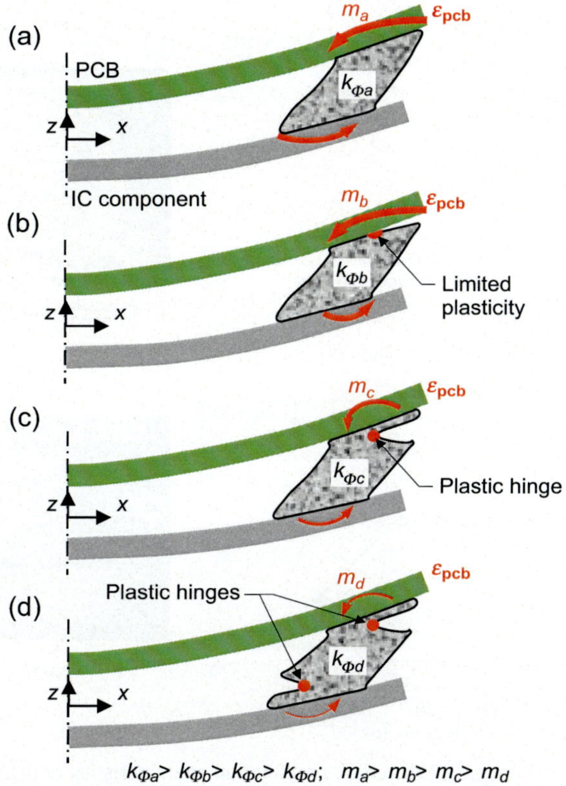

Figure 9.15 Flexural compliance of a solder joint: (a) no crack; (b) crack at PCB end in IMC; (c) crack at PCB end in solder; and (d) cracks at both ends in solder.

rotational stiffness of solder joint k_{Φ}. For the condition of fixed displacement, the magnitude of m is proportionate to k_{Φ}. Figure 9.15 illustrates the rotational stiffness of solder joints under the conditions of (a) no crack, (b) single crack in IMC, (c) single crack in bulk solder and (d) dual cracks in bulk solder. The rotational stiffness of the solder joint is inversely proportional to the opening of the crack face. Compared to a single crack in the bulk solder, a crack in the brittle IMC results in less opening of the crack faces because of limited plasticity at the crack tip. The rotational stiffness of the solder joint therefore decreases monotonically from the condition of (a) to the condition of (d); the same is true for the moment m (or σ). The term $\Delta\sigma/\sigma\Delta a$ in the bracket of Eqn (9.4) therefore increases from condition (a) to condition (d); hence, $\Delta K/\Delta a$ decreases from condition (a) to condition (d). Thus, the crack growth rate is higher in a brittle crack (e.g. a and b of Figure 9.14) than in a ductile crack (e.g. c and d of Figure 9.14).

Detailed crack growth analysis for a particular SAC101_OSP solder joint and a particular SAC305_ENIG solder joint are presented in Figure 9.16. Both solder joints exhibited some degree of mixed-mode fracture. Figure 9.16(b) shows a crack that

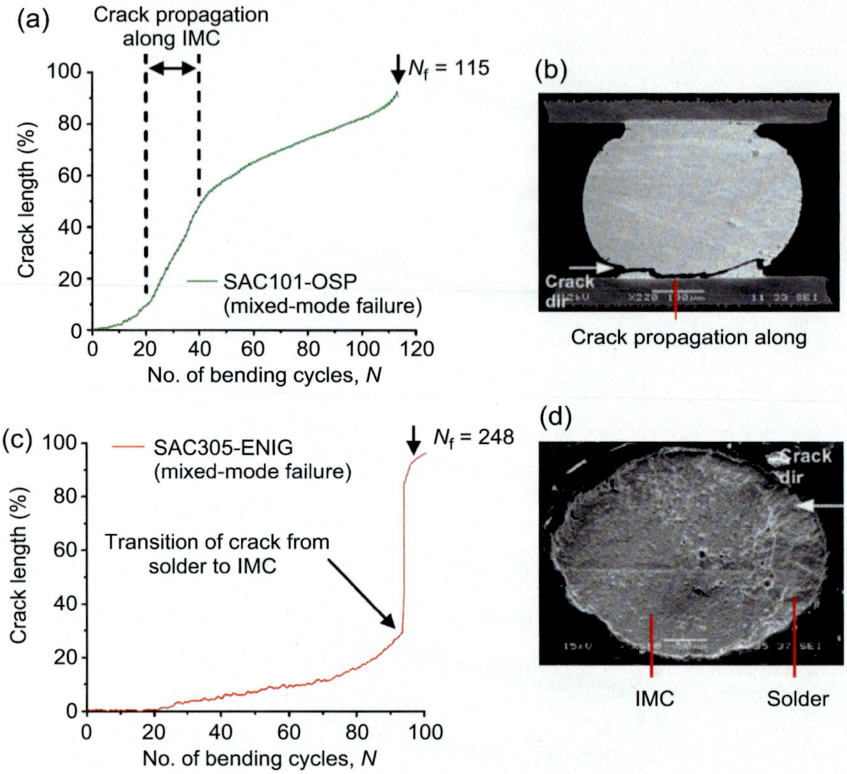

Figure 9.16 Detailed analysis of the crack propagation characteristics of (a) SAC101_OSP and (b) SAC305_ENIG solder joints.

started in the bulk solder of SAC101_OSP solder joint, then grew along the IMC layer for a short distance, and then returned to the solder before final failure. The corresponding crack growth curve in Figure 9.16(a) reflected this behaviour, where the marked section of the curve between $N/N_f = 0.17$ and 0.37, corresponding to crack propagation in the IMC, shows a rapid increase in crack length from 20% to 40%. Figure 9.16(d) shows a mixed-mode fractograph of an SAC305_ENIG solder joint. The crack initiated in and grew in the bulk solder for a short distance. Propagation of the crack into the IMC region at $N/N_f = 0.95$ resulted in a sudden catastrophic failure, as indicated by the almost vertical segment of the crack growth curve in Figure 9.16(c).

The above analysis suggests that the fatigue life of a solder joint experiencing mixed-mode is dominated by crack propagation within the bulk solder. This agrees with the earlier conclusion that the crack growth rate is higher in a brittle crack than in a ductile crack. The fatigue life of a solder joint can be improved significantly if a crack that has initiated in the bulk solder can be prevented from transiting into the IMC.

9.5 Three-dimensional fracture mechanics modelling of the crack front

Finite element analysis was performed by Wang et al. (2008) to provide insights into the observed evolution of the crack front, as depicted in Figure 9.5, and the observed characteristic of crack propagation, as depicted in Figure 9.13.

9.5.1 The finite element model

The commercial finite element software Abaqus was used to perform the analysis. A global-local modelling technique was used to allow detailed modelling of the crack front in the corner solder joint. In a nutshell, global modelling aims to obtain a good quality of displacement for the global model, especially those around the subject of interest. Details that do not affect the quality of the displacement information are ignored in the global model. In the consecutive modelling, only the subject of interest and the immediate boundary that carries the displacement information from the global model are modelled; the displacement information becomes the boundary condition for the local modelling. Details that affect the quality of the intended information are included in the local model. An important consideration in selecting the immediate boundary is that the accuracy of the displacement information obtained from the global model must not be compromised by the use of a coarse model.

The global model of the PCB assembly and the local model of the corner joint – the subject of interest – are shown in Figure 9.17. The global model is made of a quarter symmetry of the PCB assembly. The PCB, the IC component and the solder joints were modelled with relatively coarse mesh; the solder joints were modelled simply as cylinders. To enhance the quality of the global displacement from the model, the solder joints were assigned with the appropriate elastic–plastic constitutive property, and the model was assumed to experience large geometrical deformation. The PCB assembly experiences predominantly fundamental-mode deformation in BLDST and uniform curvature deformation in HSCBT. As far as the solder joint is concerned, the parameter that is of paramount importance is the PCB strain, whereas the exact mode of deformation of the PCB assembly is of much less importance. The global model was subjected to gravitational loading, which induced a fundamental-mode deformation. The magnitude of the gravitational acceleration was artificially adjusted so that the strain on the PCB reproduced that experienced by the PCB assembly in BLDST or HSCBT. The crack in the corner solder joint was not modelled in the global model.

The local model (or submodel) consisted of the corner solder joint and segments of PCB and component that carried the global displacement information. A pre-existing crack was introduced into the bulk solder on the PCB side. For this study, the crack front was modelled to propagate along a plane slightly above the copper pad and within the solder material. To facilitate the modelling of the two-dimensional crack front, the crack front was modelled using a tube constructed of rings of second-order hexahedral elements, as shown in Figure 9.18. The crack tip was made up of the collapsed edges of a ring of second-order hexahedral elements. The mid-nodes of these collapsed

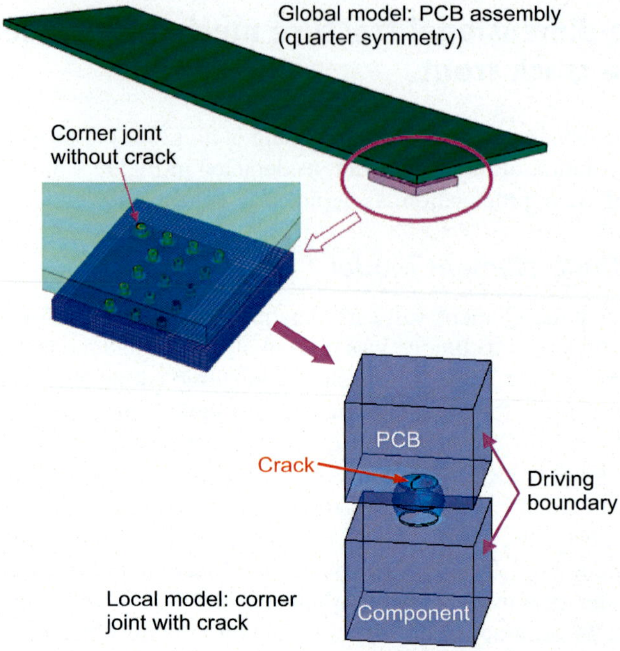

Figure 9.17 Global-local model of the PCB assembly and the corner solder joint.

elements were translated to a quarter length of the element so as to induce a singular strain field around the crack tip. The three-dimensional tube was embedded into a sea of second-order tetrahedral elements with compatible shape function.

9.5.2 Evolution of the crack front

To avoid mesh sensitivity and to improve the accuracy of the calculation of the three-dimensional J-integral, Abaqus uses volume integration over a finite domain surrounding the crack instead of area integration. The J-integrals are extracted along the crack front in the direction normal to the crack front. These are expressed in terms of angular degree, as shown in Figure 9.19, where zero degree is the direction parallel to the length of the PCB. The J-integrals along the crack front for crack lengths from 10% to 50% of the pad diameter are presented. Notice that the value of the J-integral is the highest near the middle of the crack front (zero degree) and recedes towards the edges. The higher rate of crack growth for the lower angular orientation is responsible for the observed crack front changing from concave to convex with increasing crack length, as shown in Figure 9.5.

9.5.3 Rate of crack growth with crack length

Figure 9.20 shows the values of the J-integral extracted at the middle of the crack front as a function of the crack length. It was observed that the value of the J-integral

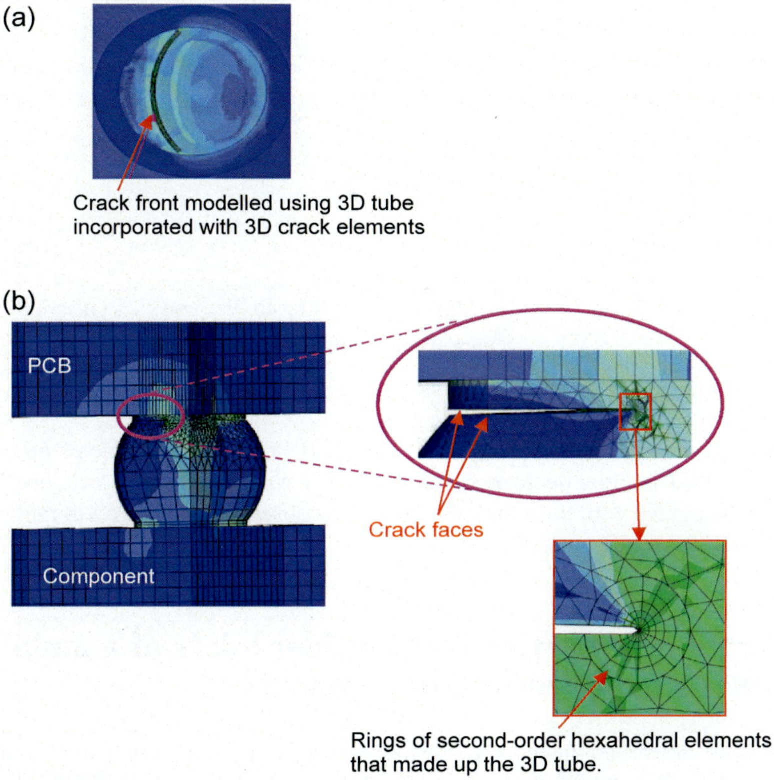

Figure 9.18 Crack front in the solder joint: (a) top view and (b) side view with enlargements of the crack and the three-dimensional tube.

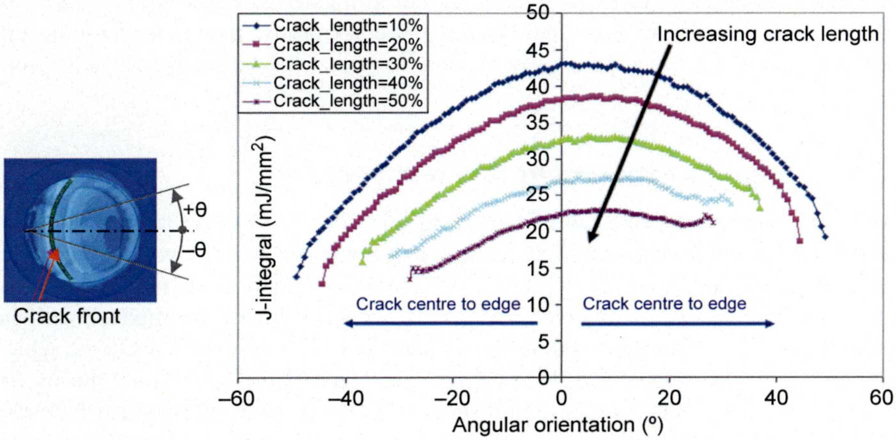

Figure 9.19 J-integral along the crack front.

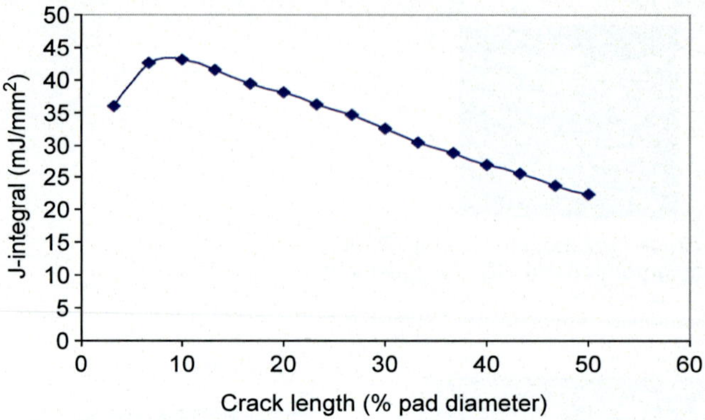

Figure 9.20 J-integral versus the crack length.

increases with increasing crack length up to about 10% of the pad diameter, after which the value of the J-integral decreases almost linearly with increasing crack length. This characteristic agrees with both the experimentally observed crack growth rates shown in Figure 9.13(d) and Eqn (9.4).

9.6 Crack propagation in the solder joints of a mobile phone experiencing drop impact

We have thus far learned much about the nature of crack propagation in ductile and brittle solder joints when the PCB assembly is subjected to BLDST and HSCBT. The PCB assembly experiences regulated bending waveforms in both loading conditions – a bending waveform of constant amplitude and frequency in the case of HSCBT and a bending waveform of receding amplitude at an almost-constant frequency in the case of BLDST. In reality, the bending waveform experienced by the PCB assembly in a portable electronic device (PED) subjected to drop impact is much more complex, as discussed in Chapter 6. With an understanding of crack propagation in solder joints under complex loading, we hope to provide more insights for robust design against drop-impact.

9.6.1 Design of experiment and test vehicle

The PCB assembly of a selected PED is replaced with a test board that is cut to identical dimensions as the original PCB, as shown in Figure 9.21. The layout of the solder joints on the test board (and the simulated IC component) is identical to those used for HSCBT with the test structure given in Figure 9.2. The test board is completed with the same four-wire resistance measurement circuits used in the previous sections, which allow in situ monitoring of the electrical resistance across the solder joint during the drop impact of the PED. *In situ* monitoring of the board strain is provided by strain gauges mounted on the reverse face of the test board.

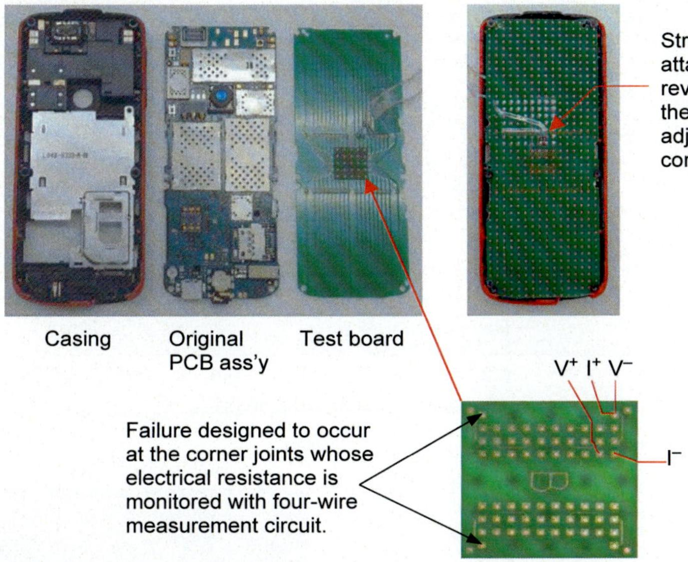

Figure 9.21 PED with a test board used for the drop test.

The PED with the test board is dropped from a height of 1 m, impacting at its bottom edge at an orientation of 45° against a steel target. The crack growth characteristics of two solder joint material systems, SnPb_ENIG and SAC101_OSP solder joints, are investigated here.

9.6.2 Results and analysis of SnPb_ENIG solder joints

The PED assembled using the test board with SnPb_ENIG solder joints was subjected to 24 consecutive drop impacts. Figure 9.22(a) shows the characteristics of crack opening, expressed as a percentage of the metal pad area, in a particular corner solder joint as a function of time for consecutive drops of the PED. The consecutive drop number is indicated in a bracket next to the drop-response. Note that not all of the responses from the 24 drop impacts are presented. The responses of some of the drops in which there is insignificant crack growth – between drops 1 and 5, for example – are neglected. That is, the horizontal axis of the figure represents the incremental time, not the absolute time. There are two jumps in the registered crack opening for each drop impact of the PED. These correspond to the ratcheting discussed in Chapter 6. The pseudo 'jumps' in the registered crack opening are attributable to the jumps in the electrical resistance due to the elastic opening, and closure of the crack surfaces corresponds to the bending of the test board. The true crack area is given by the stabilised crack opening. It appears that a crack of approximately 10% (of the pad area) is initiated by the first drop impact. The crack area increases progressively with subsequent drops. Figure 9.22(b) shows the progressively increasing crack area, computed from the residual resistance, with an increasing number of drops. The characteristic resembles a bilinear behaviour – a

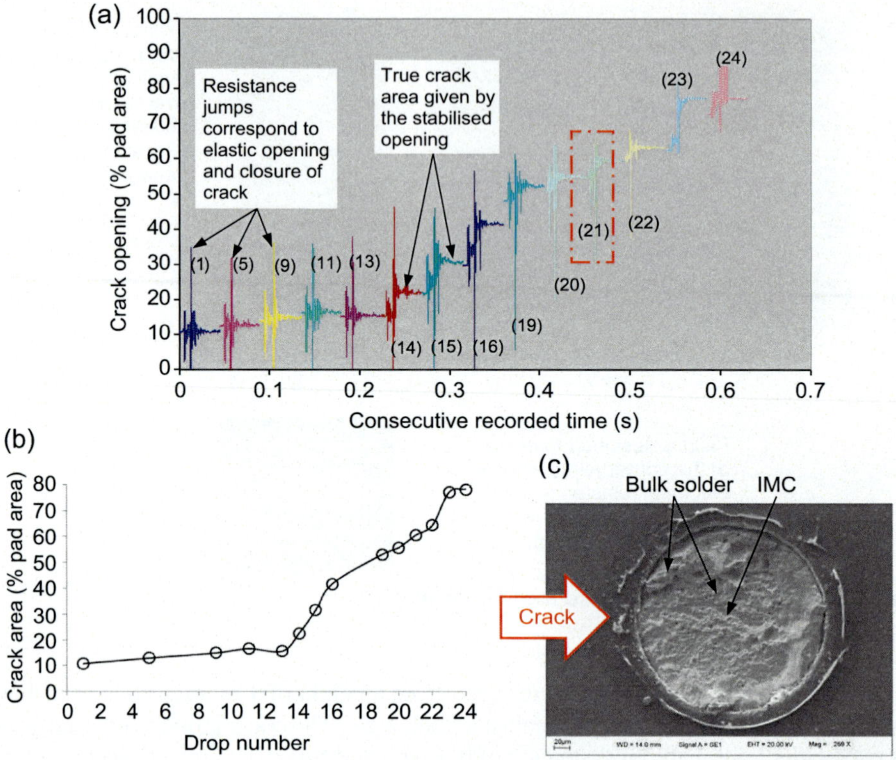

Figure 9.22 Crack growth characteristics of an SnPb_ENIG solder joint in the drop impact of a PED: (a) time history; (b) drop history; and (c) fractograph.

gradual growth rate at approximately 0.6% pad area per drop up to drop 13, followed by a rapid growth rate at approximately 6% crack area per drop. This characteristic is very different from that observed in a PCB assembly tested under HSCBT. Figure 9.22(c) shows the fractograph of the solder joint on the board side, revealing a mixture of ductile and brittle fracture modes with substantial residual solder along the fringes of the pad. However, it provides no insight on the observed bilinear crack growth characteristic.

Figure 9.23 shows the board strain and the crack opening plotted on the same time scale for drop 21. The board strain was measured with a strain gauge attached to the face opposite to the IC component and adjacent to the corner solder joint. Upon impact of the PED, the test board underwent a downward deflection, predominantly in the fundamental mode. This is illustrated in the sketches below the plots. The positive curvature resulted in compressive strain at the top of the test board and was picked up by the strain gauge, which showed a frequency of approximately 800 Hz ①. A higher flexural mode that travelled through the test board was also picked up by the strain gauge. The higher mode (at an approximate frequency of 3500 Hz) resulted in a further increase in the local curvature of the test board at which the strain gauge was attached, producing increased compressive strain ②.

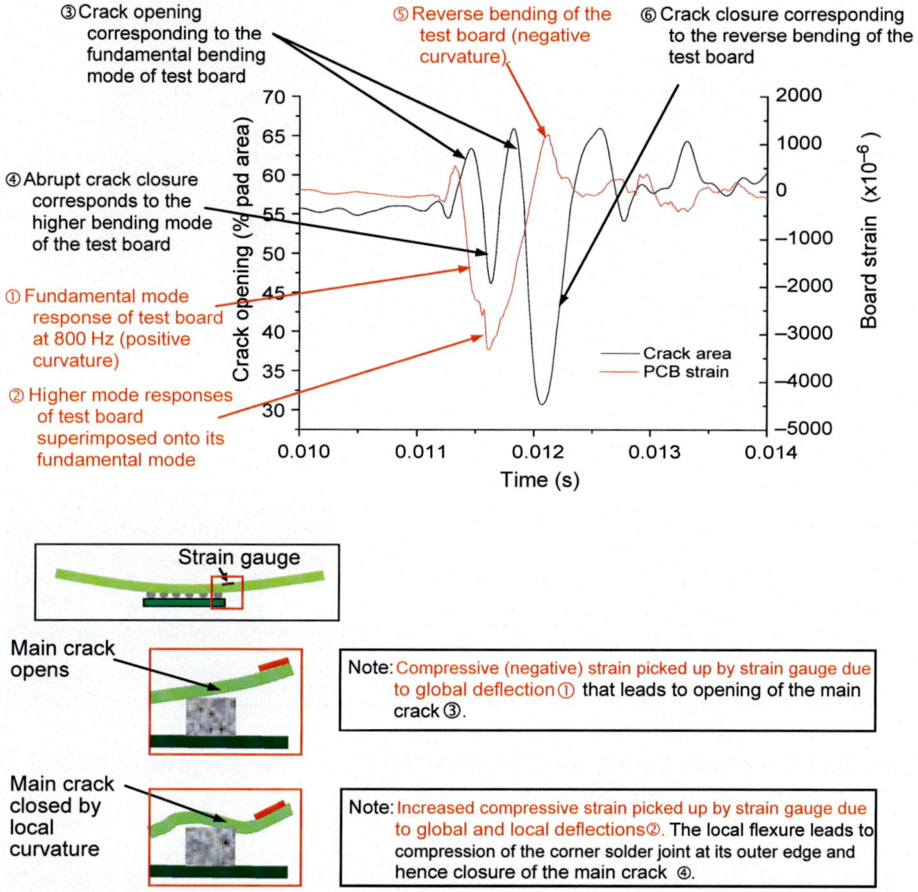

Figure 9.23 Board strain versus crack growth characteristics of an SnPb_ENIG solder joint at drop 21.

The global deflection of the test board – the fundamental mode – led to stretching of the corner solder joint at its outer edge and opened up the main crack ③, as illustrated in the second sketch. By contrast, the higher flexural mode resulted in momentary closure of the main crack ④, despite registering a compressive strain at the strain gauge, as illustrated in the third sketch. The crack returned to the opening once the higher flexural mode passed the corner solder joint. The reverse global bending of the test board resulted in positive strain being registered at the strain gauge ⑤ and a corresponding closure of the crack ⑥.

9.6.3 Results and analysis of SAC101_OSP solder joints

The test board with the SAC101_OSP solder joint was condition-aged at laboratory ambient temperatures for a couple of weeks before being assembled onto the PED

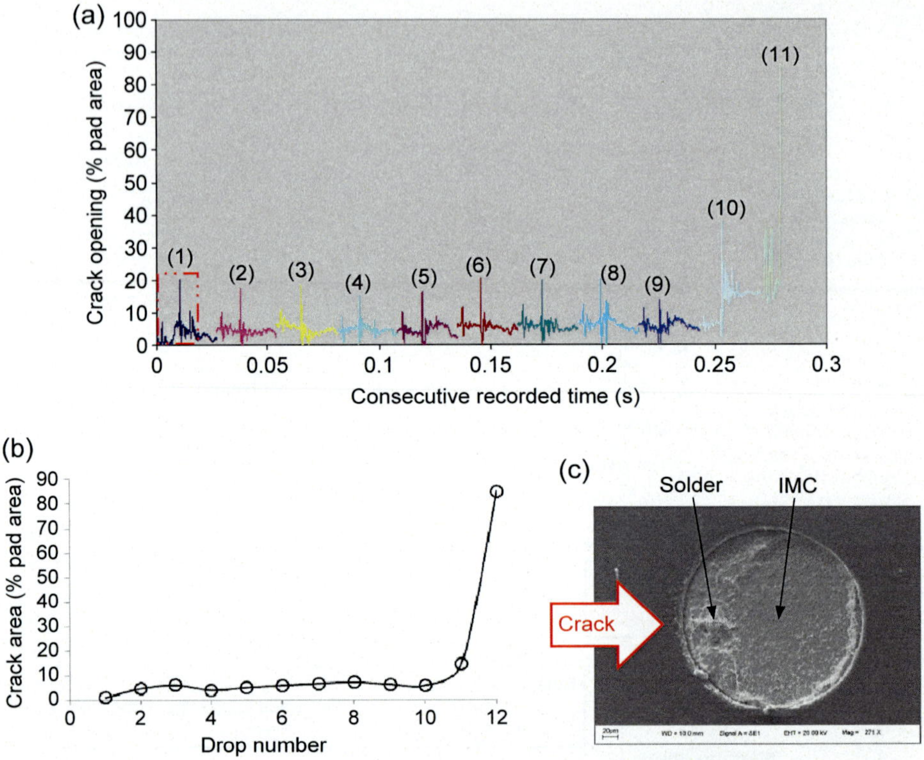

Figure 9.24 Crack growth characteristic of an SAC101_OSP solder joint in drop impact of PED: (a) time history; (b) drop history; and (c) fractograph.

and subjected to 11 consecutive drop impacts. Figure 9.24(a) shows the crack opening with time for 11 consecutive drops. Figure 9.24(b) shows characteristics of crack area, computed from the residual electrical resistance, with an increased number of drops. A period of barely discernible (approximately 6% of pad area) crack growth was observed between drop 1 and drop 10; this was followed by distinct crack growth (to 14%) after drop 10 and an abrupt failure during drop 11. The crack growth characteristics are different from that observed in HSCBT. Figure 9.24(c) shows the fractograph of the fracture surface on the test board, with distinct zones of ductile and brittle fractures; the crack propagated through the ductile zone before reaching the brittle zone. Thus, much of the life of the solder joint was spent propagating the crack in the ductile bulk solder. The solder joint underwent a near-immediate fracture once the crack path moved into the IMC layer. This observation is in agreement with that in HSCBT of the same solder joint.

Figure 9.25 shows the impact force, board strain and the crack opening plotted on the same time scale for drop 1. The impact of the PED happened at 2.5 ms and resulted in a near half-sine impact force pulse that induced a transient response of the test board – a negative strain (positive curvature) ① followed by a positive strain

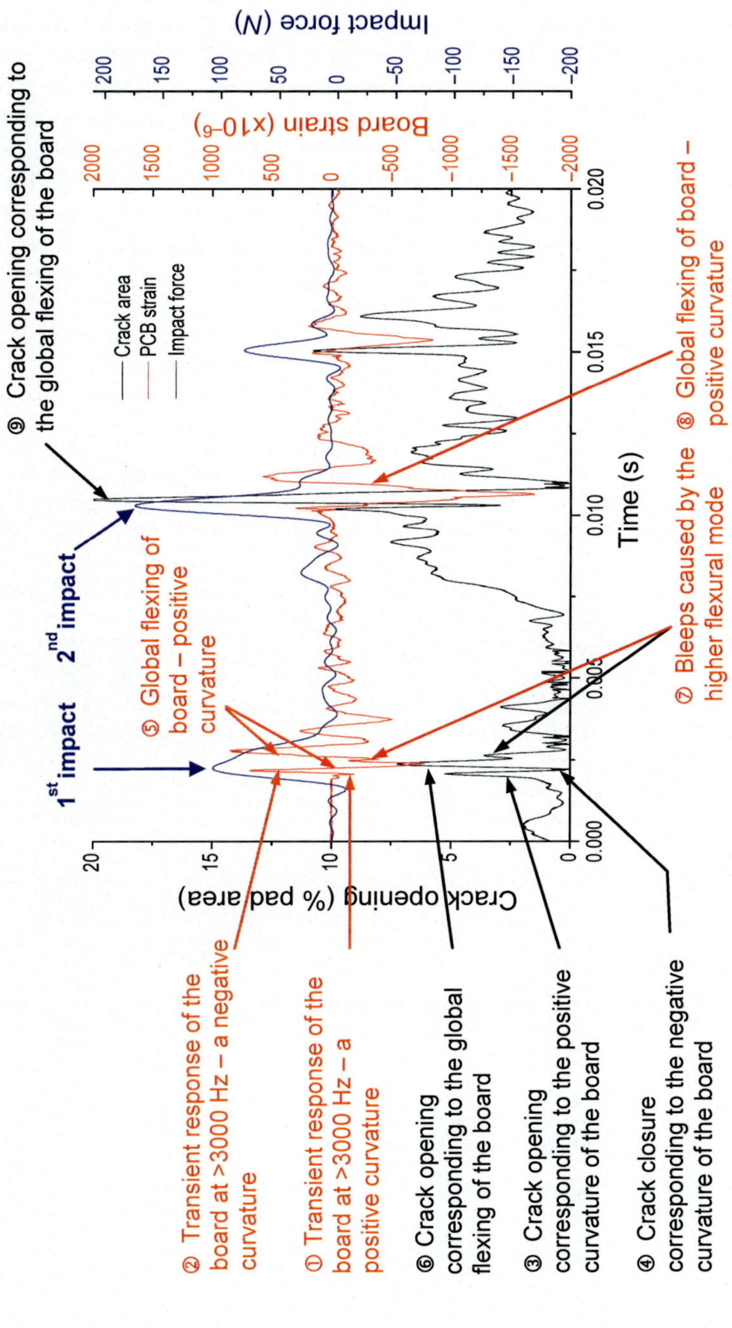

Figure 9.25 Impact force, board strain and crack growth characteristics of an SAC101_OSP solder joint at drop 1.

(negative curvature) ② at a rather high frequency. This resulted in transient opening ③ and closure ④ of the corner solder joint. The transient response of the test board was followed by global flexing (fundamental frequency) ⑤ (positive curvature) and a corresponding crack opening ⑥. The presence of the higher frequency modes led to bleeps in the board strain and the crack opening ⑦. The 45° impact of the PED at its bottom edge yielded rigid body rotation and led to a second impact of the PED at its top edge. The second impact occurred at 10.5 ms and registered a higher magnitude of impact force, most likely due to higher rigidity, as evident by the higher frequency of impact at an almost identical magnitude of impulse as the first impact. The larger magnitude of the impact force with a frequency that was closer to the first harmonic of the test board induced a larger response of the test board ⑧ and a corresponding higher crack opening ⑨.

References

Caers, J. W.-S.-L. (2010). A study of crack propagation in Pb-free solder joints under drop impact. *IEEE Transactions on Electronics Packaging Manufacturing, 33*, 84–90.

Darveaux, R. (2002). Effects of simulation methodology on solder joint crack growth correlation and faituge life prediction. *Transactions of the ASME, Journal of Electronic Packaging, 124*, 147.

Seah, S. W. (2008). Fatigue crack propagation behavior of lead-free solder joints under high-strain-rate cyclic loading. *Scripta Materialia, 59*, 1239–1242.

Wang, J. S. (2008). Fracture mechanics study of fatigue crack growth in solder joints under drop impact. In *58th Electronic components and technology conference* (pp. 1601–1605).

Zheng, D. C. (1996). Analysis of electrical resistance monitoring of PCMCIA interconnection failures. In *46th Electronic components and technology conference* (pp. 808–814) (Orlando, FL.).

Dynamic deformation of a printed circuit board in drop-shock

10

10.1 Introduction

The experimental drop-shock testing of board-level solder joints was reported in Chapter 7. This chapter provides insight into the dynamics of this event through simple analytical equations. This section presents a brief introduction to the numerical schemes used by commercial finite element (FE) code in modelling transient dynamics and an overview of the numerical modelling of board-level drop impact/shock.

10.1.1 A brief introduction to numerical schemes for transient dynamics

Computational transient dynamic modelling has a long history in the fields of aerospace and defense engineering. The flight dynamics of aircraft and ballistic projectiles were modelled extensively, even before the commercial FE codes became available. Explicit hydro-code was used extensively in the analysis of the penetration of projectiles against armour targets in the 1980s. Computational modelling of the drop impact of portable electronic products was first reported in the 1990s (Wu, Song, Yeh, & Wyatt, 1998) and became more common in the 2000s (Kim, 2001; Zhu, 2003), with focus on analysing the mechanical robustness of product housings and the liquid crystal display panels. The electronic assembly community began the computational modelling of the transient dynamics of printed circuit board (PCB) assemblies in the early 2000s, before the advent of the board-level shock-test standard JESD22-B111 in 2003 to study the nature of stresses in solder joints under various board-level test conditions (Hirata, 2001; Qiang, 2002; Sogo, 2001; Wang, 2008; Zhu, 2001) and to understand the physics of failure of the board-level solder joints (Wong, 2002). The activities in transient dynamic modelling of PCBs increased rapidly after the board-level shock-test standard JESD22-B111 was established. Based on published articles, however, it appears that transient dynamic modelling remains relatively unfamiliar to the electronics assembly community, especially for boundary conditions. A brief introduction is presented in this section.

The numerical scheme for describing the transient dynamics of an engineering system is expressed in the form of Newton's second law of motion:

$$\mathbf{M}\ddot{\mathbf{u}} + \mathbf{C}\dot{\mathbf{u}} + \mathbf{K}\mathbf{u} = \mathbf{F} \tag{10.1}$$

where $\mathbf{F}$ is the external force. A transient solution is needed when the magnitudes of the inertial force $\mathbf{M}\ddot{\mathbf{u}}$ and the damping force $\mathbf{C}\dot{\mathbf{u}}$ are significant compared to the structural deformation force $\mathbf{K}\mathbf{u}$. In general, the inertia force $\mathbf{M}\ddot{\mathbf{u}}$ becomes significant

Robust Design of Microelectronics Assemblies Against Mechanical Shock, Temperature and Moisture
http://dx.doi.org/10.1016/B978-1-84569-528-6.00010-1

when the time frequency of the applied force **F** is higher than the resonant frequency of the structure.

Numerical methods discretise the time domain and seek to satisfy dynamic equilibrium at a discrete time. There are generally two numerical methods for solving transient dynamic problems: the direct integration method and the mode-superposition method. In the direct method, Eqn (10.1) is solved in its original form via time integration. Different assumptions have led to a number of time-integration schemes, which can be classified into two main schemes: explicit schemes and implicit schemes.

10.1.1.1 Explicit schemes

An explicit scheme seeks to solve for acceleration from the equation of motion during each time increment:

$$\ddot{\mathbf{u}}_t = \mathbf{M}^{-1}\left(\mathbf{F}_t - \mathbf{K}\mathbf{u}_t - \mathbf{C}\dot{\mathbf{u}}_{t-\Delta t/2}\right) \tag{10.2}$$

Having evaluated $\ddot{\mathbf{u}}_t$, the velocity at $t + \Delta t/2$ and the displacement at $t + \Delta t/2$ are updated using the following relationships:

$$\dot{\mathbf{u}}_{t+\Delta t/2} = \dot{\mathbf{u}}_{t-\Delta t/2} + \ddot{\mathbf{u}}_t \Delta t \tag{10.3}$$

$$\mathbf{u}_{t+\Delta t} = \mathbf{u}_t + \dot{\mathbf{u}}_{t-\Delta t}\Delta t/2 \tag{10.4}$$

and the process continues. The starting of time marching for acceleration $\ddot{\mathbf{u}}_t$ at $t = 0$ requires knowledge of $\dot{\mathbf{u}}_{0-\Delta t/2}$, which may be evaluated from $\dot{\mathbf{u}}_0$ and $\ddot{\mathbf{u}}_0$ using the relationship:

$$\dot{\mathbf{u}}_{0-\Delta t/2} = \dot{\mathbf{u}}_0 - \frac{\Delta t}{2}\ddot{\mathbf{u}}_0 \tag{10.5}$$

By reducing **M** to a diagonal matrix, Eqn (10.2) becomes uncoupled and no iteration is needed during time integration, even for nonlinear problems, thus making the explicit time-marching scheme extremely efficient for nonlinear problems. The tradeoff is its relatively poor stability. A bounded solution is obtained only when the time increment Δt is less than the stable time increment $\Delta t_{\min}$. If $\Delta t \geq \Delta t_{\min}$, then the solution will be unstable and oscillations will occur in the model space. The stable time increment is defined by

$$\Delta t_{\min} = \frac{2}{\omega_{\max}}\left(\sqrt{1+\xi^2} - \xi\right) \tag{10.6}$$

where $\omega_{\max}$ and ξ are the maximum resonant frequency and the damping ratio, respectively, of the system. A conservative estimate of the stable time increment is given by

$$\Delta t_{\min} = \frac{L_e}{c_d} \tag{10.7}$$

where L_e is the characteristic length of the smallest FE in the model and c_d is the dilatational wave speed of the material, which is given by

$$c_d = \sqrt{\frac{K + 4G/3}{\rho}} \tag{10.8}$$

where K, G and ρ are the bulk modulus, shear modulus and density respectively, of the material. Physically, the stable time increment is approximately equal to the time for an elastic wave to travel the smallest elemental length in the model.

The following conclusions can be made for the explicit scheme:

- Acceleration is the field variable, and it is the natural boundary condition.
- It is best suited for a highly nonlinear or highly transient event, such as product-level drop-impact, which requires small time increments to resolve the interaction of the product with the target.

10.1.1.2 Implicit schemes

An implicit scheme seeks to solve for displacement:

$$\mathbf{K}\mathbf{u}_{t+\Delta t} = \mathbf{F}_{t+\Delta t} - \mathbf{M}\ddot{\mathbf{u}}_{t+\Delta t} - \mathbf{C}\dot{\mathbf{u}}_{t+\Delta t} \tag{10.9}$$

or

$$\mathbf{K}\Delta\mathbf{u} = \mathbf{F}_{t+\Delta t} - \mathbf{M}\ddot{\mathbf{u}}_{t+\Delta t} - \mathbf{C}\dot{\mathbf{u}}_{t+\Delta t} - \mathbf{R}_t \tag{10.10}$$

where $\mathbf{R}_t = \mathbf{K}\mathbf{u}_t$ is the internal force.

One of the most commonly used numerical methods for solving Eqn (10.10) is the Newmark-β method. The key feature of Newmark's method is the use of two parameters, γ and β, to linear-interpolate acceleration between the time interval, Δt; that is,

$$\begin{aligned} \ddot{\mathbf{u}}_\gamma &= (1-\gamma)\ddot{\mathbf{u}}_t + \gamma\ddot{\mathbf{u}}_{t+\Delta t} \\ \ddot{\mathbf{u}}_\beta &= (1-2\beta)\ddot{\mathbf{u}}_t + 2\beta\ddot{\mathbf{u}}_{t+\Delta t} \end{aligned} \tag{10.11}$$

By varying the parameter γ between 0 and 1 and the parameter β between 0 and $1/2$, the accelerations, $\ddot{\mathbf{u}}_\gamma$ and $\ddot{\mathbf{u}}_\beta$, are made to take different proportionate weights of acceleration at t and $t + \Delta t$. The velocity and displacement are given, respectively, by

$$\begin{aligned} \dot{\mathbf{u}}_{t+\Delta t} &= \dot{\mathbf{u}}_t + \ddot{\mathbf{u}}_\gamma \Delta t \\ \mathbf{u}_{t+\Delta t} &= \mathbf{u}_t + \dot{\mathbf{u}}_t\Delta t + \frac{1}{2}\ddot{\mathbf{u}}_\beta \Delta t^2 \end{aligned} \tag{10.12}$$

Substituting Eqn (10.11) into Eqn (10.12) and after some rearrangement gives

$$\begin{aligned}\dot{\mathbf{u}}_{t+\Delta t} &= \dot{\mathbf{u}}_t + (1-\gamma)\ddot{\mathbf{u}}_t\Delta t + \gamma\ddot{\mathbf{u}}_{t+\Delta t}\Delta t \\ \ddot{\mathbf{u}}_{t+\Delta t} &= \frac{\Delta\mathbf{u}}{\beta\Delta t^2} - \frac{\dot{\mathbf{u}}_t}{\beta\Delta t} - \frac{(1-2\beta)\ddot{\mathbf{u}}_t}{\beta}\end{aligned} \tag{10.13}$$

Substituting the second equation of Eqn (10.13) into its first equation gives

$$\dot{\mathbf{u}}_{t+\Delta t} = \frac{\gamma\Delta\mathbf{u}}{\beta\Delta t} + \left(1-\frac{\gamma}{\beta}\right)\dot{\mathbf{u}}_t + \left(1-\frac{\gamma}{2\beta}\right)\ddot{\mathbf{u}}_t\Delta t \tag{10.14}$$

Note that the velocity, $\dot{\mathbf{u}}_{t+\Delta t}$, and the acceleration, $\ddot{\mathbf{u}}_{t+\Delta t}$, are now expressed as a function of $\Delta\mathbf{u}$. Substituting these into Eqn (10.10) gives

$$\hat{\mathbf{K}}\Delta\mathbf{u} = \hat{\mathbf{F}}_{t+\Delta t} \tag{10.15}$$

where

$$\begin{aligned}\hat{\mathbf{K}} &= \mathbf{K} + \frac{1}{\beta\Delta t^2}\mathbf{M}_{t+\Delta t} + \frac{\gamma}{\beta\Delta t}\mathbf{C}_{t+\Delta t} \\ \hat{\mathbf{F}}_{t+\Delta t} &= \mathbf{F}_{t+\Delta t} + \mathbf{M}\left(\frac{1}{\beta\Delta t^2}\mathbf{u}_t + \frac{1}{\beta\Delta t}\dot{\mathbf{u}}_t + \frac{1-2\beta}{2\beta}\ddot{\mathbf{u}}_t\right) \\ &\quad + \mathbf{C}\left(\frac{\gamma}{\beta\Delta t}\mathbf{u}_t + \frac{\gamma-\beta}{\beta}\dot{\mathbf{u}}_t + \frac{\Delta t(\gamma-2\beta)}{2\beta}\ddot{\mathbf{u}}_t\right)\end{aligned} \tag{10.15a}$$

and the process continues.

The inverse operation of the $\hat{\mathbf{K}}$ matrix is computationally taxing. Because of the unknown matrix of $\mathbf{K}$ at $t + \Delta t$ for nonlinear problems, iteration is needed within each time increment. An implicit scheme therefore requires much longer computational time per time increment compared with the explicit scheme. However, the equation is unconditionally stable when γ and β satisfy the equation

$$2\beta \geq \gamma \geq \frac{1}{2} \tag{10.16}$$

For zero damping, the Newmark method is conditionally stable if

$$\gamma \geq \frac{1}{2},\ \ \beta \leq \frac{1}{2},\ \ \Delta t \leq \frac{1}{\omega_{\max}\sqrt{\gamma/2-\beta}} \tag{10.17}$$

Because of the unconditional stability, the Newmark method is inherently robust. However, a sufficiently small time-step must be used, say $\Delta t = 1/(10f_{max})$, to ensure there is adequate resolution of the event. Precautions must be taken for errors associated with 'numerical damping' and 'period elongation' when $\gamma \geq 1/2$, and the indefinite oscillation of the higher modes if excited.

The following conclusions can be made for the implicit scheme:

- Displacement is the basic field variable ⇒ the acceleration boundary condition must be expressed into displacement boundary condition: $\mathbf{u}_b = \iint \ddot{\mathbf{u}}_b dt$.
- The scheme is best suited for a moderate transient event, such as board-level drop-shock in which both the frequencies of the applied shock pulse (≈500 Hz) and the response of the PCB assembly (≈200 Hz) are relatively low, permitting the use of relatively larger time increments.

10.1.1.3 Mode superposition

By expressing the global displacement into a summation of its eigenvectors:

$$\mathbf{u}(t) = \mathbf{\Phi Q}(t) = \sum_{i=1}^{n} \boldsymbol{\phi}_i q_i(t) \tag{10.18}$$

where

$$\mathbf{\Phi} = [\boldsymbol{\phi}_1, \quad \boldsymbol{\phi}_2, \quad .... \quad \boldsymbol{\phi}_n], \; \mathbf{Q}(t) = \begin{bmatrix} q_1(t) \\ q_2(t) \\ ... \\ q_n(t) \end{bmatrix} \tag{10.19}$$

and $\boldsymbol{\phi}_i$ and $q_i(t)$ are the eigenvector and the corresponding participating factor of i mode, the differential equation of motion constituting of infinite degree of freedom

$$\mathbf{M}\ddot{\mathbf{u}}(t) + \mathbf{K}\mathbf{u}(t) = \mathbf{F}(t) \tag{10.20}$$

is transformed into one with n degrees of freedom:

$$\ddot{\mathbf{Q}}(t) + \boldsymbol{\omega}^2\mathbf{Q}(t) = \mathbf{\Phi}^{\mathrm{T}}\mathbf{F}(t) \tag{10.21}$$

where

$$\boldsymbol{\omega}^2 = \begin{bmatrix} \omega_1^2 & & & \\ & \omega_2^2 & & \\ & & ... & \\ & & & \omega_n^2 \end{bmatrix} \tag{10.21a}$$

Equation (10.21) degenerates into n-differential equations, each a differential equation describing a single degree of freedom (SDOF) system with angular frequency ω_i:

$$\ddot{q}_i(t) + \omega_i^2 q_i(t) = \boldsymbol{\phi}_i^{\mathrm{T}} \mathbf{F}(t) \tag{10.22}$$

Thus, the task of analysing the dynamic response $\mathbf{u}(t)$ of a system is reduced to one of (1) evaluating the eigenvalues $\boldsymbol{\omega}$ and eigenvectors $\boldsymbol{\Phi}$ of the system followed by (2) evaluating the participating factor $\mathbf{Q}(t)$ of the eigenvectors.

The mode-superposition method is the most efficient scheme for a linear system, especially if one is interested in investigating the response of a system under different load conditions.

10.1.2 A brief overview of numerical modelling of a PCB assembly subjected to impact/shock

Before the board-level drop-shock standard JESD22-B111 was established, the evaluation of the robustness of board-level solder joints was typically performed by mounting the PCB assembly to a rigid fixture, which was then dropped from a specified height against a rigid base. The shock interaction between the impacting bodies was not specified but was left to the inertia and the rigidity of the impacting bodies. The highly transient event and the non-linear contact interaction between the impacting bodies are best modelled using the explicit scheme.

Zhu (2001), Sogo et al. (Sogo, 2001), Hirata (2001), Qiang et al. (Qiang, 2002) and Wang et al. (Wang, 2002) were among the pioneers in modelling the drop impact of a PCB assembly (suspended in a fixture) using explicit codes. The force–time interaction between the impacting bodies was modelled by Sogo (2001) and Wang (2002), who also performed detailed submodelling analysis of the outermost solder joints that experienced the maximum magnitude of stress and identified the maximum stress to be along the axial direction of the solder joint, namely the peeling stress. Circumventing the complexity of contact modelling, Qiang et al. (Qiang, 2002) and Wong et al. (Wong, 2002) performed velocity impact modelling of the suspended PCB assembly using the implicit algorithm; they arrived independently at the conclusion that the flexural response of the PCB was dominated by its fundamental mode. Modelling the solder joints using beam elements, Wong et al. (Wong, 2002) also reported that the bending stress in the outermost solder joint was much higher in magnitude than the axial and shear stresses.

The JESD22-B111 test standard prescribes a half-sine acceleration shock for the impact interaction that has greatly simplified the modelling. The problem is reduced from one of drop impact to one of acceleration-shock. The prescription of an acceleration shock has eliminated the need to model the non-linear contact interaction between the impacting bodies. Instead, the prescribed acceleration shock becomes the boundary condition for the impacting bodies.

When the fundamental frequency of the test fixture is significantly higher than that of the acceleration shock, the acceleration–time profile that is prescribed between the impacting bodies may be assumed to be transmitted to the regions of the PCB that are joined to the connectors – henceforth referred to as PCB supports – without loss or distortion. That is, the test structure behaves as a rigid body. In such a case, the dynamics of the board-level drop impact can be analysed simply through application of the acceleration input directly at the PCB supports.

Tee et al. (Tee, 2004) performed drop-shock modelling of a PCB assembly subjected to a half-sine shock pulse using an explicit algorithm in which the shock pulse at the PCB supports was defined using an acceleration boundary condition – the natural boundary condition for the explicit algorithm. In view of the relatively low frequency of the prescribed acceleration shock, the JEDEC BLDST can be more efficiently modelled using implicit scheme for a linear system. The use of an implicit algorithm for acceleration-shock boundary condition was demonstrated by Wong (2005), in which the acceleration boundary condition was replaced with the displacement boundary condition; the numerical results were used to validate the analytical solutions, which too used the displacement boundary condition. Instead of replacing the acceleration boundary condition with a displacement boundary condition, Yeh and Lai (2006) demonstrated the 'support excitation' method, using the commercial ANSYS solver.

Computational modelling is capable of solving problems with complex geometry and nonlinearity. It allows users to solve complex problems without an adequate understanding of the underlying physics/mechanics, which is not at all a good practice. The following sections attempt to provide fundamental understanding through analytical analysis of board-level drop impact using increasingly sophisticated models: a spring-mass model with no damping, a spring-mass model with damping, a beam model with no damping and a beam model with damping.

10.2 Vibration of a test board in the JESD22-B111 drop-shock test

The frequency and the damping ratio of the PCB board in a PED are found to be around 400 Hz and 0.1, respectively. The fibre strain response of the test board in a JESD22-B111 was discussed in Chapter 7. The vibration frequencies and damping ratios were found to decrease with bending cycles; they range between 160 and 200 Hz and 0.013 to 0.035, respectively, as shown in Figure 10.1.

Figure 10.2(a) plots the vibration frequency against the fibre strain of the test board using the data from Figure 10.1(b); however, the first data point, which is prone to contamination by the higher frequency components, is ignored. It was observed that the frequency of the test board increases with increasing fibre strain which is larger than 0.6×10^{-3}. The increased vibration frequency reflects the increased stiffening of the test board due to the increased presence of membrane strains with increased bending of the board, whose lateral displacements are constrained by the four bolted supports. The initial clearance between the bolts and holes on the board provides for unconstrained lateral displacement for small strains, which is believed to be responsible for the independence of its resonant frequency with fibre strain for magnitudes less than 0.6×10^{-3}.

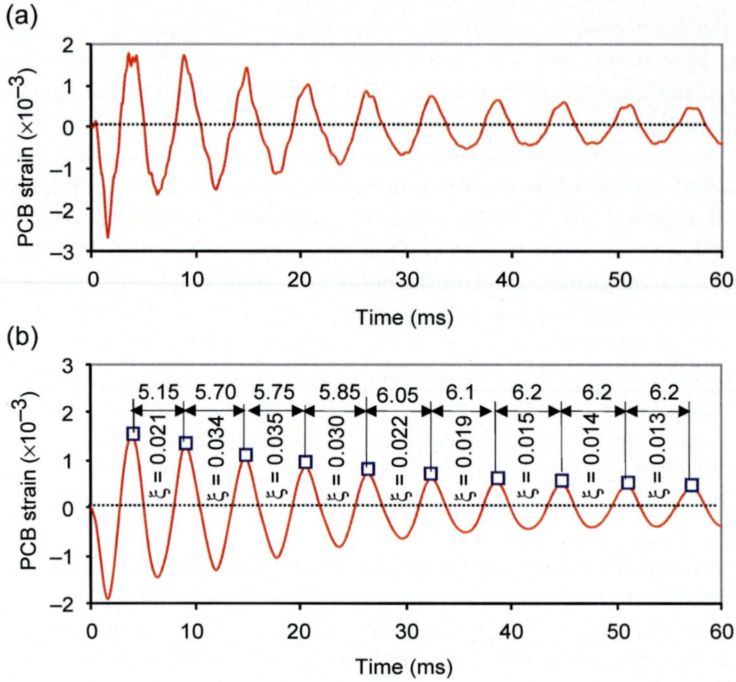

Figure 10.1 The (a) unfiltered and (b) filtered strain responses of test board in a JEDEC JESD22-B111 shock test.

The damping ratio may be expressed in terms of Rayleigh constants, α and β, as $\zeta = \alpha/2\omega + \omega\beta/2$. The first term is associated with mass damping (or viscous damping) – the damping of the entire system in a viscous medium – and is inversely proportional to the vibrating frequency. The second term is associated with structural damping (or hysteretic damping) – damping due to micro-slippage within the structure of the system – and is proportional to the vibrating frequency. Figure 10.2(b) shows damping ratio against vibrating frequency of the test board. The linear increase of damping ratio with frequency suggests hysteretic damping as the dominant physics for the board-level shock test.

10.3 Analytical solutions for a spring-mass system subjected to half-sine shock

10.3.1 An undamped spring-mass system

If only the fundamental modal response of the PCB is of interest, the PCB can be modelled as a single spring-mass system. This is depicted in Figure 10.3, where m and k represent the mass and the flexural stiffness, respectively, of the spring; Y and y are the absolute displacement of the support and the displacement of the mass, respectively; and $w = y - Y$ is the deflection of the spring.

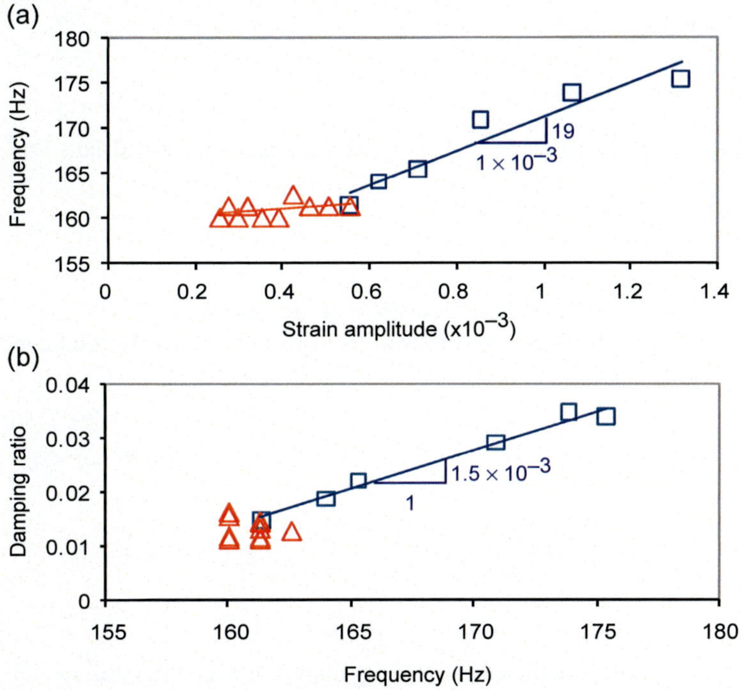

Figure 10.2 JESD22-B111 shock test. (a) Frequency versus strain; (b) damping ratio versus frequency.

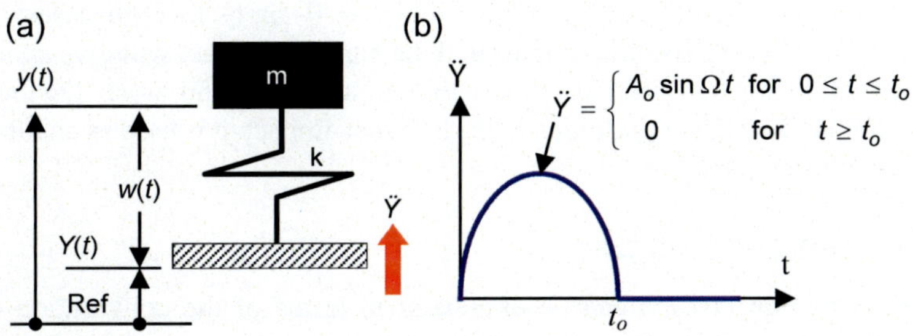

Figure 10.3 (a) PCB modelled as a single spring-mass system. (b) Input acceleration pulse.

The support is subjected to an acceleration $\ddot{Y}(t)$ described by a half-sine pulse of magnitude A_o and duration t_o (frequency $\Omega = \pi/t_o$), followed by an extended duration of nil acceleration; that is

$$\ddot{Y}(t) = \begin{cases} A_o \sin \Omega t & \text{for } 0 \leq t \leq t_o \\ 0 & \text{for } t \geq t_o \end{cases} \tag{10.23}$$

The equation of motion is given by dynamic equilibrium between the inertial force $m\ddot{y}$ and the elastic force kw:

$$m\ddot{y}(t) + kw(t) = 0 \tag{10.24}$$

The coordinate relation gives $y = Y + w$, which when substituted into Eqn (10.24) gives

$$\ddot{w}(t) + \omega^2 w(t) = -\ddot{Y}(t) \tag{10.25}$$

where $\omega^2 = \sqrt{k/m}$ is the natural frequency of the system.

The general base displacement function that satisfies the acceleration condition of Eqn (10.23) is given by

$$Y(t) = \begin{cases} Y_o + \left(\dot{Y}_o + \dfrac{A_o}{\Omega}\right)t - \dfrac{A_o}{\Omega^2}\sin\Omega t & \text{for } 0 \le t \le t_o \\ Y_o + \left(\dot{Y}_o + \dfrac{A_o}{\Omega}\right)t_o + \left(\dot{Y}_o + \dfrac{2A_o}{\Omega}\right)(t - t_o) & \text{for } t \ge t_o \end{cases} \tag{10.26}$$

Acceleration shock is induced via velocity impact; the required differential velocity may be derived either through a gravitational fall against a stationery target, as in drop impact, or simply through striking by a moving object as in hammer-impact. In the case of drop impact, the initial conditions are $y_o = Y_o = 0$ and $\dot{y}_o = \dot{Y}_o = -V_{\text{imp}}$. In the case of hammer-impact, the initial conditions are $y_o = Y_o = 0$ and $\dot{y}_o = \dot{Y}_o = 0$. The initial conditions $w_o = \dot{w}_o = 0$ apply to both cases. For the same acceleration shock, the drop impact and the hammer-impact will give an identical solution of deflection $w(t)$ and its time differentials, $\dot{w}(t)$ and $\ddot{w}(t)$. Owing to the relatively simple boundary conditions of the hammer-impact, it is used as an illustration in the rest of this chapter.

10.3.1.1 Deflection-time

The solution to Eqn (10.25) can be expressed in terms of the convolution integral (Singiresu, 1995):

$$w(t) = \begin{cases} w_o \cos\omega t + \dfrac{\dot{w}_o}{\omega}\sin\omega t - \dfrac{1}{\omega}\displaystyle\int_0^t \ddot{Y}(\tau)\sin\omega(t-\tau)\,d\tau & \text{for } 0 \le t \le t_o \\ w(t_o)\cos\omega(t - t_o) + \dfrac{\dot{w}(t_o)}{\omega}\sin\omega(t - t_o) & \text{for } t \ge t_o \end{cases} \tag{10.27}$$

$0 \leq t \leq t_o$

For the period $0 \leq t \leq t_o$, the acceleration of the support is prescribed by $\ddot{Y}(t) = A_o \sin \Omega t$. Substituting this function of $\ddot{Y}(t)$ into Eqn (10.27) together with the initial condition $w_o = \dot{w}_o = 0$, after some manipulation, gives the deflection of the mass as

$$\frac{\omega^2}{A_o} w(t) = \frac{-R_\omega}{1 - R_\omega^2} (\sin \omega t - R_\omega \sin \Omega t) \tag{10.28}$$

where $R_\omega = \omega/\Omega$. The deflection of the mass is a strong function of the frequency ratio, R_ω; $w(t) \to 0$ as $R_\omega \to 0$ and $w(t) \to -\ddot{Y}(t)/\omega^2$ as $R_\omega \to \infty$.

The acceleration of the mass is given by $\ddot{y}(t) = \ddot{w}(t) + \ddot{Y}(t)$. It may be evaluated by differentiating Eqn (10.28) twice with time:

$$\frac{\ddot{w}(t)}{A_o} = \frac{R_\omega}{1 - R_\omega^2} \left(-\frac{1}{R_\omega} \sin \Omega t + \sin \omega t \right) \tag{10.29}$$

and adding to it $\ddot{Y}(t)/A_o$:

$$\frac{\ddot{y}(t)}{A_o} = \frac{\ddot{w}(t)}{A_o} + \frac{\ddot{Y}(t)}{A_o} = \frac{R_\omega}{1 - R_\omega^2} (-R_\omega \sin \Omega t + \sin \omega t) \tag{10.30}$$

Comparing Eqn (10.30) with Eqn (10.28) yields

$$\ddot{y}(t) = -\omega^2 w(t) \tag{10.31}$$

$t \geq t_o$

For the period $t \geq t_o$, the acceleration of the support is prescribed by $\ddot{Y} = 0$. The system is now under free-vibration and the deflection of the spring is described by the second equation of Eqn (10.27). The initial conditions, $w(t_o)$ and $\dot{w}(t_o)$, are evaluated by substituting $t = t_o$ into Eqn (10.28) and its differentia, leading to

$$w(t_o) = \frac{-A_o R_\omega}{\omega^2 (1 - R_\omega^2)} \sin \pi R_\omega, \quad \dot{w}_o(t_o) = \frac{-A_o \Omega R_\omega^2 (1 + \cos \pi R_\omega)}{\omega^2 (1 - R_\omega^2)} \tag{10.32}$$

Substituting Eqn (10.32) into the second equation of Eqn (10.27), after some manipulation, gives the deflection of the mass as

$$\frac{\omega^2}{A_o} w(t) = \frac{-2R_\omega}{1 - R_\omega^2} \sin\left(\omega t - \frac{\pi R_\omega}{2} \right) \cos\left(\frac{\pi R_\omega}{2} \right) \tag{10.33}$$

The acceleration of the mass is given by $\ddot{y}(t) = \ddot{w}(t)$, which can be shown readily to be satisfying Eqn (10.31).

$0 \le t \le \infty$

The consolidated deflection-time relation of the mass is

$$\frac{\omega^2}{A_o} w(t) = \begin{cases} \dfrac{-R_\omega}{1-R_\omega^2}(\sin \omega t - R_\omega \sin \Omega t) & \text{for } 0 \le t \le t_o \\ \dfrac{-2R_\omega}{1-R_\omega^2} \sin\left(\omega t - \dfrac{\pi R_\omega}{2}\right)\cos\left(\dfrac{\pi R_\omega}{2}\right) & \text{for } t \ge t_o \end{cases} \tag{10.34}$$

Putting $\tau = t/t_o$ and noting that $\Omega = \pi/t_o$ gives $\Omega t = \pi\tau$ and $\omega t = \pi R_\omega \tau$. Substituting these into the above equations, we obtain the deflection of the mass expressed in normalised time, τ, as

$$\begin{aligned} \frac{\omega^2}{A_o} w(\tau) &= TR(R_\omega, \tau) \\ &= \begin{cases} \dfrac{-R_\omega}{1-R_\omega^2}(\sin \pi R_\omega \tau - R_\omega \sin \pi\tau) & \text{for } 0 \le \tau \le 1 \\ \dfrac{-2R_\omega}{1-R_\omega^2} \sin\left(\pi R_\omega \tau - \dfrac{\pi R_\omega}{2}\right)\cos\left(\dfrac{\pi R_\omega}{2}\right) & \text{for } \tau \ge 1 \end{cases} \end{aligned} \tag{10.35}$$

The normalised deflection-time responses of the mass for $R_\omega = 0.2, 2, 3, 4$ are shown in Figure 10.4(a) and the following points are noted:

1. An excessively high-frequency impulse (a very low R_ω) incites little response in the normalised response $\omega^2 w/A_o$, attributed to the small magnitude of ω^2; physically, there is inadequate time for the mass to response for $\tau \le 1$; that is, $y(\tau) \approx 0$, which implies

$$\frac{\omega^2}{A_o} w(\tau) \approx \frac{\omega^2}{A_o} Y(\tau) = R_\omega^2 \left[\pi\tau\left(\frac{\Omega \dot{Y}_o}{A_o} + 1\right) - \sin \pi\tau\right] \quad \text{for } 0 \le \tau \le 1 \tag{10.36}$$

   and $w(\tau) \to 0$ for $R_\omega \ll 1$.
2. The magnitude of normalised deflection first increases and then decreases with increasing R_ω;
3. There is virtually no deflection at $R_\omega = 3$ for $\tau > 1$, which could be traced to the second equation of Eqn (10.35). Setting $w(\tau) = 0$ for $\tau > 1$ gives $\cos(\pi R_\omega/2) = 0$, which gives $R_\omega = 3, 5, 7\ldots$; the condition $R_\omega = 1$ renders the second equation of Eqn (10.35) indeterminate and is excluded. That is, any odd integer of R_ω, except for $R_\omega = 1$, will return a null response of $w(\tau)$ for $\tau > 1$. Physically, this implies that the spring is under no deflection and the mass undergoes a rigid-body motion that is synchronised with the support.

The normalised deflection-time responses of the mass for $R_\omega = 3, 5, 7, 9$ are shown in Figure 10.4(b). At $R_\omega = 3$, the first equation of Eqn (10.35) is reduced to $\omega^2 w(t)/A_o = -1.5\sin^3(\pi\tau)$, giving an inverted bell shape with the highest magnitude. The area enclosed by the four curves and the normalised time axis are identical at $\tau = 2/\pi$.

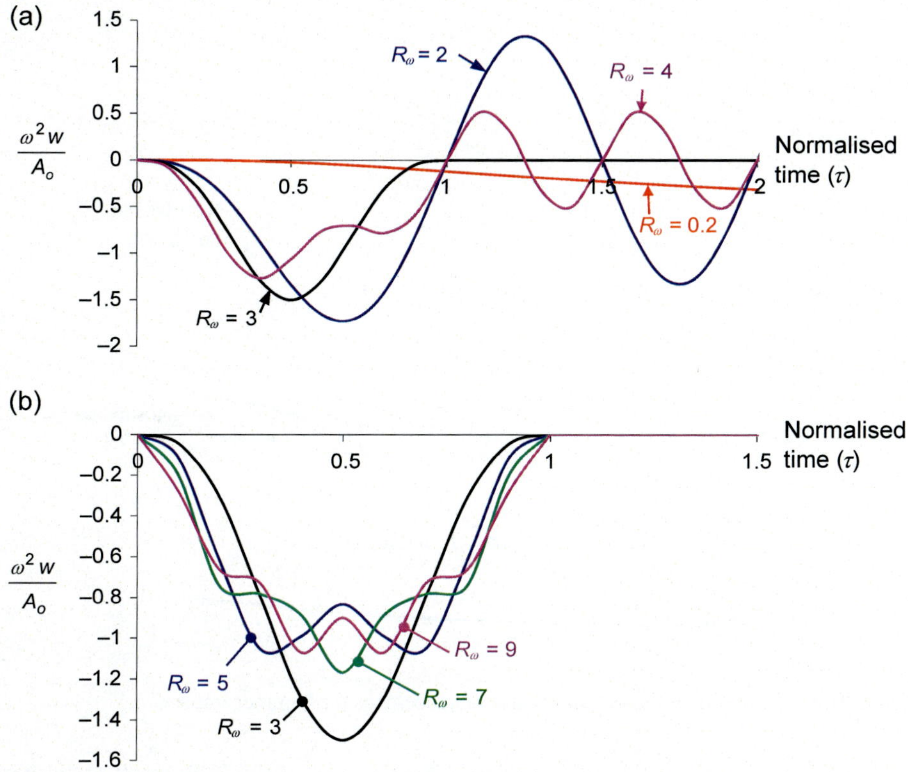

Figure 10.4 Normalised deflection-time response for selected R_ω. (a) $R_\omega = 0.2,2,3,4$; (b) $R_\omega = 3,5,7,9$.

10.3.1.2 Deflection versus base displacement

The base displacement of Eqn (10.26) may be expressed in terms of normalised time τ:

$$\frac{\omega^2}{A_o}Y(\tau) = \begin{cases} \dfrac{\omega^2 Y_o}{A_o} + \left(\dfrac{\omega \dot{Y}_o}{A_o} + R_\omega\right)\pi R_\omega \tau - R_\omega^2 \sin \pi\tau & \text{for } 0 \le \tau \le 1 \\ \dfrac{\omega^2 Y_o}{A_o} + \left(\dfrac{\omega \dot{Y}_o}{A_o} + R_\omega\right)\pi R_\omega + \left(\dfrac{\omega \dot{Y}_o}{A_o} + 2R_\omega\right)(\tau - 1)\pi R_\omega & \text{for } \tau \ge 1 \end{cases} \tag{10.37}$$

The case of hammer-impact is illustrated. Substituting the initial condition $Y_o = \dot{Y}_o = 0$ into Eqn (10.37) gives

$$\frac{\omega^2}{A_o}Y(\tau) = \begin{cases} R_\omega^2(\pi\tau - \sin \pi\tau) & \text{for } 0 \le \tau \le 1 \\ \pi R_\omega^2(2\tau - 1) & \text{for } \tau \ge 1 \end{cases} \tag{10.38}$$

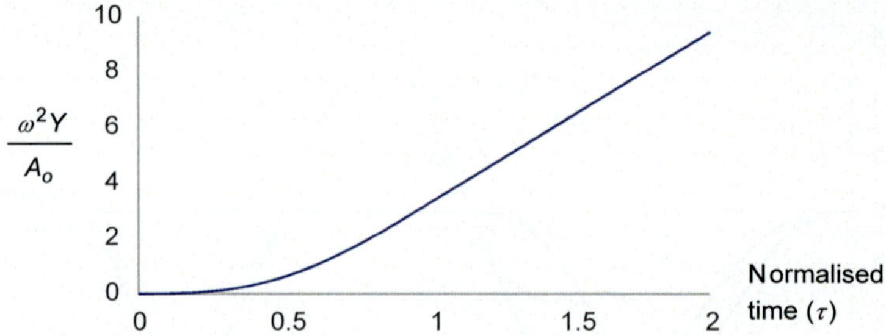

Figure 10.5 Displacement of support due to hammer impact.

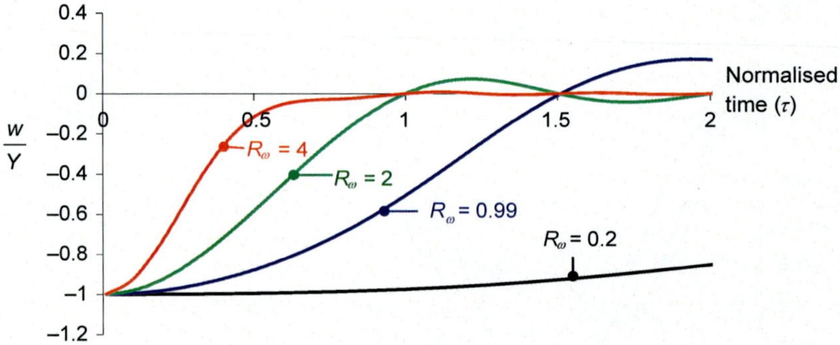

Figure 10.6 The ratios of deflection to base displacement for hammer impact.

Figure 10.5 shows the displacement of the support increases monotonically with the normalised time.

The ratio of the deflection of the spring to the displacement of the support is given by

$$\frac{w(\tau)}{Y(\tau)} = \begin{cases} \dfrac{1}{R_\omega(1-R_{\omega)}^2)}\left(\dfrac{R_\omega \sin \pi\tau - \sin \pi R_\omega \tau}{\pi\tau - \sin \pi\tau}\right) & \text{for } 0 \le \tau \le 1 \\ \dfrac{-2}{1-R_\omega^2} \; \dfrac{\sin\left(\pi R_\omega \tau - \dfrac{\pi R_\omega}{2}\right)\cos\left(\dfrac{\pi R_\omega}{2}\right)}{\pi R_\omega (2\tau - 1)} & \text{for } \tau \ge 1 \end{cases} \tag{10.39}$$

Figure 10.6 shows the ratios for $R_\omega = 0.2$, 0.99, 2 and 4. It is noted that the magnitude of the ratio decreases nearly monotonically with increasing τ. Physically, this is borne by the simple fact that the displacement of the support increases monotonically while the deflection of the spring oscillates between a fixed amplitude.

10.3.1.3 Maxima spectra of deflection

The maxima of deflection may be evaluated through finding the stationary point of Eqn (10.35); that is,

$$
\frac{\omega^2}{A_o}\frac{\mathrm{d}w}{\mathrm{d}\tau} = \begin{cases} \dfrac{-R_\omega^2}{1-R_\omega^2}(\cos \pi R_\omega \tau - \cos \pi\tau) = 0 & \text{for } 0 \le \tau \le 1 \\[2ex] \dfrac{-2R_\omega^2}{1-R_\omega^2}\cos\left(\pi R_\omega \tau - \dfrac{\pi R_\omega}{2}\right)\cos\left(\dfrac{\pi R_\omega}{2}\right) = 0 & \text{for } \tau \ge 1 \end{cases} \tag{10.40}
$$

which gives

$$
\begin{aligned} &-2\sin\left(\frac{\pi\tau(R_\omega+1)}{2}\right)\sin\left(\frac{\pi\tau(R_\omega-1)}{2}\right) = 0 \quad \text{for } 0 \le \tau \le 1 \\ &\cos\left(\pi R_\omega\tau - \frac{\pi R_\omega}{2}\right)\cos\left(\frac{\pi R_\omega}{2}\right) = 0 \quad \text{for } \tau \ge 1 \end{aligned} \tag{10.41}
$$

The normalised time τ_m that corresponds to the maxima of deflection is given by (Tsai, Yeh, Lai, & Chen, 2007)

$$
\tau_m = \begin{cases} \dfrac{2p}{R_\omega \pm 1}, & p = 1,2,3..., \quad 0 \le \tau_m \le 1 \\[2ex] \dfrac{R_\omega + n}{2R_\omega}, & n = 1,3,5..., \quad \tau_m \ge 1 \end{cases} \tag{10.42}
$$

Substituting Eqn (10.42) into Eqn (10.35) gives

$$
\frac{\omega^2}{A_o}w_{\max} = \begin{cases} \dfrac{-R_\omega}{1-R_\omega^2}\left[\sin\left(\dfrac{2\pi R_\omega p}{R_\omega \pm 1}\right) - R_\omega \sin\left(\dfrac{2\pi p}{R_\omega \pm 1}\right)\right], & p = 1,2,3..., \quad 0 \le \dfrac{2p}{R_\omega \pm 1} \le 1 \\[2ex] \dfrac{-2R_\omega}{1-R_\omega^2}\cos\left(\dfrac{\pi R_\omega}{2}\right)\sin\left(\dfrac{n\pi}{2}\right), & n = 1,3,5..., \quad \dfrac{R_\omega + n}{2R_\omega} \ge 1 \end{cases} \tag{10.43}
$$

The valid values of R_ω are given by

$$
\left.\begin{aligned} R_\omega &\ge 2p - 1 \\ R_\omega &\ge 2p + 1 \end{aligned}\right\} \quad p = 1,2,3... \qquad R_\omega \le n, \quad n = 1,3,5... \tag{10.44}
$$

Substituting Eqn (10.44) into Eqn (10.43) and noting that $\sin(n\pi/2) = 1$ for $n = 1, 3, 5...$, gives

$$\frac{\omega^2}{A_o}w_{\max} = \begin{cases} w_{m1} = \dfrac{-R_\omega}{1-R_\omega^2}\left[\sin\left(\dfrac{2\pi R_\omega p}{R_\omega+1}\right) - R_\omega \sin\left(\dfrac{2\pi p}{R_\omega+1}\right)\right], & p = 1,2,3\ldots, \quad R_\omega \geq 2p-1 \\ w_{m2} = \dfrac{-R_\omega}{1-R_\omega^2}\left[\sin\left(\dfrac{2\pi R_\omega p}{R_\omega-1}\right) - R_\omega \sin\left(\dfrac{2\pi p}{R_\omega-1}\right)\right], & p = 1,2,3\ldots, \quad R_\omega \geq 2p+1 \\ w_{m3} = \dfrac{-2R_\omega}{1-R_\omega^2}\cos\left(\dfrac{\pi R_\omega}{2}\right), & R_\omega \leq 1,3,5\ldots. \end{cases} \tag{10.45}$$

The characteristic of w_{m3} and the characteristics of w_{m1} and w_{m2} for $p = 1$ and $p = 2$ are shown in Figure 10.7. The dashed lines for w_{m1} and w_{m2} indicate the range of R_ω in which the functions are invalid for failing the constraint $R_m \geq 2p - 1$ and $R_m \geq 2p + 1$, respectively. The following points are noted:

- The functions w_{m1} and w_{m2} are invalid for $R_\omega < 1$; that is, only the function w_{m3} is valid for $R_\omega < 1$.
- Within the R_ω range wherein w_{m1} and w_{m2} are both valid, $|w_{m1}| \geq |w_{m2}|$; thus, the function w_{m2} is redundant in the discussion of maxima and will not be further investigated.

The functions $|w_{m3}|$ and the function $|w_{m1}|_p$ for $p = 1{-}5$ are shown in Figure 10.8. The dashed lines show the R_ω range in which the functions are invalid; the solid lines show the R_ω range in which the functions are valid; and the thick solid lines show the

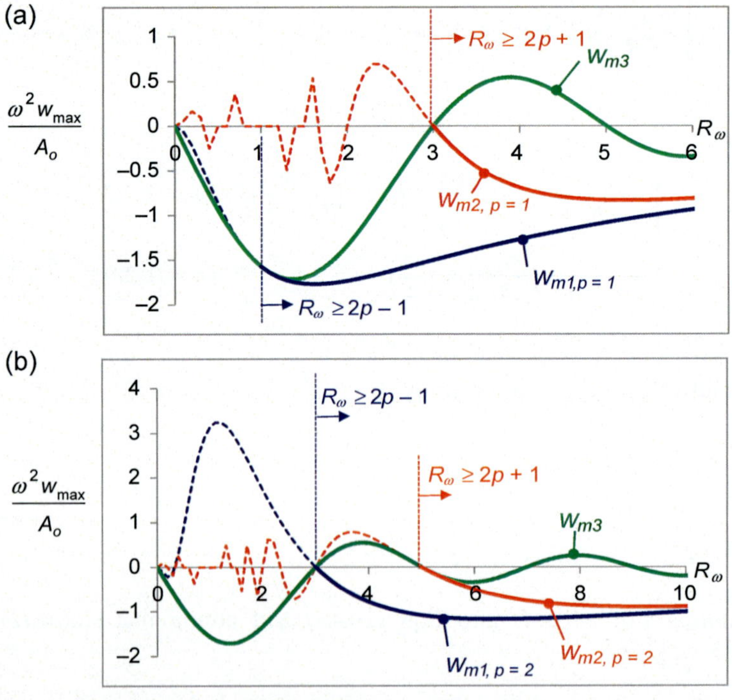

Figure 10.7 Characteristics of the maxima function *s* for normalised deflection: (a) $p = 1$; (b) $p = 2$.

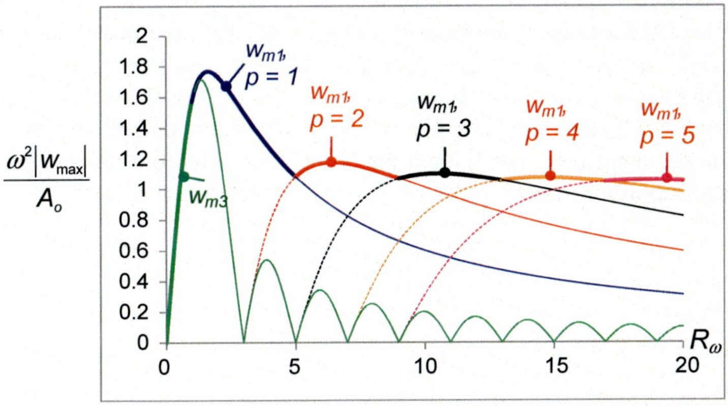

Figure 10.8 Characteristics of the absolute maxima function s for normalised deflection for $p = 1$ to $p = 5$.

R_ω range in which the functions are not only valid but have the maximum magnitude across the R_ω domain. The following points are noted:

- $|w_{m3}| \geq |w_{m1}|$ only for $R_\omega < 1$; $|w_{m1}| \geq |w_{m3}|$ for $R_\omega > 1$.
- The maxima of $|w_{m1}|_p$ occurs only over a limited range of R_ω; for example, $|w_{m1}|_{p=1}$ has the highest magnitude for $1 \leq R_\omega \leq 5$ while $|w_{m1}|_{p=2}$ has the highest magnitude for $5 \leq R_\omega \leq 9$. The range of R_ω for which $|w_{m1}|_p$ is a maximum may be described mathematically as $(4p - 3) \leq R_w \leq (4p + 1)$.

The function w_{m1} may be written more concisely as $\frac{R_\omega}{1-R_\omega}\sin\left(\frac{2p\pi}{1+R_\omega}\right)$ (Mindlin, 1946). Eqn (10.45) may now be written as

$$\frac{\omega^2}{A_o}|w_{\max}| = TR_{\max}(R_\omega)$$

$$= \begin{cases} \dfrac{2R_\omega}{1-R_\omega^2}\cos\left(\dfrac{\pi R_\omega}{2}\right) \text{ at } \tau_m = \dfrac{1+R_\omega}{2R_\omega} & \text{for } R_\omega < 1 \\ \dfrac{-R_\omega}{1-R_\omega}\sin\left(\dfrac{2p\pi}{1+R_\omega}\right) \text{ at } \tau_m = \dfrac{2p}{1+R_\omega} & \text{for } (4p-3) \leq R_{w,p} \leq (4p+1),\ p = 1,2,3... \end{cases} \quad (10.46)$$

It is further noted that:

- The maximum response of $\frac{\omega^2|w_{\max}|}{A_o}$ is 1.768, corresponding to $R_w = 1.620$.
- The function $\frac{\omega^2|w_{\max}|}{A_o}$ tends to 1 as $R_\omega \to \infty$. Mathematically, $\left.\frac{\omega^2|w_{\max}|}{A_o}\right|_{\omega\to\infty} \to 1$ implies $|w_{\max}|_{\omega\to\infty} \to \left.\frac{A_o}{\omega^2}\right|_{\omega\to\infty} \to 0$. Physically, this implies the spring-mass system translates as a rigid body when the frequency of the applied acceleration shock is much lower than that of the system, which is intuitive.

When $R_\omega \ll 1$

When the excitation frequency is significantly higher than the resonant frequency of the system, the mass has negligible response during the time of the acceleration impulse. Using the case of hammer-impact of the support as an illustration, the displacement and velocity of the mass is given by $y(t_o) = \dot{y}(t_o) \approx 0$. If it is further assumed that $w(t_o)$ is also negligibly small, then the problem is reduced to the condition of velocity impact upon the support. The velocity of the deflection is given by

$$\dot{w}(t_o) \approx -\dot{Y}(t_o) = -\int_0^{t_o} \ddot{Y}\mathrm{d}t \tag{10.47}$$

where $\int_0^{t_o} \ddot{Y}\mathrm{d}t$ is the area under the acceleration—time curve, referred to as the acceleration impulse. The velocity impact sets up a harmonic vibration in the spring-mass given by

$$w(t) \approx \frac{\dot{w}(t_o)}{\omega} \sin \omega(t - t_o) \tag{10.48}$$

The maximum magnitude of deflection is given by

$$|w_{\max}| \approx \left|\frac{\dot{w}(t_o)}{\omega}\right| = \left|\frac{\int_0^{t_o} \ddot{Y}(t)\mathrm{d}t}{\omega}\right| \tag{10.49}$$

That is, for small values of R_ω, $|w_{\max}|$ is linearly proportional to the acceleration impulse and independent of the actual description of the acceleration function $\ddot{Y}(t)$. Therefore, a square acceleration impulse of magnitude A_o and duration t_o gives the same maximum deflection as a triangular acceleration impulse of amplitude A_o and duration $2t_o$.

For the case of a half-sine acceleration impulse, $\ddot{Y} = A_o \sin \Omega t$; the velocity of the support at $t = t_o$ is given by $\dot{Y}(t_o) = \int_0^{t_o} \ddot{Y}\mathrm{d}t = \frac{2A_o}{\Omega}$, which upon substituting into Eqn (10.49) gives

$$|w_{\max}| \approx \frac{2A_o}{\omega\Omega} = \frac{2A_o t_o}{\pi\omega} \tag{10.50}$$

Hence, the maximum magnitude of deflection is proportional to the amplitude and the duration of the half-sine acceleration, but it is inversely proportional to the frequency of the spring-mass system. Equation (10.50) may be rewritten as

$$\frac{\omega^2}{A_o}|w_{\max}| = TR_{\max}(R_\omega) \approx 2R_\omega \tag{10.51}$$

That is, the maximum deflection of the spring increases linearly with R_ω at a gradient of 2. The first equation of Eqn (10.46), $TR_{\max}(R_\omega) = \frac{2R_\omega}{1-R_\omega^2}\cos\left(\frac{\pi R_\omega}{2}\right)$, has been evaluated numerically and has been found, through regression, to be

$$\frac{\omega^2}{A_o}|w_{\max}| = TR_{\max}(R_\omega) \approx \begin{cases} 1.996R_\omega & \text{for } 0 \leq R_\omega \leq 0.1 \\ 1.895R_\omega & \text{for } 0 \leq R_\omega \leq 0.5 \end{cases} \tag{10.52}$$

which has validated Eqn (10.51).

10.3.2 A damped spring-mass system

A schematic of a damped spring-mass system is depicted in Figure 10.9, where a block of damping material is introduced between the base and the mass. The equation of motion is given by the dynamic equilibrium between the inertial force $m\ddot{y}$, the damping force $c\dot{y}$ and the elastic force kw; thus, we have,

$$m\ddot{y}(t) + c\dot{w}(t) + kw(t) = 0 \tag{10.53}$$

Using the coordinate relation, $y(t) = w(t) + Y(t)$, Eqn (10.53) is expressed in terms of deflection as

$$\ddot{w}(t) + 2\zeta\omega\dot{w}(t) + \omega^2 w(t) = -\ddot{Y}(t) \tag{10.54}$$

where $\omega = k/m$ and $\zeta = c/(2m\omega)$ are the un-damped natural frequency and the damping ratio of the system, respectively; and $\ddot{Y}(t) = \begin{cases} A_o \sin \Omega t & \text{for } 0 \leq t \leq t_o \\ 0 & \text{for } t \geq t_o \end{cases}$

10.3.2.1 Deflection-time

The solution to Eqn (10.54) can be expressed in terms of the convolution integral:

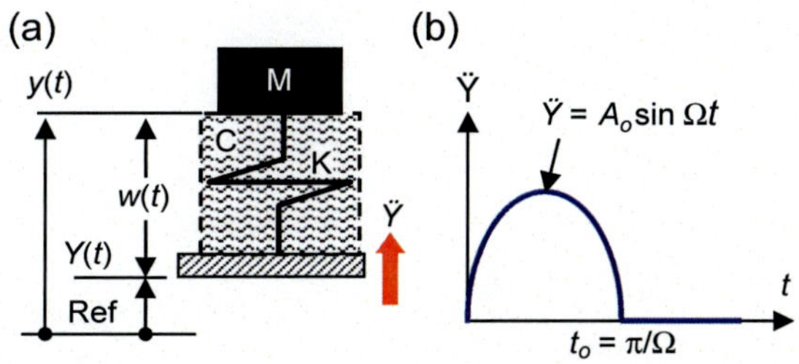

Figure 10.9 (a) PCB modelled as a single spring-damper-mass system. (b) Input acceleration pulse.

$$w(t)=\begin{cases} e^{-\zeta\omega t}\left[w(0)\cos\omega_d t+\dfrac{\dot{w}(0)+\zeta\omega w(0)}{\omega_d}\sin\omega_d t\right]-\dfrac{1}{\omega_d}\displaystyle\int_o^t \ddot{Y}(t)\cdot e^{-\zeta\omega(t-\tau)}\sin\omega_d(t-\tau)d\tau & \text{for } 0\le t\le t_o \\ e^{-\zeta\omega(t-t_o)}\left[w(t_o)\cos\omega_d(t-t_o)+\dfrac{\dot{w}(t_o)+\zeta\omega w(t_o)}{\omega_d}\sin\omega_d(t-t_o)\right] & \text{for } t\ge t_o \end{cases} \tag{10.55}$$

where $\omega_d = \sqrt{1-\zeta^2}\omega$ is the damped frequency of the system.

$0 \le t \le t_o$

For the period $0 \le t \le t_o$, the acceleration of the PCB support is prescribed by $\ddot{Y}(t) = A_o \sin \Omega t$. Substituting $\ddot{Y}(t)$ into the first equation of Eqn (10.55) plus the initial condition $w_o = \dot{w}_o = 0$, after some algebraic manipulations, gives the deflection of the mass as

$$\frac{\omega^2}{A_o}w(t) = \frac{-R_\omega e^{-\zeta\omega t}}{\left(1-R_\omega^2\right)^2+(2\zeta R_\omega)^2}TRD\_1(R_\omega,\zeta,t) \tag{10.56}$$

where

$$\begin{aligned} TRD\_1(R_\omega,\zeta,t) = {} & 2\zeta R_\omega^2\left(\cos\omega_d t - e^{\zeta\omega t}\cos\Omega t\right)+\frac{1-R_\omega^2+2\zeta^2R_\omega^2}{\sqrt{1-\zeta^2}}\sin\omega_d t \\ & - R_\omega\left(1-R_\omega^2\right)e^{\zeta\omega t}\sin\Omega t \end{aligned} \tag{10.56a}$$

$t \ge t_o$

For the period $t \ge t_o$, the system is under free vibration. The initial condition $w(t_o)$ and $\dot{w}(t_o)$ can be obtained by substituting $t = t_o$ into Eqn (10.56) and its differentiation, leading to

$$\frac{\omega^2}{A_o}w(t_o) = \frac{-R_\omega e^{-\zeta\omega t_o}}{\left(1-R_\omega^2\right)^2+(2\zeta R_\omega)^2}TRD\_1(R_\omega,\zeta,t_o) \tag{10.57}$$

and

$$\begin{aligned} \frac{\omega}{A_o}\dot{w}(t_o) = {} & -\frac{R_\omega e^{-\zeta\omega t_o}}{\left(1-R_\omega^2\right)^2+(2\zeta R_\omega)^2}\left[\left(1-R_\omega^2\right)\cos\omega_d t_o - \frac{\zeta\left(1-R_\omega^2\right)}{\sqrt{1-\zeta^2}}\sin\omega_d t_o\right] \\ & - \frac{R_\omega\left(1-R_\omega^2\right)}{\left(1-R_\omega^2\right)^2+(2\zeta R_\omega)^2} \end{aligned} \tag{10.58}$$

Substituting the above into the second equation of Eqn (10.55) and after some algebraic manipulation, we obtain the deflection of the mass as

$$\frac{\omega^2}{A_o}w(t) = \frac{-R_\omega e^{-\zeta\omega t}}{\left(1-R_{(\omega)}^2\right)^2+(2\zeta R_\omega)^2}TRD\_2(R_\omega,\zeta,t) \tag{10.59}$$

where

$$\begin{aligned}TRD\_2(R_\omega,\zeta,t) = {} & 2\zeta R_\omega^2\left[\cos\omega_d t + e^{\zeta\pi R_\omega}\cos\left(\omega_d t - \pi\sqrt{1-\zeta^2}R_\omega\right)\right]\\ & + \frac{1-R_\omega^2+2\zeta^2R_\omega^2}{\sqrt{1-\zeta^2}}\left[\sin\omega_d t + e^{\zeta\pi R_\omega}\sin\left(\omega_d t - \pi\sqrt{1-\zeta^2}R_\omega\right)\right]\end{aligned} \tag{10.59a}$$

$0 \le t \le \infty$

The consolidated deflection−time relation of the mass is

$$\begin{aligned}\frac{\omega^2}{A_o}w(t) &= TRD(R_\omega,\zeta,t)\\ &= \frac{-R_\omega e^{-\zeta\omega t}}{\left(1-R_{(\omega)}^2\right)^2+(2\zeta R_\omega)^2}\begin{Bmatrix} TRD\_1(R_\omega,\zeta,t) & \text{for } 0\le t\le t_o\\ TRD\_2(R_\omega,\zeta,t) & \text{for } t\ge t_o\end{Bmatrix}\end{aligned} \tag{10.60}$$

Putting $\Omega t = \pi\tau$, $\omega t = \pi R_\omega\tau$, $\omega_d t = \sqrt{1-\zeta^2}\,\pi R_\omega\tau = \mu\tau$ and $\tau = t/t_o$ gives the deflection of the mass expressed in the normalised time, τ, as

$$\begin{aligned}\frac{\omega^2}{A_o}w(\tau) &= TRD(R_\omega,\zeta,\tau)\\ &= \frac{-R_\omega e^{-\zeta\pi R_\omega\tau}}{\left(1-R_{(\omega)}^2\right)^2+(2\zeta R_\omega)^2}\begin{Bmatrix} TRD\_1(\zeta,R_\omega,\tau) & \text{for } 0\le \tau\le 1\\ TRD\_2(\zeta,R_\omega,\tau) & \text{for } \tau\ge 1\end{Bmatrix}\end{aligned} \tag{10.61}$$

where

$$\begin{aligned}TRD\_1(\zeta,R_\omega,\tau) &= 2\zeta R_\omega^2\cos\mu\tau + \frac{1-R_\omega^2+2\zeta^2R_\omega^2}{\sqrt{1-\zeta^2}}\sin\mu\tau\\ &\quad - R_\omega e^{\zeta\pi R_\omega\tau}\left[2\zeta R_\omega\cos\pi\tau + \left(1-R_{(\omega)}^2\right)\sin\pi\tau\right]\\ TRD\_2(\zeta,R_\omega,\tau) &= 2\zeta R_\omega^2\left[\cos\mu\tau + e^{\zeta\pi R_\omega}\cos\mu(\tau-1)\right]\\ &\quad + \frac{1-R_\omega^2+2\zeta^2R_\omega^2}{\sqrt{1-\zeta^2}}\left[\sin\mu\tau + e^{\zeta\pi R_\omega}\sin\mu(\tau-1)\right]\end{aligned} \tag{10.61a}$$

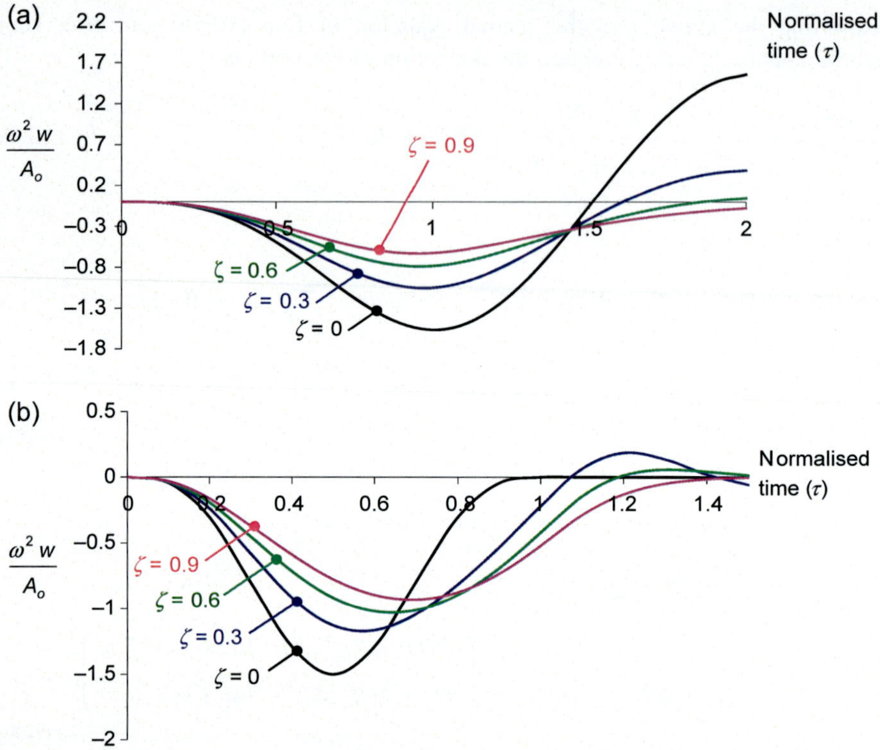

Figure 10.10 Normalised deflection-time response of a damped spring-mass system for (a) $R_\omega = 0.99$ and (b) $R_\omega = 3$.

It is noted that Eqn (10.61) is reduced to Eqn (10.35) when $\zeta = 0$. Furthermore, it can be shown that $\ddot{y}(t) = -\omega^2 w(t)$.

The normalised deflection-time response of the mass for $R_\omega = 0.99$ and $R_\omega = 3$ for $\zeta = 0$, 0.3, 0.6, 0.9 are illustrated in Figure 10.10(a) and (b), respectively. It is noted that:

1. Damping reduces the magnitude of deflection, as expected.
2. Damping shifts the time at which the maximum deflection occurs: to the left for $R_\omega \leq 1.2$ and to the right for $R_\omega \geq 1.2$.

10.3.2.2 Maxima spectra of deflection

There is no simple analytical equation for describing the maxima of the normalised deflection of a damped spring-mass system. Nevertheless, the maxima of the normalised deflection for any value of R_ω described by Eqn (10.61) may be sought numerically. The maxima spectra are shown in Figure 10.11. We shall designate the maxima function as $TRD_{\max}(R_\omega, \zeta)$. It is noted that with increasing damping ratio (1) the

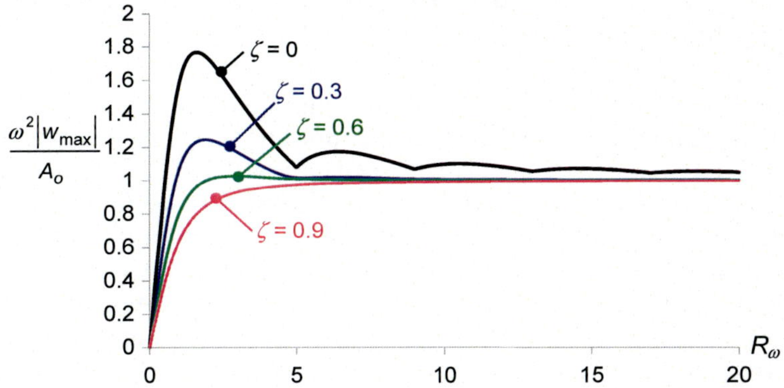

Figure 10.11 Maxima spectra of a damped spring-mass system.

magnitude of the maximum response decreases, as expected and (2) the frequency ratio R_ω at which the maximum response occurs increases.

When $R_\omega \ll 1$

Again, we use the case of hammer-impact of the support as an example, which may be approximated as a velocity impact for $R_\omega \ll 1$. The harmonic vibration of the damped spring-mass system is given by

$$w(t) \approx e^{-\zeta\omega(t-t_o)} \frac{\dot{w}(t_o)}{\omega_d} \sin \omega_d (t - t_o) \tag{10.62}$$

Differentiating Eqn (10.62) with respect to time gives the maximum deflection as

$$\frac{\omega^2}{A_o} |w_{\max}| = \left| \frac{\omega \dot{Y}(t_o)}{A_o} \exp\left[-\frac{\zeta \cos^{-1}(\zeta)}{\sqrt{1-\zeta^2}} \right] \right| \tag{10.63}$$

Again, $|w_{\max}|$ is linearly proportional to the acceleration impulse and independent of the actual acceleration shock. For the case of a half-sine acceleration shock, substituting $\dot{Y}(t_o) = 2A_o/\Omega$ into the above equation gives

$$\frac{\omega^2}{A_o} |w_{\max}| = TRD_{\max}(R_\omega, \zeta) \approx k(\zeta) R_\omega \tag{10.64}$$

where

$$k(\zeta) = 2 \exp\left[-\frac{\zeta \cos^{-1}(\zeta)}{\sqrt{1-\zeta^2}} \right] \tag{10.64a}$$

Thus, for small values of R_ω, $|w_{\max}|$ for half-sine acceleration, shock increases linearly with R_ω, whose gradient decreases with increasing damping ratio (see Figure 10.11).

10.4 Analytical solutions for a beam/plate subjected to half-sine shock

In general, integrated circuit (IC) components that are mounted on a test board are significantly smaller in size than the test board itself; hence, the test board can be modelled as a beam/plate of equivalent resonant frequency. The vehicle used for this analysis consists of a rectangular strip of PCB that is supported at two locations along its full width such that it experiences symmetric bending about its mid-length when subjected to mechanical shock applied to the two supports. Compared to the asymmetrical deformation of the four-point supported test board as specified in the JESD22-B111 test method, the two-edge supported test board design offers greater ease of analysis of the experimental data. In any case, the analysis performed here will be applicable qualitatively to a four-point supported test board.

10.4.1 The equation of motion

Figure 10.12 shows an elemental length of a beam of thickness h that is in dynamic equilibrium. The equation of motion in the vertical direction is given by

$$\overline{m}\,\frac{\partial^2 y}{\partial t^2} + c\,\frac{\partial w}{\partial t} - \frac{\partial V}{\partial x} = F \tag{10.65}$$

where $\overline{m} = \rho h$ and c are the mass and damping coefficient, respectively, of the beam per unit cross-sectional area; V is the sectional force per unit width; and F is the distributed force. The condition of moment equilibrium and the moment–curvature relation gives, respectively,

$$V = -\frac{\mathrm{d}M}{\mathrm{d}x} \tag{10.66}$$

$$M = D\frac{\mathrm{d}^2 w}{\mathrm{d}x^2} \tag{10.67}$$

Figure 10.12 Equilibrium of a beam element.

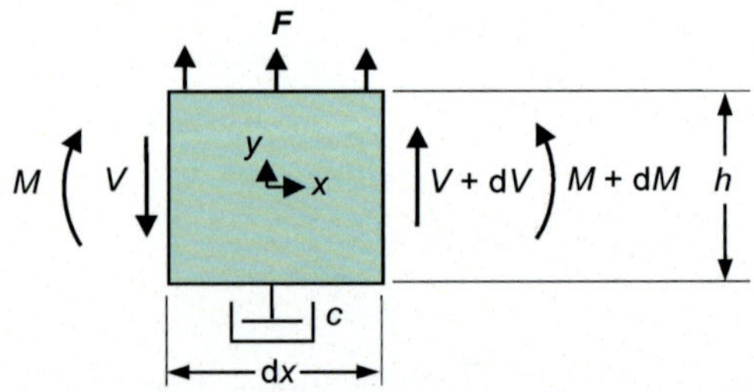

where M is bending moment per unit width and D is the flexural rigidity of the beam per unit width. Substituting these two relations into Eqn (10.65) gives the equation of motion as

$$\ddot{y} + \lambda^2 w^{IV} + 2\zeta\omega\dot{w} = F \tag{10.68}$$

where $\zeta = c/2\overline{m}\omega$; and $\lambda = \sqrt{D/\overline{m}}$ is the characteristic constant of the beam.

10.4.2 An undamped beam

The equation of motion of a test board modelled as an undamped beam subjected to a half-sine acceleration shock applied to its two supported ends as shown in Figure 10.13 is given by

$$\ddot{y}(x,t) + \lambda^2 w^{IV}(x,t) = 0 \tag{10.69}$$

Using the coordinate relation, $y(x, t) = w(x, t) + Y(t)$, Eqn (10.69) can be expressed in terms of deflection as

$$\ddot{w}(x,t) + \lambda^2 w^{IV}(x,t) = -\ddot{Y}(t) \tag{10.70}$$

where $\ddot{Y}(t) = \begin{cases} A_o \sin \Omega t & \text{for } 0 \leq t \leq t_o \\ 0 & \text{for } t \geq t_o \end{cases}$

10.4.2.1 Deflection-time

The dynamic deflection of the beam may be described by a linear superposition of the dynamics of individual deformation response as:

$$w(x,t) = \sum_{n}^{\infty} w_n(x,t) \tag{10.71}$$

where the individual modal deflection is given by

$$w_n(x,t) = W_n(x)q_n(t) \tag{10.72}$$

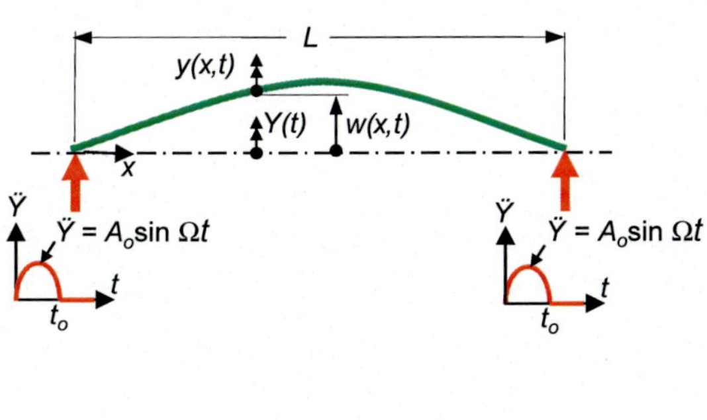

Figure 10.13 Test board modelled as a beam subjected to half-sine acceleration at end supports.

where $W_n(x)$ is the nth modal deflection function (or eigenfunction), and $q_n(t)$ is the corresponding participating factor. Substituting Eqns (10.71) and (10.72) into Eqn (10.70) gives

$$\sum_{n=1}^{\infty} W_n(x)\ddot{q}_n(t) + \sum_{n=1}^{\infty} \omega_n^2 W_n(x) q_n(t) = -\ddot{Y}(t) \tag{10.73}$$

Equation (10.73) can be simplified by first multiplying it with $W_m(x)$ followed by integration over the length of the board to give

$$\sum_{n=1}^{\infty} \ddot{q}_n(t) \int_0^L W_m(x) W_n(x)\mathrm{d}x + \sum_{n=1}^{\infty} \omega_n^2 q_n(t) \int_0^L W_m(x) W_n(x)\mathrm{d}x = -\int_0^L W_m(x)\ddot{Y}(t)\mathrm{d}x \tag{10.74}$$

and then applying the orthogonality principle

$$\int_0^L W_n(x)\cdot W_m(x)\mathrm{d}x = 0 \quad n \neq m \tag{10.75}$$

to reduce the equation to

$$\ddot{q}_n(t) + \omega_n^2 q_n(t) = -\frac{\int_0^L W_n(x)\mathrm{d}x}{\int_0^L W_n^2(x)\mathrm{d}x}\ddot{Y}(t) \tag{10.76}$$

The eigenfunction $W_n(x)$ can be shown to satisfy the homogeneous equation

$$W_n^{IV}(x) - \beta_n^4 W_n(x) = 0 \tag{10.77}$$

where $\beta_n = \sqrt{\omega_n/\lambda}$, and ω_n is the n-modal frequency of the board. The general solution is given by

$$W(x) = C_1 \cos \beta x + C_2 \sin \beta x + C_3 \cosh \beta x + C_4 \sinh \beta x \tag{10.78}$$

where the coefficients are defined by the support conditions of the beam. For simplicity, the case of the simply supported test board that deflects symmetrically about the mid-length is analysed, while not sacrificing the physical insights. The eigenfunction is given by

$$W_n(x) = \sin \beta_n x, \quad \beta_n = \frac{n\pi}{L}, \quad n = 1, 3, 5... \tag{10.79}$$

where L is the span of the test board between the two supports. It is understood that $n = 1, 3, 5...$ in the following equations and hence will not be specifically stated. Substituting Eqn (10.79) into Eqn (10.76) leads to

$$\ddot{q}_n(t) + \omega_n^2 q_n(t) = -\frac{4}{n\pi}\ddot{Y}(t) \tag{10.80}$$

Comparing Eqn (10.80) with Eqn (10.25), it becomes clear that the solution for $q_n(\tau)$ takes the form of Eqn (10.35); that is,

$$\frac{\omega_n^2}{A_o} q_n(\tau) = \frac{4}{n\pi} TR(R_{\omega_n}, \tau) \tag{10.81}$$

where $R_{\omega_n} = \omega_n/\Omega$. Using the relationship $\omega_n = \lambda\beta_n^2 = \lambda(n\pi/L)^2$, the nth modal frequency ω_n may be expressed in terms of the fundamental frequency ω_1 as

$$\omega_n = n^2\omega_1 \tag{10.82}$$

Substituting Eqn (10.82) into Eqn (10.81), we have

$$\frac{\omega_1^2}{A_o} q_n(\tau) = \frac{4}{n^5\pi} TR(R_{\omega_n}, \tau) \tag{10.83}$$

The modal deflection response is therefore

$$\frac{\omega_1^2}{A_o} w_n(x, \tau) = \frac{4}{n^5\pi} TR(R_{\omega_n}, \tau) \sin \beta_n x \tag{10.84}$$

and the deflection of the board modelled as a beam is given by

$$\frac{\omega_1^2}{A_o} w(x, \tau) = \frac{\omega_1^2}{A_o} \sum_{n=1,3,5...}^{\infty} w_n(x, \tau) = \sum_{n=1,3,5...}^{\infty} \left[\frac{4}{n^5\pi} TR(R_{\omega_n}, \tau) \sin \beta_n x\right] \tag{10.85}$$

10.4.2.2 Maxima spectra of deflection

The maximum of the modal deflection occurs at $x = L/2$ and takes the form:

$$\frac{\omega_1^2}{A_o} \left|w_{n,\max}\right| = \frac{4}{n^5\pi} TR_{\max}(R_{\omega_n}) \tag{10.86}$$

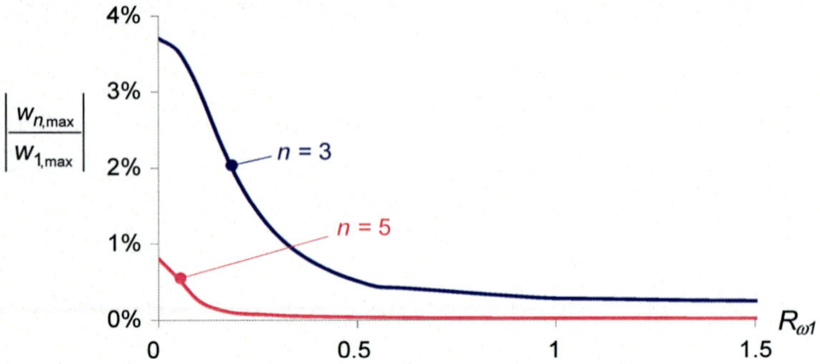

Figure 10.14 Ratios of higher mode to the fundamental mode: deflection.

where the function $TR_{\max}(R_{\omega_n})$ has been defined in Eqn (10.46), simply replacing R_ω with R_{ω_n}.

The ratio of the higher mode to the fundamental mode response is given by

$$\left|\frac{w_{n,\max}}{w_{1,\max}}\right| = \frac{1}{n^5}\frac{TR_{\max}(R_{\omega_n})}{TR_{\max}(R_{\omega_1})} \tag{10.87}$$

Equation (10.87) is bounded at two limiting values of R_{ω_1}: $R_{\omega_1} \to 0$ and $R_{\omega_1} \to \infty$. At small values of R_{ω_1} and R_{ω_n}, Eqn (10.51) gives $TR_{\max}(R_{\omega_1}) \approx 2R_{\omega_1}$ and $TR_{\max}(n^2 R_{\omega_1}) \approx 2n^2 R_{\omega_1}$; thus,

$$\left|\frac{w_{n,\max}}{w_{1,\max}}\right|_{R_{\omega_1}\to 0} \approx \frac{1}{n^3} \tag{10.88}$$

At large values of R_{ω_1} and R_{ω_n}, Eqn (10.46) gives $TR_{\max}(R_{\omega_1}) \approx 1$ and $TR_{\max}(R_{\omega_n}) \approx 1$; thus,

$$\left|\frac{w_{n,\max}}{w_{1,\max}}\right|_{R_{\omega_1}\to \infty} \approx \frac{1}{n^5} \tag{10.89}$$

The function $\left|\frac{w_{n,\max}}{w_{1,\max}}\right|$ for $n = 3$ and 5, respectively, are shown in Figure 10.14. It becomes obvious that the higher modes have relatively little contribution in the deflection of the beam; and the contribution of the higher mode decreases with increasing frequency ratio of R_{ω_1}. For all practical purposes, the higher mode deflection response can be ignored.

It is noted that $|w_{\max}| \neq \sum_n^\infty |w_{n,\max}|$ because the maximum of individual mode does not occur at the same time and at the same location. There is no simple analytical equation for $|w_{\max}|$. Nevertheless, because of the dominance of the fundamental mode, the maximum of the board deflection can be approximated by the maximum of the fundamental modal deflection; that is,

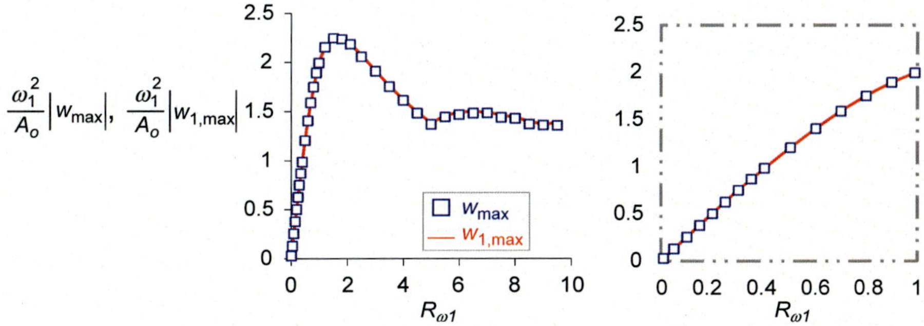

Figure 10.15 Comparison of w_{max} and $w_{1,max}$.

$$\frac{\omega_1^2}{A_o}|w_{max}| \approx \frac{\omega_1^2}{A_o}|w_{1,max}| = \frac{4}{\pi} TR_{max}(R_{\omega_1}) \tag{10.90}$$

The goodness of Eqn (10.90) for estimating the maximum board deflection is illustrated in Figure 10.15 (the figure to the right shows the expanded view for $0 \le R_{\omega_1} < 1$), in which the data for $\frac{\omega_1^2}{A_o}|w_{max}|$ have been computed numerically.

10.4.2.3 Fibre strain-time

The simple bending relationship for the beam is given by

$$\frac{M_n(x,\tau)}{D} = -\frac{2\varepsilon_n(x,\tau)}{h} = \frac{\partial^2 w_n(x,\tau)}{\partial x^2} \tag{10.91}$$

Substituting Eqn (10.84) into the above and noting that $\frac{\partial^2}{\partial x^2}\sin\beta_n x = -\left(\frac{n\pi}{L}\right)^2 \sin\beta_n x$ gives

$$-\frac{\omega_1^2 L^2}{A_o D} M_n(x,\tau) = \frac{2\omega_1^2 L^2}{A_o h}\varepsilon_n(x,\tau) = \frac{4\pi}{n^3} TR(R_{\omega_n},\tau)\sin\beta_n x \tag{10.92}$$

The fibre strain of the board modelled as a beam is given by

$$\frac{2\omega_1^2 L^2}{A_o h}\varepsilon(x,\tau) = \frac{2\omega_1^2 L^2}{A_o h}\sum_{n=1,3,5\ldots}^{\infty}\varepsilon_n(x,\tau) = \sum_{n=1,3,5\ldots}^{\infty}\left[\frac{4\pi}{n^3} TR(R_{\omega_n},\tau)\sin\beta_n x\right] \tag{10.93}$$

10.4.2.4 Maxima spectra of fibre strain

The maxima of the modal bending moment and fibre strain occur at $x = L/2$ and take the form:

$$\frac{\omega_1^2 L^2}{A_o D}|M_{n,max}| = \frac{2\omega_1^2 L^2}{A_o h}|\varepsilon_{n,max}| = \frac{4\pi}{n^3} TR_{max}(R_{\omega_n}) \tag{10.94}$$

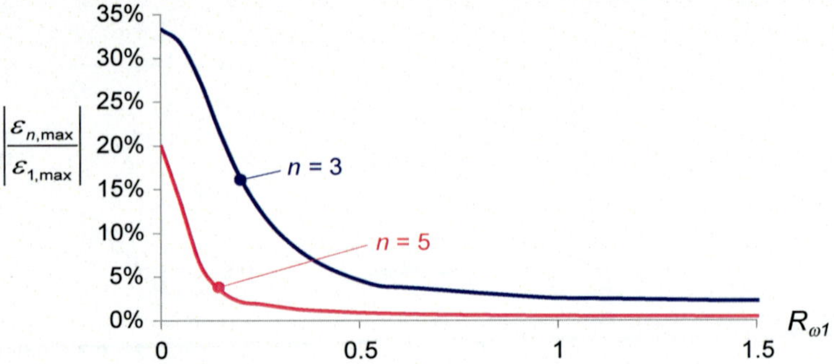

Figure 10.16 Ratios of the higher mode to the fundamental mode: fibre strain.

The ratio of the higher mode to the fundamental mode response is given by

$$\left|\frac{\varepsilon_{n,\max}}{\varepsilon_{1,\max}}\right| = \frac{1}{n^3}\frac{TR_{\max}(R_{\omega_n})}{TR_{\max}(R_{\omega_1})} \tag{10.95}$$

The maximum magnitude of the ratio is:

$$\left|\frac{\varepsilon_{n,\max}}{\varepsilon_{1,\max}}\right|_{R_{\omega_1}\to 0} \approx \frac{1}{n} \tag{10.96}$$

The function $\left|\frac{\varepsilon_{n,\max}}{\varepsilon_{1,\max}}\right|$ for $n = 3$ and 5, respectively, are shown in Figure 10.16. It becomes obvious that the higher modes have a significant contribution to the fibre strain of the beam, but the contribution recedes rapidly with increasing R_{ω_1} and becomes negligible for $R_{\omega_1} > 0.3$.

Again, the maximum board strain is not the simple summation of the maxima of the individual modal strain; that is, $|\varepsilon_{\max}| \neq \sum_{n=1,3,5}^{\infty}|\varepsilon_{n,\max}|$. Nevertheless, due to the relative dominance of the fundamental mode for $R_{\omega_1} > 0.3$, the maximum board strain is approximated by the maximum fundamental modal strain; that is,

$$\frac{2\omega_1^2L^2}{A_oh}|\varepsilon_{\max}| \approx \frac{2\omega_1^2L^2}{A_oh}|\varepsilon_{1,\max}| = 4\pi\cdot TR_{\max}(R_{\omega_1}) \tag{10.97}$$

For small values of R_ω, the function $TR_{\max}$ (R_ω) is proportional to the magnitude of the acceleration impulse, independent of the actual function of the acceleration shock; and

$$\frac{2\omega_1^2L^2}{A_oh}|\varepsilon_{\max}| \approx 8\pi R_{\omega_1} \tag{10.98}$$

The goodness of Eqn (10.97) is illustrated in Figure 10.17 (the figure to the right shows the expanded view for $0 \leq R_{\omega_1} < 1$). The deviation appears to be much more subdued than suggested in Figure 10.16. This is attributed to the fact that

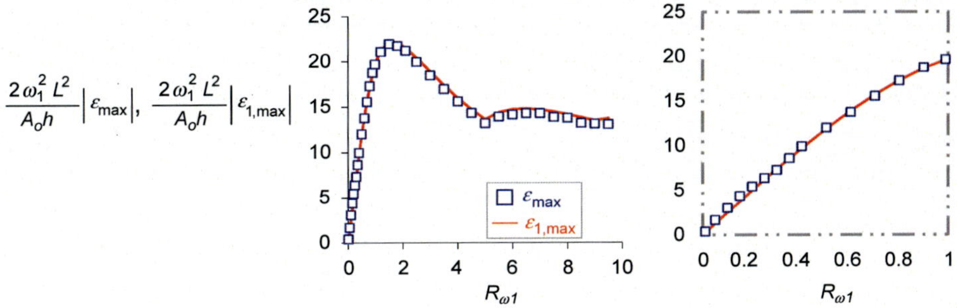

Figure 10.17 Comparison of ε_{max} and $\varepsilon_{1,max}$.

Figure 10.16 is expressed as a percentage difference, while the absolute magnitude of $\varepsilon_{1,max}$ is relatively small for $R_{\omega_1} < 0.3$. Equation (10.97) is thus a good approximation for the maximum of the fibre strain. The relative insensitivity of the fibre strain to the higher deformation mode endorses the practice of using fibre strain as a measurement of the severity of the loading experienced by a PCB assembly.

10.4.2.5 Acceleration-time

Differentiating the modal deflection function twice with respect to time gives the modal w-acceleration function below:

$$\frac{\ddot{w}_n(x,\tau)}{A_o} = \begin{cases} \frac{-4}{n\pi}[TR(R_{\omega_n},\tau) + \sin \pi\tau]\sin \beta_n x & \text{for } 0 \le \tau \le 1 \\ \frac{-4}{n\pi} TR(R_{\omega_n},\tau)\sin \beta_n x & \text{for } \tau \ge 1 \end{cases} \tag{10.99}$$

The modal acceleration of the board modelled as a beam is given by

$$\frac{\ddot{y}_n(x,\tau)}{A_o} = \frac{\ddot{w}_n(x,\tau)}{A_o} + \frac{\ddot{Y}(\tau)}{A_o}$$
$$= \begin{cases} \frac{-4}{n\pi}[TR_n(R_{\omega_n},\tau) + \sin \pi\tau]\sin \beta_n x + \sin \pi\tau & \text{for } 0 \le \tau \le 1 \\ \frac{-4}{n\pi} TR_n(R_{\omega_n},\tau)\sin \beta_n x & \text{for } \tau \ge 1 \end{cases} \tag{10.100}$$

A common mistake in evaluating the y-acceleration of the board is in assuming that $\ddot{y}(x,\tau) = \sum_{n=1,3,5\ldots}^{\infty} \ddot{y}_n(x,\tau)$, which would grossly overestimate the magnitude of y-acceleration for $0 \le \tau \le 1$ due to the repeated summation of $\ddot{Y}(\tau)$. The correct procedure is to evaluate the resultant w-acceleration of the board using $\ddot{w}(x,\tau) = \sum_{n=1,3,5\ldots}^{\infty} \ddot{w}_n(x,\tau)$ followed by addition of $\ddot{Y}(\tau)$; that is,

Table 10.1 Parameters for the FE and the analytical models

Beam					Input pulse	
L (mm)	h (mm)	E (GPa)	ρ (g/cm^2)	ω_1 (rad/s)	A_o (g)	Ω (rad/s)
100	1	24	2	986	1000	3142

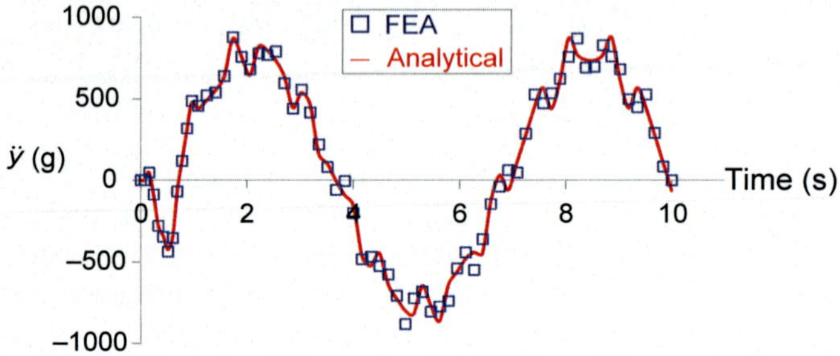

Figure 10.18 Absolute acceleration at the mid-length of the board: comparison between analytical and FEA.

$$\frac{\ddot{y}(x,\tau)}{A_o} = \begin{cases} \dfrac{\ddot{w}(x,\tau)}{A_o} + \sin \pi\tau & \text{for } 0 \le \tau \le 1 \\ \dfrac{\ddot{w}(x,\tau)}{A_o} & \text{for } \tau \ge 1 \end{cases} \tag{10.101}$$

Equation (10.101) is benchmarked against FE analysis. The parameters for the validation vehicle are shown in Table 10.1. Figure 10.18 compares the absolute acceleration of the validation vehicle at its mid-length evaluated by Eqn (10.101) and by finite element analysis (FEA) using implicit scheme in which the initial conditions $y_o = \dot{y}_o = Y_o = \dot{Y}_o = 0$ and the displacement boundary conditions

$$Y(t) = \begin{cases} \dfrac{A_o t}{\Omega} - \dfrac{A_o}{\Omega^2} \sin \Omega t & \text{for } 0 \le t \le t_o \\ \dfrac{A_o t_o}{\Omega} + \dfrac{2A_o}{\Omega}(t - t_o) & \text{for } t \ge t_o \end{cases}$$

are prescribed. The two results match up almost perfectly.

10.4.2.6 Maxima spectra of acceleration

The maximum of the modal w-acceleration occurs at $x = L/2$ and takes the form:

$$\frac{\ddot{w}_{n,\max}}{A_o} = \begin{cases} \dfrac{4}{n\pi} \max[TR_n(R_{\omega_n}, \tau) + \sin \pi\tau] & \text{for } 0 \le \tau \le 1 \\ \dfrac{4}{n\pi} TR_{\max}(R_{\omega_n}) & \text{for } \tau \ge 1 \end{cases} \tag{10.102}$$

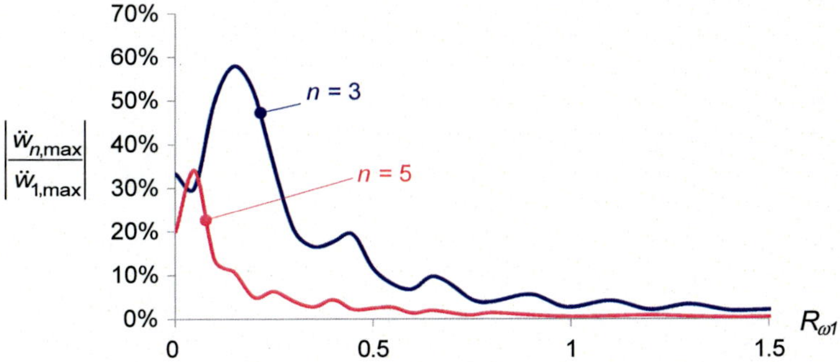

Figure 10.19 Ratios of the higher mode to the fundamental mode: acceleration.

The ratios $\left|\frac{\ddot{w}_{n,\max}}{\ddot{w}_{1,\max}}\right|$ for $n = 3$ and 5 have been evaluated numerically; these are shown in Figure 10.19. Unlike the deflection or fibre strain, the ratio does not decrease monotonically with increasing R_{ω_1}. The maximum ratio of 58% for $n = 3\%$ and 34% for $n = 5$ occurs at $R_{\omega_1} = 0.15$ and $R_{\omega_1} = 0.05$, respectively. The strong presence of the higher modes is responsible for the 'noisy' acceleration response observed in the experimental measurements.

Despite the significant contribution of the higher model accelerations, it is nevertheless tempting to approximate the maximum board acceleration to the maximum fundamental modal acceleration; that is,

$$\frac{|\ddot{y}_{\max}|}{A_o} \approx \frac{|\ddot{y}_{1,\max}|}{A_o} = \begin{cases} \mathrm{Max}\left\{\frac{4}{\pi}[TR(R_{\omega_1}, \tau) + \sin \pi\tau] + \sin \pi\tau\right\} & \text{for } 0 \leq \tau \leq 1 \\ \frac{4}{\pi} TR_{\max}(R_{\omega_1}) & \text{for } \tau \geq 1 \end{cases} \tag{10.103}$$

If the last term in the first equation, $\sin \pi\tau$, had been $\frac{4}{\pi}\sin \pi\tau$, then Eqn (10.103) may be reduced to

$$\frac{|\ddot{y}_{\max}|}{A_o} \approx \frac{|\ddot{y}_{1,\max}|}{A_o} \approx \frac{4}{\pi} TR_{\max}(R_{\omega_1}) \tag{10.104}$$

We recall in the discussions of the maxima for a spring-mass system that the maxima for the period $0 \leq \tau \leq 1$ corresponds to $R_{\omega_1} > 1$. We shall therefore expect Eqn (10.104) to have overpredicted the magnitude of acceleration, approximately by the magnitude of $(4 - \pi)/\pi \approx 0.27$ for $R_{\omega_1} > 1$. This is shown in Figure 10.20, where a persistence deviation of $\approx 0.3/A_o$ is observed for $R_{\omega_1} > 1$. The observed deviation for $R_{\omega_1} < 1$ is due to the contributions of the higher modal acceleration, which has been shown to be more pronounced at small magnitudes of R_{ω_1}.

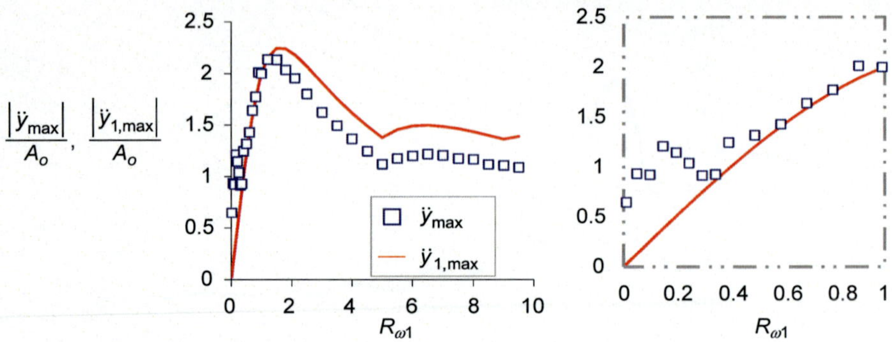

Figure 10.20 Comparison of $\ddot{y}_{max}$ and $\ddot{y}_{1,max}$.

10.4.3 A damped beam

The equation of motion of a PCB modelled as a beam subjected to half-sine acceleration shock taking into account of damping is given by

$$\lambda^2 w^{IV}(x,t) + \ddot{w}(x,t) + 2\zeta\omega\dot{w}(x,t) = \begin{cases} -A_o \sin \Omega t & \text{for } 0 \le t \le t_o \\ 0 & \text{for } t \ge t_o \end{cases} \tag{10.105}$$

Following the procedures in Section 10.4.2 leads to the following equation for the modal participating factor:

$$\ddot{q}_n(t) + 2\zeta_n\omega_n\dot{q}_n(t) + \omega_n^2 q_n(t) = \begin{cases} -\dfrac{4}{n\pi}A_o \sin \Omega t & \text{for } 0 \le t \le t_o \\ 0 & \text{for } t \ge t_o \end{cases} \tag{10.106}$$

Comparing Eqn (10.106) with Eqn (10.54) suggests the solution to be

$$\frac{q_n(\tau)}{A_o}\omega_n^2 = \frac{4}{n\pi} TRD(R_{\omega_n}, \zeta_n, \tau); \quad n = 1, 3, 5\ldots \tag{10.107}$$

where the function $TRD(R_{\omega_n}, \zeta_n, \tau)$ has been defined in Eqn (10.61). The modal deflection and the modal fibre strain of the board modelled as a beam with damping are simply

$$\begin{aligned} \frac{\omega_1^2}{A_o} w_n(x,\tau) &= \frac{4}{n^5\pi} TRD(R_{\omega_n}, \zeta_n, \tau) \sin \beta_n x \\ \frac{2\omega_1^2 L^2}{A_o h} \varepsilon_n(x,\tau) &= \frac{4\pi}{n^3} TRD(R_{\omega_n}, \zeta_n, \tau) \sin \beta_n x \end{aligned} \tag{10.108}$$

The corresponding maxima are

$$\begin{aligned}\frac{\omega_1^2}{A_o} w_{n,\max} &= \frac{4}{n^5 \pi} TRD_{\max}(R_{\omega_n}, \zeta_n) \\ \frac{2\omega_1^2 L^2}{A_o h} \varepsilon_{n,\max} &= \frac{4\pi}{n^3} TRD_{\max}(R_{\omega_n}, \zeta_n)\end{aligned} \tag{10.109}$$

From Eqn (10.64), the function $TRD_{\max}(R_{\omega_n}, \zeta_n)$ may be approximated by a linear function of R_{ω_n} for small values of R_{ω_n}; that is,

$$TRD_{\max}(R_{\omega_n}, \zeta_n) \approx k(\zeta_n) R_{\omega_n} \tag{10.110}$$

where

$$k(\zeta_n) = 2 \exp\left[-\frac{\zeta_n \cos^{-1}(\zeta_n)}{\sqrt{1-\zeta_n^2}} \right] \tag{10.110a}$$

Thus,

$$\begin{aligned}\frac{\omega_1^2}{A_o} w_{n,\max} &\approx \frac{4}{n^5 \pi} k(\zeta_n) R_{\omega_n} \\ \frac{2\omega_1^2 L^2}{A_o h} \varepsilon_{n,\max} &\approx \frac{4\pi}{n^3} k(\zeta_n) R_{\omega_n}\end{aligned} \tag{10.111}$$

10.4.4 An undamped plate

The PCB is more appropriately modelled as a plate. Figure 10.21 depicts a plate subjected to a half-sine acceleration input pulse along its four edges that are simply supported. The assumption of the plate being simply supported along its edges instead of four corners is not ideal but is necessary for analytical simplicity.

10.4.4.1 Equation of motion

The equation of motion is given by

$$\ddot{z}(x, y, t) + \lambda^2 \nabla^2 \nabla^2 w(x, y, t) = 0 \tag{10.112}$$

or for a half-sine acceleration at the supports:

$$\ddot{w}(x, y, t) + \lambda^2 \nabla^2 \nabla^2 w(x, y, t) = \begin{cases} -A_o \sin \Omega t & \text{for } 0 \le t \le t_o \\ 0 & \text{for } t \ge t_o \end{cases} \tag{10.113}$$

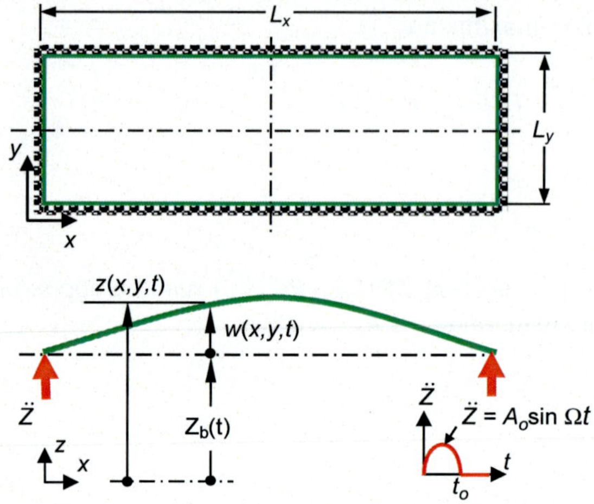

Figure 10.21 Plate model and the coordinate system (simply supported on four edges).

where $\lambda = \sqrt{D/\overline{m}}$, $D = \frac{Eh^3}{12(1-\nu^2)}$, $\overline{m} = \rho h$; and

$$\nabla^2 \nabla^2 w = \frac{\partial^4 w}{\partial x^4} + 2\frac{\partial^4 w}{\partial x^2 \partial y^2} + \frac{\partial^4 w}{\partial y^4} \tag{10.113a}$$

10.4.4.2 Deflection-time

Describing the deflection as a linear superposition of eigenfunction:

$$w(x, y, t) = \sum_{n=1}^{\infty} \sum_{m=1}^{\infty} w_{nm}(x, y, t) \tag{10.114}$$

where

$$w_{nm}(x, y, t) = W_n(x) W_m(y) q_{nm}(t) \tag{10.115}$$

and following the procedures derived for the beam model, the eigenfunctions of a rectangular plate simply supported at its four edges are given by Eqn (10.116). The associated modal frequency is given by Eqn (10.117):

$$\begin{aligned} W_n(x) &= \sin \beta_n x, \quad \beta_n = \frac{n\pi}{L_x} \\ W_m(y) &= \sin \alpha_m y, \quad \alpha_m = \frac{m\pi}{L_y} \end{aligned}, \quad n = 1, 3, 5 \text{ and } m = 1, 3, 5\ldots \tag{10.116}$$

$$\omega_{nm} = \lambda(\beta_n^2 + \alpha_m^2) = \frac{\lambda\pi^2}{L_x^2}(n^2 + r_a^2 m^2), \text{ where } r_a = L_x/L_y \tag{10.117}$$

For reasons of brevity, the definitions $n = 1, 3, 5\ldots$ and $m = 1, 3, 5\ldots$ will be hidden in the following discussions. Following the procedures described for the beam model, the modal participating factor can be shown to be given by

$$\frac{q_{nm}(t)}{A_o}\omega_{nm}^2 = \frac{16}{nm\pi^2}TR(R_{\omega_{mn}}, \tau) \tag{10.118}$$

From Eqn (10.117), the frequency ω_{nm} may be expressed in terms of the fundamental frequency as

$$\omega_{nm} = \left(\frac{n^2 + r_a^2 m^2}{1 + r_a^2}\right)\omega_{11} \tag{10.119}$$

Substituting into Eqn (10.118) and then into Eqn (10.115) gives the modal deflection as

$$\frac{\omega_{11}^2}{A_o}w_{nm}(x, y, \tau) = \left(\frac{1 + r_a^2}{n^2 + r_a^2 m^2}\right)^2 \frac{16}{\pi^2 nm}TR(R_{\omega_{nm}}, \tau)\sin\beta_n x \cdot \sin\alpha_m y \tag{10.120}$$

The deflection of the board modelled as a plate is given by

$$\frac{\omega_{11}^2}{A_o}w(x, \tau) = \frac{\omega_{11}^2}{A_o}\sum_{n=1,3,5\ldots}^{\infty}\sum_{m=1,3,5\ldots}^{\infty}w_{nm}(x, y, \tau) \tag{10.121}$$

10.4.4.3 Maxima spectra of deflection

The maximum of the modal deflection occurs at ($x = a/2$, $y = b/2$) and takes the form:

$$\frac{\omega_1^2}{A_o}\left|w_{nm,\max}\right| = \left(\frac{1 + r_a^2}{n^2 + r_a^2 m^2}\right)^2 \frac{16}{\pi^2 nm}TR_{\max}(R_{\omega_{nm}}) \tag{10.122}$$

where the function $TR_{\max}(R_{\omega_{nm}})$ has been defined in Eqn (10.46). The ratio of the higher mode to the fundamental mode response is given by

$$\left|\frac{w_{nm,\max}}{w_{11,\max}}\right| = \left(\frac{1 + r_a^2}{n^2 + r_a^2 m^2}\right)^2 \frac{1}{nm}\frac{TR_{\max}(R_{\omega_{nm}})}{TR_{\max}(R_{\omega_{11}})} \tag{10.123}$$

The maximum magnitude of the ratio is:

$$\left|\frac{w_{nm,\max}}{w_{11,\max}}\right|_{R_{\omega_{11}}\to 0} \approx \left(\frac{1+r_a^2}{n^2+r_a^2m^2}\right)\frac{1}{nm} \tag{10.124}$$

In view of the dominance of the fundamental mode, the maximum board deflection can be approximated by the maximum fundamental modal deflection; that is,

$$\frac{\omega_{11}^2}{A_o}|w_{\max}| \approx \frac{\omega_{11}^2}{A_o}|w_{11,\max}| = \frac{16}{\pi^2}TR_{\max}(R_{\omega_{11}}) \tag{10.125}$$

The goodness of Eqn (10.125) for estimating the maximum board deflection is illustrated in Figure 10.22 (the figure to the right shows the expanded view for $0 \le R_{\omega_1} < 1$).

10.4.4.4 Fibre strain-time

The simple bending relationship

$$\frac{2\varepsilon_{x_i,nm}(x,y,\tau)}{h} = -\frac{d^2w_{nm}(x,y,\tau)}{dx_i^2}, \quad x_i = x,y \tag{10.126}$$

gives the fibre strain in the board in the two principal directions. These are:

$$\begin{aligned}\frac{2\omega_{11}^2L_x^2}{A_oh}\varepsilon_{x,nm}(x,y,\tau) &= (n\pi)^2\frac{\omega_{11}^2}{A_o}w_{nm}(x,y,\tau)\\ \frac{2\omega_{11}^2L_y^2}{A_oh}\varepsilon_{y,nm}(x,y,\tau) &= (m\pi)^2\frac{\omega_{11}^2}{A_o}w_{nm}(x,y,\tau)\end{aligned} \tag{10.127}$$

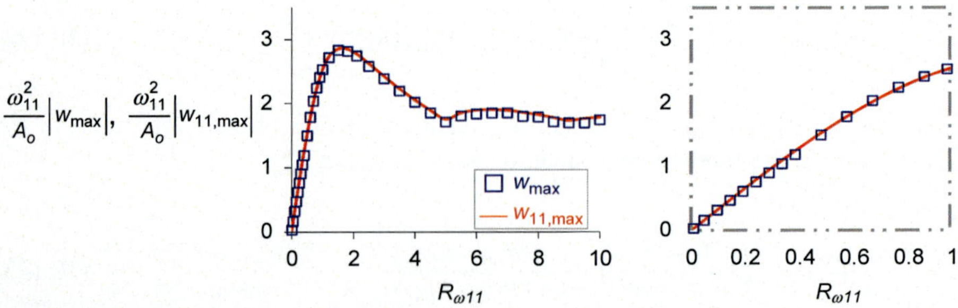

Figure 10.22 Comparison of $w_{\max}$ and $w_{11,\max}$.

The simple bending relationship

$$\frac{M_{xx,nm}(x,y,\tau)}{D} = -\left(\frac{\partial^2}{\partial x^2} + \nu\frac{\partial^2}{\partial y^2}\right)w_{nm}(x,y,\tau) \tag{10.128}$$

gives the bending moment in the board in the two principle directions as

$$\begin{aligned}\frac{\omega_{11}^2 L_x^2}{A_o D} M_{xx,nm}(x,y,\tau) &= \left[n^2 + \nu(r_a m)^2\right]\frac{\pi^2\omega_{11}^2}{A_o} w_{nm}(x,y,\tau)\\ \frac{\omega_{11}^2 L_y^2}{A_o D} M_{yy,nm}(x,y,\tau) &= \left[m^2 + \nu(n/r_a)^2\right]\frac{\pi^2\omega_{11}^2}{A_o} w_{nm}(x,y,\tau)\end{aligned} \tag{10.129}$$

10.4.4.5 *Maxima spectra of fibre strain*

The ratio of the higher mode to the fundamental mode response of fibre strain is given by

$$\begin{aligned}\left|\frac{\varepsilon_{x,nm,\max}}{\varepsilon_{x,11,\max}}\right| &= \left(\frac{1+r_a^2}{n^2+r_a^2 m^2}\right)^2 \frac{n}{m}\frac{TR_{\max}(R_{\omega_{nm}})}{TR_{\max}(R_{\omega_{11}})}\\ \left|\frac{\varepsilon_{y,nm,\max}}{\varepsilon_{y,11,\max}}\right| &= \left(\frac{1+r_a^2}{n^2+r_a^2 m^2}\right)^2 \frac{m}{n}\frac{TR_{\max}(R_{\omega_{nm}})}{TR_{\max}(R_{\omega_{11}})}\end{aligned} \tag{10.130}$$

The maximum magnitude of the ratio is

$$\left|\frac{\varepsilon_{x,nm,\max}}{\varepsilon_{x,11,\max}}\right|_{R_{\omega_{11}}\to 0} \approx \left(\frac{1+r_a^2}{n^2+r_a^2 m^2}\right)\frac{n}{m} \tag{10.131}$$

The functions $\left|\frac{\varepsilon_{x,nm,\max}}{\varepsilon_{x,11,\max}}\right|$ for $\{n,m\} = \{3,1\}$ and $\{3,3\}$, respectively, are shown in Figure 10.23. It is noted that the ratio is significantly higher than that of a beam model. The frequency ratio $R_{\omega_{11}}$ is approximately 0.2 for the JEDEC JESD22-B111 test standard, which suggests a contribution of more than 40% from the next higher mode of deflection. It is however worth noting that the frequency ratios R_{ω_1} for most commercial portable electronics are higher than 0.4 rendering a much less contribution from the higher modal strain.

For $R_{\omega_1} > 0.5$, the maximum board strain can be approximated by the maximum fundamental modal strain; that is

$$\frac{2\omega_{11}^2 L_x^2}{A_o h}\left|\varepsilon_{x,\max}\right| \approx \frac{2\omega_{11}^2 L_x^2}{A_o h}\left|\varepsilon_{x,11,\max}\right| = 16\cdot TR_{\max}(R_{\omega_{11}}) \tag{10.132}$$

The goodness of Eqn (10.132) for estimating the maximum board deflection is illustrated in Figure 10.24 (the figure to the right shows the expanded view for $0 \le R_{\omega_{11}} < 1$).

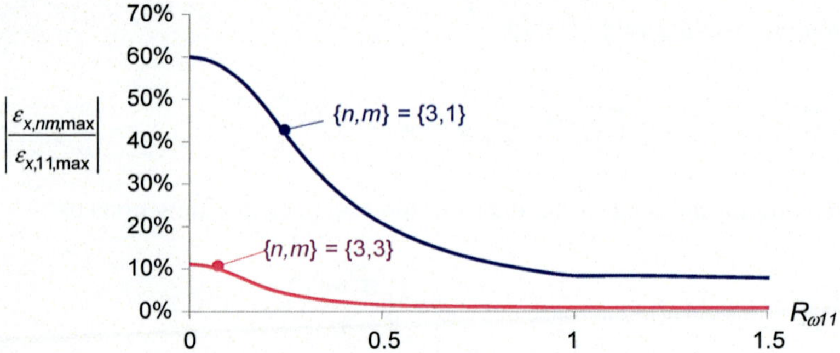

Figure 10.23 Ratios of higher mode to the fundamental mode $-\varepsilon_x$.

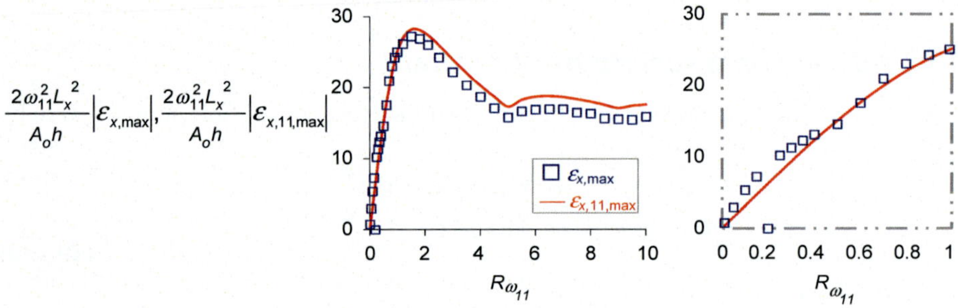

Figure 10.24 Comparison of $\varepsilon_{x,\max}$ and $\varepsilon_{x,11,\max}$.

10.4.4.6 Acceleration-time

Differentiating the modal deflection function twice with respect to time gives the modal *w*-acceleration function as

$$\frac{\ddot{w}_{nm}(x,y,\tau)}{A_o} = \begin{cases} -\dfrac{16}{\pi^2 nm}[TR(R_{\omega_{nm}},\tau) + \sin \pi\tau] \sin \beta_n x \cdot \sin \alpha_m y & \text{for } 0 \leq \tau \leq 1 \\ -\dfrac{16}{\pi^2 nm} TR(R_{\omega_{nm}},\tau) \sin \beta_n x \cdot \sin \alpha_m y & \text{for } \tau \geq 1 \end{cases} \tag{10.133}$$

The absolute acceleration of the board is simply $\ddot{z}(x,\tau) = \ddot{w}(x,\tau) + \ddot{Y}(\tau)$, which gives

$$\frac{\ddot{z}(x,\tau)}{A_o} = \begin{cases} \dfrac{\ddot{w}_{nm}(x,y,\tau)}{A_o} + \sin \pi\tau & \text{for } 0 \leq \tau \leq 1 \\ \dfrac{\ddot{w}_{nm}(x,y,\tau)}{A_o} & \text{for } \tau \geq 1 \end{cases} \tag{10.134}$$

Table 10.2 Parameters for the FE and the analytical models

Plate						Input pulse	
a (mm)	b (mm)	h (mm)	E (GPa)	ρ (g/cm^2)	ω_1 (rad/s)	A_o (g)	Ω (rad/s)
100	100	1	24	2	2073	1000	3142

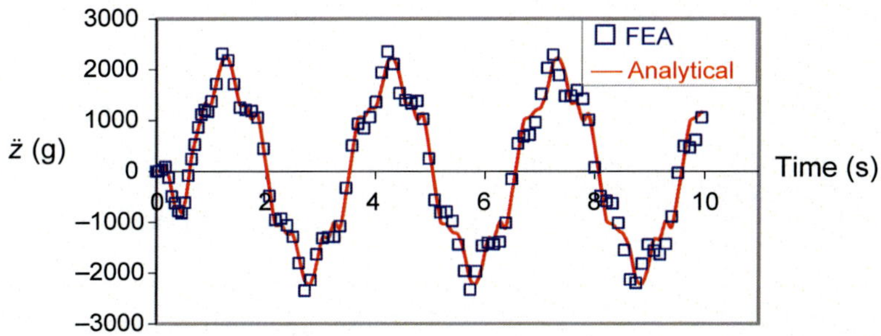

Figure 10.25 Absolute acceleration at the middle of the board – comparison between analytical and FEA.

Equation (10.134) is benchmarked against FEA. The parameters for the validation vehicle are given in Table 10.2. Figure 10.25 compares the absolute acceleration of the validation vehicle at its midpoint evaluated by FEA and using Eqn (10.134). The two results matched up almost perfectly.

10.5 Analysing the effects of imperfect half-sine acceleration shock

10.5.1 Distortion of half-sine acceleration pulse

The half-sine excitation specified in the JEDEC board level test standard JESD22-B111 is established experimentally through manipulating the fall height of the test structure and the stiffness of the shock generation pad. If the shock generation is non-dissipative, then the acceleration of the test structure is given by $\ddot{Y}(t) = V_o \Omega \sin \Omega t$, where V_o is the impacting velocity. To limit the acceleration within the half period of $t_o = \Omega/\pi$, the shock generation pad has to be made of materials of extremely high damping. In doing so, the acceleration-time profile of the fixture will not be a perfect half-sine, but a profile that has a distortion that skews towards the acceleration axis. Therefore, it is in practice impossible to introduce an ideal half-sine acceleration shock to the PCB assembly. The effects of a distorted acceleration shock on the resultant deflection and fibre strain of a PCB modelled as a beam simply supported at two ends will be investigated in this section.

A distorted acceleration shock may be expressed as a Fourier series of odd function

$$A(t) = \begin{cases} A_o \sin \Omega t + \sum_{i=2}^{\infty} b_i \sin(i\Omega t) & \text{for } 0 \le t \le t_o \\ 0 & \text{for } t \ge t_o \end{cases} \tag{10.135}$$

where

$$b_i = \frac{2}{t_o} \int_0^{t_o} f(x) \sin(i\Omega t) \mathrm{d}x \tag{10.135a}$$

Equation (10.135) consists of an ideal half-sine excitation function $A_p(t) = \begin{cases} A_o \sin \Omega t & \text{for } 0 \le t \le t_o \\ 0 & \text{for } t \ge t_o \end{cases}$ and some higher order harmonic functions $A_d(t) = \begin{cases} \sum_{i=2}^{\infty} b_i \sin(i\Omega t) & \text{for } 0 \le t \le t_o \\ 0 & \text{for } t \ge t_o \end{cases}$. These higher order harmonic functions are responsible for the distortion from the ideal half-sine excitation and will henceforth be referred to as deviatory functions.

The case of a distorted excitation that is expressible using a single deviatory function is used as an illustration:

$$A(t) = \begin{cases} A_o \sin \Omega t + b_2 \sin(2\Omega t) & \text{for } 0 \le t \le t_o \\ 0 & \text{for } t \ge t_o \end{cases} \tag{10.136}$$

Figure 10.26 shows the ideal excitation, $A_p(t) = \begin{cases} A_o \sin \Omega t & \text{for } 0 \le t \le t_o \\ 0 & \text{for } t \ge t_o \end{cases}$, the deviatory function, $A_d(t) = \begin{cases} b_2 \sin(2\Omega t) & \text{for } 0 \le t \le t_o \\ 0 & \text{for } t \ge t_o \end{cases}$, and the distorted accelerated shock, $A(t) = A_p(t) + A_d(t)$, which is skewed to the left. The coefficient b_2 has the ratio $b_2 = 0.5A_o$. The resultant amplitude of the distorted excitation is 30% higher

Figure 10.26 Ideal and distorted excitations for the case of $b_2 = 0.5A_o$.

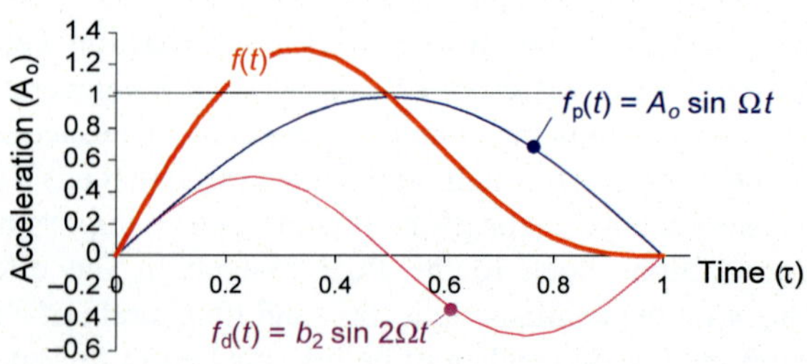

than that of the ideal excitation while the two have equal acceleration impulse $\int_0^{t_o} \ddot{Y}(t)\mathrm{d}t$. To facilitate evaluation of the responses of the PCB using solutions developed in the previous section for half-sine acceleration shock, the deviatory function is decomposed into two half-sine functions; that is

$$A_d(t) = \left\{\begin{array}{ll} b_2 \sin 2\Omega t & \text{for } 0 \leq t \leq \frac{t_o}{2} \\ 0 & \text{for } t \geq \frac{t_o}{2} \end{array}\right\} + \left\{\begin{array}{ll} 0 & \text{for } t \leq \frac{t_o}{2} \\ -b_2 \sin 2\Omega\left(t - \frac{t_o}{2}\right) & \text{for } \frac{t_o}{2} \leq t \leq t_o \\ 0 & \text{for } t \geq t_o \end{array}\right\} \tag{10.137}$$

the last is indeed a translated half-sine function. Substituting the above into Eqn (10.136) gives

$$A(t) = \left\{\begin{array}{ll} A_o \sin \Omega t & \text{for } 0 \leq t \leq t_o \\ 0 & \text{for } t \geq t_o \end{array}\right\} + \left\{\begin{array}{ll} b_2 \sin 2\Omega t & \text{for } 0 \leq t \leq \frac{t_o}{2} \\ 0 & \text{for } t \geq \frac{t_o}{2} \end{array}\right\}$$
$$- \left\{\begin{array}{ll} 0 & \text{for } t \leq \frac{t_o}{2} \\ -b_2 \sin 2\Omega\left(t - \frac{t_o}{2}\right) & \text{for } \frac{t_o}{2} \leq t \leq t_o \\ 0 & \text{for } t \geq t_o \end{array}\right\} \tag{10.138}$$

From Eqn (10.108), the modal strain-time responses of PCB, modelled as a beam, corresponding to the three half-sine excitations are, respectively,

$$\frac{2\omega_1^2 L^2}{A_o h}\varepsilon_n(x,\tau) = \frac{4\pi}{n^3} TRD(R_{\omega_n}, \zeta_n, \tau)\sin \beta_n x \tag{10.139}$$

$$\frac{2\omega_1^2 L^2}{b_2 h}\varepsilon_n(x,\tau) = \frac{4\pi}{n^3} TRD\left(\frac{R_{\omega_n}}{2}, \zeta_n, \tau_2\right)\sin \beta_n x \tag{10.140}$$

and

$$\frac{2\omega_1^2 L^2}{b_2 h}\varepsilon_n(x,\tau - 1/2) = \left\{\begin{array}{ll} 0 & \text{for } \tau \leq 1/2 \\ \frac{4\pi}{n^3} TRD\left(\frac{R_{\omega_n}}{2}, \zeta_n, \tau_3\right)\sin \beta_n x & \text{for } \tau \geq 1/2 \end{array}\right. \tag{10.141}$$

where $\tau_2 = t/(t_o/2) = 2\tau$ and $\tau_3 = (t - t_o/2)/(t_o/2) = 2\tau - 1$. The sum of these three responses gives the modal strain response of the PCB. The strain response of a PCB at its mid-length, $x = L/2$, to the excitation $b_2/A_o = 0.5$ is illustrated in Figure 10.27 for (a) $R_{\omega_1} = 0.2$, (b) $R_{\omega_1} = 1.5$ and (c) $R_{\omega_1} = 5$. In view of the dominance of the fundamental mode, only the response of the fundamental mode is presented; and damping is also ignored owing to the observed relatively small magnitude of damping observed in practice. For ease of discussions, we shall henceforth refer to the response to the ideal excitation $A_p(t)$ as the ideal response $\varepsilon_p(t)$, the response to the deviatory function $A_d(t)$ as deviatory response $\varepsilon_d(t)$; and that to the resultant distorted excitation $A(t)$ as the distorted response $\varepsilon(t)$. It is noted that:

- For the case of $R_{\omega_1} = 0.2$, the two deviatory responses, $\varepsilon_{d1}(\tau)$ and $\varepsilon_{d2}(\tau)$, corresponding to Eqns (10.140) and (10.141), respectively, are almost equal and opposite at any time τ. As a consequence, the distorted response is almost identical as the ideal response; that is, $\varepsilon(\tau) \approx \varepsilon_p(\tau)$.
- For the case of $R_{\omega_1} = 1.5$, the two deviatory responses are offset in time, leading to a 20% increase in magnitude of the resultant response; that is $\varepsilon_{\max} \approx 1.2\ \varepsilon_{p,\max}$.
- For the case of $R_{\omega_1} = 5$, the offset in time for the two deviatory responses increases and $\varepsilon_{\max} \approx 1.7\ \varepsilon_{p,\max}$.

The case of $R_{\omega_1} = 0.2$ represents the conditions of JESD22-B111 test standard. It has been highlighted in Section 10.4.2 that at small values of R_ω the function $TR_{\max}(R_\omega)$ is proportional to the magnitude of the acceleration impulse, independent of the actual function of the acceleration shock. The near identical response of $\varepsilon(\tau)$ and $\varepsilon_p(\tau)$ for the case of $R_{\omega_1} = 0.2$ is thus not surprising. Therefore, the test condition of JESD22-B111 test standard is indeed insensitive to the distortion of the acceleration shock profile. For a particular deign of test board, the JEDEC test is reproducible at different test site so long as the acceleration impulse is reproduced. However, Eqn (10.98), which is repeated below,

$$\frac{2\omega_1^2 L^2}{A_o h}|\varepsilon_{\max}| \approx 8\pi R_{\omega_1} \tag{10.98}$$

suggests that test boards of different resonant frequencies will experience different magnitude of fibre strain when subjected to identical acceleration shock. The frequency of a test board depends on the assembled IC components. It is hence impractical for a test standard to prescribe both the acceleration shock and the board fibre strain as the imposing test parameters. Among these two conditions, the board strain is obviously more relevant.

Deviatory functions that are made of even harmonics, that is $i = 2, 4, 6\ldots$ as in the above illustration, will result in a skewed excitation, and a reduced enclosed area for the same amplitude and duration as the ideal excitation. By contrast, deviatory functions that are made of odd harmonics will result in a symmetric excitation and either a reduced or an enlarged enclosed area. While such distortion is uncommon, it is

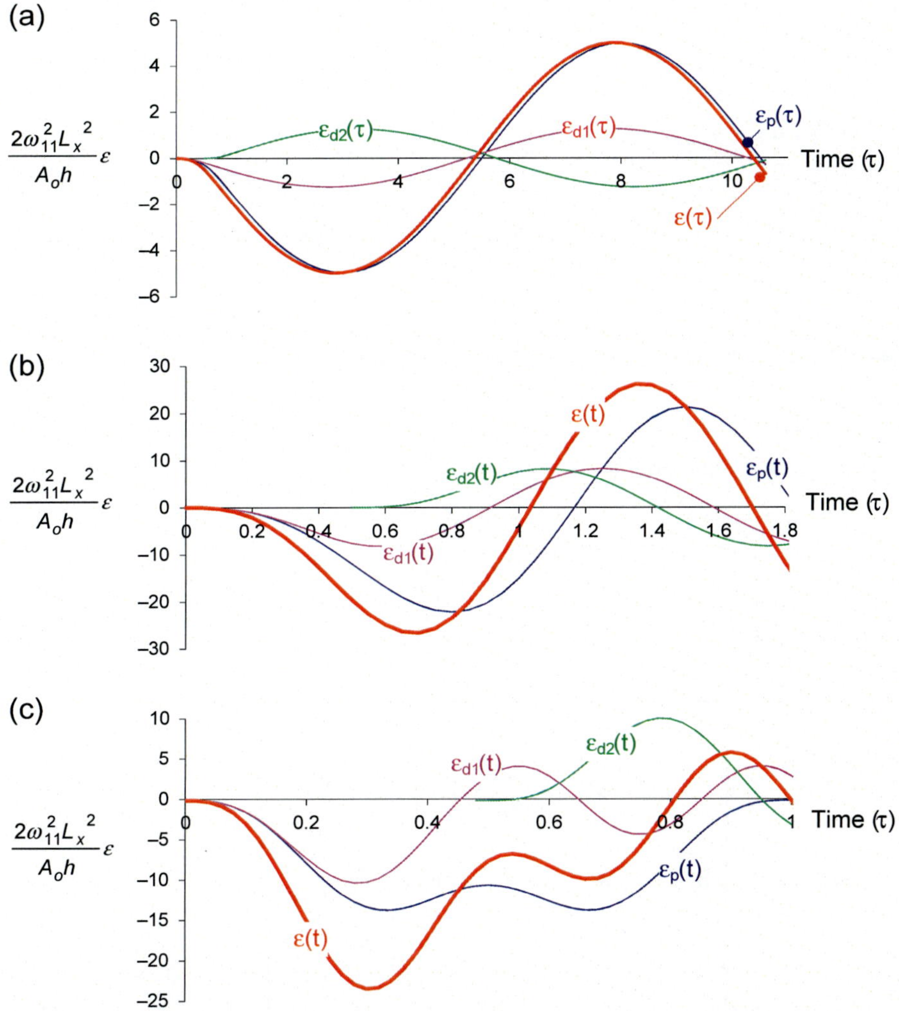

Figure 10.27 Responses to the distorted half-sine excitation of Eqn (10.138) for the case of $b_2/A_o = 0.5$, $n = 1$, $\zeta_n = 0$ and $x = L/2$ for (a) $R_{\omega_1} = 0.2$, (b) $R_{\omega_1} = 1.5$ and (c) $R_{\omega_1} = 5$.

nevertheless analysed for completeness. As an illustration, a similar analysis is performed for the excitation $f(t) = \begin{cases} A_o \sin \Omega t + b_3 \sin(3\Omega t) & \text{for } 0 \le t \le t_o \\ 0 & \text{for } t \ge t_o \end{cases}$, where $b_3/A_o = 0.5$ is shown in Figure 10.28. The amplitude of the distorted excitation is 8% higher than that of the ideal excitation while it can be shown that the fractional increase in acceleration impulse is given by $\frac{b_3/A_o}{3} = 17\%$. The case of $b_3/A_o = -0.5$ will lead to a reduction in the acceleration impulse of similar magnitude.

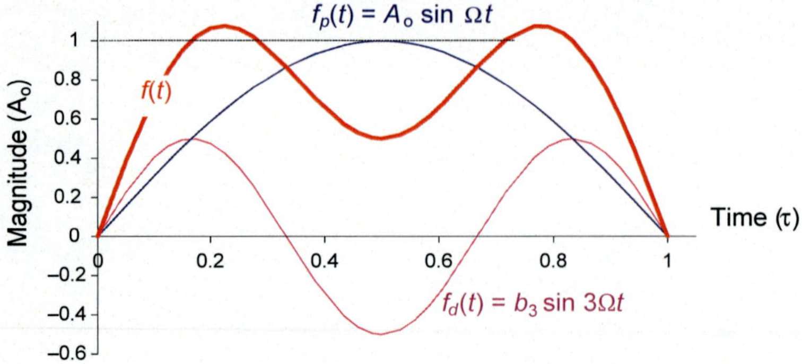

Figure 10.28 Ideal and distorted excitations for the case of $b_3 = 0.5A_o$.

The deviatory function is decomposed into

$$f_d(t) = \begin{Bmatrix} b_3 \sin 3\Omega t & \text{for } 0 \leq t \leq \dfrac{t_o}{3} \\ 0 & \text{for } t \geq \dfrac{t_o}{3} \end{Bmatrix} + \begin{Bmatrix} 0 & \text{for } t \leq \dfrac{t_o}{3} \\ -b_3 \sin 3\Omega\left(t - \dfrac{t_o}{3}\right) & \text{for } \dfrac{t_o}{3} \leq t \leq \dfrac{2t_o}{3} \\ 0 & \text{for } t \geq \dfrac{2t_o}{3} \end{Bmatrix} + \begin{Bmatrix} 0 & \text{for } t \leq \dfrac{2t_o}{3} \\ b_3 \sin 3\Omega\left(t - \dfrac{2t_o}{3}\right) & \text{for } \dfrac{2t_o}{3} \leq t \leq t_o \\ 0 & \text{for } t \geq t_o \end{Bmatrix} \quad (10.142)$$

The corresponding strain-responses to the deviatory functions are

$$\frac{2\omega_1^2 L^2}{b_3 h}\varepsilon_n(x,\tau) = \frac{4\pi}{n^3} TRD\left(\frac{R_{\omega_n}}{2}, \zeta_n, \tau_2\right)\sin\beta_n x \quad (10.143)$$

$$\frac{2\omega_1^2 L^2}{b_3 h}\varepsilon_n(x,\tau-1/2) = \begin{cases} 0 & \text{for } \tau \leq 1/3 \\ \dfrac{4\pi}{n^3} TRD\left(\dfrac{R_{\omega_n}}{2},\zeta_n,\tau_3\right)\sin\beta_n x & \text{for } \tau \geq 1/3 \end{cases} \tag{10.144}$$

$$\frac{2\omega_1^2 L^2}{b_3 h}\varepsilon_n(x,\tau-1/2) = \begin{cases} 0 & \text{for } \tau \leq 2/3 \\ \dfrac{4\pi}{n^3} TRD\left(\dfrac{R_{\omega_n}}{2},\zeta_n,\tau_4\right)\sin\beta_n x & \text{for } \tau \geq 2/3 \end{cases} \tag{10.145}$$

where $\tau_2 = t/(t_o/3) = 3\tau$, $\tau_3 = (t - t_o/3)/(t_o/3) = 3\tau - 1$, $\tau_4 = (t - 2t_o/3)/(t_o/3) = 3\tau - 2$. The responses for the cases of $b_3/A_o = 0.5$, $n = 1$, $\zeta_n = 0$ and $x = L/2$ are illustrated in Figure 10.29 for (a) $R_{\omega_1} = 0.2$, (b) $R_{\omega_1} = 1.5$ and (c) $R_{\omega_1} = 5$. It is noted that (a) $\varepsilon_{\max} \approx 1.7\ \varepsilon_{p,\max}$ for $R_{\omega_1} = 5.0$, which is similar to the case of $b_2/A_o = 0.5$; (b) $\varepsilon_{\max} \approx 0.89\ \varepsilon_{p,\max}$ for $R_{\omega_1} = 1.5$ – a 11% reduction in amplitude; and (c) $\varepsilon_{\max} \approx 1.16\ \varepsilon_{p,\max}$ for $R_{\omega_1} = 0.2$, which is similar to the increase in the acceleration impulse.

10.5.2 Acceleration spikes

A less common form of distortion is the contamination of a half-sine excitation by a high-frequency spike, which can happen in a poorly designed or maintained shock equipment. Figure 10.30 gives an illustration of a half-sine spike $A_s(t)$ of amplitude A_s and duration t_s that superimposes onto a half-sine excitation after a delay of t_d. The frequency of the spike is given by $\Omega_s = \pi/t_s$. Expressing t_s as a fraction of t_o; that is $t_s = t_o/s$, then $\Omega_s = s\Omega$. The acceleration excitation may be described mathematically as

$$\begin{aligned} A(t) &= \begin{Bmatrix} A_o \sin o\Omega t & \text{for } 0 \leq t \leq t_o \\ 0 & \text{for } t \geq t_o \end{Bmatrix} \\ &+ \begin{Bmatrix} 0 & \text{for } t \leq t_d \\ A_s \sin s\Omega(t - t_d) & \text{for } t_d \leq t \leq (t_d + t_s) \\ 0 & \text{for } t \geq (t_d + t_s) \end{Bmatrix} \end{aligned} \tag{10.146}$$

The response of a board may be sought through the use of superposition. The case of a board modelled as an un-damped spring-mass system is illustrated. The maxima response function of the spring-mass system to a perfect half-sine acceleration shock has been presented in Eqn (10.46). The frequency ratio of the elastic system to the acceleration-spike is given by $R_{\omega,s} = \omega/\Omega_s = R_\omega/s$. Substituting this into

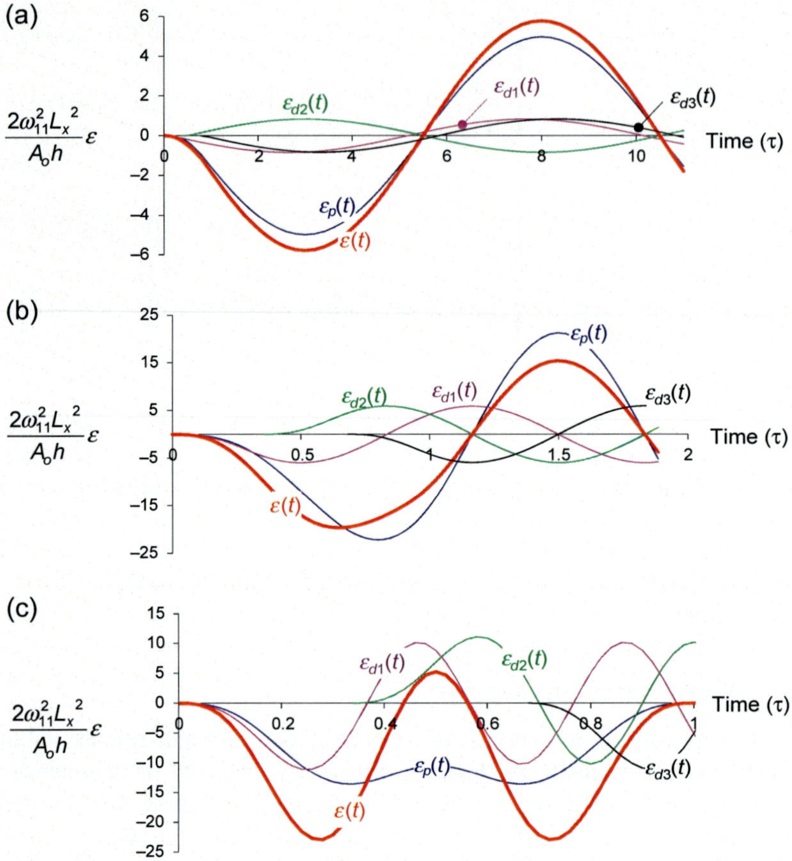

Figure 10.29 Responses to the distorted half-sine excitation of $b_3/A_o = 0.5$ for (a) $R_{\omega_1} = 0.2$, (b) $R_{\omega_1} = 1.5$ and (c) $R_{\omega_1} = 5$.

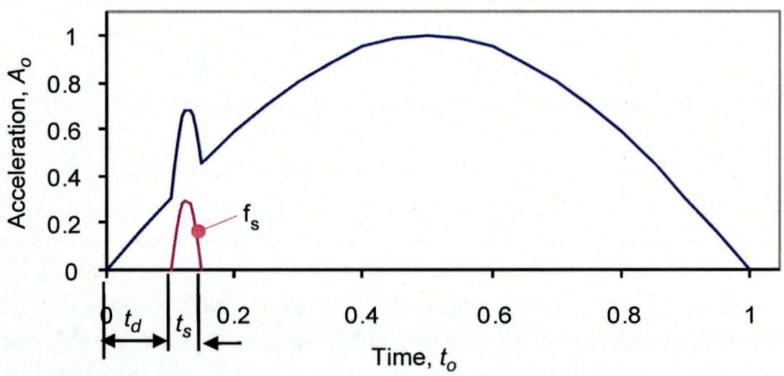

Figure 10.30 Half-sine excitation superimposed with a spike.

Eqn (10.46) gives the maxima response function of the spring-mass system to a half-sine acceleration-spike described by Eqn (10.146) as

$$TR_{\max,s}(R_{\omega,s}) = \begin{cases} \dfrac{2R_{\omega,s}}{1-R_{\omega,s}^2}\cos\left(\dfrac{\pi R_{\omega,s}}{2}\right) & \text{at } t_{m,s} = t_d + \left(\dfrac{1+R_{\omega,s}}{2R_{\omega,s}}\right)\dfrac{t_o}{s} \text{ for } R_{\omega,s} < 1 \\ \dfrac{-R_{\omega,s}}{1-R_{\omega,s}}\sin\left(\dfrac{2p\pi}{1+R_{\omega,s}}\right) & \text{at } t_{m,s} = t_d + \left(\dfrac{2p}{1+R_{\omega,s}}\right)\dfrac{t_o}{s} \text{ for } (4p-3) \le R_{\omega,s} \le (4p+1),\ p = 1,2,3... \end{cases} \tag{10.147}$$

The case of $A_s = A_o$, $t_d = 0$, $t_s = t_o/20$ (or $R_{\omega,s} = R_\omega/20$) is illustrated for $0 \le R_\omega \le 1$. The maximum responses and the corresponding occurring times are shown graphically in Figure 10.31. It is observed that:

- The response due to the spike is merely 5% the responses of the main excitation. The magnitude of 5% can be rationalised using Eqn (10.51), which suggests $TR_{\max} \approx 2R_\omega$ for $R_\omega \ll 1$. The fact that $R_{\omega,s} = R_\omega/20$ then leads naturally to the conclusion that $TR_{\max,s} = 5\%\ TR_{\max,p}$.
- The occurrence time for the maxima of the spike, $t_{m,s}$ leads the excitation, $t_{m,p}$ by a constant margin, which may be evaluated with the aid of Eqn (10.147), giving the difference between the two occurrence time to be

$$\Delta t_m = \frac{1+R_\omega}{2R_\omega}t_o - \frac{t_o}{s}\left(\frac{1+R_\omega/s}{2R_\omega/s}\right) = \frac{t_o}{2}\left(1-\frac{1}{s}\right) \tag{10.148}$$

Substituting $s = 20$ into the above equation gives $\Delta t_m = 0.475t_o$.

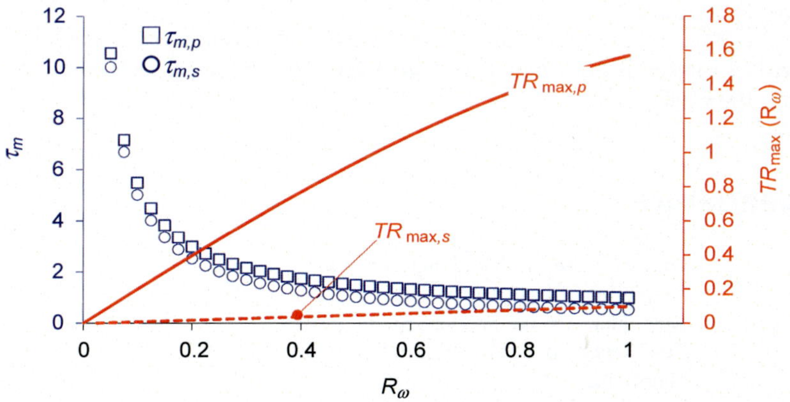

Figure 10.31 Maximum responses and the corresponding times of excitation and spike for $A_s = A_o$, $t_d = 0$, $t_s = t_o/20$.

References

Hirata, I. (2001). Drop-simulation of electronic boards mounted with CSPs. In *1st International conference for electronic packaging* (pp. 422–426).

Kim, J.-Y. (2001). Characterization of wafer level package for mobile phone application. In *51st electronic components and technology conference* (pp. 1398–1401).

Mindlin, R. (1946). Dynamics of package cushioning. *Bell System Technical Journal, 24*, 424–426.

Qiang, Y. (2002). Dynamic behaviour of electronics package and impact reliability of BGA solder joints. In *Intersociety conf. on thermal phenomena* (pp. 953–960).

Singiresu, S. (1995). *Mechanical vibration* (3rd ed.). Addition-Wesley.

Sogo, T. H. (2001). Estimation of fall impact strength for BGA solder joint. In *1st International conference for electronic packaging* (pp. 369–373).

Tee, T. (2004). Advanced experimental and simulation techniques for analysis of dynamic responses during drop impact. In *54th Electronic component technology conference* (pp. 1088–1094).

Tsai, T.-Y., Yeh, C.-L., Lai, Y.-S., & Chen, R.-S. (2007). Response spectra analysis for undamped structural systems subjected to half-sine impact acceleration pulses. *Microelectronics Reliability, 47*, 1239–1245.

Wang, J. (2002). Modeling solder joint reliability of BGA packages subject to drop impact loading using submodeling. In *Abaqus user's conference* (pp. 1–11).

Wang, J. S. (2008). Fracture mechanics study of fatigue crack growth in solder joints under drop impact. In *58th Electron. compon. technol. conf.* (pp. 1601–1605).

Wong, E. H. (2002). Drop impact test – mechanics and physics of failure. In *4th Electronic packaging technology conference* (pp. 327–333).

Wong, E. (2005). Dynamics of board level drop impact. *Journal of Electronic Packaging, 127*, 200–207.

Wu, J., Song, G., Yeh, C. P., & Wyatt, K. (1998). *Drop/impact simulation and test validation of telecommunication products* (pp. 330–336).

Yeh, C.-L., & Lai, Y.-S. (2006). Support excitation scheme for transient analysis of JEDEC board-level drop test. *Microelectronics Reliability, 46*, 626–636.

Zhu, L. (2001). Submodeling technique for BGA reliability analysis of CSP packaging subjected to an impact loading. In *Int. electron. packag. tech. conf., (pp. IPACK2001-15873). Kauai, HI.*

Zhu, L. (2003). Modeling technique for reliability assessment of portable electronic product subject to drop impact. In *53rd electronic component technology conference* (pp. 100–104).

Nomenclature

a, b, L	Length, width of plate; length of beam
A_o	Amplitude of base acceleration
A_p	Area under the half-sine excitation pulse
b_1, b_i	Coefficients of ideal half-sine excitation; coefficient of deviatory function of *i* harmonic
d_s	Diameter of solder joint assuming circular in cross-section
D	Flexural stiffness

$\overline{D}_e$	Flexural compliance of the test board assembly
E, I, ρ, ν	Modulus, moment of inertia, density, Poisson ratio
h	Thickness of test board
h_s	Height of solder joint assuming cylindrical in shape
k_a, k_s, k_Φ	Equivalent axial, shear and flexural stiffness of the solder joints, respectively
L	Span of test board between supports
M	Bending moment
p	Pitch between neighbouring solder joints
q_n	The *n*th modal participating factor (or generalised displacement)
R_{ω_n}	The ratio of the *n*th modal frequency of the vibrating body to the excitation frequency of the support, $R_{\omega_n} = \omega_n/\Omega$
t_o	Half duration of base acceleration
$w, w_n, w_{n,\max}$	Deflection, modal deflection, maximum modal deflection
$y, y_n, y_{n,\max}$	Displacement, modal displacement, maximum modal displacement of board
$Y(t)$	Absolute displacement of support
W_n	The *n*th eigenfunction
β_n	The *n*th harmonic eigenfunction
$\varepsilon, \varepsilon_n, \varepsilon_{n,\max}$	The fibre strain, modal fibre strain, maximum fibre strain
λ	Characteristic constant of test board modelled as beam/plate given by $\lambda = \omega_n/\beta_n^2$
λ_x	In-plane compliance of the test board assembly
τ	Normalised time, $\tau = t/t_o$
ω_n	The *n*th angular modal frequency of vibrating body
ω_{dn}	Damped *n*th modal angular resonant frequency of test board
Ω	Angular frequency of base acceleration
ζn	The *n*th modal damping ratio

Mathematical symbols

$$TR(R_\omega,\tau) = \begin{cases} \dfrac{-R_\omega}{1-R_\omega^2}(\sin \pi R_\omega \tau - R_\omega \sin \pi\tau) & \text{for } 0 \le \tau \le 1 \\[2ex] \dfrac{-2R_\omega}{1-R_\omega^2}\sin\left(\pi R_\omega \tau - \dfrac{\pi R_\omega}{2}\right)\cos\left(\dfrac{\pi R_\omega}{2}\right) & \text{for } \tau \ge 1 \end{cases}$$

$$TR_{\max}(R_\omega) = \begin{cases} \dfrac{2R_\omega}{1-R_\omega^2}\cos\left(\dfrac{\pi R_\omega}{2}\right) & \text{for } R_\omega < 1 \\[2ex] \dfrac{-R_\omega}{1-R_\omega}\sin\left(\dfrac{2p\pi}{1+\mathrm{R}_\omega}\right) & \text{for } (4p-3) \le R_{w,p} \le (4p+1),\ p = 1,2,3... \end{cases}$$

$$TRD(R_\omega,\zeta,\tau) = \frac{-R_\omega e^{-\zeta\pi R_\omega \tau}}{\left(1-R_\omega^2\right)^2 + (2\zeta R_\omega)^2}\left\{\begin{array}{ll} TRD\_1(R_\omega,\zeta,\tau) & \text{for } 0 \le \tau \le 1 \\ TRD\_2(R_\omega,\zeta,\tau) & \text{for } \tau \ge 1 \end{array}\right\}$$

$$TRD\_1(R_\omega, \zeta_n, \tau) = 2\zeta R_\omega^2 \cos \mu\tau + \frac{1 - R_\omega^2 + 2\zeta^2 R_\omega^2}{\sqrt{1-\zeta^2}} \sin \mu\tau - R_\omega e^{\zeta\pi R_\omega \tau}\left[2\zeta R_\omega \cos \pi\tau + \left(1 - R_\omega^2\right) \sin \pi\tau\right]$$

$$TRD\_2(R_\omega, \zeta_n, \tau) = 2\zeta R_\omega^2 \left[\cos \mu\tau + e^{\zeta\pi R_\omega} \cos \mu(\tau - 1)\right] + \frac{1 - R_\omega^2 + 2\zeta^2 R_\omega^2}{\sqrt{1-\zeta^2}} \left[\sin \mu\tau + e^{\zeta\pi R_\omega} \sin \mu(\tau - 1)\right]$$

$$\omega_{dn} = \sqrt{1-\zeta^2}\,\omega_n, \quad \mu_n = \sqrt{1-\zeta^2}\,\pi R_{\omega_n}$$

Stresses in solder joints due to the bending deformation of printed circuit boards in microelectronics assemblies

11

11.1 Introduction

Flexural deformation of a printed circuit board (PCB) assembly may be caused by inertia, such as drop impact of portable electronic devices (PEDs), and may be simulated by subsystem test methods such as acceleration shock and high-speed cyclic-bending testing, etc. Flexural deformation of a PCB assembly may also be caused by mechanically applied bending, such as the bending of a PED in the rear pocket of individuals who are sitting, the bending of a PCB assembly during key pressing or during insertion into connectors or during handling. Flexural deformation of a PCB assembly leads to differential flexural deformation of the IC component and the PCB that must be accommodated by solder joints resulting in their eventual fatigue failure. The fatigue resistance equations of solder joints have been established in Chapter 8. The chapter analytically derives the driving forces for the fatigue failure of solder joints, which is essential for their robust design against bending of PCB assembly.

Compared to the case of thermally induced stress, there are very few published studies on the theoretical analysis of trilayer structure subjected to mechanically induced bending. Earlier research (Hovgaard, 1934; Troelsch, 1934) focused on lap shearing of a bonded joint assuming that only shearing stresses are present. The significance of the peeling stress was first highlighted by Goland and Reissner (Goland, 1944) in their classical analysis of a lap-shear specimen. Two extreme cases were analysed: a non-compliant bonding layer and a highly compliant bonding layer. In the former, the members are treated as two-dimensional (2-D) elastic bodies and the stress field analysed using stress functions. The highest magnitude of stress is concentrated in a small region near the edge, and the magnitude of the peeling stress is significantly higher than that of the shearing stress. In the latter, the bond layer is modelled as a distribution of shear and compression springs, and the problems are analysed as beams on elastic foundation. The stresses spread out more along the length of the bonding and the peeling and shearing stresses are of similar magnitudes.

The first stress analysis in the interconnections of a PCB assembly in bending is attributed to Suhir (1988), who analysed the stress in the leads of a leaded IC component mounted to a PCB that is subjected to symmetrically applied moments to the PCB. The leads are modelled as a distribution of axial springs and the problem is treated as

Robust Design of Microelectronics Assemblies Against Mechanical Shock, Temperature and Moisture
http://dx.doi.org/10.1016/B978-1-84569-528-6.00011-3

beams on elastic foundation. However, the assumption of the lead as being capable of undergoing only stretching and compression is not satisfactory as the leads, being a thin element with curvilinear geometry, must necessarily also undergo flexural deformation. Following the derivation procedures of Suhir (1988), Wong (2007) has incorporated flexural compliance to the discrete element and extended the solution to asymmetric bending. With much reduced height than a lead, a solder joint in a ball grid array packaging experiences significant shearing. Treating the bonding layer as a relatively compliant layer with negligible longitudinal stiffness and the outer layers as a combination of beams and 2-D elastic bodies, Wong and Wong (2008) presented the stresses in the bonding layer for two cases: a continuous bonding layer as in underfilled solder joints and a discrete array of solder joints.

In practice, mechanical bending of the PCB assembly in a PED can occur while it is in an operating condition, when the PCB assembly also experiences heating. The combined loadings of mismatched thermal expansion and mechanical bending are analysed. The derivations of analytical equations, the validation of results and use of derived equations for robust design analysis are presented in this chapter.

11.2 The fundamentals

11.2.1 The simple theory of composite beams

The analysis of multilayer beams is not new, and the theory of composite beam subjected to simple bending is well established. When a multilayer structure undergoes bending that is applied across its entire cross-section, the theory of composite beam assumes that its cross-section remains plane and perpendicular to the longitudinal axis of the structure as shown in Figure 11.1. In this case, elements in the beam experience no shearing but only stretching, whose magnitude is defined by its geometrical distance from the neutral axis, z, independent of the structural composition of the beam; that is,

$$\varepsilon_x = -\frac{z}{\rho} \tag{11.1}$$

Figure 11.1 Composite beam: plane remains plane and perpendicular to the longitudinal axis.

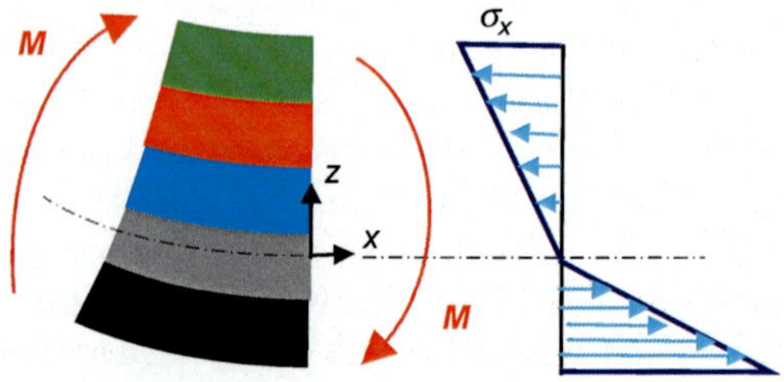

where ρ is the radius of curvature. For the case of plane stress, the accompanying bending stress is given by

$$\sigma_x = E_i \varepsilon_x = -\frac{E_i z}{\rho}; \quad i = 1, 2, ...n \tag{11.2}$$

where n is the number of materials constituting the composite beam. The position of the neutral axis is defined by

$$\sum_{i=1}^{n} \int_{\text{Area}i} \sigma_x \text{d}A_i = \sum_{i=1}^{n} E_i \int_{\text{Area}i} z\text{d}A_i = 0 \tag{11.3}$$

The moment–curvature relation is given by

$$M = -\sum_{i=1}^{n} \int_{\text{Area}} z(\sigma_x \text{d}A) = \frac{1}{\rho} \sum_{i=1}^{n} E_i \int_{\text{Area}} z^2 \text{d}A_i = \frac{1}{\rho} \sum E_i I_{iG} \tag{11.4}$$

where $I_{i,G}$ is the second moment of area of member #i about the neutral axis of the composite beam and $\sum E_i I_{iG}$ is the flexural rigidity of the composite beam. Thus, the flexural rigidity of a composite beam is given by the sum of the flexural rigidity of the individual member, with respect to the neutral axis of the composite beam. Combining Eqns (11.2) and (11.4) gives

$$\frac{M}{\sum E_i I_{iG}} = \frac{1}{\rho} = -\frac{\varepsilon_x}{z} = -\frac{\sigma_{xi}}{E_i z} \tag{11.5}$$

Equation (11.5) gives the relation between the applied moment (across the entire cross-section of a composite beam) and the fibre stress at distance z from the neutral axis of the composite beam.

11.2.2 Mechanical blending of a bilayer (material) structure

The above simple theory of composite beams does not apply when a bending moment is not applied across the entire cross-section of a beam. This is the case for the bending of a PCB assembly when the bending moment is applied only across the cross-section of the PCB but not the IC component nor the solder joints. Planes do not remain planes, and the members experience both shearing and stretching; on top of these, the discrete solder joints also experience bending (Wong & Wong, 2008). The bending of a two-layer structure is explained in this section as a precursor to the following analysis of a trilayer structure. See Chapter 2.1 for more fundamental analysis.

Referring to Figure 11.2, member #2 of the structure is subjected to an applied +ve moment of M_2 at the free edge ($x = L$); the bilayer structure is assumed to acquire a positive curvature resulting in the development of an interfacial normal stress, f_a,

Figure 11.2 Pure bending of a bilayer beam: deformed state.

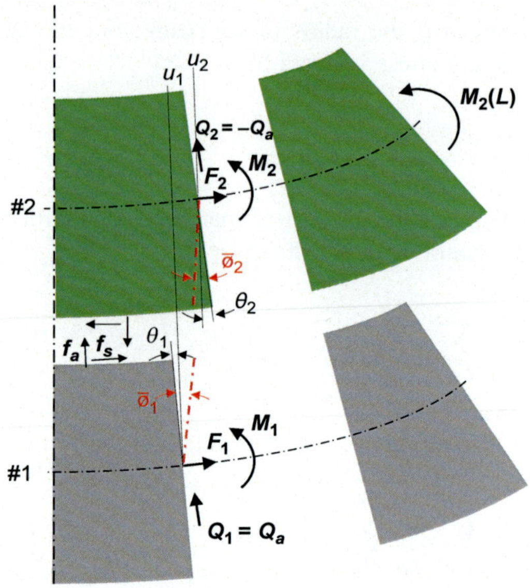

so as to enforce z-compatibility. The neutral axis of member #2 is also assumed to acquire a positive x-direction displacement relative to that of member #1, resulting in the development of interfacial shear stress f_s so as to enforce x-compatibility. The cross-sections of the members acquire moments, M, shear force, Q, and longitudinal force, F. Due to bending, the entire cross-section undergoes a rotation, θ, about the y-axis; and due to shearing, half of the section undergoes a rotation, $\overline{\phi}$, about the y-axis.

The force−displacement relations

$$\lambda_z f_a = w_2 - w_1 \tag{11.6}$$

$$\kappa_s \frac{\mathrm{d}f_s}{\mathrm{d}x} = \alpha_{21}\Delta T - (\lambda_{x1} + \lambda_{x2})F + \frac{1}{2}\left(h_1\frac{\mathrm{d}\theta_1}{\mathrm{d}x} + h_2\frac{\mathrm{d}\theta_2}{\mathrm{d}x}\right) \tag{11.7}$$

$$\frac{\mathrm{d}^3 w_{bi}}{\mathrm{d}x^3} = \frac{\mathrm{d}^2\theta_i}{\mathrm{d}x^2} = \frac{1}{D_i}\left(\frac{f_s h_i}{2} - Q_{ai}\right) \tag{11.8}$$

and the differential equations of interfacial stresses

$$\frac{\mathrm{d}^2 f_s}{\mathrm{d}x^2} - \beta^2 f_s = \frac{\overline{D}_{21}}{\kappa_s} Q_a \tag{11.9}$$

$$\frac{\mathrm{d}^4 f_a}{\mathrm{d}x^4} - \frac{\kappa_\phi}{\lambda_z}\frac{\mathrm{d}^2 f_a}{\mathrm{d}x^2} + \frac{\overline{D}_e}{\lambda_z} f_a = \frac{\overline{D}_{21}}{\lambda_z}\frac{\mathrm{d}f_s}{\mathrm{d}x} - \frac{\kappa_{21}}{\lambda_z}\frac{\mathrm{d}^3 f_s}{\mathrm{d}x^3} \tag{11.10}$$

where, $\beta^2 = \lambda_x/\kappa_s$, $\lambda_x = \lambda_{x1} + \lambda_{x2} + \lambda_{x\theta} \approx \sum_{i=1}^{2} \frac{4}{E_i h_i}$, $\lambda_{xi} \approx \frac{1}{E_i h_i}$, $\kappa_s = \kappa_{s1} + \kappa_{s2}$, $\kappa_{si} \approx \frac{h_i}{8G_i}$, $\lambda_z = \lambda_{z1} + \lambda_{z2}$, $\lambda_{zi} \approx \frac{3h_i}{8E_i}$, $\overline{D}_{21} = \frac{1}{2}\left(\frac{h_2}{D_2} - \frac{h_1}{D_1}\right)$, $\overline{D}_e = \frac{1}{D_2} + \frac{1}{D_1}$, $\kappa_{21} = \frac{1}{8}\left(\frac{1}{G_2} - \frac{1}{G_1}\right)$ and $\kappa_\phi = \kappa_{\phi 1} + \kappa_{\phi 2}$, $\kappa_{\phi i} = \frac{3}{4G_i h_i}$ have been derived in Section 2.3.2 and are reproduced above.

11.2.2.1 *Interfacial shear stress, $f_s(x)$*

We shall, for simplicity, assume for the moment that the term containing the sectional force, Q_a, can be ignored. The differential equation, Eqn (11.9), is reduced to

$$\frac{d^2 f_s}{dx^2} - \beta^2 f_s = 0 \tag{11.11}$$

The symmetry boundary condition gives $df_s(0)/dx = 0$, which is satisfied by

$$f_s(x) = A_s \frac{\cosh \beta x}{\sinh \beta L} \tag{11.12}$$

Substituting the boundary conditions: $F(L) = 0$, $d\theta_1(L)/dx = 0$ and $d\theta_2(L)/dx = 1/\rho_2(L)$ into Eqn (11.7) gives

$$\kappa_s \frac{df_s(L)}{dx} = \alpha_{21}\Delta T + \frac{h_2}{2\rho_2(L)} = \alpha_{21}\Delta T - \varepsilon_2(L) \tag{11.13}$$

Differentiating Eqn (11.12) wrt x gives

$$\frac{df_s(x)}{dx} = A_s\beta \frac{\sinh \beta x}{\sinh \beta L} \tag{11.14}$$

Put $x = L$ and equate with Eqn (11.13) to give

$$A_s = \frac{\alpha_{21}\Delta T - \varepsilon_2(L)}{\beta \kappa_s} \tag{11.15}$$

11.2.2.2 *Interfacial normal stress, $f_a(x)$*

For simplicity's sake, we shall ignore the shear compliances κ_ϕ and κ_{21}. Equation (11.10) is reduced to

$$\frac{d^4 f_a}{dx^4} + \frac{\overline{D}_e}{\lambda_z} f_a = \frac{\overline{D}_{21}}{\lambda_z} \frac{df_s}{dx} \tag{11.16}$$

Differentiating Eqn (11.9) wrt x and rearranging gives

$$\frac{df_s}{dx} = \frac{1}{\beta^2}\frac{d^3 f_s}{dx^3} + \frac{\overline{D}_{21}}{\beta^2 \kappa_s} f_a \tag{11.17}$$

Differentiating Eqn (11.12) thrice wrt x and substituting into Eqn (11.17) yields

$$\frac{df_s}{dx} = A_s \beta \frac{\sinh \beta x}{\sinh \beta L} + \frac{\overline{D}_{21}}{\lambda_x} f_a \tag{11.18}$$

Now, substituting Eqn (11.18) into Eqn (11.16), we have

$$\frac{d^4 f_a}{dx^4} + 4\alpha_z^4 f_a = \frac{\overline{D}_{21}\beta A_s}{\lambda_z}\frac{\sinh \beta x}{\sinh \beta L} \tag{11.19}$$

where $4\alpha_z^4 = \frac{\overline{D}_e - \overline{D}_{21}^2/\lambda_x}{\lambda_z}$.

For reasons of simplicity, ignore shear deformation such that $w = w_b$; the four boundary conditions for $f_a(x)$ are

- the equilibrium of z-directional force: $\int_0^L f_a(x)dx = 0$;
- the symmetry of f_a: $\frac{df_a(0)}{dx} = 0$;
- the curvatures at the free edge: $\frac{d^2 w_1(L)}{dx^2} = 0$ and $\frac{d^2 w_2(L)}{dx^2} = \frac{1}{\rho_2(L)}$; substituting with the differential of Eqn (11.6) gives $\lambda_z \frac{d^2 f_a(L)}{dx^2} = \frac{1}{\rho_2(L)}$;
- from Eqn (11.8) and noting that $Q_{ai}(L) = 0$: $\frac{d^3 w_1(L)}{dx^3} = \frac{1}{D_1}\frac{f_s h_1}{2}$; $\frac{d^3 w_2(L)}{dx^3} = \frac{1}{D_2}\frac{f_s h_2}{2}$; substituting w_i''' with the differentiation of Eqn (11.6) gives $\lambda_z \frac{d^3 f_a(L)}{dx^3} = \overline{D}_{21} f_s(L)$; and substituting $f_s(L)$ with Eqn (11.12) gives $\lambda_z \frac{d^3 f_a(L)}{dx^3} \approx \overline{D}_{21} A_s$.

Solving Eqn (11.19) and together with the four boundary conditions gives

$$\begin{aligned} f_a(x) = {} & C_b(\chi_1 \xi_{1x} + \chi_2 \xi_{2x}) + C_s\Big\{e^{\alpha_z(x-L)}[C_1 \sin \alpha_z(x-L) \\ & + C_2 \cos \alpha_z(x-L)] + e^{\beta(x-L)}\Big\} \end{aligned} \tag{11.20}$$

where C_b is associated with mechanical bending and is given by

$$C_b = \frac{1}{\lambda_z \alpha_z^2 \rho_2(L)} \approx \frac{2}{\sqrt{\lambda_z \overline{D}_e}\rho_2(L)} = -\frac{4\varepsilon_2(L)}{\sqrt{\lambda_z \overline{D}_e} h_2} \tag{11.20a}$$

and C_s is associated with both thermal expansion and mechanical bending and is given by

$$C_s = \frac{\overline{D}_{21}[\alpha_{21}\Delta T - \varepsilon_2(L)]}{\lambda_z \kappa_s (4\alpha_z^4 + \beta^4)} \tag{11.20b}$$

and

$$\chi_i = \frac{\xi_{3L} \mp \xi_{4L}}{\xi_{5L}}; i = 1, 2$$

$$\begin{aligned} \xi_{1x} &= \cos(\alpha_z x)\cosh(\alpha_z x), \xi_{2x} = \sin(\alpha_z x)\sinh(\alpha_z x), \xi_{3L} \\ &= \cos(\alpha_z L)\sinh(\alpha_z L) \\ \xi_{4L} &= \sin(\alpha_z L)\cosh(\alpha_z L), \xi_{5L} = \sin(2\alpha_z L) + \sinh(2\alpha_z L) \\ C_1 &= -\frac{\beta^2}{2\alpha_z^2}, C_2 = C_1 - \frac{2\alpha_z}{\beta} \end{aligned} \tag{11.20c}$$

Ignoring the thermal strain and putting $\alpha_z = \beta$ and $h_1 = h_2 = h$ gives $\frac{C_s}{C_b} = \frac{3(E_2 - E_1)}{20(E_2 + E_1)}$. In practice, E_1 and E_2 are of the same order of magnitude; therefore, $C_s \ll C_b$ and Eqn (11.20) is reduced to

$$f_a(x) = C_b(\chi_1 \xi_{1x} + \chi_2 \xi_{2x}) \tag{11.21}$$

Asymmetric bending

Equation (11.21) is strictly valid for symmetric bending. Figure 11.3 shows the asymmetric bending of a PCB assembly modelled as a bilayer assembly. The assembly experiences a positive moment M_L and M_R and accompanied shear forces Q_L and Q_R on the left and the right edges, respectively, of the PCB. The corresponding curvatures are ρ_L and ρ_R, respectively. The interface between the two layers is assumed to experience only normal stress, f_a, but no shear stress.

The equilibrium of member #1 gives

$$\int_{-L}^{L} f_a(x)\,dx = 0 \text{ and } \int_{-L}^{L} f_a(x)x\,dx = 0 \tag{11.22}$$

and the equilibrium of member #2 gives

$$Q_R - Q_L - \int_{-L}^{L} f_a(x)\,dx = 0 \text{ and } M_L - M_R - (Q_R + Q_L)L + \int_{-L}^{L} f_a(x)x\,dx = 0 \tag{11.23}$$

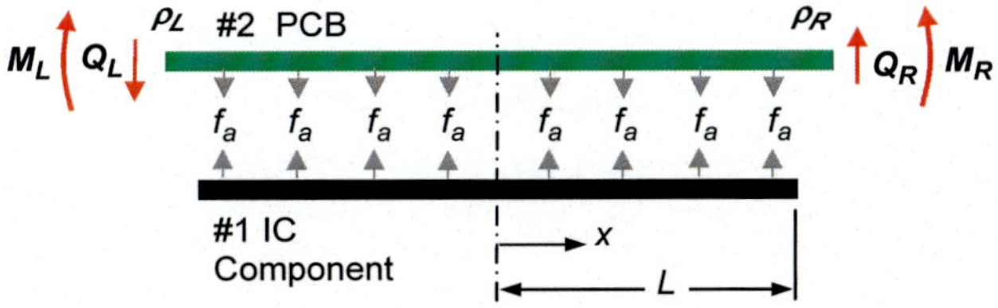

Figure 11.3 Asymmetric bending of PCB assembly assuming no interfacial shearing stress.

Substituting Eqn (11.22) into Eqn (11.23) gives

$$Q_L = Q_R = \frac{M_L - M_R}{2L} \tag{11.24}$$

Ignoring the shearing stress, Eqn (11.16) is reduced to

$$\frac{d^4 f_a}{dx^4} + \frac{\overline{D}_e}{\lambda_z} f_a = 0 \tag{11.25}$$

The general solution is

$$f_a(x) = B_1\xi_{1x} + B_2\xi_{2x} + B_3\xi_{3x} + B_4\xi_{4x} \tag{11.26}$$

where B_1 to B_4 are constants and

$$\begin{aligned} \xi_{1x} &= \cos(\alpha_z x)\cosh(\alpha_z x), \ \xi_{2x} = \sin(\alpha_z x)\sinh(\alpha_z x), \\ \xi_{3x} &= \cos(\alpha_z x)\sinh(\alpha_z x), \ \xi_{4x} = \sin(\alpha_z x)\cosh(\alpha_z x) \end{aligned} \tag{11.26a}$$

Applying the asymmetric boundary conditions:

- Moment, $M_i = D_i \frac{d^2 w_i}{dx^2}$, and noting that

$$M_2(-L) = M_L,\ M_1(-L) = 0 \Rightarrow \frac{M_L}{D_2} = \frac{d^2[w_2(-L) - w_1(-L)]}{dx^2} = \frac{\lambda_z d^2 f_a(-L)}{dx^2}$$

$$M_2(L) = M_R,\ M_1(L) = 0 \Rightarrow \frac{M_R}{D_2} = \frac{d^2[w_2(L) - w_1(L)]}{dx^2} = \frac{\lambda_z d^2 f_a(L)}{dx^2}$$

- Shear force, $Q_i = -D_i \frac{d^3 w_i}{dx^3}$, and noting that

$$Q_2(-L) = Q_L, Q_1(-L) = 0 \Rightarrow -\frac{Q_L}{D_2} = \frac{d^3[w_2(-L) - w_1(-L)]}{dx^3} = \frac{\lambda_z d^3 f_a(-L)}{dx^3}$$

$$Q_2(L) = Q_R, Q_1(L) = 0 \Rightarrow -\frac{Q_R}{D_2} = \frac{d^3[w_2(L) - w_1(L)]}{dx^3} = \frac{\lambda_z d^3 f_a(L)}{dx^3}$$

the coefficients in Eqn (11.26) have been evaluated as

$$\begin{aligned} B_1 &= \frac{M_L + M_R}{2\lambda_z \alpha_z^2 D_2}\chi_1, \ B_2 = \frac{M_L + M_R}{2\lambda_z \alpha_z^2 D_2}\chi_2, \\ B_3 &= \frac{M_L - M_R}{2\lambda_z \alpha_z^2 D_2}\chi_3, \ B_4 = \frac{M_L - M_R}{2\lambda_z \alpha_z^2 D_2}\chi_4 \end{aligned} \tag{11.27}$$

where

$$\chi_i = \frac{\xi_{3L} \mp \xi_{4L}}{\xi_{5L}} \text{ for } i = 1, 2$$

$$\chi_3 = \frac{\xi_{1L} - \xi_{2L} - \xi_{3L}/(\alpha_z L)}{\xi_{6L}}, \quad \chi_4 = \frac{\xi_{1L} + \xi_{2L} - \xi_{4L}/(\alpha_z L)}{\xi_{6L}} \tag{11.27a}$$

$$\xi_{5L} = \sin(2\alpha_z L) + \sinh(2\alpha_z L), \quad \xi_{6L} = \sin(2\alpha_z L) - \sinh(2\alpha_z L)$$

Substituting Eqn (11.27) into Eqn (11.26) gives

$$f_a(x) = \frac{M_L + M_R}{2\lambda_z \alpha_z^2 D_2}(\chi_1 \xi_{1x} + \chi_2 \xi_{2x}) + \frac{M_L - M_R}{2\lambda_z \alpha_z^2 D_2}(\chi_3 \xi_{3x} + \chi_4 \xi_{4x}) \tag{11.28}$$

For the case of symmetric bending, $M_L = M_R$, the second term in Eqn (11.28) disappears:

$$f_{a,\text{sym}}(x) = \frac{1}{\lambda_z \alpha_z^2 \rho_R}(\chi_1 \xi_{1x} + \chi_2 \xi_{2x}) \tag{11.29}$$

For the case of antisymmetric bending, $M_L = -M_R$, the first term in Eqn (11.28) disappears:

$$f_{a,\text{asym}}(x) = -\frac{1}{\lambda_z \alpha_z^2 \rho_R}(\chi_3 \xi_{3x} + \chi_4 \xi_{4x}) \tag{11.30}$$

The maximum distributed stress typically occurs at $x = L$; thus,

$$f_{a,\max} = \frac{M_L + M_R}{4\lambda_z \alpha_z^2 D_2}\psi_1(\alpha_z L) + \frac{M_L - M_R}{4\lambda_z \alpha_z^2 D_2}\psi_2(\alpha_z L) \tag{11.31}$$

where,

$$\psi_1(\alpha_z L) = -\frac{\xi_{6L}}{\xi_{5L}}$$

$$\psi_2(\alpha_z L) = \frac{\cosh(2\alpha_z L) - \cos(2\alpha_z L) - \alpha_z L \xi_{5L}}{\alpha_z L \xi_{6L}} \tag{11.32}$$

Figure 11.4 shows plots of $\psi_1(\alpha_z L)$ and $\psi_2(\alpha_z L)$, which shows fast asymptotic convergence of $\psi_1(\alpha_z L)$ and a very slow asymptotic convergence of $\psi_1(\alpha_z L)$ toward the value of 1 with increasing $\alpha_z L$.

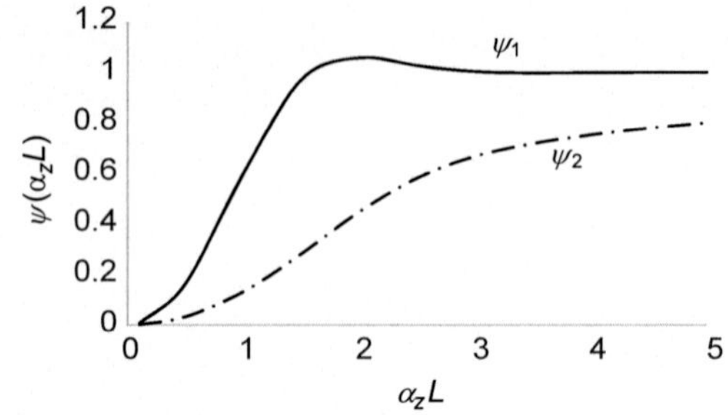

Figure 11.4 Functions $\psi_1(\alpha_z L)$ and $\psi_2(\alpha_z L)$.

11.2.2.3 Updating of shear stress, $f_s(x)$

Integrating Eqn (11.21) gives the sectional force $Q_a(x)$ as

$$Q_a(x) = -\int_L^x C_b(\chi_1 \xi_{1x} + \chi_2 \xi_{2x})\mathrm{d}x = \frac{C_b}{\alpha_z}\left(\frac{\xi_{4L}\xi_{3x} - \xi_{3L}\xi_{4x}}{\xi_{5L}}\right) \tag{11.33}$$

Substituting Eqn (11.33) into Eqn (11.9) and using the boundary conditions: $f_s'(0) = 0$ and $f_s'(L) = \frac{\alpha_{21}\Delta T - \varepsilon_2(L)}{\kappa_s}$, we have

$$f_s(x) \approx \left[A_s - \frac{C_b\overline{D}_{21}\left(2\alpha_z^2 + \beta^2\right)}{2\beta\kappa_s\left(4\alpha_z^4 + \beta^4\right)}\right]\frac{\cosh \beta x}{\sinh \beta L} + \frac{C_b\overline{D}_{21}}{\alpha_z\kappa_s\left(4\alpha_z^4 + \beta^4\right)} \times \frac{\xi_{3L}\left(\beta^2\xi_{4x} + 2\alpha_z^2\xi_{3x}\right) + \xi_{4L}\left(2\alpha_z^2\xi_{4x} - \beta^2\xi_{3x}\right)}{\xi_{5L}} \tag{11.34}$$

Ignoring thermal strains, and putting $\alpha_z = \beta$, $h_1 = h_2 = h$ and $\nu_1 = \nu_2 = \nu$ gives $\frac{C_b\overline{D}_{21}(2\alpha_z^2+\beta^2)}{2\beta\kappa_s(4\alpha_z^4+\beta^4)} \approx \frac{192(1+\nu)^2(E_1 - E_2)}{5(E_1+E_2)}A_s$, which is significant compared to A_s. This is also true for the last term in Eqn (11.34).

11.3 Bending of a PCB assembly that has a continuous bonding layer

Most bilayer structures are formed through bonding. If the bonding layer is relatively thin, then the assembly may be treated as a bilayer structure and the solution described in Section 11.2 applies. If the bonding layer is relatively thick, then the assembly has to be treated as a trilayer assembly. With increased fragility of the solder joints due to reducing feature size, it has become a common practice for the PCB assembly in the PEDs to have the solder joints underfilled with polymer to form a continuous bonding later. The bonding is at risk of delamination and the IC component of fracturing during drop impact.

11.3.1 Derivations

Figure 11.5 shows the deformed state of a trilayer assembly due to an externally applied moment at the free edge of member #2. The two outer layers are assumed to be capable of flexural and shear deformation while the bonding layer is assumed to have significantly higher stretch and flexural compliances than the two outer layers. The bonding layer experiences a linear variation of transverse stress over its thickness. The transverse stress is decomposed into a uniform component f_a and a linear component f_b. The transverse stress on the top surface of the bonding layer has the magnitude $f_a + f_b$, while that on the bottom surface has the magnitude $f_a - f_b$.

The force–deformation relation

$$\frac{d^2\theta_i}{dx^2} = \frac{1}{D_i}\left(\frac{f_s h_i}{2} - Q_i\right) = \frac{1}{D_i}\left(\frac{h_i + h_3}{2} f_s \mp Q_a\right) \quad (11.35)$$

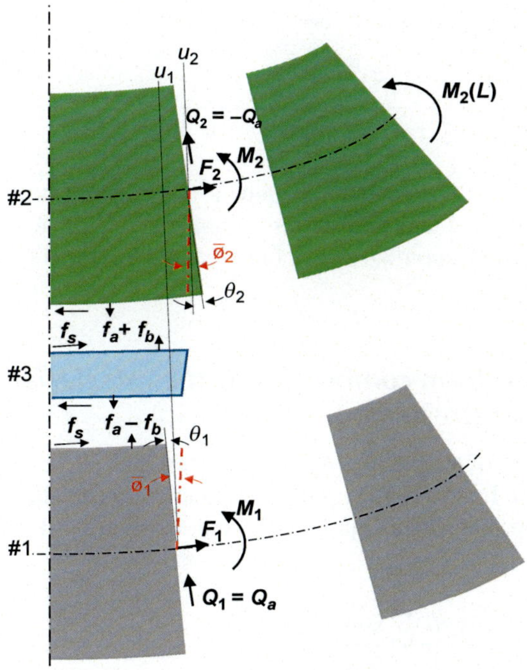

Figure 11.5 Pure bending of a trilayer beam: deformed state.

and the differential equations for the interfacial stresses

$$\frac{d^2 f_s}{dx^2} - \beta^2 f_s = \frac{\overline{D}_{21}}{\kappa_s} Q_a \tag{11.9}$$

$$\frac{d^4 f_a}{dx^4} + \frac{\overline{D}_e}{\lambda_z} = \frac{\overline{D}_{21}^*}{\lambda_z} \frac{df_s}{dx} \tag{11.36}$$

where, $\beta^2 = \lambda_x/\kappa_s$, $\lambda_x = \lambda_{x1} + \lambda_{x2} + \lambda_{x\theta}^* \approx \sum_{i=1}^{2} \frac{1}{E_i h_i}\left(4 + \frac{3h_3}{h_i}\right)$, $\kappa_s = \sum_{i=1}^{2} \kappa_{si}^* + \frac{h_3}{G_3}$, $\kappa_{si}^* \approx \frac{h_i \mp 2h_3}{8G_i}$, $\overline{D}_{21}^* = \frac{1}{2}\left(\frac{h_2+h_3}{D_2} - \frac{h_1+h_3}{D_1}\right)$, $\lambda_z = \sum_{i=1}^{2} \lambda_{zi} + \frac{h_3}{E_3}$ and $\lambda_{zi} = \frac{3h_i}{8E_i}$ have been derived in Section 4.2.2 and are reproduced above.

11.3.1.1 *Interfacial shear stress, $f_s(x)$*

The interfacial shear stress is identical to that in a bilayer structure, which is given by Eqn (11.34). However, the equation does not satisfy the boundary condition $f_s(L) = 0$. To enforce the boundary condition, the shear stress is assumed to take the form

$$f_s(x) = A_s'\left(\frac{\cosh \beta x}{\sinh \beta L} - \coth \beta L \frac{\cosh n\beta x}{\cosh n\beta L}\right) \tag{11.37}$$

and the magnitude of A_s' is taken to be

$$A_s' \approx A_s - \frac{C_b \overline{D}_{21}\left(2\alpha_z^2 + \beta^2\right)}{2\beta\kappa_s\left(4\alpha_z^4 + \beta^4\right)} + \frac{\alpha_z C_b \overline{D}_{21}}{\kappa_s\left(4\alpha_z^4 + \beta^4\right)} \tag{11.38}$$

If the ratio h/D of the two outer members are of similar magnitude such that $\overline{D}_{21} \approx 0$, then the second and third terms diminish and $A_s' \approx A_s$. The function for n has been evaluated in Section 2.4.4 as a function of βh_3 and is reproduced below:

$$n \approx 4(\beta h_3)^{-1.37} \tag{11.39}$$

11.3.1.2 *The uniform component of the interfacial normal stress, $f_a(x)$*

We shall, for reasons of simplicity, assume the interfacial shear stress $f_s(x)$ to take the form of Eqn (11.12). Following the same manipulations as detailed in Section 11.2 for the bilayer structure, we obtain Eqn (11.18); and substituting it into Eqn (11.36) this gives

$$\frac{d^4 f_a}{dx^4} + 4\alpha_z^{*4} f_a = \frac{A_s \beta \overline{D}_{21}^*}{\lambda_z} \frac{\sinh \beta x}{\sinh \beta L} \tag{11.40}$$

where $\alpha_z^* = \sqrt[4]{\frac{\overline{D}_e - \overline{D}_{21}^* \overline{D}_{21}/\lambda_x}{4\lambda_z}}$. The solution to Eqn (11.40) is identical to that of Eqn (11.20), which may be reduced to

$$f_a(x) \approx C_b^*(\chi_1\xi_{1x} + \chi_2\xi_{2x}) \tag{11.41}$$

where,

$$C_b^* = \frac{1}{\lambda_z \alpha_z^{*2} \rho_2(L)} = -\frac{2\varepsilon_2(L)}{\lambda_z \alpha_z^{*2} h_2} \approx -\frac{4\varepsilon_2(L)}{\sqrt{\lambda_z \overline{D}_e} h_2} \tag{11.42}$$

11.3.1.3 The linear component of the interfacial normal stress, $f_b(x)$

Figure 11.6 shows an element $dx \cdot dz$ that is in equilibrium. The z-equilibrium of the element gives

$$d\sigma_z dx + d\tau_{xz} dz = 0 \tag{11.43}$$

which, upon dividing by $dx \cdot dz$ yields

$$\frac{d\sigma_z}{dz} + \frac{d\tau_{xz}}{dx} = 0 \tag{11.44}$$

Thus, the presence of a shear stress τ_{xz} that varies along the x-direction must be accompanied by a normal stress σ_z that varies along the z-direction. Substituting τ_{xz} with f_s and considering the fact that f_s in the bonding layer does not vary along its thickness, then Eqn (11.44) is reduced to

$$\frac{\Delta\sigma_z}{\Delta z} = -\frac{df_s}{dx} \tag{11.45}$$

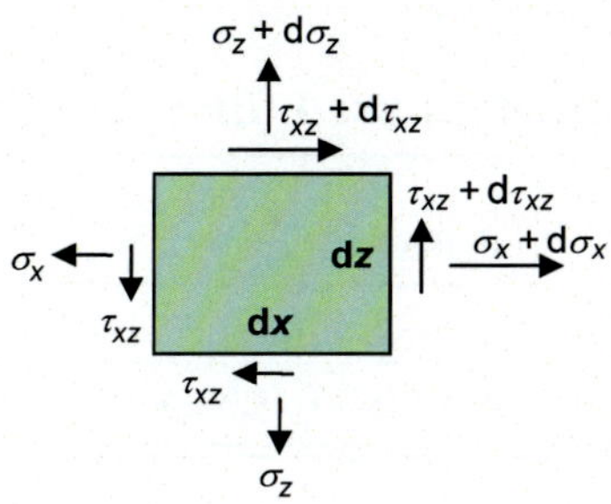

Figure 11.6 An element in equilibrium.

Referring to Figure 11.5, $\sigma_z = f_a + f_b$ at $z = h_3/2$ and $\sigma_z = f_a - f_b$ at $z = -h_3/2$, which give $\Delta\sigma_z = 2f_b$ and $\Delta z = h_3$. Substituting these relations of $\Delta\sigma_z$ and Δz into Eqn (11.45) gives

$$f_b = -\frac{h_3}{2}\frac{\mathrm{d}f_s}{\mathrm{d}x} \tag{11.46}$$

Differentiating Eqn (11.37) wrt x and substituting into Eqn (11.46) gives

$$f_b = -\frac{h_3 A_s' \beta}{2}\left(\frac{\sinh \beta x}{\sinh \beta L} - n \coth \beta L \frac{\sinh n\beta x}{\cosh n\beta L}\right) \tag{11.47}$$

11.3.2 Validations

The analytical solutions are validated against finite element analysis (FEA) in this section. The basic dimensions and material properties for the models used in the validations are identical to those used in Section 2.4.6 and these are shown in Table 11.1. The compliances and the characteristic constants are given in Table 11.2. The units for individual parameters are listed below the tables.

The finite element (FE) model is shown in Figure 11.7(a), which shows a symmetric half of the PCB assembly. The PCB and the IC component are modelled using 50 × 10 quadratic plane elements while the sandwiched layer is modelled using 50 × 5

Table 11.1 **Basic parameters used for validation analyses**

Component	Dimensions		Material properties			Diff. Temp.
	L	h	E	G	α	ΔT
#1, IC device		1	24	9.2	5	
#2, Substrate	5	1.25	24	9.2	15	100
#3, Adhesive		0.5	4.9	1.9	0	

Note: L, h, (mm); E, G (GPa), α (10^{-6}/°C), ΔT (°C). The elastic modulus of IC package as provided is an estimation of the composite property for silicon and moulding compound.

Table 11.2 **Derived parameters used for validation**

	Compliances			Characteristic constants	
Source	κ_s	λ_x	λ_z	β	α_z
Wong	2.91×10^{-4}	4.49×10^{-4}	1.32×10^{-4}	1.24	1.17

Note: λ_x (N^{-1} mm), λ_z, κ_s (N^{-1} mm^3), α_z, β(mm^{-1}).

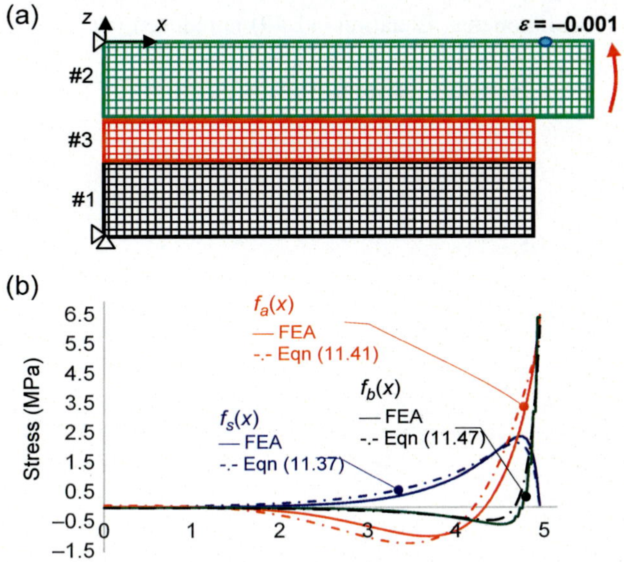

Figure 11.7 (a) FE model, (b) comparison of interfacial stresses.

quadratic plane elements. The plane elements in the sandwiched layer are defined with orthotropic material properties wherein the lateral elastic moduli (E_x & E_y) and the out-of-plane shear moduli (G_{xy} & G_{yz}) have been assigned a negligible value compared to its transverse modulus (E_z) and its in-plane shear modulus (G_{xz}).

The FE model is prescribed with a symmetric strain of −0.001 on the top fibre at the free edge of the PCB, corresponding to a positive radius of curvature of 500 mm. The interfacial shear stress, the uniform component and the linear component of the interfacial normal stresses are plotted in Figure 11.7(b). It is observed that the results from the analytical equations compared very well with the FE results.

11.3.3 Robust design analysis

11.3.3.1 Interfacial delamination

The delamination of the interfaces is a major concern for sandwich structures in service. Between the shearing and the peeling stresses, the latter is the main driving force for interfacial delamination. The maximum peeling stress is given by

$$\sigma_{\text{peeling,max}} = f_a(L) + f_b(L) \tag{11.48}$$

Substituting $x = L$ into Eqn (11.41) gives

$$f_a(L) = -\frac{2\varepsilon_2(L)}{\lambda_z \alpha_z^{*2} h_2}\left(-\frac{\xi_{6L}}{\xi_{5L}}\right) \tag{11.49}$$

The trigonometric function approaches unity for $\alpha_z L > 2$, which is indeed the case for most practical IC components. Equation (11.49) is reduced to

$$f_a(L) = -\frac{2\varepsilon_2(L)}{\lambda_z \alpha_z^{*2} h_2} \approx -\frac{4\varepsilon_2(L)}{\sqrt{\lambda_z \overline{D}_e} h_2} \tag{11.50}$$

Substituting $x = L$ into Eqn (11.47) gives

$$\begin{aligned} f_b(L) &= \frac{A_s' \beta h_3}{2}(n-1) \\ &\approx \frac{A_s' \beta h_3 n}{2} \quad \text{for} \quad n \gg 1 \end{aligned} \tag{11.51}$$

Substituting $n = 4(\beta h_3)^{-\gamma}$ from Eqn (11.39), where $\gamma = 1.37$, into the above and for $n \gg 1$,

$$f_b(L) \approx \frac{2A_s'}{(\beta h_3)^{\gamma-1}} \tag{11.52}$$

For the two outer members with similar magnitude of h/D, $A_s' \approx A_s$ and Eqn (11.52) may be reduced to

$$f_b(L) \approx \frac{2A_s}{(\beta h_3)^{\gamma-1}} = \frac{2[\alpha_{21}\Delta T - \varepsilon_2(L)]}{\kappa_s \beta^{\gamma} h_3^{\gamma-1}} = \frac{2[\alpha_{21}\Delta T - \varepsilon_2(L)]}{\lambda_x^{\gamma/2} \kappa_s^{1-\gamma/2} h_3^{\gamma-1}} \tag{11.53}$$

Substituting Eqns (11.50) and (11.53) into Eqn (11.48) gives

$$\sigma_{\text{peeling,max}} \approx \frac{2\alpha_{21}\Delta T}{\lambda_x^{\gamma/2} \kappa_s^{1-\gamma/2} h_3^{\gamma-1}} - 4\varepsilon_2(L)\left(\frac{1}{\sqrt{\lambda_z \overline{D}_e} h_2} + \frac{1}{2\lambda_x^{\gamma/2} \kappa_s^{1-\gamma/2} h_3^{\gamma-1}}\right) \tag{11.54}$$

Equation (11.54) suggests that the magnitude of the peeling stress may be reduced by increasing the flexural, the stretch, the shear and the transverse compliances of the structure. This can be achieved by reducing the moduli of the constituting materials, by reducing the thicknesses of the two outer layers and by increasing the thickness of the bonding layer.

11.3.3.2 Fracturing of the outer members

Compared to the PCB, the IC component is relatively fragile, especially if it is made of bare chip. Excessive bending of the IC component may lead to its fracture. The longitudinal stress, σ_{xi}, in the outer members has been given in Section 2.4.7 of Chapter 2 and is reproduced below:

$$\sigma_{xi} = \frac{2(3Z-1)}{h_i} F(x) + \frac{6(2Z-1)}{h_i^2} \int_L^x Q_i \mathrm{d}x, \quad i = 1, 2 \tag{11.55}$$

where $Z = z/h_i$ is the normalised coordinate and

$$\begin{aligned} Q_i &= Q_a \mp Q_b \\ Q_b &= \frac{h_3 f_s}{2} \end{aligned} \tag{11.55a}$$

Substituting Eqn (11.55a) and the relation $F(x) = -\int_L^x f_s \mathrm{d}x$ into Eqn (11.55) gives

$$\sigma_{xi} = \left[\frac{2(3Z-1)}{h_i} \pm \frac{3h_3(2Z-1)}{h_i^2}\right] F(x) + \frac{6(2Z-1)}{h_i^2} \int_L^x Q_a \mathrm{d}x, \quad i = 1,2 \tag{11.56}$$

The function Q_a has been given in Eqn (11.33) and its definite integral is given by

$$\int_L^x Q_a \mathrm{d}x = C_b^* \frac{2\xi_{3L}(\xi_{1x} - \xi_{2x}) + 2\xi_{4L}(\xi_{1x} + \xi_{2x}) - \xi_{5L}}{4\alpha_z^{*2} \xi_{5L}} \tag{11.56a}$$

The maximum in-plane stress occurs at the middle of the member ($x = 0$); thus,

$$\sigma_{xi} = \left[\frac{2(3Z-1)}{h_i} \pm \frac{3h_3(2Z-1)}{h_i^2}\right] F(0) + \frac{6(2Z-1)}{h_i^2} \int_L^0 Q_a \mathrm{d}x, \quad i = 1,2 \tag{11.57}$$

Using the simple expression of $f_s(x) = A_s \frac{\cosh \beta x}{\sinh \beta L}$, the surface traction at $x = 0$ is given by

$$\begin{aligned} F(0) &= -\int_L^0 f_x \mathrm{d}x = \frac{\alpha_{21}\Delta T - \varepsilon_2(L)}{\lambda_x}\left(1 - \frac{1}{\sinh \beta L}\right) \\ &\approx \frac{\alpha_{21}\Delta T - \varepsilon_2(L)}{\lambda_x} \quad \text{for } \beta L > 3 \end{aligned} \tag{11.57a}$$

while the integral in the equation is given by

$$\begin{aligned} \int_L^0 Q_a \mathrm{d}x &= C_b^* \frac{2\xi_{3L} + 2\xi_{4L} - \xi_{5L}}{4\alpha_z^{*2}\xi_{5L}} \\ &\approx \frac{-C_b}{4\alpha_z^2} \quad \text{for } \alpha_z L > 2 \\ &= \frac{2\varepsilon_2(L)}{\overline{D}_e h_2} \quad \text{for } \alpha_z L > 2 \end{aligned} \tag{11.57b}$$

The maximum in-plane stress in member #1 becomes

$$\sigma_{x1} = \left[\frac{2(3Z-1)}{h_1} + \frac{3h_3(2Z-1)}{h_1^2}\right]\frac{\alpha_{21}\Delta T - \varepsilon_2(L)}{\lambda_x} + \frac{12(2Z-1)\varepsilon_2(L)}{\overline{D}_e h_1^2 h_2}, \quad i = 1,2 \tag{11.58}$$

For a positive curvature (or a negative $\varepsilon_2(L)$), the maximum in-plane stress in member #1 may occur at either $Z = 0$ (the exposed surface) or $Z = 1$ (the interface with the bonding layer) depending on the dominance of either the bending stress or the interfacial shear stress. Substituting $Z = 0$ and 1, respectively, into Eqn (11.58) gives

$$\begin{aligned} \sigma_{x1,\max}\big|_{Z=0} &\approx -\left(\frac{2}{h_1} + \frac{3h_3}{h_1^2}\right)\frac{\alpha_{21}\Delta T - \varepsilon_2(L)}{\lambda_x} - \frac{12\varepsilon_2(L)}{\overline{D}_e h_1^2 h_2} \\ \sigma_{x1,\max}\big|_{Z=1} &\approx \left(\frac{4}{h_1} + \frac{3h_3}{h_1^2}\right)\frac{\alpha_{21}\Delta T - \varepsilon_2(L)}{\lambda_x} + \frac{12\varepsilon_2(L)}{\overline{D}_e h_1^2 h_2} \end{aligned} \tag{11.59}$$

Assuming $h_1 = h_2 = h$ and ignoring $\alpha_{21}\Delta T$, the ratio of the first term to the second term in the two equations in Eqn (11.59) is, respectively, $\frac{2h+3h_3}{4h+3h_3}$ and 1; that is, $\sigma_{x1,\max}\big|_{Z=1} = 0$. This suggests that maximum peeling stress is more likely to occur at $Z = 0$ for positive $\alpha_{21}\Delta T$. Thus,

$$\sigma_{x1,\max} \approx -\left(\frac{2}{h_1} + \frac{3h_3}{h_1^2}\right)\frac{\alpha_{21}\Delta T - \varepsilon_2(L)}{\lambda_x} - \frac{12\varepsilon_2(L)}{\overline{D}_e h_1^2 h_2} \tag{11.60}$$

The first term is attributed to f_s while the second term to f_a. The ratio $\frac{2h+3h_3}{4h+3h_3}$ suggests that the second term has a larger magnitude than the first term and the ratio decreases with increasing h_3. The risk of tensile fracture of the IC component in the presence of mechanical bending can be reduced by decreasing the elastic modulus of the IC component and the PCB, by reducing h_2, and increasing h_1 and h_3.

11.4 Bending of a PCB assembly that has a discrete bonding layer

Underfilling of solder joints is an expensive operation, and the practice is typically reserved for PEDs. The solder joints in the PCB assembly in almost all other applications are not underfilled. The discrete solder joints are vulnerable to mechanical bending.

11.4.1 Derivations

A discrete member is acted upon by a pair of shear force F_s and a pair of axial force F_a. The pair of shear force introduces a shear couple on the discrete member, which is countered by an anti-symmetric pair of moment m_s. On top of this, in the presence of relative rotation Φ between the two outer layers, the discrete member will be acted upon by a symmetric pair of moment m_Φ.

11.4.1.1 The average axial and shear stresses in a discrete member

Figure 11.8 shows the shear force F_s, the axial forces F_a and the shear-couple-induced moment M_s acting on a discrete member. The magnitudes of F_s and F_a can be estimated from the interfacial shear stress $f_s(x)$ and the uniform component of the interfacial normal stress $f_a(x)$ evaluated in the last section. Assuming a uniform distributed discrete member with a spacing pitch of $p \times q$, the axial force F_a and shear force F_s, in a discrete member that is at distance Γ from the mid-length of a trilayer assembly can be obtained through definite integration of $f_a(x)$ and $f_s(x)$ over a pitch, p, between the limit $(\Gamma - p/2)$ to $(\Gamma + p/2)$ and for a width of q.

The magnitudes of the forces are

$$F_s(\Gamma) = q \int_{\Gamma-p/2}^{\Gamma+p/2} f_s(x)\mathrm{d}x \tag{11.61}$$

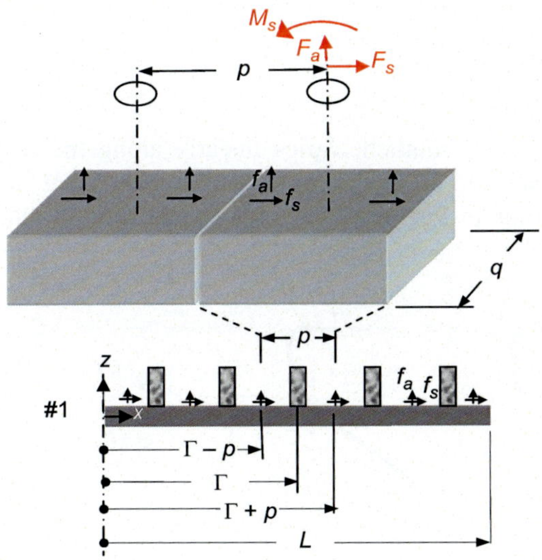

Figure 11.8 Integration of interfacial stresses into forces over an area equivalent to that covered by a single discrete member.

$$F_a(\Gamma) = q \int_{\Gamma-p/2}^{\Gamma+p/2} f_a(x)\mathrm{d}x \tag{11.62}$$

Assuming the discrete member to be uniform in its cross-section with area A_3, the average shear and the average axial stresses in the member are given, respectively, by

$$\overline{\tau}(\Gamma) = \frac{q}{A_3} \int_{\Gamma-p/2}^{\Gamma+p/2} f_s(x)\mathrm{d}x = \frac{qA_{s'}}{\beta A_3}\left[\frac{\sinh \beta x}{\sinh \beta L} - \frac{\coth \beta L}{n}\frac{\sinh n\beta x}{\cosh n\beta L}\right]_{\Gamma-p/2}^{\Gamma+p/2} \tag{11.63}$$

$$\overline{\sigma}_a(\Gamma) = \frac{q}{A_3} \int_{\Gamma-p/2}^{\Gamma+p/2} f_a(x)\mathrm{d}x = \frac{qC_b^*}{A_3\alpha_z^*}\left[\frac{-\xi_{4L}\xi_{3x} + \xi_{3L}\xi_{4x}}{\xi_{5L}}\right]_{\Gamma-p/2}^{\Gamma+p/2} \tag{11.64}$$

11.4.1.2 The bending moment on a discrete member

An individual discrete member is acted upon by two pairs of moments. This is shown in Figure 11.9. The moments at the ends of the discrete member interfacing with member #2 (PCB) and member #1 (IC component) are designated as M_{b2} and M_{b1}, respectively, where the subscript b suggests the discrete member is treated as a beam.

The first pair of moments is induced by the shear couple – we shall refer to this pair as shear-couple moment – given by

$$M_s(\Gamma) = \frac{h_3 F_s(\Gamma)}{2} \tag{11.65}$$

The magnitude of this moment varies linearly along the length of the discrete element – from a positive M_s at the top end to a negative M_s at the bottom end of the member (referring to Figure 11.9). Taking the end of the discrete member

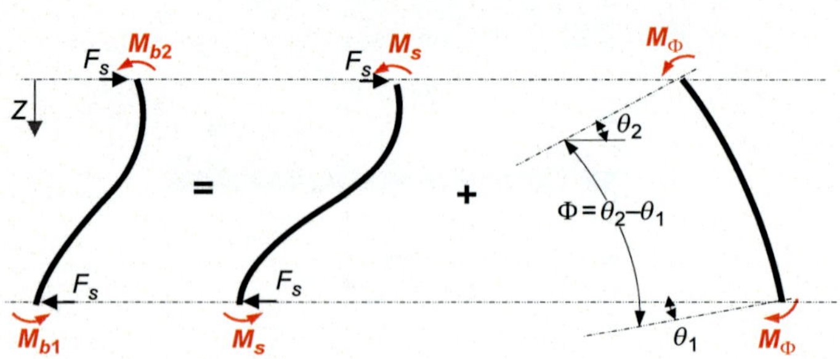

Figure 11.9 Components of bending moment.

interfacing with member #2 (PCB) as the origin, the shear-couple moment along the length of the discrete member is given by

$$M_s(\Gamma, Z) = M_s(\Gamma) - F_s(\Gamma)h_3 Z = F_s(\Gamma)h_3\left(\frac{1}{2} - Z\right) \tag{11.66}$$

where $0 \leq Z \leq 1$.

The second pair of moments is induced by the change in the inclusive angle of the two outer members, $\theta_{21} = \theta_2 - \theta_1$; we shall refer to this pair as differential rotational moment. The magnitude of the moment is given by

$$M_\Phi(\Gamma) = D_3\theta_{21}(\Gamma) \tag{11.67}$$

where D_3 is the flexural rigidity of the discrete element. From Eqn (11.35), we obtain

$$\begin{aligned}\frac{d^2\theta_2}{dx^2} &= \frac{1}{D_2}\frac{dM_2}{dx} = \frac{1}{D_2}\left[\frac{f_s(h_1 + h_3)}{2} + Q_a\right]\\ \frac{d^2\theta_1}{dx^2} &= \frac{1}{D_1}\frac{dM_1}{dx} = \frac{1}{D_1}\left[\frac{f_s(h_2 + h_3)}{2} - Q_a\right]\end{aligned} \tag{11.68}$$

The differential is given by

$$\frac{d^2\theta_{21}(x)}{dx^2} = \overline{D}_{21}^{*} f_s(x) + \overline{D}_e Q_a(x) \tag{11.69}$$

The shear stress $f_s(x)$ and the sectional shear force $Q_a(x)$ are assumed to take the simple forms

$$f_s(x) = A_s \frac{\cosh \beta x}{\sinh \beta L} \tag{11.12}$$

and

$$Q_a(x) = \frac{C_b}{\alpha_z}\left(\frac{\xi_{4L}\xi_{3x} - \xi_{3L}\xi_{4x}}{\xi_{5L}}\right) \tag{11.33}$$

Substituting the above into Eqn (11.69) and together with the boundary conditions:

- $\theta_{21}(0) = \theta_2(0) - \theta_1(0) = 0$, gradient continuity at $x = 0$,
- $\frac{d\theta_{21}(L)}{dx} = \frac{d\theta_2(L)}{dx} - \frac{d\theta_1(L)}{dx} = \frac{1}{\rho_2(L)} = -\frac{2\varepsilon_2(L)}{h_2}$, curvature at free edge

the differential rotation for individual discrete element at distance Γ away from the mid-plane of the PCB assembly has been evaluated to be

$$\theta_{21}(\Gamma) \approx \frac{2}{\alpha_z \rho_2(L)} \left(\frac{\xi_{3L}\xi_{3\Gamma} + \xi_{4L}\xi_{4\Gamma}}{\xi_{5L}} \right) \tag{11.70}$$

The first term in Eqn (11.70) is attributed to the sectional force Q_a, and the second term is attributed to the interfacial shear stress f_s. In general, the contribution of the interfacial shear stress to differential rotation is relatively small and may be ignored. The differential rotational moment is simply

$$M_\Phi(\Gamma) \approx D_3\theta_{21}(\Gamma) \approx \frac{2D_3}{\alpha_z \rho_2(L)} \times \left\{ \frac{\xi_{4L}[\xi_{4\Gamma} - \alpha_z \Gamma(\xi_{1L} + \xi_{2L})] + \xi_{3L}[\xi_{3\Gamma} - \alpha_z \Gamma(\xi_{1L} - \xi_{2L})]}{\xi_{5L}} + \frac{\alpha_z \Gamma}{2} \right\} \tag{11.71}$$

Unlike the shear-couple moment whose magnitude varies along the length of a discrete element, the magnitude of the differential rotational moment does not vary along the length of a discrete element.

The total moment along the length of the discrete element is

$$M(\Gamma, Z) = M_\Phi(\Gamma) + M_s(\Gamma, Z) = M_\Phi(\Gamma) + F_s(\Gamma)h_3\left(\frac{1}{2} - Z\right) \tag{11.72}$$

The moments at the two ends of the discrete member are, respectively,

$$\begin{aligned} M_{b2}(\Gamma) &= M_\Phi(\Gamma) + \frac{F_s(\Gamma)h_3}{2} \\ M_{b1}(\Gamma) &= M_\Phi(\Gamma) - \frac{F_s(\Gamma)h_3}{2} \end{aligned} \tag{11.73}$$

Thus, the end of the discrete member joining the PCB experiences a higher magnitude of bending moment than the end joining the IC component.

11.4.1.3 Bending and longitudinal stresses

The bending stress on the outer fibre of a cylindrical discrete element, assuming plane remains plane, is simply

$$\bar{\sigma}_b(\Gamma, Z) = \frac{M(\Gamma, Z)d_3}{2I_3} = \left[M_\Phi(\Gamma) + F_s(\Gamma)h_3\left(\frac{1}{2} - Z\right) \right] \frac{d_3}{2I_3} \tag{11.74}$$

where d_3 and I_3 are the diameter and the second moment of area of the discrete member. The longitudinal stress on the outer fibre of a discrete member is the sum of the axial stress, and the bending stress given by

$$\overline{\sigma}_L(\Gamma, Z) = \overline{\sigma}_a(\Gamma) + \left[M_\Phi(\Gamma) + F_s(\Gamma)h_3\left(\frac{1}{2} - Z\right)\right]\frac{d_3}{2I_3} \quad (11.75)$$

11.4.1.4 The shear compliance of discrete members

Compared to a continuous layer, a discrete member has an additional freedom of rotation, which leads to relative x-displacement between the two ends of the member as shown in Figure 11.10. This contributes to additional shear compliance.

The curvature of the discrete member due to the shear-couple moment is given by

$$\frac{d^2x_3}{dz^2} = \frac{M_s(\Gamma, z)}{D_3} \quad (11.76)$$

where $D_3 = E_3I_3$ is the flexural stiffness of the discrete member. The relative x-displacement between the two ends of the member, Δu_{b3}, is given by

$$\Delta u_{b3}(\Gamma) = \frac{1}{D_3}b\int_0^{h_3}\int_0^{z} M_s(\Gamma, \zeta)\mathrm{d}\zeta\mathrm{d}z = \frac{h_3^2}{D_3}\int_0^{1}\int_0^{Z} M_s(\Gamma, \zeta)\mathrm{d}\zeta\mathrm{d}Z = \frac{F_s(\Gamma)h_3^3}{12D_3} \quad (11.77)$$

The shear compliance due to the flexural deformation of the discrete member is defined as $\frac{\Delta u_{b3}}{\overline{f}_s}$, where $\overline{f}_s$ is the average shear stress over the area of integration. Putting $F_s = \overline{f}_s pq$, the shear compliance is

$$\kappa_{s3\vartheta} = \frac{\Delta u_{b3}}{\overline{f}_s} = \frac{pqh_3^3}{12D_3} \quad (11.78)$$

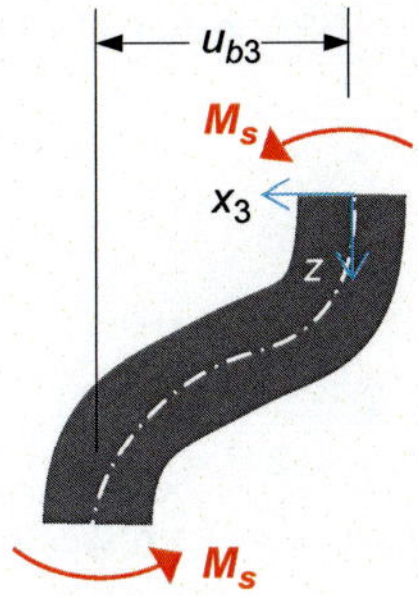

Figure 11.10 Relative x-displacement between the two ends of a discrete member due to moment M_s.

Therefore, the total shear compliance of a PCB assembly with a discrete sandwiched layer is

$$\kappa_s = \kappa_{s1} + \kappa_{s2} + \kappa^*_{s3} \tag{11.79}$$

where $\kappa^*_{s3} = \kappa_{s3} + \kappa_{s\vartheta}$

11.4.2 Validations

Figure 11.11(a) shows a symmetric half of the PCB assembly. The IC component, the PCB and the solder joints are modelled using Timoshenko beam elements that are capable of shear deformation. To account for the thickness of the IC component and the PCB, the mid-planes of the beam elements for IC component and the PCB were offset away from the interconnects by half their respective thicknesses. The basic dimensions and material properties for the models used in the validations are identical to those given in Table 11.1. The diameter of the cylindrical joints is 0.5 mm, and the joints are spaced uniformly with a pitch of 1 × 1 mm. A stretch modulus of 25 GPa is assigned to the solder joints. A symmetric strain of magnitude −0.001 is induced on

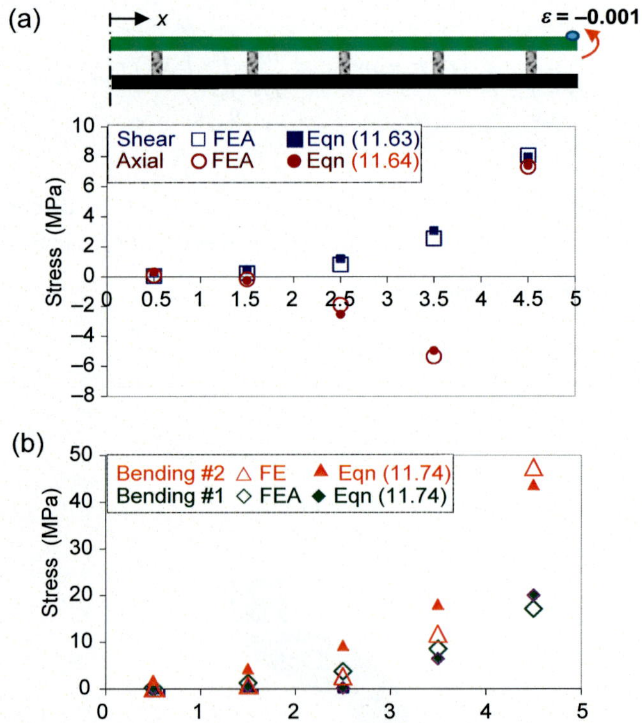

Figure 11.11 (a) FE model, (b) comparisons of stresses in the discrete members.

the top fibre at the free edges of the PCB. The shear, axial and bending stresses in the discrete elements obtained from the analytical equations are compared against those obtained from FEA in Figure 11.11(b). The solder joints experience similar magnitude of shear and axial stresses but significantly higher magnitude of bending stresses; the end of the solder joint joining the PCB experiences much higher stress than the end joining the IC component, as suggested in Eqn (11.73). Overall, the commuted stresses from the analytical equations agree very well with those from the FEA.

11.4.3 Robust design analysis

Among the discrete members, the one that is near the edge experiences the highest magnitude of the axial and bending stresses, and hence, the highest magnitude of the longitudinal stress. The end of the critical discrete member joining the PCB experiences a significantly higher magnitude of the longitudinal stress than the other end; the maximum magnitude of longitudinal stress in the discrete members is therefore given by

$$\overline{\sigma}_{L,\max} = \overline{\sigma}_a(L-p/2) + \left[M_\Phi(L) + F_s(L-p/2)\frac{h_3}{2}\right]\frac{d_3}{2I_3} \tag{11.80}$$

where,

$$F_s(L-p/2) \approx \frac{qA_s}{\beta}\left[\frac{\sinh \beta x}{\sinh \beta L}\right]_{L-p}^{L} = \frac{qA_s}{\beta}f(\beta L, \beta p), \tag{11.80a}$$

$$\overline{\sigma}_a(L-p/2) \approx \frac{C_b q}{\alpha_z A_3}\left[\frac{-\xi_{4L}\xi_{3x} + \xi_{3L}\xi_{4x}}{\xi_{5L}}\right]_{L-p}^{L} = \frac{4q\alpha_z}{\overline{D}_e A_3 \rho(L)} g(\alpha_z L, \alpha_z p), \tag{11.80b}$$

$$m_\Phi(L) \approx \frac{2D_3}{\alpha_z \rho_2(L)} h(\alpha_z L) \tag{11.80c}$$

where A_s is already given in Eqn (11.15) and

$$f(\beta L, \beta p) = 1 - \frac{\sinh \beta(L-p)}{\sinh \beta L}$$

$$g(\alpha_z L, \alpha_z p) = \frac{\xi_{4L}\xi_{3(L-p)} - \xi_{3L}\xi_{4(L-p)}}{\xi_{5L}}$$

$$h(\alpha_z L) = \frac{\xi_{4L}[\xi_{4L} - \alpha_z L(\xi_{1L} + \xi_{2L})] + \xi_{3L}[\xi_{3L} - \alpha_z L(\xi_{1L} - \xi_{2L})]}{\xi_{5L}} + \frac{\alpha_z L}{2} \tag{11.80d}$$

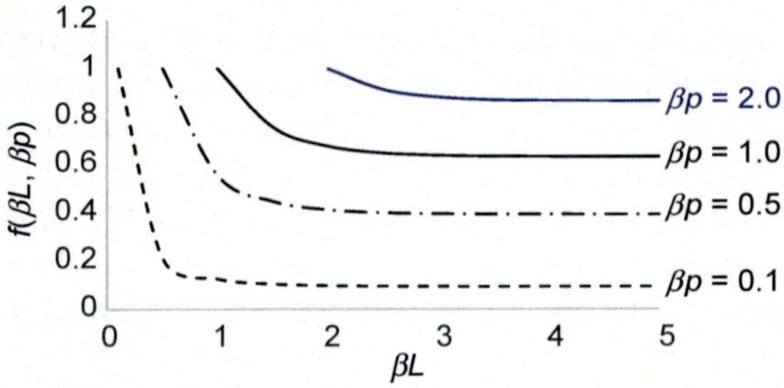

Figure 11.12 Function $f(\beta L, \beta p)$ versus βL.

The function $f(\beta L, \beta p)$ is plotted against βL in Figure 11.12. It is noted that the magnitude of $f(\beta L, \beta p)$ decreases monotonically with increasing magnitude of βL and the function reaches saturation for $\beta L > 1.5$ and $p \ll L$; this implies that increasing the length of the IC component beyond $\beta L > 1.5$ does not increase the maximum magnitude of F_s in the solder joints. The saturated magnitude of $f(\beta L, \beta p)$ may be approximated by

$$f_{\text{sat}}(\beta p) = 1 - e^{-\beta p} \text{ for } \beta L > 1.5 \text{ and } p \ll L \tag{11.81}$$

The function $g(\alpha_z L, \alpha_z p)$ is plotted against $\alpha_z L$ in Figure 11.13. It is noted that the magnitude of $g(\alpha_z L, \alpha_z p)$ increases with increasing magnitude of $\alpha_z L$ and generally reaches saturation for $\alpha_z L > 2$ and $p \ll L$; this implies that increasing the length of the IC component beyond $\alpha_z L > 2$ does not increase the maximum magnitude of $\overline{\sigma}_a$ in the solder joints. It is further noted that the saturated magnitude of $g(\alpha_z L, \alpha_z p)$ does not increase monotonically with increasing $\alpha_z p$ but decreases after reaching a peak at $\alpha_z p \approx 0.8$. The decreasing magnitude of $\overline{\sigma}_a$ with increasing pitch is unintuitive and is attributed to the averaging of f_a, which rises with a steep gradient near the edge, over increasingly large distance as the pitch increases. The saturated magnitude of $g(\alpha_z L, \alpha_z p)$ near and after $\alpha_z p = 0.8$ is deemed unreliable. Figure 11.14 shows the plot of the saturated magnitude of $g(\alpha_z L, \alpha_z p)$ for $0.01 \leq \alpha_z p \leq 0.5$; the curve can be described with a simple parabolic function:

$$g_{\text{sat}}(\alpha_z p) = -0.3837(\alpha_z p)^2 + 0.4816\alpha_z p \text{ for } \alpha_z L > 2 \text{ and } p \ll L \tag{11.82}$$

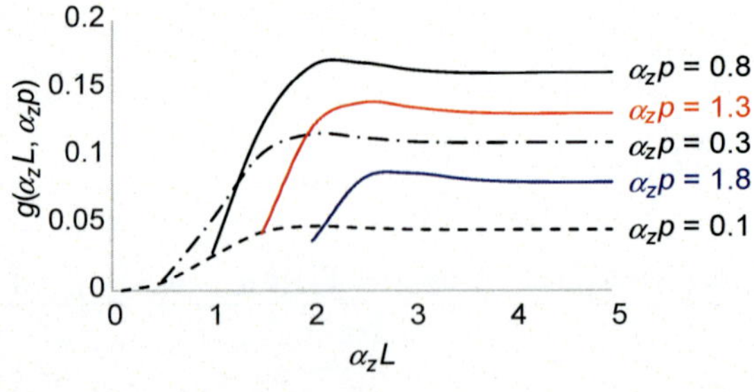

Figure 11.13 Function $g(\alpha_z L, \alpha_z p)$ versus $\alpha_z L$.

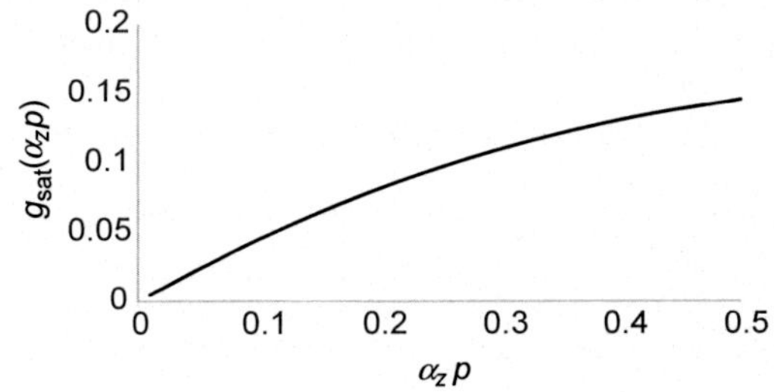

Figure 11.14 Function $g_{sat}(\alpha_z p)$.

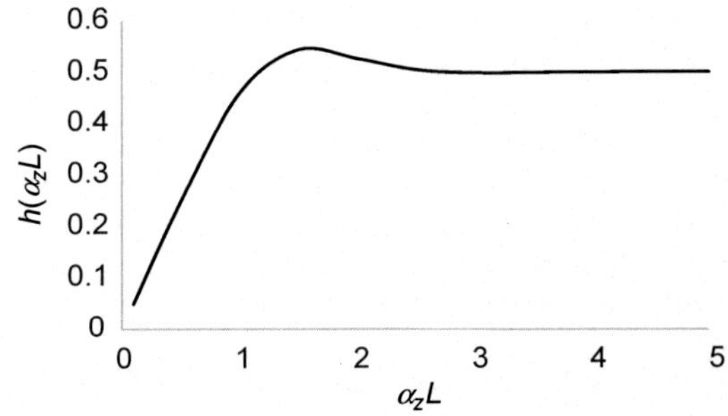

Figure 11.15 Function $h(\alpha_z L)$.

The function $h(\alpha_z L)$ is plotted against $\alpha_z L$ in Figure 11.15. It is noted that the magnitude of $h(\alpha_z L)$ increases initially with increasing magnitude of $\alpha_z L$ but reaches the saturated value of 0.5 for $\alpha_z L > 2$.

Substituting the above back into Eqn (11.80) gives

$$\overline{\sigma}_{L,\max} \approx \frac{4q\alpha_z}{\overline{D}_e A_3 \rho(L)} g_{\text{sat}}(p) + \frac{E_3 d_3}{2\alpha_z \rho_2(L)} + \frac{qA_s h_3 d_3}{4\beta I_3}\left(1 - e^{-\beta p}\right) \tag{11.83}$$

The magnitude of $\overline{\sigma}_{L,\max}$ can be reduced by reducing the diameter and the elastic modulus of the discrete member, by increasing the flexural compliance and x-compliance of the system, and by reducing the shear compliance of the system.

References

Goland, M. (1944). Stresses in cemented joints. *ASME Journal of Applied Mechanics, 11*, A17–A27.

Hovgaard, W. (1934). The stress distribution in longitudinal welds and adjoining structures. *Journal of Mathematics and Physics, 13*, 195–248.

Suhir, E. (1988). On a paradoxical phenomenon related to beams on elastic foundation: could external compliant leads reduce the strength of a surface-mounted device? *Journal of Applied Mechanics, 55*, 818.

Troelsch, H. (1934). Distribution of shear in welded connections. *Transactions of ASCE, 99*, 409–436.

Wong, E.-W. (2007). Analytical solutions for interconnect stress in board level drop impact. *IEEE Advanced Packaging, 30*, 654–664.

Wong, E., & Wong, C. (2008). Trilayer structures subjected to combined temperature and mechanical loadings. *IEEE Transactions on Components and Packaging Technology, 31*, 790–800.

Nomenclature

X_i	Subscript $i = 1, 2, 3$ for IC package, PCB and interconnect, respectively
A_3, d_3, I_3	Cross-sectional area, diameter, second moment of area of a single discrete interconnect
L, h_i	Half-length and thickness of the section of PCB assembly of interest
D_i, E_i, G_i, I_i	Flexural rigidity, elastic modulus, shear modulus, second moment of inertia
$\overline{D}_e, \overline{D}_{21}$	Effective flexural compliance of members #1 and #2; differential flexural compliance between member #2 and #1
f_a, f_b, f_s	Interfacial stresses: uniform transverse stress, linear transverse stress, shear stress
F_a, F_s	Forces in a discrete element: axial force, shear force
$\sigma_x, \sigma_b, \sigma_z, \tau_{xz}$	Stresses along the thickness of outer members: x-directional stress, x-directional bending stress, z-directional stress, shear stress
F_i	x-direction sectional traction (per unit y-width) acting along the neutral axis of member i
M_i	Sectional moment (per unit y-width)
m_s	Shear-couple moment in a discrete element
p	Pitch between neighbouring interconnects
Q_i	Sectional shear force (per unit y-width)
u	x-direction displacement of the neutral axis due to stretching
Δu	Differential x-direction displacement
w_b, w_s	z-direction displacement of the neutral axis due to bending, shearing
α_z, β	Characteristic constants for z-direction deformation, shear deformation
α_{ij}	Differential coefficient of thermal expansion between members #i and #j
$\kappa_{si}, \kappa_{s3\vartheta}, \kappa_s$	Shear compliances (shear displacement between the neutral axis and the interface per unit f_s for $i = 1, 2$ and shear displacement between the two interfaces for member #3): of member i, of discrete member #3 due to flexing, of the structure between the neutral axis of member #1 and member #2
$\kappa_{\phi i}, \kappa_{\phi}$	Shear compliance (shear-induced rotation per unit Q_a): of member #i, of the system
κ_{21}	Differential shear compliance (shear-induced rotation per unit f_s) between member #2 and #1
$\lambda_{xi}, \lambda_{x\theta}, \lambda_x$	Stretch compliances (displacement per unit traction force): of the neutral axis of member #i, due to rotation of members, of the system

λ_{zi}, λ_z — Transverse compliance (displacement between the neutral axis and the interface per unit f_a): of member #i; of the system (between the neutral axis of member #1 and member #2)

$\theta, \phi, \overline{\phi}$ — Rotation of neutral axis: due to bending, due to shearing; mean shear strain between shear surface and neutral axis

$\overline{\sigma}_a, \overline{\tau}, \overline{\sigma}_b$ — Average stresses in a discrete interconnect: axial stress, shear stress, bending stress

ΔT — Temperature change

Γ — Distance of a discrete interconnect from the mid-plane of the structure

Mathematical symbols

$$A_s = \frac{\alpha_{21}\Delta T}{\beta\kappa_s}$$

$$\alpha_z = \sqrt[4]{\frac{\overline{D}_e}{4\lambda_z}} \approx \begin{cases} \sqrt[4]{\dfrac{\overline{D}_e - \overline{D}_{21}^2/\lambda_x}{4\lambda_z}} & \text{for bilayer structure} \\ \sqrt[4]{\dfrac{\overline{D}_e - \overline{D}_{21}^*\overline{D}_{21}/\lambda_x}{4\lambda_z}} & \text{for trilayer structure} \end{cases}$$

$$\beta^2 = \lambda_x/\kappa_s$$

$$C_s = \begin{cases} \dfrac{\overline{D}_{21}\alpha_{21}\Delta T}{\lambda_z\kappa_s(4\alpha_z^4 + \beta^4)} & \text{for bilayer structure} \\ \dfrac{\overline{D}_{21}^*\alpha_{21}\Delta T}{\lambda_z\kappa_s(4\alpha_z^4 + \beta^4)} & \text{for trilayer structure} \end{cases}$$

$$\overline{D}_e = \frac{1}{D_2} + \frac{1}{D_1}$$

$$D_i = \tfrac{E_i h_i^3}{12},\ i = 1, 2, \text{ (plane stress; substitutes } E_i \text{ with } E_i/(1-\nu_i^2) \text{ for plane strain)}$$

$$\begin{matrix} \overline{D}_{21} \\ \overline{D}_{21}^* \end{matrix} = \begin{cases} \dfrac{1}{2}\left(\dfrac{h_2}{D_2} - \dfrac{h_1}{D_1}\right) \\ \dfrac{1}{2}\left(\dfrac{h_2 + h_3}{D_2} - \dfrac{h_1 + h_3}{D_1}\right) \end{cases}$$

$$\kappa_s \approx \begin{cases} \kappa_{s1} + \kappa_{s2} & \text{for bilayer structure} \\ \kappa_{s1}^* + \kappa_{s2}^* + \kappa_{s3} & \text{for trilayer structure with continuous sandwiched layer} \\ \kappa_{s1}^* + \kappa_{s2}^* + \kappa_{s3} + \kappa_{s3\vartheta} & \text{for trilayer structure with discrete sandwiched layer} \end{cases}$$

$$\begin{matrix} \kappa_{si} \\ \kappa_{si}^* \end{matrix} = \begin{cases} \dfrac{h_i}{8G_i} + \dfrac{Q_{ai}}{2G_i f_s} \approx \dfrac{h_i}{8G_i} \\ \dfrac{1}{8G_i}(h_i \mp 2h_3) + \dfrac{Q_a}{2G_i f_s} \approx \dfrac{1}{8G_i}(h_i \mp 2h_3) \end{cases}, \quad i = 1, 2$$

$$\kappa_{s3} = \begin{cases} \dfrac{h_3}{G_3} & \text{for continuous sandwiched layer} \\ \dfrac{pqh_3}{G_3 A_3} & \text{smeared property for discrete sandwiched layer} \end{cases}$$

$$\kappa_{s3\vartheta} = \frac{h_3^3 pq}{12 E_3 I_3}$$

$$\kappa_\phi = \kappa_{\phi 1} + \kappa_{\phi 2}, \quad \kappa_{\phi i} = \frac{3}{4G_i h_i}$$

$$\kappa_{21} = \frac{1}{2}\left(\frac{1}{G_2} - \frac{1}{G_1}\right)$$

$$\lambda_x = \begin{cases} \lambda_{x1} + \lambda_{x2} + \lambda_{x\theta} \approx \displaystyle\sum_{i=1}^{2} \frac{4}{E_i h_i} & \text{for bilayer structure} \\ \lambda_{x1} + \lambda_{x2} + \lambda_{x\theta}^* \approx \displaystyle\sum_{i=1}^{2} \frac{1}{E_i h_i}\left(4 + \frac{3h_3}{h_i}\right) & \text{for trilayer structure} \end{cases}$$

$$\lambda_{xi} = \frac{1}{E_i F}\left[\frac{F}{h_i} - \nu_i\left(\frac{h_i}{8}\frac{\mathrm{d}f_s}{\mathrm{d}x} + \frac{1}{2}f_a\right)\right] \approx \frac{1}{E_i h_i},$$

$i = 1, 2$, (plane stress; substitutes E_i with $E_i/(1-\nu_i^2)$ for plane strain)

$$\begin{matrix} \lambda_{x\theta} \\ \lambda_{x\theta}^* \end{matrix} = \begin{cases} \dfrac{1}{4}\left(\dfrac{h_1^2}{D_1} + \dfrac{h_2^2}{D_2}\right) \\ \dfrac{1}{4}\left(\dfrac{h_1^2 + h_1 h_3}{D_1} + \dfrac{h_2^2 + h_2 h_3}{D_2}\right) \end{cases}, \quad i = 1, 2$$

$$\lambda_z = \lambda_{z1} + \lambda_{z2} + \lambda_3$$

$$\lambda_{zi} = \frac{h}{Ef_a}\left[\left(\frac{11h}{192}\frac{\mathrm{d}f_s}{\mathrm{d}x} + \frac{13}{32}f_a\right) + \nu\left(\frac{5F}{4h} + \frac{3}{2h^2}\int_L^x Q_a \mathrm{d}x\right)\right] \approx \frac{13h_i}{32E_i},$$

$i = 1, 2$ (plane stress; substitutes E_i with $E_i/(1-\nu_i^2)$ for plane strain)

$$\lambda_{z3} = \begin{cases} \dfrac{h_3}{E_3} & \text{for continuous sandwiched layer} \\[2ex] \dfrac{pqh_3}{E_3A_3} & \text{smeared property for discrete sandwiched layer} \end{cases}$$

$$\chi_1 = \frac{\xi_{3L} - \xi_{4L}}{\xi_{5L}}, \quad \chi_2 = \frac{\xi_{3L} + \xi_{4L}}{\xi_{5L}}, \quad \chi_3 = \frac{\xi_{1L} - \xi_{2L} - \xi_{3L}/(\alpha_z L)}{\xi_{6L}},$$

$$\chi_4 = \frac{\xi_{1L} + \xi_{2L} - \xi_{4L}/(\alpha_z L)}{\xi_{6L}}$$

$$\xi_{1x} = \cos(\alpha_z x)\cosh(\alpha_z x), \quad \xi_{2x} = \sin(\alpha_z x)\sinh(\alpha_z x), \quad \xi_{3x} = \cos(\alpha_z x)\sinh(\alpha_z x),$$

$$\xi_{4x} = \sin(\alpha_z x)\cosh(\alpha_z x), \quad \xi_{5L} = \sin(2\alpha_z L) + \sinh(2\alpha_z L),$$

$$\xi_{6L} = \sin(2\alpha_z L) - \sinh(2\alpha_z L)$$

Rate-dependent stress–strain properties of solders

12

12.1 Introduction

12.1.1 The fundamentals

A *stress wave* is a form of acoustic wave that travels at finite velocity in a solid. Any applied stress will induce disequilibrium, leading to particles moving and adjusting themselves to the disequilibrium stress. This particle adjustment is propagated at a definite speed. At a low strain rate (relative to the propagation velocity and dimensions of the object), the stress wave will make multiple reflections and eventually attain equilibrium in the solid and a quasi-static condition may be assumed. At a high strain rate, there is insufficient time for stress equilibrium to be achieved within the medium, so the stress wave phenomenon occurs.

Plastic flow is a kinetic process. Although it is often convenient to think of a polycrystalline solid as having a well-defined yield strength, below which it does not flow and above which the flow is rapid, this is true only at absolute zero temperature. In general, the stress at which to induce further deformation to a solid depends on the strain history of the solid, the applied strain rate and the temperature (Armstrong & Walley, 2008; Field, 1994, 2004; Fallansbee, 1985; Gray, 2003). Plastic deformation in metals occurs through transportation of dislocation. Finite time is needed for the transportation; thus, transportation of dislocation in the solder joints is inhibited at high strain rate resulting in the suppression of plastic deformation. A full description of the constitutive properties of metals must include strain rate and temperature.

Under an isothermal condition, the flow stress of a material is a function of its strain and strain rate. The constitutive properties have assumed many forms (Armstrong & Walley, 2008); principally, the exponential equation

$$\dot{\varepsilon} = a\,\exp(\beta\sigma + \chi\varepsilon) \tag{12.1}$$

is credited to Prandtl (Prandtl, 1928), where a, β, and χ are experimental constants; assuming $\dot{\varepsilon}$ to be independent of ε, Eqn (12.1) may be reduced to the hyperbolic sine equation:

$$\dot{\varepsilon} = C\,\sinh(\beta\sigma) \tag{12.2}$$

the Johnson–Cook equation (Johnson & Cook, 1985):

$$\sigma = (a + B'\varepsilon^n)(1 + C\,\log\dot{\varepsilon}) \tag{12.3}$$

Robust Design of Microelectronics Assemblies Against Mechanical Shock, Temperature and Moisture
http://dx.doi.org/10.1016/B978-1-84569-528-6.00012-5

the power-law equation (Zener & Hollomon, 1944; Alder & Philips, 1954; Baraya, 1965):

$$\sigma = \sigma_o \dot{\varepsilon}^{\alpha} \tag{12.4}$$

where $\sigma_o = S_o \varepsilon^n$ for ductile material (Ramberg & Osgood, 1943); and the power-law equation in the form of shear stress–shear strain-shear strain rate (Nemat-Nasser, Li, & Isaacs, 1994).

$$\tau = \tau_o \left[1 + \frac{\gamma}{\gamma_o}\right]^n \left[\frac{\dot{\gamma}}{\dot{\gamma}_o}\right]^m \tag{12.5}$$

12.1.2 Strain rates in solder joints

Many engineering fields such as impact metal forming, crash worthiness of automobiles, birdstrike on aircraft, meteorite impact on satellites, and penetration by projectiles, etc. have substantial experience in robust design at high strain rate, including characterising and using the rate-dependent constitutive properties of metals. Microelectronic packaging could tap into this wealth of knowledge.

The solder joint within a portable electronic device that is experiencing drop impact does not, in general, experience direct stress from the impact. Rather, the mechanical shock induces flexural deformation of the printed circuit board (PCB); and the solder joints deform plastically to accommodate the imposed strain. This results in a significantly lower strain rate in the solder joint than would be under direct impact.

The frequencies of response of the PCB in a number of portable electronic devices subjected to 1 m of drop impact have been found in Chapter 6 to be between 300 Hz and 700 Hz. The magnitude of the registered strain on the PCB varies widely between devices and also between the orientation of impact but typically ranges from 1000 to 3000 microstrain. Using the same vehicle described in Section 7.1, the plastic strain in the outer solder joint when the PCB is subjected to a static bending strain of 3000 microstrain has been modelled using finite element analysis, illustrated in Figure 12.1. The bulk of the solder joint experiences up to 0.002 plastic strain. A ring of elements at the top and bottom faces of the solder joint experiences a high concentration of plastic strain with magnitude up to 0.072. Assuming sinusoidal deflection, the corresponding strain rate for the plastic strain in the solder joint can be estimated using the simple relation $\dot{\varepsilon} = 2\pi f \varepsilon$, where f is the frequency of flexing. Assuming a frequency of 100 Hz, the corresponding strain rates for the strain of 0.002 and 0.072 have been evaluated to be 1.25 and 45 s^{-1}, respectively. For practical reasons, it is estimated that the range of strain rate that would be of interest falls between 0.1 s^{-1} and 250 s^{-1}. It is worth noting that a ring of material along the edges of the solder joint would experience infinite strain rate because of geometrical discontinuity; however, the volume of these materials is insignificant with respect to any effect on the global stiffness of the solder joint.

The quasi-static constitutive properties of solders are readily available in handbooks (Benjamin, 1989) or from the suppliers of solders. There is also no lack of reports on the constitutive properties of eutectic 63Sn37Pb solder at strain rates at which creep

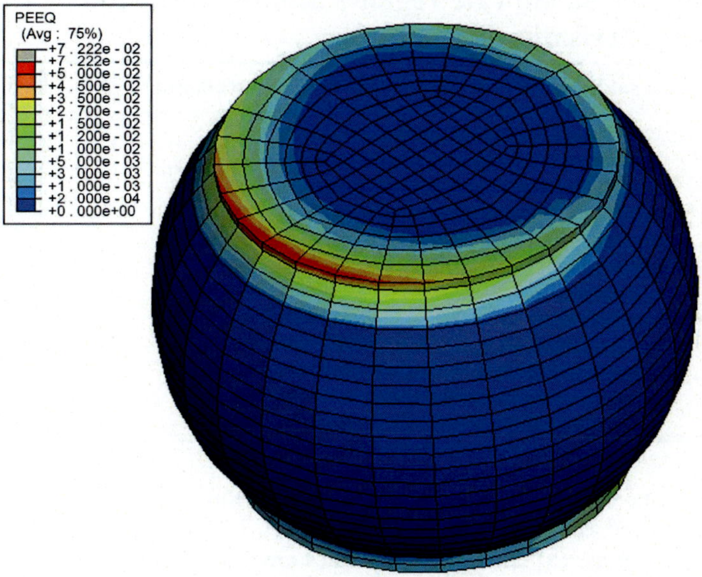

Figure 12.1 Typical plastic strain distribution in a solder joint.

mechanisms are active (Kawashima, 1992; Shi, 2002; Shohji, 2004; Solomon, 1990; Wray, 1973). However, there are far fewer reports on the constitutive properties of solder at high strain rate. Early work by Clyens and Campbell (Clyens & Campbell, 1974) reported the stress–strain characteristic of the 62Sn38Pb solder over the strain range 200–4000 s^{-1} in shear. More recently, Wang and Yi (Wang & Yi, 2002) reported compressive constitutive properties of the 63Sn37Pb solder over the range 340–1150 s^{-1} using the split Hopkinson pressure bar (SHPB) technique. However, the reported stress–strain curves show dramatic strain softening after the yield that has cast doubt on the quality of the data. Siviour et al. (Siviour, 2005) published compressive stress–strain data for tin-based solders, including 63SnPb, 96.5Sn3.5Ag, 95.5Sn3.8Ag0.7Cu, 96.5Sn3.0Ag0.5Cu and 96.2Sn2.5Ag0.8Cu0.5Sb, using the SHPB technique at strain rates between 500 s^{-1} and 3000 s^{-1} and for the temperature range between −40 °C and 60 °C. Nevertheless, the strain rates from the above references are deemed to be too high for solder joints in drop-impact conditions.

12.2 An overview of the experimental techniques for high strain rate characterisation

A summary of the experimental technique for characterizing the stress–strain property of materials at high strain rate has been presented by Gray (Gray, 2003). A historical account of the development of the Hopkinson bar and the related techniques is given by Field (2004).

Table 12.1 Strain rate regimes and experimental characterisation techniques

Strain rate regime	Experimental technique
Low rate $\dot{\varepsilon} < 0.1\ s^{-1}$	Standard test procedures using conventional mechanical tester
Medium rate $0.1\ s^{-1} \leq \dot{\varepsilon} \leq 200\ s^{-1}$	• Mechanical tester with ultra capacity • Cam plastometer • Drop weight
High rate $200\ s^{-1} \leq \dot{\varepsilon} \leq 10^5\ s^{-1}$	• Hopkinson pressure bar • Taylor rod impact
Very high rate $\dot{\varepsilon} \geq 10^5\ s^{-1}$	Flyer plate impact

The strain rate used in the mechanical characterisation of materials may be conveniently categorised into four regimes (Fallansbee, 1985) as given in Table 12.1 in which the appropriate experimental techniques are also listed.

12.2.1 Conventional mechanical tester

Mechanical testers are typically screw or servo-hydraulic driven and offer flexibility in force and displacement control. They are routinely used to characterise the stress–strain response of materials at strain rate less than $0.1\ s^{-1}$, but can be used to achieve higher strain rate.

In view of the relative ease of holding specimens in compression, it is much easier to perform compression than tension tests at high strain rate. Compression is achieved by positioning the specimen, typically uniform in cross-section, on a fixed anvil while another anvil mounted on a crosshead is travelled at a controlled speed toward the fixed anvil. The displacement of the crosshead $u(t)$ is measured using a linear velocity displacement transducer while the compressive force $F(t)$ is measured through a load cell. Care is needed to ensure free expansion of the two ends of the specimen that are in contact with the anvils such that the condition of uniaxial load and uniform cross-sectional stress prevail.

In the ideal case, a uniform stress and strain develops over the cross-section $A(t)$ and along the length $l(t)$ of the test specimen. The true stress $\sigma(t)$ in the specimen is given by:

$$\sigma(t) = \frac{F(t)}{A(t)} \tag{12.6}$$

while the incremental true strain in the specimen is given by:

$$d\varepsilon(t) = \frac{dl(t)}{l(t)} \tag{12.7}$$

Integrating Eqn (12.7) gives the true strain in the specimen as:

$$\varepsilon(t) = \int_{l_o}^{l} \frac{dl}{l} = \ln(l) - \ln(l_o) = \ln\frac{l(t)}{l_o} \tag{12.8}$$

where l_o is the initial length of the specimen.

The rate of change of true strain, referred to as the instantaneous strain rate or the true strain rate, experienced by the test specimen is given by:

$$\dot{\varepsilon}(t) = \frac{d\varepsilon(t)}{dt} = \frac{1}{l(t)}\left(\frac{dl(t)}{dt}\right) = \frac{V(t)}{l(t)} \tag{12.9}$$

where $V = dl/dt$ is the relative velocity at which the two ends of the anvils on which the two faces of the test specimen rest are moving toward each other.

In practice, the crosshead speed of the mechanical tester is typically set at a constant speed V_o; this gives:

$$\dot{\varepsilon}|_{V=V_o} = \frac{V_o}{l_o - V_o t} \tag{12.10}$$

Therefore, specimens tested using conventional mechanical testers typically experience monotonically increasing strain rate with time and hence with displacement. Moreover, an initial gap between the compression anvil and the test specimen must be catered to allow for the acceleration of the crosshead from rest to the desired test speed of V_o.

For reasons of easy reading, the term *true* will be omitted in the subsequent text when describing true stress, true strain and true strain rate.

12.2.1.1 The upper limit of mechanical testers

The practically attainable strain rate of a mechanical tester is limited by the inertia of its machine elements. Most conventional mechanical testers have a maximum crosshead speed of a few millimetres per second. Using a specimen of a few millimetres length, it is possible to achieve a strain rate of a couple of strains per second. However, this is way below the strain rates in the solder joint of a portable electronic device subjected to drop impact. Specially designed testing machines, typically equipped with high-capacity servo-hydraulic valves and high-speed control and data-acquisition instrumentation, can be used during compression testing to achieve strain rates as high as 200 s^{-1} (Gray, 2003). However, such specialised machines are extremely expensive and rare; care must also be taken to ensure that the accuracy of the measured load and displacement are not affected by the dynamics of the machine structure and the measuring instruments.

12.2.2 Split Hopkinson pressure bar (SHPB)

The Hopkinson bar technique is named after Bertram Hopkinson who, in 1914, studied the shape and evolution of stress pulses as a function of time as they propagated down

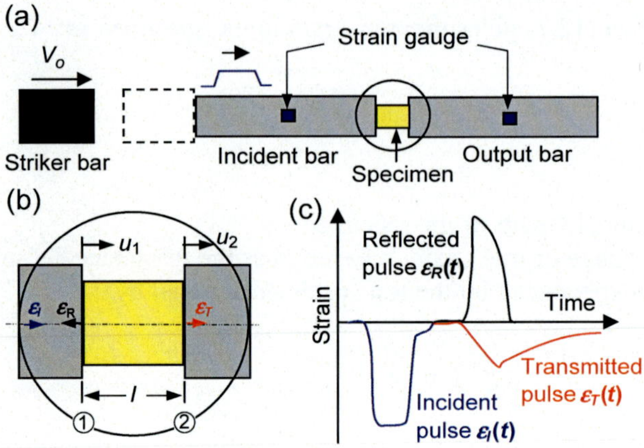

Figure 12.2 (a) Schematic of SHPB; (b) enlarged view of specimen; and (c) strain wave across interfaces.

a long rod. Building on the experience of Hopkinson and Davies (1948) and Kolsky (1949) had independently devised the split Hopkinson pressure bar (SHPB) technique that is now commonly used for characterizing the dynamic stress–strain response of materials. Readers are advised to refer to the work of Field (Field, 2004) for a detailed narration of the historical development of material testing at high strain rate. A schematic of the SHPB technique is shown in Figure 12.2. A specimen is sandwiched between an incident bar and an output bar; a striker bar of similar diameter as the incident and the output bars strikes the incident bar squarely at a given velocity V_o and sets up a stress wave propagating down the incident bar, the specimen and the output bars. The stress–strain response of the specimen can be evaluated from measurements of the incident, the reflected, and the transmitted strain-time pulse in the incident and output bars; the strain-time pulse is measured at some distances from the specimen so that the incident and the reflected pulses are distinguishable.

If the incident bar, the output bar and the test specimen are of similar cross-section, and ignoring radial inertial, the stress wave in the bars and test specimen may be described using the one-dimensional (1-D) wave equation:

$$\frac{\partial^2 u}{\partial t^2} = c_L^2 \frac{\partial^2 u}{\partial x^2} \tag{12.11}$$

where u is the displacement of particles on a longitudinal medium along the longitudinal direction; $c_L = \sqrt{(E/\rho)}$ is the propagation velocity of the stress wave; and E and ρ are the elastic modulus and density of the medium, respectively. The general solution to Eqn (12.11) is in the form:

$$u = f(x - c_L t) + g(x + c_L t) \tag{12.12}$$

where $f(x - c_L t)$ and $g(x + c_L t)$ describes a wave that propagates in the +ve and in the −ve x-direction, respectively. Differentiating Eqn (12.12) with respect to x and t, respectively, gives strain and particle velocity:

$$\varepsilon = \frac{\partial u}{\partial x} = f'(x - c_L t) + g'(x + c_L t)$$
$$V = \frac{\partial u}{\partial t} = -c_L.f'(x - c_L t) + c_L.g'(x + c_L t) \tag{12.13}$$

The velocity and strain are related by the constant c_L; that is,

$$V = \begin{cases} -c_L \varepsilon & \text{for stress wave propagating in the } +\text{ve } x\text{-direction} \\ c_L \varepsilon & \text{for stress wave propagating in the } -\text{ve } x\text{-direction} \end{cases}$$

Continuity of strain at interface ① of the input bar gives:

$$\varepsilon_{1,I} = \varepsilon_I(t) + \varepsilon_R(t) \tag{12.14}$$

The particle velocity in the input bar at interface ① is given by:

$$V_{1,I} = c_{L,I}[-\varepsilon_I(t) + \varepsilon_R(t)] \tag{12.15}$$

and the stress in the input bar at interface ① is simply:

$$\sigma_{1,I} = E_I[\varepsilon_I(t) + \varepsilon_R(t)] \tag{12.16}$$

In practice, the cross-sectional area of the specimen A_s is usually marginally smaller than that of the incident bar A_I; the conditions of force equilibrium and continuity at interface ① give the stress and the particle velocity in the specimen as:

$$\sigma_{1,s} = \sigma_{1,I}\frac{A_I}{A_s}, \; V_{1,s} = V_{1,I} \tag{12.17}$$

Note that $\sigma_{1,s}$ may exceed the yield strength of the specimen.

Performing similar analysis at interface ② yields:

$$\varepsilon_{2,T} = \varepsilon_T(t)$$
$$\sigma_{2,T} = E_T \varepsilon_T(t), \quad \sigma_{2,s} = \sigma_{2,T}\frac{A_T}{A_s} \tag{12.18}$$
$$V_{2,T} = -c_{L,T}.\varepsilon_T(t), \quad V_{2,s} = V_{2,T}$$

The specimen does not experience the same magnitude of stress along its length at time t. The average stress in the specimen is given by:

$$\sigma_{ave}(t) = \frac{1}{2}\left[\sigma_{1,s}(t) + \sigma_{2,s}(t)\right] = \frac{E_I A_I[\varepsilon_I(t) + \varepsilon_R(t)] + E_T A_T \varepsilon_T(t)}{2A_s(t)} \qquad (12.19)$$

Similarly, strain rate in the specimen varies along its length. From Eqn (12.9), the mean strain rate in the specimen is:

$$\bar{\dot{\varepsilon}}(t) = \frac{V_{2,s}(t) - V_{1,s}(t)}{l(t)} = \frac{c_{L,I}[\varepsilon_I(t) - \varepsilon_R(t)] - c_{L,T}\varepsilon_T(t)}{l(t)} \qquad (12.20)$$

The mean strain in the specimen is simply:

$$\bar{\varepsilon}(t) = \int_0^t \bar{\dot{\varepsilon}}(\tau)\mathrm{d}\tau \qquad (12.21)$$

Assuming that after an initial ringing up period the specimen reaches force equilibrium, then $\sigma_{1,I}A_I = \sigma_{2,T}A_T$, which implies that $E_I A_I[\varepsilon_I(t) + \varepsilon_R(t)] = E_T A_T \varepsilon_T(t)$. If the input and the output bars are made of identical material and have identical cross-sections, then $\varepsilon_T(t) = \varepsilon_I(t) + \varepsilon_R(t)$. Substituting this into Eqn (12.20) gives:

$$\dot{\varepsilon}(t) = \frac{2c_L\varepsilon_R(t)}{l} \qquad (12.22)$$

From Eqn (12.18), the stress in the specimen can be taken from the transmitted bar as:

$$\sigma_s(t) = E_T\varepsilon_T(t)\frac{A_T}{A_s(t)} \qquad (12.23)$$

12.2.2.1 The lower limit of SHPB

The absence of machine structure and measuring instruments in the SHPB technique allows measurement at high strain rate and is limited only by the radial inertia of the test bars due to the Poisson ratio effect. The upper limit of strain rate for the SHPB technique is approximately 10^{-5} s^{-1} for a material that has high sound speed (Gray, 2003). However, as far as the solders for drop-impact applications are concerned, it is the lower limit of the SHPB technique that is of particular interest. The strain rate of interest in the solder joints of portable electronic devices subjected to drop-impact ranges from less than 1 s^{-1} to a couple of hundreds of strains per second. Even the upper range of the strain rate of interest stretches the lower limit of the SHPB technique, which will be analysed in detail below.

Referring to Figure 12.2, upon striking the impact bar a compressive stress wave is initiated in the striker bar at its impact end, which propagates toward the free end; upon reaching the free end, the stress wave turns into a tension wave and propagates back toward the impact end. Upon reaching the impact end, the striker bar rebounds and breaks contact with the incident bar. The duration of the incident pulse is given by:

$$t_{\text{incident}} = \frac{2L_{\text{st}}}{c_{\text{L,st}}} \tag{12.24}$$

where L_{st} is the length of the striker bar and $c_{\text{L,st}}$ is the velocity of the longitudinal wave in the striker bar. Assuming the strain rate in the specimen does not change with time (this condition suggests that the incident strain pulse is of constant magnitude with time), then

$$t_{\text{incident}} = \frac{\varepsilon}{\dot{\varepsilon}} \tag{12.25}$$

and substituting this into Eqn (12.4) gives:

$$L_{\text{st}} = \frac{c_{\text{L,st}}\varepsilon}{2\dot{\varepsilon}} \tag{12.26}$$

The striker bar is typically made of steel the longitudinal wave velocity of which is $c_{\text{L,st}} \approx 5000$ m/s. For a specimen strain and strain rate of $\varepsilon = 0.1$ and $\dot{\varepsilon} = 1000\ \text{s}^{-1}$, respectively, the required length of the striker bar is $L_{\text{st}} = 0.25$ m, which is about the practical limit before the striker bar may experience buckling. Reducing the strain rate to say $100\ \text{s}^{-1}$ requires a striker bar of 2.5 m in length, which is impossible.

12.2.3 Drop-weight technique

There is a gap of strain rate between the upper limit of the conventional mechanical tester and the lower limit of the SHPB technique. The strain rate experienced by solder joints in the drop impact of portable electronic devices happens to fall within this gap. While this regime of strain rate can be managed using an ultra-capacity mechanical tester or a cam plastometer (Table 12.1), these items require heavy capital investment. By contrast, a drop-weight tester is relatively economical to set up.

The drop-weight technique uses a falling weight to introduce compressive load to the specimen. A typical drop-weight testing set-up is illustrated in Figure 12.3. The compression load on the specimen is measured using a load cell placed at the bottom of the specimen. Given the desired initial strain rate (or engineering strain rate), $\dot{\varepsilon}_{\text{o}}$, and the initial length of the specimen l_{o}, the initial velocity of the drop weight V_{o} is:

$$V_{\text{o}} = \dot{\varepsilon}_{\text{o}} l_{\text{o}} \tag{12.27}$$

Assuming equilibrium of forces at the top and the bottom faces of the specimen, the force that retards the drop weight is the same as that measured by the load cell.

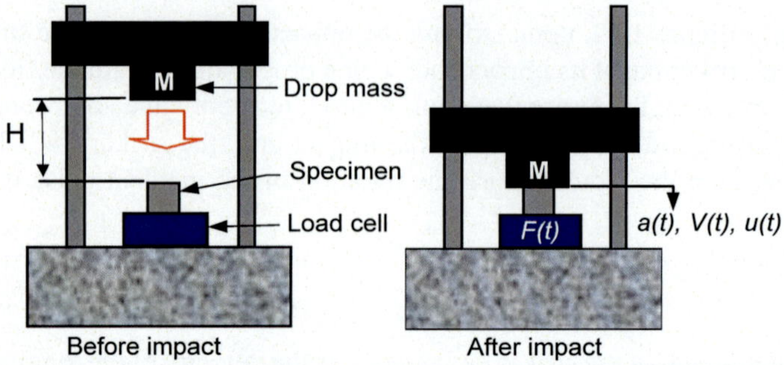

Figure 12.3 Schematic of drop-weight test technique.

Assuming negligible frictional force, the acceleration, and hence the velocity and the displacement, of the drop weight can be evaluated:

$$V(t) - V_o = \frac{1}{M}\int_o^t F(\tau)\mathrm{d}\tau$$

$$u(t) - 0 = \int_o^t V(\tau)\mathrm{d}\tau \qquad (12.28)$$

Assuming the drop weight and the load cell to be perfectly rigid, then the instantaneous length of the specimen $l(t)$ is given by:

$$l(t) = l_o - u(t) \qquad (12.29)$$

Alternatively, the drop-weight displacement may be measured directly using a laser linear measurement instrument.

The drop-weight technique suffers from the risk of 'amplification of load' due to the inertia of the load cell. Modeling the load cell as a spring-mass system, the response of the load cell to a half-sine impact force is identical to that due to a half-sine shock, the solution of which has been given in Chapter 10. Figure 12.4 shows the response spectra of a load cell of resonant frequency ω impacted upon by a half-sine force pulse of $F = F_o \sin \Omega t$. Depending on the frequency ratio, the load cell may register a very different magnitude of load F_d from the actual applied load F_o. An ideal load cell is one whose natural frequency is at least five times that of the highest frequency component of the impact pulse. Accompanying the 'amplification' is the ringing of the load due to the load cell being set into a harmonic motion.

All compression test methods suffer from the barrelling of a specimen caused by the frictional contacts at the two ends of specimen and buckling of specimen. While the elimination of barrelling requires a specimen of large length-to-diameter aspect ratio, the reverse is true for buckling.

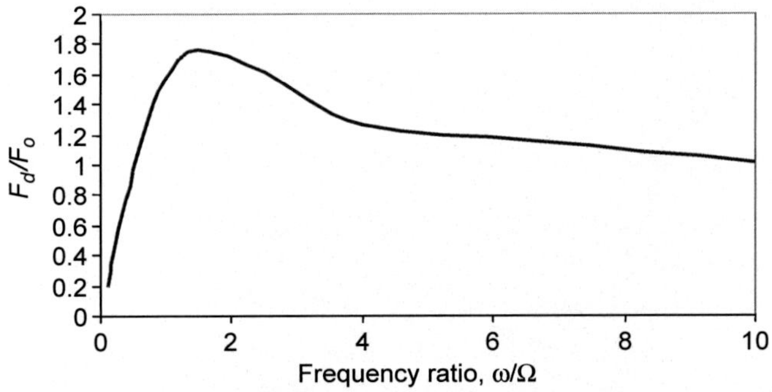

Figure 12.4 Dynamic amplification of load cell.

12.3 Characterisations of solders using drop-weight test

12.3.1 Materials and experimental set-up

The stress–strain properties of four tin-based solders are characterised at eight engineering strain rates and these are listed in Table 12.2. Two specimens were tested at each strain rate. The engineering strain rates $\dot{\varepsilon}_o$ between 0.005 s^{-1} and 6.0 s^{-1} were obtained using the standard Instron Micro-Force Tester, the crosshead of which was driven at constant speed. For the engineering strain rate of 6.0 s^{-1}, a level was

Table 12.2 Test matrix for solders

Solders		Initial strain rate, $\dot{\varepsilon}_o$ (s^{-1})	Test methods	Drop weight
Compositions	**Designation**			
63.0% Sn, 37.0% Pb	Sn37Pb	0.005	Mech. tester	N.A.
98.5% Sn–1.0% Ag–0.1% Cu	SAC101	0.05	Mech. tester	N.A.
95.5% Sn–3.0% Ag–0.5% Cu	SAC305	1.0	Mech. tester	N.A.
96.5% Sn–3.5% Ag	Sn3.5Ag	6.0	Mech. tester (with leverage)	N.A.
		65	Drop weight	55 kg
		100	Drop weight	55 kg
		200	Drop weight	45 kg
		300	Drop weight	45 kg

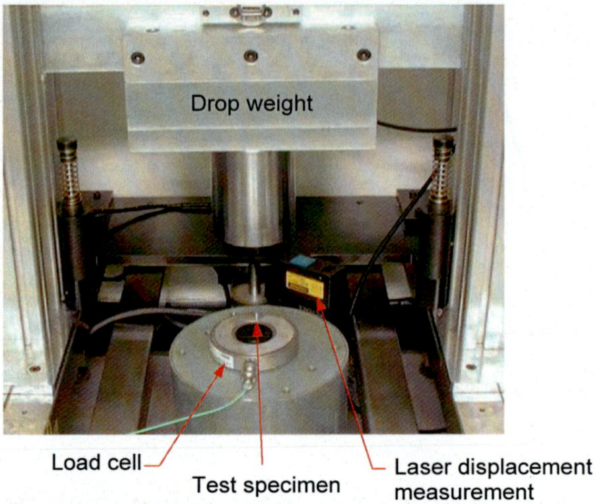

Figure 12.5 Drop-weight tester and test set-up.

used to amplify the crosshead speed by six times. The engineering strain rates $\dot{\varepsilon}_0$ between 65 s^{-1} and 300 s^{-1} were characterised using a drop-weight tester that was specially developed for this purpose. The specimens used in the test were 4 mm in diameter and in length. Test specimens were machined from extruded bars for Sn37 Pb, Sn-3.5Ag and SAC305 solder alloys, and from cast bars for SAC101 solder alloy.

A picture of the drop-weight tester is shown in Figure 12.5. The drop-weight tester consists of two parallel frames along which a horizontal structure attached with a weight of desired mass slides vertically. The drop weight is elevated using a motorised drive; at the desired height, the weight is released and falls squarely onto the specimen, compressing it between the drop weight and a load cell (Model 9081A from Kistler that has a resonant frequency of 50 kHz). The load cell is precompressed and is attached rigidly to an anvil of substantial mass to prevent the rigid body motion of the load cell during impact. The anvil is isolated from the test frame of the test apparatus to eliminate transmission of the impact wave to the test frame. The effect of contact friction is minimised through the application of graphite lubricant across the two contact surfaces of the specimen. The drop height is varied to achieve the desired engineering strain rate. Drop weights of 55, 55, 45 and 45 kg are used for $\dot{\varepsilon}_0 = 65$, 100, 200 and 300 s^{-1}, respectively. The higher drop weights for the lower strain rates are necessary to ensure that there is sufficient kinetic energy to induce the required strain in the specimen. For ease of mounting, the drop weight is offset from the two vertical frames. A laser distance measurement device (model LK-G87 from Keyence that has a resolution of 0.2 μm and dynamic data rate of 20 μs) is attached to the anvil to measure the displacement of the drop weight relative to the anvil. Unfortunately, laser measurements were only taken for strain rates of 65 and 100 s^{-1}.

12.3.2 The basic characteristics

12.3.2.1 Force–velocity–time characteristics

Mechanical tester

It is inevitable that the surfaces of the compression anvil and the specimen are not perfectly parallel to each other; a small pre-compression is applied to the specimen to take up this deviation. Figure 12.6 shows the force–velocity–time characteristics of Sn37 Pb solder obtained using a standard mechanical tester at $\dot{\varepsilon}_o = 1\ \mathrm{s}^{-1}$. A compressive force is registered as positive and a downward velocity of the drop weight is registered as positive. The crosshead is accelerated to the programed speed almost instantaneously followed inevitably by transitory oscillation before settling at the programed speed. The initial rapid rise in compressive force is attributable to the elastic deformations of the test specimen, the load cell and the test structure. This is followed by a more gradual linear increase in the magnitude of compressive force with time, which is the result of a combination of (1) strain hardening, (2) increasing diameter of the specimen $A(t) = A_o l_o/(l_o - V_o t)$ and (3) increasing strain rate $\dot{\varepsilon}(t) = V_o/(l_o - V_o t)$.

Drop-weight tester

Figure 12.7 shows the force–time characteristic for Sn37Pb solder for $\dot{\varepsilon}_o = 65\ \mathrm{s}^{-1}$, which shows ① an initial rapid rise in the contact force to $t \approx 0.5$ ms, followed by ② a period of sustained and gradual increase in contact force to a peak force at $t = 12$ ms, then by ③ a rapid descent in contact force until nil magnitude at $t = 14.3$ ms, and finally followed by ④ a period of nil force before another ⑤ impulse of force is registered between 16 ms $\leq t \leq$ 18 ms. Physically, response ① is the combination of the establishment of full area contact between the anvil and the test specimen and the elastic deformation of test specimen; response ② is associated with increasing diameter of test specimen accompanied by strain hardening; response ③ is associated with elastic unloading of the specimen; response ④ is associated with rebound and airborne of the drop weight; and response ④ is due to the secondary impact of the drop weight against the load cell after the rebound.

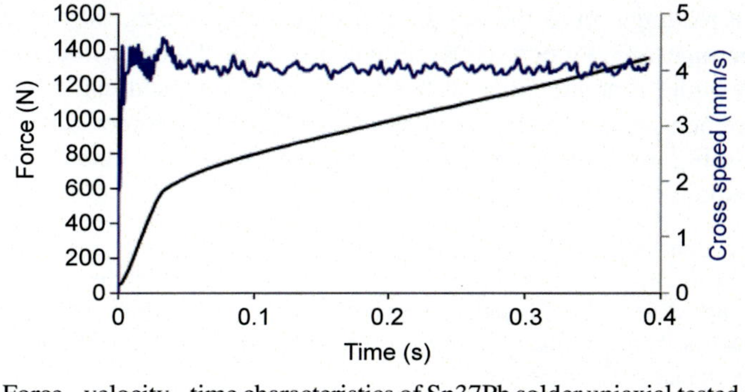

Figure 12.6 Force–velocity–time characteristics of Sn37Pb solder uniaxial tested at $\dot{\varepsilon}_o = 1\ \mathrm{s}^{-1}$.

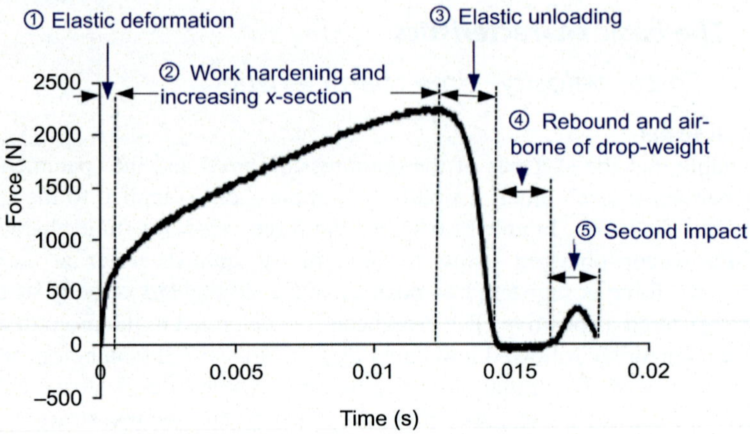

Figure 12.7 Force–time characteristics of the Sn37Pb solder for $\dot{\varepsilon}_o = 65\ s^{-1}$.

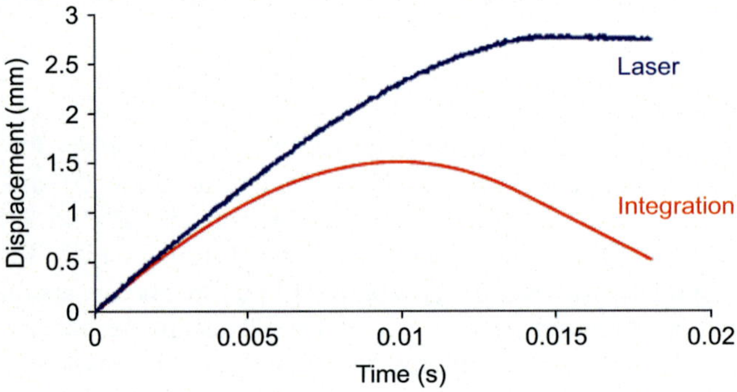

Figure 12.8 Comparison of the displacement measurements using direct integration and laser measurement instrument for Sn37Pb solder for $\dot{\varepsilon}_o = 65\ s^{-1}$.

Figure 12.8 shows the displacement–time characteristics of Sn37Pb solder at $\dot{\varepsilon}_o = 65\ s^{-1}$ for a specimen obtained using laser measurement (u_{laser}) and by integration (u_{int}) using Eqn (12.28), respectively. The two measurement methods give similar displacement readings up to the duration of 3 ms, after which, they diverge and the laser measurement gives consistently higher readings than the integration method. The velocity–time characteristics of the dead weight obtained using integration and laser measurement, respectively, are superimposed onto the force–time measurement and shown in Figure 12.9. Note that the compressive force is taken as positive. The following responses are envisaged:

- There are two instances when there is no impact force: at $t = 0$ and when the drop weight is airborne; the gradient of the velocity–time curves should be zero at these two instances. The first condition, $dV(t = 0)/dt$, is satisfied by both the V–t curves; however, the airborne condition is satisfied only by the V–t curve using integration while that by laser measurement shows a marginal delay in response.

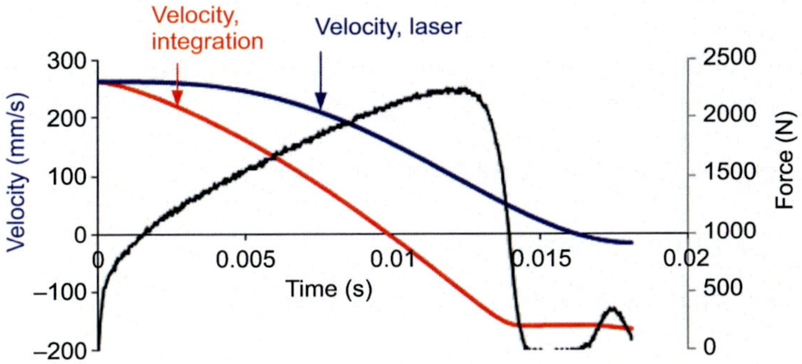

Figure 12.9 Force–velocity–time characteristics of Sn37Pb solder for $\dot{\varepsilon}_o = 65\ \mathrm{s}^{-1}$.

- The drop weight is expected to come to a full stop ($V = 0$) after the instance of peak load but before the instance of elastic unloading; this occurrence has been marginally delayed for the data using laser measurement and marginally advanced for the data using integration.
- Physically, the drop weight should have acquired a small magnitude of upward (negative) velocity during the initial rebound. This is reflected in the data using laser measurement. By contrast, the rebound velocity obtained using integration appears to be grossly excessive.

Overall, it appears that the laser measurements give more reliable displacement data.

The difference in displacement measurements between the laser and the integration methods translates to the difference in the measured strain in the test specimen. Mathematically, this is given by:

$$\Delta\varepsilon = \varepsilon_{\text{integration}} - \varepsilon_{\text{laser}} = \ln\left(\frac{l_o - u_{\text{integration}}}{l_o - u_{\text{laser}}}\right) \tag{12.30}$$

Figure 12.10 shows plots of deviation in strain, $\Delta\varepsilon$, versus the strain obtained using laser measurement, $\varepsilon_{\text{laser}}$, for the Sn37Pb solder at $\dot{\varepsilon}_o = 65\ \mathrm{s}^{-1}$ and $\dot{\varepsilon}_o = 100\ \mathrm{s}^{-1}$.

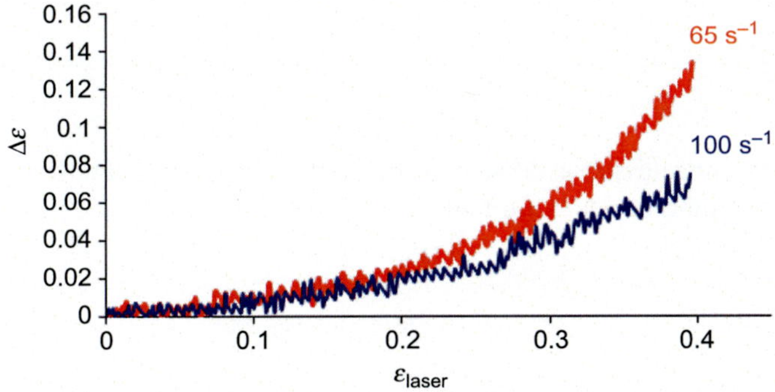

Figure 12.10 Deviation in the measured strain for Sn37Pb solder at $\dot{\varepsilon}_o = 65\ \mathrm{s}^{-1}$ and $\dot{\varepsilon}_o = 100\ \mathrm{s}^{-1}$.

Note that the deviation increases with increasing strain. Nevertheless, the absolute deviation is only 0.027 at $\varepsilon_{\text{laser}} = 0.2$ and 0.06 at $\varepsilon_{\text{laser}} = 0.3$ for $\dot{\varepsilon}_o = 65\ \text{s}^{-1}$; and the magnitude of deviation decreases with increasing strain rate. For all practical purposes, the deviation is deemed negligible. For this reason and for the reason of maintaining consistency between the force and the displacement measurements, the integration method is used in the subsequent analyses.

12.3.2.2 Strain rate characteristics

Differentiating strain rate with respect to strain gives:

$$\frac{\mathrm{d}\dot{\varepsilon}}{\mathrm{d}\varepsilon} = \frac{\mathrm{d}\left(Vl^{-1}\right)}{\mathrm{d}\varepsilon} = \frac{\mathrm{d}V}{l\mathrm{d}\varepsilon} + V\frac{\mathrm{d}l^{-1}}{\mathrm{d}\varepsilon} \tag{12.31}$$

Denoting compressive strain as positive, the instantaneous length of the specimen is given by:

$$l = l_o e^{-\varepsilon} \tag{12.32}$$

Substituting this into Eqn (12.31) gives:

$$\frac{\mathrm{d}\dot{\varepsilon}}{\mathrm{d}\varepsilon} = \frac{e^{\varepsilon}}{l_o}\left(\frac{\mathrm{d}V}{\mathrm{d}\varepsilon} + V\right) = \frac{1}{l_o}\frac{\mathrm{d}}{\mathrm{d}\varepsilon}\left(e^{\varepsilon}V\right) \tag{12.33}$$

Integrating Eqn (12.33) w.r.t. strain and together with the initial condition $\dot{\varepsilon}_o = V_o/l_o$ give the equation for strain rate as:

$$\dot{\varepsilon} = \frac{e^{\varepsilon}V}{l_o} \tag{12.34}$$

The ratio of the instantaneous (true) strain rate to the initial (engineering) strain rate is given by:

$$\frac{\dot{\varepsilon}}{\dot{\varepsilon}_o} = \frac{e^{\varepsilon}V}{V_o} \tag{12.35}$$

Fractional strain rate deviation $\Delta\dot{\varepsilon}$ may be defined as the fractional deviation of the true strain rate $\dot{\varepsilon}$ from the initial strain rate $\dot{\varepsilon}_o$:

$$\Delta\dot{\varepsilon} = \frac{\dot{\varepsilon}}{\dot{\varepsilon}_o} - 1 \tag{12.36}$$

Mechanical tester

Substituting $V = V_o$ into Eqn (12.35) gives:

$$\dot{\varepsilon} = \dot{\varepsilon}_o e^{\varepsilon} \tag{12.37}$$

That is, the strain rate in the test specimen increases exponentially with strain. This is illustrated in Figure 12.11 using the experimental data of Sn37Pb solder at $\dot{\varepsilon}_o = 1\ \text{s}^{-1}$. The fractional strain rate deviation is given by:

$$\Delta\dot{\varepsilon} = e^{\varepsilon} - 1 \tag{12.38}$$

Thus, the fractional deviation of the true strain rate from the initial strain rate increases with increasing strain. At $\varepsilon = 0.5$, $\Delta\dot{\varepsilon} = 65\%$, which is very significant.

Drop-weight tester

Figure 12.12 shows the experimental velocity versus strain characteristic of the Sn37Pb solder at $\dot{\varepsilon}_o = 100\ \text{s}^{-1}$ for $0 \le \varepsilon \le 0.5$, which can be described by the parabolic equation:

$$V = -201\varepsilon^2 - 151\varepsilon + 403 \tag{12.39}$$

Also shown in the figure are the experimental strain rate characteristics of the solder, which suggests a non-monotonic variation of strain rate – increasing strain rate from $\varepsilon = 0$ to peak strain rate at $\varepsilon \approx 0.4$, followed by decreasing strain rate. The exact instance of strain that corresponds to maximum strain rate can be evaluated mathematically by substituting Eqn (12.39) into Eqn (12.33) to give:

$$\frac{d\dot{\varepsilon}}{d\varepsilon} = \frac{e^{\varepsilon}}{l_o}\left(-201\varepsilon^2 - 553\varepsilon + 252\right) \tag{12.40}$$

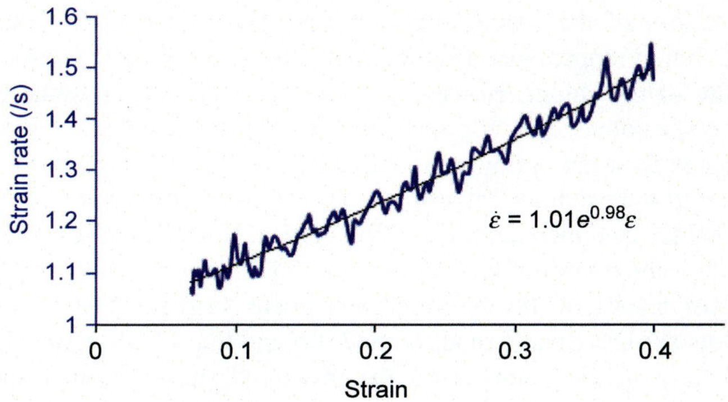

Figure 12.11 $\dot{\varepsilon}$ versus ε for SnPb solder at $\dot{\varepsilon}_o = 1\ \text{s}^{-1}$.

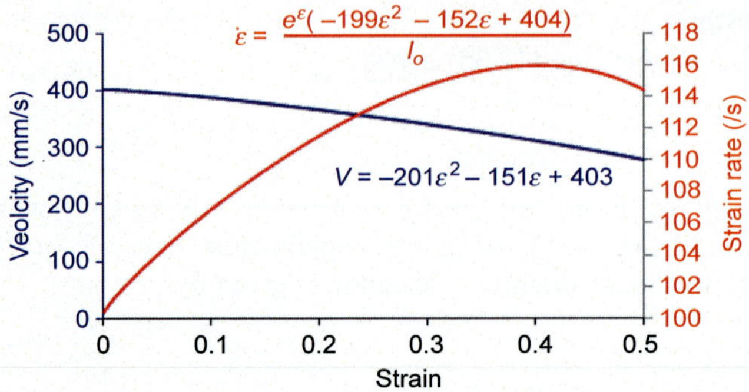

Figure 12.12 Experimental velocity and strain rate versus strain and the fitting equations for SnPb solder at $\dot{\varepsilon}_o = 100\ s^{-1}$.

and setting $d\dot{\varepsilon}/d\varepsilon = 0$ to arrive at $\varepsilon = 0.40$. The mathematical equation for strain rate can be evaluated by substituting Eqn (12.39) into Eqn (12.34) giving:

$$\dot{\varepsilon} = \frac{e^{\varepsilon}\left(-201\varepsilon^2 - 151\varepsilon + 403\right)}{l_o} \tag{12.41}$$

The coefficients $\{-201, -151, 403\}$ agree remarkably well with the coefficients $\{-199, -152, 404\}$ obtained from least-squares fitting of the test data as shown in Figure 12.12. The maximum fractional strain rate deviation is 16%.

12.3.2.3 Visual appearance of test specimens

Figure 12.13 shows the visual appearances of deformed specimens of Sn37Pb, SAC101, Sn3.5Ag and SAC305 solder alloys. In general, the specimens suffer very limited bellowing; therefore, the conditions of uniaxial stress and uniform cross-sectional stress/strain can be assumed for these specimens. The circumferential surfaces of the solder materials show rather distinct characteristics. A wrinkled pattern is observed along the circumferential surface of the Sn37Pb solder specimen (Figure 12.13(a)), probably due to the collapse of the oxide layer. Interestingly, the SAC101 solder specimen displayed 3-D lumpy appearance (Figure 12.13(b)), possibly a reflection of the large grainy structure of the solder. However, cracks are observed on some Sn3.5Ag and SAC305 solder specimens. Sn3.5Ag solder specimens typically exhibit shallow circumferential cracks on their top or bottom surface (Figure 12.13(c) and (d)) that do not typically propagate to their cylindrical surface. The crack is attributed to contact friction. The SAC305 solder specimen tends to exhibit vertical crack along its cylindrical surface (Figure 12.13(e)) and grows inward in the circumferential direction (Figure 12.13(f)). The presence of cracks in the deformed specimens suggests that the Sn3.5Ag and SAC305 solders are less ductile than the Sn37Pb and the SAC101 solders. Inevitably, cracking will have a direct impact on the measured stress–strain responses of the specimens.

(a) (b)
(c) (d)
(e) (f)

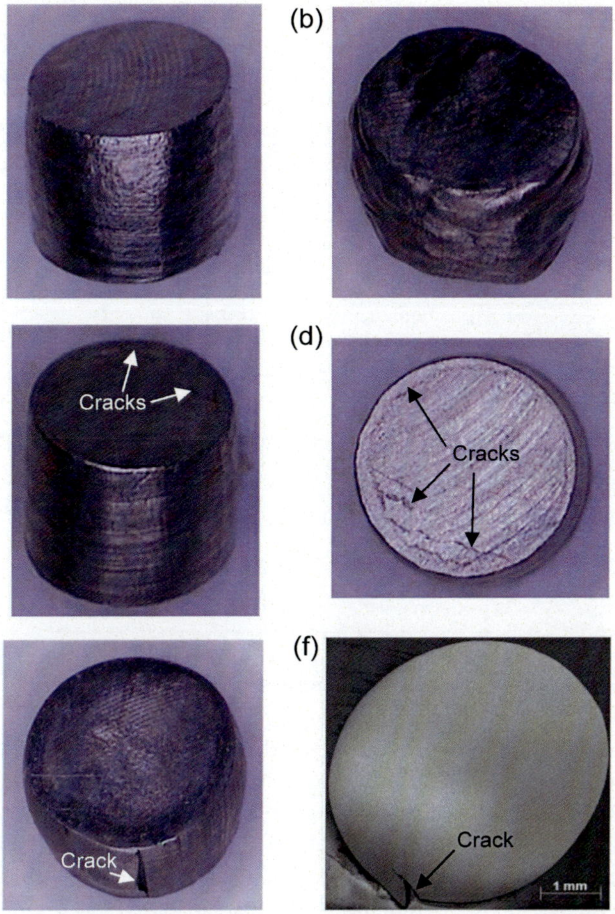

Figure 12.13 Distinct characteristics of deformed solder specimens: (a) Sn37Pb, (b) SAC101, (c) and (d) Sn3.5Ag, and (e) and (f) SAC305.

12.4 The stress–strain characteristics of solders

12.4.1 Expressed as non-iso-strain rate

The stress–strain characteristics of Sn37Pb, SAC101, Sn3.5Ag and SAC305 solders are shown in Figure 12.14 identified by the initial strain rate $\dot{\varepsilon}_o = V_o/l_o$. It is noted that the elastic response of the solders is unable to be reliably measured using this method in view of the significant elastic deformation of the test structure including the load cell. For practical reasons, the terms *strain* and *plastic strain* are interchangeable in the following discussions.

High-frequency resonant noises from the load cell are observed for the higher strain rate but damped off rapidly with increasing strain. A low-pass filter set at a frequency

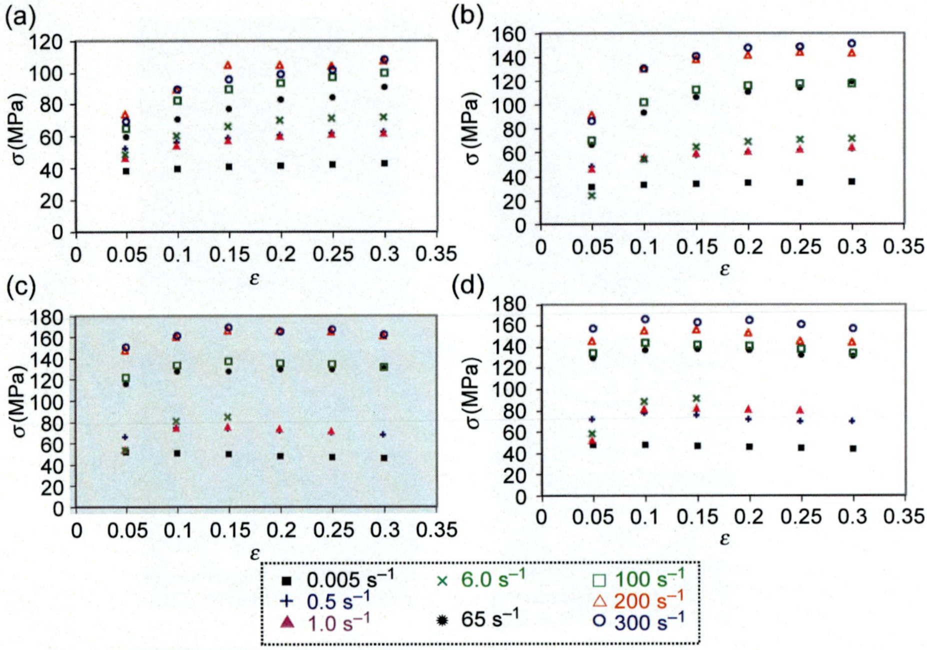

Figure 12.14 Stress–strain characteristics of (a) Sn37Pb, (b) SAC101, (c) Sn3.5Ag, and (d) SAC305 expressed in curves of initial strain rate $\dot{\varepsilon}_o = V_o/l_o$.

just below the resonant frequency of the load cell is used to filter off the noise. Nevertheless, the quality of the data at the low strain region ($|\varepsilon| < 0.05$) may have been compromised by the noises and filtering. 'Strain softening' is observed for some of the solder alloys at a very low strain rate of 0.005 s^{-1} attributed to creep. However, strain softening is observed for Sn3.5Ag and SAC305 solders for almost all strain rates. Given the evidence of microcracking in the test specimens of Sn3.5Ag and SAC305 solders, it is highly likely that the observed strain softening at higher strain rate for the two solders is an artefact of microcracking rather than true material response.

12.4.2 Expressed as iso-strain rate

It is also clear from the discussion in Section 12.3.2 that the experimental strain rate in a mechanical or a drop-weight test is a function of strain; that is, $\dot{\varepsilon}(\varepsilon)$; hence, the experimental stress–strain data are not obtained at constant strain rate. For the purposes of constructing a true rate-dependent constitutive equation, these experimental stress–strain data must be transformed to an iso-strain rate condition. The transformation relation can be established using any of the five constitutive equations from Eqn (12.1) to Eqn (12.5). Using Eqn (12.4) as an illustration and assuming σ_o independent of strain rate, the stress–strain data pair $\{\sigma_{\dot{\varepsilon}_1}, \dot{\varepsilon}_1\}$ may be transformed to $\{\sigma_{\dot{\varepsilon}_2}, \dot{\varepsilon}_2\}$ using the relation:

$$\sigma_{\dot{\varepsilon}_2} = \sigma_{\dot{\varepsilon}_1}\left(\frac{\dot{\varepsilon}_2}{\dot{\varepsilon}_1}\right)^{\alpha} \tag{12.42}$$

Designating the stress and the instantaneous strain rate under the condition of non-iso-strain rate as $\sigma_{\dot{\varepsilon}_i}(\varepsilon)$ and $\dot{\varepsilon}_i(\varepsilon)$, respectively, these may be transformed to the desired iso-true-strain rate as:

$$\sigma_{\dot{\varepsilon}_{\text{iso}}}(\varepsilon) = \sigma_{\dot{\varepsilon}_i}(\varepsilon)\left(\frac{\dot{\varepsilon}_{\text{iso}}}{\dot{\varepsilon}_i(\varepsilon)}\right)^{\alpha} \tag{12.43}$$

Choosing the initial strain rate $\dot{\varepsilon}_o$ as the desired iso-true strain rate, that is, $\dot{\varepsilon}_{\text{iso}} \rightarrow \dot{\varepsilon}_o$, and substituting with Eqn (12.35), $\dot{\varepsilon}_i/\dot{\varepsilon}_o = e^{\varepsilon}V_i/V_o$ gives:

$$\sigma_{\dot{\varepsilon}_{\text{iso}}}(\varepsilon) = \sigma_{\dot{\varepsilon}_i}(\varepsilon)\left(\frac{V_o}{e^{\varepsilon}V(\varepsilon)}\right)^{\alpha} \tag{12.44}$$

The fractional stress deviation Δ_σ is defined as:

$$\Delta_\sigma = \frac{\sigma_{\dot{\varepsilon}_i}}{\sigma_{\dot{\varepsilon}_{\text{iso}}}} - 1 = \left(\frac{\dot{\varepsilon}_i}{\dot{\varepsilon}_{\text{iso}}}\right)^{\alpha} - 1 \tag{12.45}$$

Choosing the initial strain rate $\dot{\varepsilon}_o$ as the desired iso-true-strain rate, then:

$$\Delta_\sigma = \left(\frac{e^{\varepsilon}V(\varepsilon)}{V_o}\right)^{\alpha} - 1 \tag{12.46}$$

12.4.2.1 Mechanical tester

For tests performed using a mechanical tester, $V = V_o$, the stress transformation equation, Eqn (12.44), becomes:

$$\sigma_{\dot{\varepsilon}_{\text{iso}}} = \sigma_{\dot{\varepsilon}_i} e^{-\alpha\varepsilon} \tag{12.47}$$

and the fractional stress deviation becomes

$$\Delta_\sigma = e^{\alpha\varepsilon} - 1 \tag{12.48}$$

For $\alpha\varepsilon < 0.1$, $e^{\alpha\varepsilon} \approx 1 + \alpha\varepsilon$; and

$$\Delta_\sigma \approx \alpha\varepsilon \tag{12.48a}$$

That is, the fractional stress difference Δ_σ increases almost linearly with strain with a gradient α, independent of the initial strain rate. Note that at $\varepsilon = 0.5$ and $\alpha = 0.1$, $\Delta_{\dot{\varepsilon}}$ is 65% while Δ_σ is a mere 5%; the effect of nonuniform strain rate on the true magnitude of flow stress is therefore not as alarming as it might have appeared.

12.4.2.2 Drop-weight tester

For tests performed using the drop-weight technique and for $\alpha\varepsilon < 0.1$, substituting $e^{\alpha\varepsilon} \approx 1 + \alpha\varepsilon$ into Eqn (12.46) gives:

$$\Delta_\sigma = (1 + \alpha\varepsilon)\left(\frac{V}{V_o}\right)^\alpha - 1 \tag{12.49}$$

For a drop weight of large kinetic energy, the ratio V/V_o will be less than but near unity, and so is the function $(V/V_o)^\alpha$; the fractional stress error will be bounded and fall within the range $0 \le \Delta_\sigma \le \alpha\varepsilon$. Conversely, the ratio V/V_o for a drop weight of small kinetic energy may become significantly less than unity and the fractional stress error will be unbounded and $\Delta_\sigma \le 0$. Thus, we should always guard against a drop weight of insufficient kinetic energy. This is particularly susceptible when testing at low strain rate.

The fractional stress difference Δ_σ for the Sn3.5A5 solder at the initial strain rate of $\dot{\varepsilon}_o = 65$, 100, 200 and 300 s^{-1} is shown in Figure 12.15. The magnitudes of Δ_σ are less than 4% at $\varepsilon = 0.3$ for $\dot{\varepsilon}_o = 100$, 200 and 300 s^{-1}; but note that $\Delta_\sigma = -6\%$ at $\varepsilon = 0.3$ for $\dot{\varepsilon}_o = 65\ s^{-1}$ and increased exponentially to 15% at $\varepsilon = 0.35$. In hindsight, a significantly larger mass of drop weight should have been used for the test at $\dot{\varepsilon}_o = 65\ s^{-1}$.

Using the transformation equation, Eqn (12.44), the experimental flow stresses of the solders have been mapped into iso-strain rate at $\dot{\varepsilon} = \dot{\varepsilon}_o$, and these are given in Table 12.3 for six strains and eight strain rates. Note that the flow stress for Sn3.5Ag and SAC305 solders at $\dot{\varepsilon} = 1\ s^{-1}$ and $\dot{\varepsilon} = 6\ s^{-1}$, and for SAC101 solder at $\dot{\varepsilon}_o = 6\ s^{-1}$, are missing due to unplanned experimental procedures.

12.5 The constitutive equations of solders

12.5.1 $\sigma = S_o\varepsilon^n\dot{\varepsilon}^\alpha$

Rewriting Eqn (12.4) as:

$$\sigma = \sigma_o(\varepsilon)\dot{\varepsilon}^\alpha \tag{12.50}$$

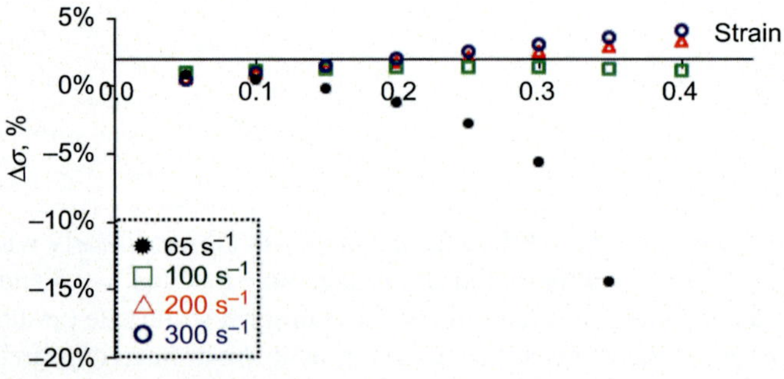

Figure 12.15 Fractional stress difference Δ_σ of Sn3.5Ag solder.

Table 12.3 **Stress–strain at iso-strain rates**

Mat	Fitted equation, Eqn (12.50) $\sigma = S_o \varepsilon^n \dot{\varepsilon}^\alpha$	ε	$\dot{\varepsilon}$ 0.005	0.5	1	6	65	100	200	300
Sn37Pb	$S_o = 84.2$	0.05	37.9	52.2	46.9	48.5	59.8	64.5	74.0	69.2
	$\alpha = 0.077$	0.10	39.6	56.6	54.6	60.4	70.3	82.1	89.5	89.5
	$n = 0.182$	0.15	40.8	59.1	57.9	66.2	77.0	89.3	104.7	96.0
	$\zeta_o = 21.1$	0.20	41.6	60.8	60.0	69.9	82.9	93.0	104.6	99.1
		0.25	42.3	61.9	61.5	71.5	84.3	97.2	104.2	101.6
		0.30	42.7	62.5	62.1	72.2	90.5	99.7	107.4	108.3
SAC101	$S_o = 96.7$	0.05	30.8	48.2	46.9	24.5	66.2	70.0	91.1	86.3
	$\alpha = 0.124$	0.10	32.8	54.3	55.8	54.7	93.2	101.8	130.8	130.6
	$n = 0.265$	0.15	33.7	57.7	59.2	64.8	106.3	112.0	137.9	140.3
	$\zeta_o = 90.3$	0.20	34.2	60.3	61.5	68.9	110.3	115.6	141.2	147.4
		0.25	34.7	62.1	63.3	71.1	114.4	117.3	143.8	148.2
		0.30	35.1	63.4	64.3		119.5	117.5	143.4	150.9

Continued

Table 12.3 Continued

Mat	Fitted equation, Eqn (12.50) $\sigma = S_o \varepsilon^n \dot{\varepsilon}^\alpha$	ε	$\dot{\varepsilon}$							
			0.005	0.5	1	6	65	100	200	300
Sn3.5Ag	$S_o = 94.2$	0.05	51.0	65.8	54.4	52.9	115.8	121.9	147.8	150.8
	$\alpha = 0.115$ $n = 0.092$	0.10	50.2	74.4	74.3	80.9	127.1	133.5	160.7	160.8
	$\zeta_o = 81.8$	0.15	49.2	73.7	75.3	84.7	127.4	136.7	166.6	169.1
		0.20	47.7	71.8	74.0		129.3	135.4	165.9	165.0
		0.25	46.3	69.5	72.1		129.0	133.8	164.9	166.7
		0.30	45.2	68.1			130.3	131.1	160.9	162.6
SAC305	$S_o = 88.8$	0.05	47.7	71.1	51.1	56.8	129.2	134.3	144.6	157.1
	$\alpha = 0.117$ $n = 0.059$	0.10	47.6	77.1	80.8	87.3	136.8	143.6	154.7	166.6
	$\zeta_o = 42.0$	0.15	47.0	74.9	81.8	90.1	138.3	141.9	155.3	163.2
		0.20	45.6	71.1	80.9		137.0	141.3	152.4	165.1
		0.25	44.4	68.7	79.3		132.4	138.1	144.5	161.4
		0.30	43.6	68.6			130.7	133.6	143.9	157.6

where

$$\sigma_o(\varepsilon) = S_o\varepsilon^n \tag{12.50a}$$

and S_o, n and α are constants, the values of which for the respective solders have been evaluated through regression by minimising $\zeta_{\ln} = \sum_i^m [\ln(\sigma_{\exp}) - \ln(\sigma_{Eq(50)})]_i^2$, where $m = 8$ (strain rate) times 6 (strain) for full sets of data, but may be smaller for some sets of data; for example, Sn3.5Ag solder has a total of 45 sets of data. The use of logarithmic data leads to optimal regression for the power-law equation. The coefficients S_o, n and α together with the coefficient of fitting error, $\zeta_o = \frac{\sum_i^m (\sigma_{\exp} - \sigma_{Eq(50)})_i^2}{m}$, are tabulated in the second column of Table 12.3. The following observations are made:

- Sn37Pb solder has the least rate sensitivity at 0.077, which agrees with the value of 0.08 reported by Clyens (Clyens & Campbell, 1974) for Sn38Pb solder.
- SAC101 solder has the highest rate sensitivity and strain hardening, which agree with the stress–strain plots shown in Figure 12.14; it also has the highest value of coefficient S_o, which, however, does not agree with the stress–strain plots shown in Figure 12.14.
- Sn3.5Ag and SAC305 solders show the least strain hardening, which could be due to microcracking of the test specimen.

12.5.2 $\sigma = S(\varepsilon, \dot{\varepsilon})\varepsilon^{n(\dot{\varepsilon})}\dot{\varepsilon}^{\alpha}$

Expressing $\sigma = S_o\varepsilon^n\dot{\varepsilon}^\alpha$ as $\ln\sigma = \ln(S_o\varepsilon^n) + \alpha\ln(\dot{\varepsilon})$, the coefficients $S_o\varepsilon^n$ and α for the respective strain have been evaluated and are tabulated in the third last and last columns of Table 12.4, which is an expansion of Table 12.3. Substituting the value of coefficient n given in Table 12.3 into $S_o\varepsilon^n$, the coefficient S_o is then evaluated and shown in the second last column of Table 12.4. The following observations are made:

- The coefficient S_o decreases with increasing the magnitude of strain as shown in Figure 12.16. The value of S_o computed at $\varepsilon = 0.05$ deviates significantly from the linear trend of the others. It is initially suspected that this is caused by the contribution of elastic strain becoming significant at small strain thus distorting the accuracy of the data. However, further analysis, which will be elaborated, suggests that the nonlinear trend can be real. Nevertheless, a linear trend $S_1(\varepsilon) = S_{o,1}(1 + a_1\varepsilon)$ is assumed for simplicity. This is shown at the bottom of the column.
- The strain rate sensitivity index α is almost independent of strain. An average value is evaluated and tabulated at the bottom of the column.

Expressing $\sigma = S_o\varepsilon^n\dot{\varepsilon}^\alpha$ as $\ln\sigma = \ln(S_o\dot{\varepsilon}^\alpha) + n\ln(\varepsilon)$, the coefficients $S_o\dot{\varepsilon}^\alpha$ and n for respective strain rates are evaluated and are tabulated in the rows following the stress data. Substituting the value of coefficient α in Table 12.3 into $S_o\dot{\varepsilon}^\alpha$, the coefficient S_o is then evaluated. The following observations are made:

- The absence of adequate data for Sn3.5Ag and SAC305 solders at $\dot{\varepsilon}_o = 1\ \text{s}^{-1}$ and $\dot{\varepsilon}_o = 6\ \text{s}^{-1}$ and SAC101 solder at $\dot{\varepsilon}_o = 6\ \text{s}^{-1}$ has affected rather significantly the accuracy of α and S_o evaluated. The affected values of α and a_o are ignored in the subsequent calculations.

Table 12.4 Stress–strain at iso-strain rates (expansion)

Mat	Fitted equation Eqn (12.51) $\sigma = S(\varepsilon, \dot{\varepsilon})\varepsilon^{n(\dot{\varepsilon})}\dot{\varepsilon}^{\alpha}$ $S_1(\varepsilon) = S_{o,1}(1+a_1\varepsilon)$ $S_2(\dot{\varepsilon}) = S_{o,2}(1+a_2 \ln\dot{\varepsilon})$ $n(\dot{\varepsilon}) = n_o(1+n_1 \ln\dot{\varepsilon})$	ε	$\dot{\varepsilon}$ 0.005	0.5	1	6	65	100	200	300	$S_1\varepsilon^n$	S_1	α
Sn37Pb	$\alpha = 0.077$	0.05	37.9	52.2	46.9	48.5	59.8	64.5	74.0	69.2	49.7	85.7	0.056
		0.10	39.6	56.6	54.6	60.4	70.3	82.1	89.5	89.5	56.8	86.4	0.073
		0.15	40.8	59.1	57.9	66.2	77.0	89.3	104.7	96.0	60.6	85.5	0.081
		0.20	41.6	60.8	60.0	69.9	82.9	93.0	104.6	99.1	62.7	84.0	0.082
		0.25	42.3	61.9	61.5	71.5	84.3	97.2	104.2	101.6	63.9	82.2	0.082
		0.30	42.7	62.5	62.1	72.2	90.5	99.7	107.4	108.3	65.2	81.2	0.086
	$S_{o,1} = 88$ $S_{o,2} = 80$ $a_1 = -0.24$	$S_2\dot{\varepsilon}^{\alpha}$	46.4	71.3	76.6	98.4	117.6	136.3	143.8	144.5		$S_{o,1}(1 + a_1 \varepsilon)$	0.077
	$a_2 = 0.031$	S_2	69.7	75.2	76.6	85.7	85.4	95.7	95.8	93.3	$S_{o,2}(1 + a_2 \ln \dot{\varepsilon})$		
	$n_o = 0.15$ $n_1 = 0.105$	n	0.069	0.102	0.157	0.225	0.225	0.238	0.209	0.232	$n_o(1 + n_1 \ln \dot{\varepsilon})$		
SAC101	$\alpha = 0.124$	0.05	30.8	48.2	46.9	24.5	66.2	70.0	91.1	86.3	44.2	97.7	0.094
		0.10	32.8	54.3	55.8	54.7	93.2	101.8	130.8	130.6	57.7	106.3	0.126
		0.15	33.7	57.7	59.2	64.8	106.3	112.0	137.9	140.3	62.2	102.8	0.131
		0.20	34.2	60.3	61.5	68.9	110.3	115.6	141.2	147.4	64.4	98.6	0.132
		0.25	34.7	62.1	63.3	71.1	114.4	117.3	143.8	148.2	65.9	95.1	0.132
		0.30	35.1	63.4	64.3		119.5	117.5	143.4	150.9	66.9	92.0	0.132
	$S_{o,1} = 105$	$S_2\dot{\varepsilon}^{\alpha}$	38.3	76.9	81.0	153.2	183.2	178.0	205.6	232.1		$S_{o,1}(1 + a_1 \varepsilon)$	0.124
	$S_{o,2} = 88$ $a_1 = -0.34$ $a_2 = 0.041$	S_2	74.1	83.8	81.0	0.488	109.0	100.3	106.3	114.1	$S_{o,2}(1 + a_2 \ln \dot{\varepsilon})$		
	$n_o = 0.18$ $n_1 = 0.117$	n	0.070	0.154	0.175	122.5	0.320	0.282	0.242	0.298	$n_o(1 + n_1 \ln \dot{\varepsilon})$		

Sn3.5Ag	$\alpha = 0.115$	0.05	51.0	65.8	54.4	52.9	115.8	121.9	147.8	150.8	69.7	91.7	0.109
		0.10	50.2	74.4	74.3	80.9	127.1	133.5	160.7	160.8	80.6	99.5	0.110
		0.15	49.2	73.7	75.3	84.7	127.4	136.7	166.6	169.1	81.2	96.6	0.115
		0.20	47.7	71.8	74.0		129.3	135.4	165.9	165.0	81.7	94.7	0.117
		0.25	46.3	69.5	72.1		129.0	133.8	164.9	166.7	80.0	90.8	0.120
		0.30	45.2	68.1			130.3	131.1	160.9	162.6	80.3	89.7	0.118
	$S_{o,1} = 98$ $S_{o,2} = 83$ $a_1 = -0.22$	$S_2\dot{\varepsilon}^{\alpha}$	42.5	72.2	98.9	208.9	141.7	143.0	177.9	177.8		$S_{o,1}(1 + a_1\, \varepsilon)$	0.115
	$a_2 = 0.017$	S_2	78.2	78.2	0.173	0.448	87.6	84.2	96.7	92.2	$S_{o,2}(1 + a_2 \ln \dot{\varepsilon})$		
	$n_o = 9.6\times10^{-4}$ $n_1 = 11.2$	n	−0.067	0.013	98.9	170.0	0.060	0.042	0.053	0.048	$n_o(1 + n_1 \ln \dot{\varepsilon})$		
SAC305	$\alpha = 0.117$	0.05	47.7	71.1	51.1	56.8	129.2	134.3	144.6	157.1	70.5	84.1	0.119
		0.10	47.6	77.1	80.8	87.3	136.8	143.6	154.7	166.6	82.7	94.7	0.115
		0.15	47.0	74.9	81.8	90.1	138.3	141.9	155.3	163.2	82.5	92.2	0.115
		0.20	45.6	71.1	80.9		137.0	141.3	152.4	165.1	82.0	90.2	0.118
		0.25	44.4	68.7	79.3		132.4	138.1	144.5	161.4	79.7	86.4	0.118
		0.30	43.6	68.6			130.7	133.6	143.9	157.6	78.6	84.4	0.117
	$S_{o,1} = 91$ $S_{o,2} = 78$ $a_1 = -0.16$	$S_2\dot{\varepsilon}^{\alpha}$	41.8	67.4	126.5	220.2	135.7	138.6	147.1	162.7		$S_{o,1}(1 + a_1\, \varepsilon)$	0.117
	$a_2 = 0.008$	S_2	77.5	73.0	0.265	0.441	83.4	81.0	79.3	83.6	$S_{o,2}(1 + a_2 \ln \dot{\varepsilon})$		
	$n = -0.014$	n	−0.051	−0.034	126.5	178.6	0.007	−0.001	−0.007	0.003	$n_o(1 + n_1 \ln \dot{\varepsilon})$		

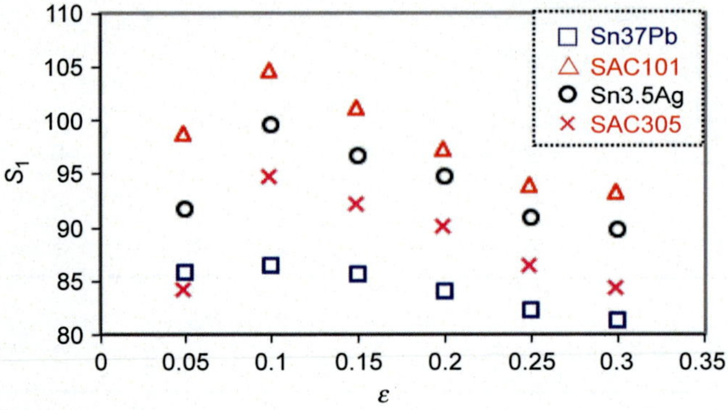

Figure 12.16 Function $S_1(\varepsilon)$.

- The coefficient S_o increases almost logarithmically with increasing strain rate; that is, $S_2(\dot{\varepsilon}) = S_{o,2}(1 + a_2 \ln \dot{\varepsilon})$. This is shown to the right of the row.
- Strain hardening index n increases almost logarithmically with increasing strain rate: $n(\dot{\varepsilon}) = n_o[1 + n_1 \ln(\dot{\varepsilon})]$. This is shown to the right of the row.
- Sn3.5Ag and SAC305 solders exhibit little to no distinguishable dependency on strain rate. For this reason, an average hardening index is computed for the respective solders.
- Sn3.5Ag and SAC305 solders exhibit a negative harden index at $\dot{\varepsilon} = 0.005\ \text{s}^{-1}$, possibly a result of creep. The SAC305 solder gives a negative hardening index even at higher strain rates (100 and 200 s^{-1}), likely to be a result of macrocrack.

The coefficients $S_{o,1}$, $S_{o,2}$, a_1, a_2, n_o, n_1 and α for respective solders are summarised in the second column of Table 12.4.

With the new insights gained from the analysis, the constitutive equation is refined to:

$$\sigma = S(\varepsilon, \dot{\varepsilon})\varepsilon^{n(\dot{\varepsilon})}\dot{\varepsilon}^{\alpha} \tag{12.51}$$

where it is suggested that:

$$S(\varepsilon, \dot{\varepsilon}) = S_o(1 + a_1\varepsilon + a_2 \ln \dot{\varepsilon}) \text{ and } n(\dot{\varepsilon}) = n_o(1 + n_1 \ln \dot{\varepsilon}) \tag{12.51a}$$

However, the fact that $S_{o,1} \neq S_{o,2}$ renders the expression $S(\varepsilon, \dot{\varepsilon}) = S_o(1 + a_1\varepsilon + a_2 \ln \dot{\varepsilon})$ invalid.

12.5.3 $\sigma = S(\varepsilon, \varepsilon', \dot{\varepsilon})\varepsilon^{n(\dot{\varepsilon})}\dot{\varepsilon}^{\alpha}$

Referring to plots of S_1 versus strain in Figure 12.16, the magnitude of S_1 is observed to take a nosedive for $\varepsilon \leq 0.10$. The linearization $S_1(\varepsilon) = S_{o,1}(1 + a_1\varepsilon)$ has significantly overestimated the magnitude of $S_1(\varepsilon = 0) = S_{o,1}$. The error from linearizing $S_1(\varepsilon)$ can be minimised by translating the reference strain from $\varepsilon = 0$ to $\varepsilon = \varepsilon'$, where $\varepsilon' \geq 0.10$. This is expressed mathematically as $S_1(\varepsilon) = S_{o,1}[1 + a_1(\varepsilon - \varepsilon')]$. The coefficient

$S_{o,1}$ can be made equal to the coefficient $S_{o,2}$ through appropriate selection of ε'. The refined constitutive equation becomes:

$$\sigma = S(\varepsilon, \varepsilon', \dot{\varepsilon})\varepsilon^{n(\dot{\varepsilon})}\dot{\varepsilon}^{\alpha} \tag{12.52}$$

where

$$S(\varepsilon, \varepsilon', \dot{\varepsilon}) = S_o[1 + a_1(\varepsilon - \varepsilon') + a_2 \ln \dot{\varepsilon}], n(\dot{\varepsilon}) = n_o[1 + n_1 \ln(\dot{\varepsilon})] \tag{12.52a}$$

The coefficients S_o, a_1, a_2, ε', α, n_o and n_1 for respective solders have been evaluated through regression by minimising $\zeta_1 = \frac{\sum_i^m (\sigma_{\text{exp}} - \sigma_{\text{eq}(52)})_i^2}{m}$. The complete sets of coefficients, including the error coefficient ζ_1 for respective solders, are tabulated in Table 12.5. The coefficients associated with Eqn (12.50) are also given for ease of comparison. On comparing the values of ζ_1 with ζ_o, it is clear that Eqn (12.52) gives a much better quality of fit than Eqn (12.50). The magnitudes of S_o for the four solders using Eqn (12.52) have much better agreement with the stress–strain plots of Figure 12.14.

12.5.4 $\sigma = S(\varepsilon, \varepsilon')\varepsilon^n\dot{\varepsilon}^{\alpha}$

The coefficients a_2 and n_1 are associated with strain rate. Their relatively small magnitudes are noted and their contribution is investigated. The magnitude of a_2 that will give rise to a unit change in S_o (or the magnitude of n_1 that will give rise to a unit change in n_o) is just $1/\ln(\dot{\varepsilon})$. For the maximum strain rate of 300 s^{-1}, this gives a value of 0.175. Since $a_2 \ll 0.175$ for the four solders, it is apparent that a_2 has little contribution to S_o and may be ignored. The function $a_o(\varepsilon, \varepsilon', \dot{\varepsilon})$ may be reduced to $S_o(\varepsilon, \varepsilon') = S_o[1 + a_1(\varepsilon - \varepsilon')]$. The contribution of n_1 to n_o ranges from 0% for $\dot{\varepsilon} = 1\ s^{-1}$ to 24% at $\dot{\varepsilon} = 300\ s^{-1}$ for the Sn37Pb solder, which is significant; however, for simplicity, the coefficient n_1 is set to null such that the hardening index n is now independent of strain rate. Equation (12.52) is reduced to:

$$\sigma = S(\varepsilon, \varepsilon')\varepsilon^n\dot{\varepsilon}^{\alpha} \tag{12.53}$$

where

$$S(\varepsilon, \varepsilon') = S_o[1 + a_1(\varepsilon - \varepsilon')] \tag{12.53a}$$

The coefficients S_o, a_1, ε', α and n for respective solder alloys are evaluated through regression by minimising $\zeta_2 = \frac{\sum_i^m (\sigma_{\text{exp}} - \sigma_{\text{eq}(53)})_i^2}{m}$, and the complete sets of coefficients are tabulated in Table 12.5. Comparing the values of ζ_2 with ζ_1, it becomes clear that Eqn (12.53) that is described with only five coefficients sacrifices little in the quality of fit compared with Eqn (12.52) that is described with seven coefficients. At the same time, it is clear from comparing the values of ζ_2 with ζ_o that Eqn (12.50), described with three coefficients, is grossly inadequate.

Table 12.5 Coefficients for $\sigma = S_0 \varepsilon^n \dot{\varepsilon}^\alpha$, $\sigma = S_0(\varepsilon, \varepsilon', \dot{\varepsilon})\varepsilon^{n(\dot{\varepsilon})}\dot{\varepsilon}^\alpha$ and $\sigma = S_0(\varepsilon, \varepsilon')\varepsilon^n \dot{\varepsilon}^\alpha$

	Coefficients for Eqn (12.50) $\sigma = S_o \varepsilon^n \dot{\varepsilon}^\alpha$							Fitting error
Solder alloys	S_o			n			α	ζ_o
Sn37Pb	84			0.18			0.077	21.1
SAC101	98			0.26			0.124	90.3
Sn3.5Ag	94			0.09			0.115	81.8
SAC305	89			0.06			0.117	42.0
Solder alloys	Coefficients for Eqn (12.52) $\sigma = S_o[1 + a_1(\varepsilon - \varepsilon') + a_2 \ln \dot{\varepsilon}]\varepsilon^{n_o(1+n_1 \ln \varepsilon)}\dot{\varepsilon}^\alpha$							Fitting error
	S_o	a_1	a_2	ε'	n_0	n_1	α	ζ_1
Sn37Pb	71	−0.58	9×10^{-5}	0.58	0.21	0.043	0.1	15.4
SAC101	57	−1.1	-6×10^{-4}	0.89	0.31	−0.037	0.13	40.8
Sn3.5Ag	87	−0.94	-4×10^{-5}	0.37	0.16	0.002	0.132	49.8
SAC305	99	−0.79	-9×10^{-4}	0.17	0.12	−0.03	0.118	15.2
Solder alloys	Coefficients for Eqn (12.53) $\sigma_f = S_o[1 + a_1(\varepsilon - \varepsilon')]\varepsilon^n \dot{\varepsilon}^\alpha$							Fitting error
	S_o	a_1		ε'	n		α	ζ_2
Sn37Pb	75	−0.58		0.59	0.242		0.086	15.8
SAC101	55	−1.1		0.83	0.262		0.148	41.5
Sn3.5Ag	86	−0.94		0.37	0.159		0.131	49.8
SAC305	97	−0.80		0.17	0.103		0.123	15.3

Comparing the coefficients of Eqn (12.50) with those of Eqns (12.52) and (12.53), it is noted that the use of the simpler Eqn (12.50) would lead to an underestimation of the rate sensitivity α and an overestimation of the stress coefficient S_o; for example, the stress coefficient of the SAC 101 solder has been overestimated by more than 70% when using Eqn (12.50). The accurate characterisation of stress coefficient is highly desirable in that solders with a low stress coefficient are more robust in drop impact.

12.6 Constant strain rate testing

The ideal test condition is one in which the strain rate experienced by a test specimen is constant with strain. The feasibility of such an ideal test condition is to be investigated.

The strain rate in a test specimen subjected to compression is given by Eqn (12.29) as: $\dot{\varepsilon}(\varepsilon) = V(\varepsilon)e^{\varepsilon}/l_o$. By introducing a compression velocity profile,

$$V(\varepsilon) = V_o e^{-\varepsilon} \tag{12.54}$$

a constant strain rate of $\dot{\varepsilon}_o = V_o/l_o$ can be achieved. The strain in the specimen at any time is given by $\varepsilon = \dot{\varepsilon}_o t$. Substituting this into Eqn (12.54) gives the velocity for constant strain rate compression as:

$$V(t) = V_o e^{-\dot{\varepsilon}_o t} \tag{12.55}$$

12.6.1 Mechanical tester

Integrating Eqn (12.55) with respect to time together with the initial condition, $u(0) = 0$, gives the required displacement time for the crosshead of a mechanical tester:

$$u(t) = l_o\left(1 - e^{-\dot{\varepsilon}_o t}\right) \tag{12.56}$$

This displacement characteristic can be easily delivered by the controller of commercial mechanical testers.

12.6.2 Drop-weight tester

The critical role of the kinetic energy of the drop weight in the fractional stress deviation Δ_σ is discussed in Section 12.4.2. The velocity of the drop weight is dictated by the desired strain rate; thus, the mass of the drop weight is the only test parameter that is free for manipulation. So far, the selection of the mass of drop weight has been arbitrary. The objective of this section is to find the optimum mass of the drop weight that will give rise to a minimum variation in strain rate in the test specimen over the range of strain of interest.

12.6.2.1 Method 1: assume a constant strain rate

The momentum relation gives:

$$F = m\dot{V} \tag{12.57}$$

Differentiating Eqn (12.55) with respect to time and substituting into Eqn (12.57) gives:

$$F = -mV_o\dot{\varepsilon}_o e^{-\dot{\varepsilon}_o t} \tag{12.58}$$

Under the condition of dynamic equilibrium, F is equilibrated to the reaction force, $-\sigma A$, from the test specimen. Substituting $\sigma = S_o\varepsilon^n\dot{\varepsilon}_o^\alpha$ from Eqn (12.50) and $A(\varepsilon) = A_o e^\varepsilon$, $V_o = \dot{\varepsilon}_o l_o$, $\dot{\varepsilon}_o t = \varepsilon$ into the reaction force gives:

$$m = \frac{S_o A_o \dot{\varepsilon}_o^{\alpha-2}}{l_o} g(\varepsilon) \tag{12.59}$$

where

$$g(\varepsilon) = e^{2\varepsilon}\varepsilon^n \tag{12.60}$$

Eqns (12.59) and (12.60) suggest a drop mass that varies with strain. Figure 12.17 plots the function $g(\varepsilon)$ for hardening index n equals to 0.1, 0.2 and 0.3 for $0 \leq \varepsilon \leq 0.5$, which shows large variations of nearly 100% for $0.05 \leq \varepsilon \leq 0.5$.

In practice, it is impossible to vary the mass of the drop weight as a function of strain; instead, a mean value of $g(\varepsilon)$,

$$\overline{g} = \frac{\int_o^\xi e^{2\varepsilon}\varepsilon^n d\varepsilon}{\xi} \tag{12.61}$$

is sought, which may be evaluated numerically. As an illustration, using $n = 0.182$ for the Sn37Pb solder, the value of $\overline{g}$ for $0 \leq \varepsilon \leq 0.5$ has been evaluated to be 1.33.

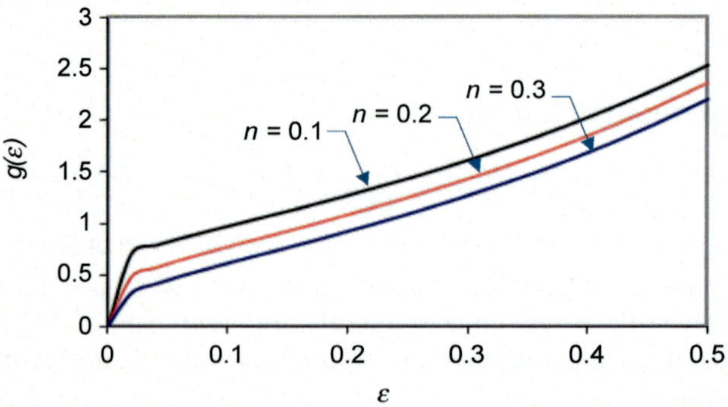

Figure 12.17 Plots of $g(\varepsilon)$

Substituting $\overline{g} = 1.33$, $A_o = 12.57\ \text{mm}^2$, $l_o = 0.004$ m, $\varepsilon_o = 100\ \text{s}^{-1}$; and $S_o = 84.15\ \text{N mm}^{-2}$, $\alpha = 0.077$ for the SnPb solder into Eqn (12.59), the optimum mass of the drop weight is calculated as $\overline{m} = 50$ kg.

By fixing the mass, the strain rate will now vary with strain. Equation (12.29) gives: $V(\varepsilon) = l_o\dot{\varepsilon}e^{-\varepsilon}$, which upon differentiation with respect to time gives:

$$\dot{V} = \frac{dV}{d\varepsilon}\dot{\varepsilon} = l_o\dot{\varepsilon}\left(-\dot{\varepsilon}e^{-\varepsilon} + e^{-\varepsilon}\frac{d\dot{\varepsilon}}{d\varepsilon}\right) = l_o e^{-\varepsilon}\left(-\dot{\varepsilon}^2 + \ddot{\varepsilon}\right) \tag{12.62}$$

Substituting Eqn (12.62) and $F = -S_o A_o \dot{\varepsilon}_o^{\alpha} e^{\varepsilon} \varepsilon^n$ into Eqn (12.57) yields:

$$\ddot{\varepsilon} - \dot{\varepsilon}^2 + \frac{S_o A_o \dot{\varepsilon}_o^{\alpha} e^{2\varepsilon} \varepsilon^n}{\overline{m} l_o} = 0 \tag{12.63}$$

which suggests a rather complex function of strain rate.

12.6.2.2 Method 2: assume a parabolic function for velocity

The variation of the function $g(\varepsilon)$ for $0 \le \varepsilon \le 0.5$ using the above method is excessively large to yield an optimum mass for the drop weight. It is suggested in Figure 12.12 that the velocity of drop weight may be described rather well using a parabolic function:

$$V = V_o\left(1 + b_1\varepsilon + b_2\varepsilon^2\right) \tag{12.64}$$

The compressive strain rate in the test specimen is given by Eqn (12.29) as $\dot{\varepsilon} = Ve^{\varepsilon}/l_o$; that is,

$$\dot{\varepsilon} = \dot{\varepsilon}_o h(\varepsilon) \tag{12.65}$$

where

$$h(\varepsilon) = e^{\varepsilon}\left(1 + b_1\varepsilon + b_2\varepsilon^2\right) \tag{12.65a}$$

The momentum relation gives:

$$F = -\sigma A = m\dot{V} = mV_o(b_1 + 2b_2\varepsilon)\dot{\varepsilon} \tag{12.66}$$

Substituting $\sigma = S_o\varepsilon^n\dot{\varepsilon}^{\alpha}$, $A(t) = A_o e^{\varepsilon}$, $V_o = \dot{\varepsilon}_o l_o$, and Eqn (12.65) into Eqn (12.66) gives:

$$m = \frac{S_o A_o \dot{\varepsilon}_o^{\alpha-2}}{l_o} f(\varepsilon) \tag{12.67}$$

where,

$$f(\varepsilon) = -\frac{e^{\varepsilon}\varepsilon^{n}[h(\varepsilon)]^{\alpha-1}}{b_1 + 2b_2\varepsilon} \tag{12.67a}$$

The objective is to minimise the variation of strain rate and mass within the strain range of interest. This can be accomplished numerically by minimising the sum of the mean-normalised variances of $h(\varepsilon)$ for $0 \leq \varepsilon \leq \xi$ and $f(\varepsilon)$ for $0.1\xi \leq \varepsilon \leq \xi$, where ξ is the upper limit of the strain range of interest; that is,

$$\text{Minimise}\left[\frac{h_{\text{variance}}}{\bar{h}}\bigg|_{0\leq\varepsilon\leq\xi} + \frac{f_{\text{variance}}}{\bar{f}}\bigg|_{0.1\xi\leq\varepsilon\leq\xi}\right] \tag{12.68}$$

Using Sn37Pb solder (assuming $n = 0.182$ and $\alpha = 0.077$) as an illustration, the optimum values of b_1 and b_2 for $\xi = 0.5$ have been evaluated to be -0.43 and -0.67, respectively. The functions $h(\varepsilon)$ and $f(\varepsilon)$ are plotted in Figure 12.18. The peak magnitude of $h(\varepsilon)$ occurs near the middle of the strain range, which is an ideal condition. The variation of $h(\varepsilon)$ is limited to 10%. Comparatively, the condition $b_1 = b_2 = 0$ will give $1.0 \leq h(\varepsilon) \leq 1.65$, a variation of 65%. The variation of $f(\varepsilon)$ is limited to 7% for $0.1\xi \leq \varepsilon \leq \xi$, which is a significant improvement over that of function $g(\varepsilon)$. The mean magnitude of function $f(\varepsilon)$ is $\bar{f} = 1.23$ compared to $\bar{g} = 1.33$. Substituting $f(\varepsilon) = 1.23$, $A_o = 12.57\ \text{mm}^2$, $l_o = 0.004$ m, $\dot{\varepsilon}_o = 100\ \text{s}^{-1}$; and $S_o = 84.15\ \text{N mm}^{-2}$, $\alpha = 0.077$ for the Sn37Pb solder into Eqn (12.67) yields the optimum mass of the drop weight to be $\bar{m} = 46$ kg.

The drop-weight mass of 55 kg was used in the characterisation experiment for the Sn37Pb solder at $\dot{\varepsilon}_o = 100\ \text{s}^{-1}$; the experimental function $h(\varepsilon)_{\text{exp}}$ has also been plotted in Figure 12.18. The peak strain rate of $116\ \text{s}^{-1}$ occurs at $\varepsilon = 0.4$, suggesting the mass of the drop weight to be marginally excessive. The values of $h(\varepsilon)_{\text{exp}}$ vary within 16%, which is a reasonably good result.

At first glance, it appears that we are in a paradoxical situation—we should know the material constants, $\{A_o, n, \alpha\}$, which are supposed to be evaluated through

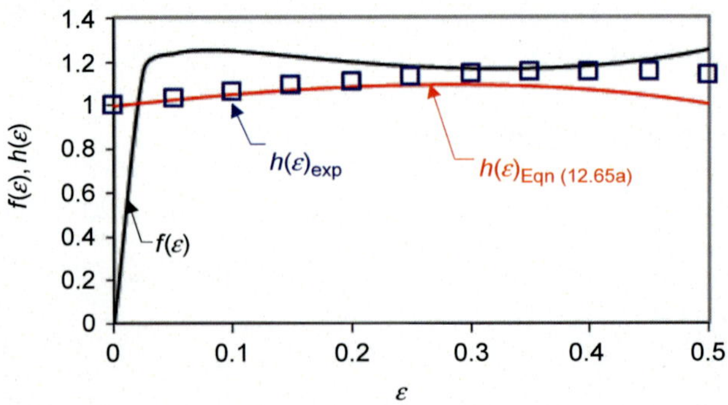

Figure 12.18 Plots of $h(\varepsilon)_{\text{Eqn (12.65)}}$ $f(\varepsilon)$, and $h(\varepsilon)_{\text{exp}}$ for Sn37Pb solder at $\dot{\varepsilon}_o = 100\ \text{s}^{-1}$.

drop-weight experiments, but before we can decide on the optimum mass of the drop weight to be used for the experiment. In reality, however, a significant deviation from the optimal mass of the drop weight will result in only marginal variations in the experimental strain rate as has been shown for the Sn37Pb solder at $\dot{\varepsilon}_0 = 100\ s^{-1}$, and hence, the material constants, $\{A_0, n, \alpha\}$. Thus, we can calculate the 'optimum' mass of the drop weight using the best-guessed material constants. In any case, it is better to err on the side of a larger mass as the stress deviation, Δ_σ, will then be bounded, while that for an under-mass drop weight is not bounded (Section 12.4.2).

References

Alder, J., & Philips, V. (1954). *Journal of the Institute of Metals, 80*, 80.

Armstrong, R., & Walley, S. (2008). High strain rate properties of metals and alloys. *International Materials Reviews, 53*, 105.

Baraya, G. J. (1965). The dynamic compression of circular cylinders of super-pure aluminium at elevated temperatures. *International Journal of Mechanical Sciences, 7*, 621.

Benjamin, D. (1989). Metals handbook. In *Properties and selection: Nonferrous alloys and pure metals* (Volume 2) ASM.

Clyens, S., & Campbell, J. (1974). Behaviour of Cu and Pb/Sn eutectic in torsion at high strain rate. *Institute of Physics Conference Series, 21*, 62.

Fallansbee, P. S. (1985). High strain rate compression testing. In *Metals handbook* (Vol. 8) ASM.

Field, J. W. (1994). Experimental methods at high rates of strain. *Journal of Physics IV*, C3–C8.

Field, J. (2004). Review of experimental techniques for high rate defomation and shock studies. *International Journal of Impact Engineering, 30*, 115–725.

Gray, G. (2003). High strain rate testing of materials. In E. Kaufmann (Ed.), *Characteriasation of materials, V1*. Wiley-Interscience.

Johnson, G., & Cook, W. (1985). Fracture characteristics of three metals subjected to various strains, strain rates, temperatures, pressures. *Engineering Fracture Mechanics, 21*, 31–48.

Kawashima, K. (1992). Strain-rate and temperature-dependent stress–strain and temperature-dependent stress–strain. *Journal of Materials Science, 27*, 6387.

Nemat-Nasser, S., Li, Y. F., & Isaacs, J. B. (1994). Experimental/computational valuation of flow stress at high strain rates with application to adiabatic shear banding. *Mechanics of Materials, 17*, 111–134.

Prandtl, L. (1928). Ein Gedankenmodell zur kinetischen Theorie der festen Körper. *Zeitschrift für Angewandte Mathematik und Mechanik, 8*, 85–106.

Ramberg, W., & Osgood, W. R. (1943). *Description of stress- strain curves by three parameters. Technical Note No. 902*. Washington, DC: National Advisory Committee for Aeronautics.

Shi, X. W. (2002). A new creep constitutive model for eutectic solder Alloy. *ASME Transactions Journal of Electronic Packaging, 124*, 85–90.

Shohji, I. (2004). Comparison of low-melting lead-free solders in tensile properties with Sn-Pb eutectic solder. *Journal of Materials Science—Materials in Electronics, 15*, 219.

Siviour, C. (2005). Mechanical properties of SnPb and lead-free solders at high rates of strain. *Journal of Physics D: Applied Physics, 38*, 4131–4139.

Solomon, H. (1990). The creep and strain rate sensitivity of a high Pb content solder with comparisons to 60Sn/40Pb solder. *Journal of Electronic Materials, 19*, 929.

Wang, B., & Yi, S. (2002). Dynamic plastic behavior of 63 wt% Sn 37 wt% Pb eutectic solder under high strain rate. *Journal of Materials Science Letters, 21*, 697–698.
Wray, P. (1973). Experimental study of the origin of strain-rate sensitivity in a common Pb-Sn Alloy. *Metallurgical Transactions, 4*, 2475.
Zener, C., & Hollomon, J. (1944). Effects of strain rate upon plastic flow of steel. *Journal of Applied Physics, 15*, 22–32.

Index

Note: Page numbers followed by "f" and "t" indicate figures and tables respectively.

Printed in the United States
By Bookmasters